Flora of Florida, Volume IV

UNIVERSITY PRESS OF FLORIDA

Florida A&M University, Tallahassee
Florida Atlantic University, Boca Raton
Florida Gulf Coast University, Ft. Myers
Florida International University, Miami
Florida State University, Tallahassee
New College of Florida, Sarasota
University of Central Florida, Orlando
University of Florida, Gainesville
University of North Florida, Jacksonville
University of South Florida, Tampa
University of West Florida, Pensacola

Flora of & Florida

VOLUME IV

DICOTYLEDONS, COMBRETACEAE THROUGH AMARANTHACEAE

Richard P. Wunderlin, Bruce F. Hansen, and Alan R. Franck

University Press of Florida

Gainesville · Tallahassee · Tampa · Boca Raton

Pensacola · Orlando · Miami · Jacksonville · Ft. Myers · Sarasota

142578

This book may be available in an electronic edition.

22 21 20 19 18 17   6 5 4 3 2 1

Flora of Florida, Volume IV: Dicotyledons, Combretaceae through Amaranthaceae
ISBN 978-0-8130-6248-8 (cloth)

The Library of Congress has catalogued the first volume in the Flora of Florida set as follows:
Wunderlin, Richard P., 1939–
Flora of Florida/Richard P. Wunderlin and Bruce F. Hansen with the collaboration of Edwin L.
Bridges and Jack B. Fisher.
p. cm.
Includes bibliographical references and indexes.
Contents: v. 1. Pteridophytes and gymnosperms.
ISBN 978-0-8130-1805-8 (cloth: alk. paper)
    1. Botany—Florida—Classification. 2. Plants—Identification. I. Hansen, Bruce F. II. Title.
QK154.W85 2000
581.9759—dc21 00-032599

The University Press of Florida is the scholarly publishing agency for the State University System
of Florida, comprising Florida A&M University, Florida Atlantic University, Florida Gulf Coast
University, Florida International University, Florida State University, New College of Florida,
University of Central Florida, University of Florida, University of North Florida, University of South
Florida, and University of West Florida.

University Press of Florida
15 Northwest 15th Street
Gainesville, FL 32611-2079
http://upress.ufl.edu

Dedicated to the memory of George Ralph Cooley (1896–1986), New York investment banker, patron of botany, personal friend of the first president of the University of South Florida, John Allen, and founder of the USF Herbarium

Contents

Acknowledgments

The facilities and collections of many herbaria were utilized in preparing this volume. The courtesies extended and the loan of specimens by the curators are gratefully appreciated. These include the Florida Museum of Natural History (FLAS), Florida State University (FSU), Harvard Herbaria (A, GH), Marie Selby Botanical Gardens (SEL), New York Botanical Garden (NY), University of Central Florida (FTU), and University of North Carolina–Chapel Hill (NCU). We are especially grateful to Kent Perkins (FLAS), Loran Anderson (FSU), and Austin Mast (FSU) for their continuous support. The treatment for the Plumbaginaceae was provided by Rani Vajravelu (FTU).

The *Flora of Florida* project has been strongly supported by the University of South Florida Institute for Systematic Botany.

Introduction

Volume 1 of the *Flora of Florida* provides background information on the physical setting, vegetation, history of botanical exploration, and systematic treatments of the pteridophytes and gymnosperms. Volumes 2 through 7 will contain the dicotyledons and volumes 8 through 10, the monocotyledons.

This volume contains the taxonomic treatments of 31 families of the dicotyledons (see table of contents).

ORGANIZATION OF THE FLORA

Taxa Included

Florida, with more than 4,300 taxa, has the third most diverse vascular plant flora of any state in the United States. The *Flora of Florida* is a treatment of all indigenous and naturalized vascular plant taxa currently known to occur in the state. Naturalized is defined as those nonindigenous taxa growing outside of cultivation and naturally reproducing. This includes plants that have escaped from cultivation as well as those that were intentionally or accidentally introduced by human activities in post-Columbian times. Taxa that have not been recently recollected and may no longer exist in the wild in Florida are formally treated both for historical completeness and on the premise that they may be rediscovered in the future.

A taxon is formally treated in this flora if (1) an herbarium specimen has been seen to document its occurrence in Florida, or (2) a specimen is cited from Florida in a monograph or revision whose treatment is considered sound.

Taxa Excluded

Literature reports of taxa attributed to Florida that are considered to be erroneous or highly questionable and therefore to be excluded from this flora are listed following the treatment for the genus, or in the case of genera not otherwise treated, at the end of the family. The reason for exclusion is given in each case. Most commonly, the taxon is excluded because it is based on a misidentified specimen(s), lack of documentation by means of a specimen, or it is based on a misapplied name, that is, a name correctly applied to a plant not found in Florida.

Systematic Arrangement

Recent studies have demonstrated that the traditional dicotyledons are paraphyletic and that the monophyletic monocotyledons are derived from within the dicotyledons. We believe that the arrangement as proposed by the Angiosperm Phylogeny Group III (Stevens, 2016) has merit and is followed in this work with slight modifications. The linear sequence of families used here essentially follows that proposed by Haston et al. (2009). For convenience, the genera and species within each family are arranged alphabetically.

Descriptions

Descriptions are based on Florida material and are given for each family, genus, species, and infraspecific taxon.

Common Names

Non-Latinized names given for the taxa are derived from published sources as well as from our own experience. No attempt is made to list all names that have been applied to a taxon, or to standardize names with a specific source, or to supply a name for species where one is not in general usage. For plants lacking a common name, the generic name may be used as is the usual practice.

Derivation of Latin Names

The derivation of the generic name and that of each specific and infraspecific epithet is given.

Synonymy

A full literature citation is given for each species, infraspecific taxon, and synonym. Synonyms listed are only those that have been cited for Florida in manuals, monographic treatments, and technical papers. Also included is the basionym and all homotypic synonyms of a name introduced into synonymy. The homotypic synonyms are listed in chronological order in a single paragraph, and the paragraphs of synonyms are put in chronological order according to the basionym of each. If the type of a taxon is a Florida collection and is known, this information is given. We do not attempt to lectotypify the numerous Florida taxa needing lectotypification in the belief that this is best left to monographers.

For families and genera, only the author and date of publication is given. Family and generic synonyms listed are those that have been used in the major publications pertinent to the Florida flora.

Citation of periodical literature conforms to that cited in *Botanico-Periodicum-Huntianum* (Lawrence et al., 1968) and *Botanico-Periodicum-Huntianum/Supplementum* (Bridson and Smith, 1991). Other literature citations conform to that cited in *Taxonomic Literature*, edition 2 (Stafleu and Cowan, 1976, et seq.). Author abbreviations are those listed in *Authors of Plant Names* (Brummitt and Powell, 1992).

Habitat

The terminology used for plant communities generally follows that of Myers and Ewel (1990), but may vary.

Distribution

The global distribution is given for each family and genus where native and naturalized. Relative abundance in Florida (ranked as common, frequent, occasional, or rare) and the distribution are given for each species and infraspecific taxon. The format for distribution of species and infraspecific taxa is: Florida; North America (continental United States, Canada, and Greenland); tropical America (West Indies, Mexico, Central America, South America); Old World (Europe, Africa, Asia, Australia, Pacific Islands). For taxa occurring in all of these areas, the phrase "nearly cosmopolitan" is used. For taxa of limited distribution in Florida, range statements by county are usually given. For taxa of wide distribution in Florida, the range is given in general terms: *panhandle*—from the Suwannee River west to Escambia County; *peninsula*—east of the Suwannee River and south of the Georgia line southward through the Florida Keys. Because of the vast floristic differences in peninsular Florida, this region is often further subdivided into northern, central, and southern regions and the keys. The northern region is east of the Suwannee River and south of the Georgia line southward through Gilchrist, Alachua, Putnam, and Flagler Counties. The central region extends from Levy, Marion, and Volusia Counties southward through Lee, Hendry, and Palm Beach Counties. The southern peninsula consists of the southernmost four counties (Collier, Broward, Monroe, and Miami-Dade). The Florida Keys consist of the chain of islands from Key Largo to the Marquesas Keys and the Dry Tortugas. Politically, they are part of Monroe County. The panhandle is subdivided into eastern, central, and western regions. The eastern region consists of the counties west of the Suwannee River west through Jefferson County, the central region extends from Leon and Wakulla Counties west through Holmes, Washington, and Bay Counties while the western region consists of the westernmost four counties (Walton, Okaloosa, Santa Rosa, and Escambia). Since the species distribution may change as new data are added, refer to the *Atlas of Florida Plants* website (http://florida.plantatlas.usf.edu/) for current information.

Endemic or Exotic Status

Endemic taxa are those whose global native distribution is confined to the political boundary of Florida. If a taxon is a non-native, the region of nativity is given. Non-native taxa are those that are known to have become part of the flora following the occupation by Europeans in the sixteenth century. Admittedly, this is an arbitrary starting point because several species are believed to have been introduced by Paleo-Indians before 1513 when La Florida was discovered by Ponce de Leon. Technically, these are considered as native. Another problem in interpretation arises when propagules arrive after 1513 by some means other than human activity (that is, hurricanes, storms, sea drift, or animals) and the species becomes established. Again technically, these are considered as non-natives. It is sometimes difficult to determine whether a widespread species is native or a non-native, and our opinion may differ from that of others.

Reproductive Season

The sexual reproductive (flowering) season for each species and infraspecific taxon is given. The reproductive seasons are broadly defined as follows: spring—March through May; summer—June through September; fall—October through November; winter—December through February. Species "flowering out of season" are sometimes encountered.

Hybrids

Named hybrids are listed along with the putative parents, relative nomenclature, and usually with a comment concerning distribution in Florida.

References

Major monographs, revisions, and other pertinent literature cited in the text, other than those cited in the nomenclature, are given at the end of the volume.

TAXONOMIC CONCEPTS

Taxonomic interpretations and nomenclature are generally in accord with recent monographs or revisions for the various groups except where it is believed that recent evidence necessitates a change. Citation of a monograph or revision in the text implies consideration of the work during the preparation of the treatment, but not necessarily acceptance. Where a difference of opinion exists among published treatments or the treatment in this work deviates from that of the reference cited, a discussion of alternative opinions is often provided.

Species, subspecies, and varieties are considered as entities with a high degree of population integrity. Color forms and minor morphotypes that occur within a species and that may be formally recognized as *forma* by other authors are accorded no formal recognition in this work.

No nomenclatural innovations are intentionally published in the *Flora*.

LITERATURE CITED

Bridson, G.D.R., and E. R. Smith. 1991. Botanico-Periodicum-Huntianum/Supplementum. Pittsburgh: Hunt Botanical Library.

Brummitt, R. K., and C. E. Powell. 1992. Authors of Plant Names. Royal Botanical Gardens, Kew. Basildon: Her Majesty's Stationery Office.

Haston, E., J. E. Richardson, P. F. Stevens, M. W. Chase, and D. J. Harris. 2009. The linear Angiosperm Phylogeny Group (LAPG) III: a linear sequence of the families in APG III. Bot. J. Linnean Soc. 161: 128–31.

Lawrence, G.H.M., A.F.G. Buchheim, G. S. Daniels, and H. Dolezal. 1968. Botanico-Periodicum-Huntianum. Pittsburgh: Hunt Botanical Library.

Myers, R. I., and J. J. Ewel, eds. 1990. Ecosystems of Florida. Gainesville: University Press of Florida.

Stafleu, F. A., and R. S. Cowan. 1976 et seq. Taxonomic Literature. Edition 2. Utrecht: Bohn, Scheltema, and Holkema.

Stevens, P. F. 2016. Angiosperm Phylogeny Website. Version. Version 13 [continuously updated]. http://www.mobot.org/MOBOT/research/APweb/.

Systematic Treatments

Keys to Major Vascular Plant Groups

1. Plant reproducing by spores .... PTERIDOPHYTES (volume I)
1. Plant reproducing by seeds.
    2. Leaves with a single midvein or with simple or sometimes dichotomously branched veins, these closely parallel and lacking secondary interconnecting cross-veinlets; seeds borne on the surface of specialized bract-scale structures aggregated into woody or fleshy cones or a single seed partly or wholly surrounded by a fleshy aril and drupelike or berrylike; perianth lacking .... GYMNOSPERMS (volume I)
    2. Leaves with parallel veins with secondary interconnecting cross-veinlets or with reticulate veins; seeds borne enclosed within specialized structures (carpels); perianth usually present.
        3. Vascular bundles occurring in a ring or in concentric cylinders; cotyledons 2; flower parts usually in other than whorls of 3 or multiples thereof; leaves usually reticulate-veined .... DICOTYLEDONS (volumes II–VII)
        3. Vascular bundles scattered (or rarely single); cotyledon 1; flower parts often in whorls of 3 or multiples thereof; leaves usually parallel-veined (sometimes with midvein only) (some plants diminutive, floating aquatics, the plant body thalloid, not differentiated into stems and leaves, rootless or with 1–few unbranched roots or plants partly or wholly submersed aquatics, the leaves, flowers, and fruits often much reduced) .... MONOCOTYLEDONS (volumes VIII–X)

Dicotyledons

COMBRETACEAE R. Br., nom. cons. 1810. COMBRETUM FAMILY

Shrubs, trees, or woody vines. Leaves alternate or opposite, simple, pinnate-veined, petiolate, stipulate or estipulate. Flowers in terminal or axillary spikes, racemes, panicles, or heads, actinomorphic, bisexual or unisexual (staminate) (plants dioecious or polygamodioecious), bracteate, bracteolate or ebracteolate; hypanthium prolonged beyond the ovary, the lower part adnate to the ovary, the upper part free; sepals 4 or 5, connate; petals 5 and free or absent; nectaries present; stamens 5–10, the filaments free, the anthers 2-locular, versatile, longitudinally dehiscent; ovary 2- to 5-carpellate, 1-loculate, the style 1. Fruit a drupe.

A family of 14 genera and about 500 species; nearly cosmopolitan.

Terminaliaceae J. St.-Hil. (1805).

Selected references: Graham (1964b); Stace (2010).

1. Leaves opposite, decussate.
    2. Tree or erect shrub; petiole with nectar glands; flowers inconspicuous, the petals ca. 1 mm long, greenish white ....**Laguncularia**
    2. Vine or scandent shrub; petiole without nectar glands; flowers showy, the petals 1–2 cm long, white to pink or red..**Combretum**
1. Leaves alternate, spiral.
    3. Flowers in dense spherical or oblong heads; fruits in a dry, conelike head ....**Conocarpus**
    3. Flowers in spikes (these sometimes reduced to a few flowers); fruit a dry or fleshy drupe.
        4. Leaves somewhat succulent; petals white ....**Lumnitzera**
        4. Leaves chartaceous; petals absent .... **Terminalia**

Combretum Loefl., nom. cons. 1758.

Woody vines. Leaves opposite, decussate, pinnate-veined, petiolate, estipulate. Flowers in terminal or axillary spikes, bisexual, bracteate, ebracteolate; sepals 5, connate; petals 5, free; stamens 10, free. Fruit a drupe.

A genus of about 250 species; North America, West Indies, Mexico, Central America, South America, Africa, Asia, and Australia. [Derived from a name applied by Pliny the Elder to a climbing plant.]

Quisqualis L. (1762).

Combretum indicum (L.) DeFilipps [Of India.] RANGOON CREEPER.

Quisqualis indica Linnaeus, Sp. Pl., ed. 2. 556. 1762. *Quisqualis pubescens* Burman f., Fl. Indica 104. 1768, nom. illegit. *Combretum indicum* (Linnaeus) DeFilipps, Useful Pl. Dominica 277. 1998.

Climbing woody vine, to 6 m; branchlets pubescent. Leaves with the blade elliptic to oblong-elliptic or slightly obovate, 4–18 cm long, 1.5–9 cm wide, the apex acuminate to subcaudate, the base obtuse to subcordate, the margin entire, the upper and lower surfaces glabrate to pubescent and with inconspicuous short-stipitate glands, the petiole 0.5–2 cm long. Flowers in a spike 1.5–7.5 cm long; bracts lanceolate to ovate-lanceolate, 6–10 mm long; lower hypanthium 3–5 mm long, the upper hypanthium 5–6.5(8) cm long, sericeous; sepals triangular to deltate, 1–3 mm long, sericeous; petals oblong, 10–15(20) mm long, white or pink, becoming red, the outer surface sericeous, the inner surface glabrous; nectary on the inner surface of the hypanthium near the base; stamens 4–8 mm long, included; style adnate to the upper hypanthium except for the distal 10–20 mm. Fruit ovate-elliptic, 2.5–4 cm long, red, 5-winged, glabrous or sparsely glandular-pubescent.

Disturbed sites. Rare; Highlands, Broward, and Miami-Dade Counties. Escaped from cultivation. Florida; West Indies, Mexico, Central America, and South America; Africa, Asia, and Australia. Native to Asia. Summer.

Conocarpus L., 1753. BUTTONWOOD

Shrubs or trees. Leaves alternate, spiral, pinnate-veined, petiolate, estipulate. Flowers in pedunculate racemes or panicles of conelike heads, unisexual (plants dioecious), bracteate, ebracteolate; sepals 5, connate; petals absent; stamens (5)10, free. Fruit a drupe, scalelike, aggregated into conelike heads; seed 1.

A genus of 2 species; North America, West Indies, Mexico, Central America, Africa, and Asia. [From the Greek *konos*, cone, and *karpos*, fruit, in reference to the conelike fruiting clusters.]

Conocarpus erectus L. [Upright, in reference to the habit.] BUTTONWOOD.

Conocarpus erectus Linnaeus, Sp. Pl. 176. 1753. *Conocarpus erectus* Linnaeus var. *arboreus* de Candolle, Prodr. 3: 16. 1828, nom. inadmiss. *Terminalia erecta* (Linnaeus) Baillon, Hist. Pl. 6: 266, 275. 1876.

Conocarpus erectus Linnaeus var. *sericeus* de Candolle, Prodr. 3: 16. 1828. *Conocarpus sericeus* (de Candolle) G. Don, Gen. Hist. 2: 662. 1832. *Conocarpus erectus* Linnaeus forma *sericeus* (de Candolle) Stace, in Harling & Sparre, Fl. Ecuador 81: 58. 2007.

Conocarpus erectus Linnaeus var. *sericeus* Grisebach, Fl. Brit. W.I. 277. 1860; non de Candolle, 1828. *Conocarpus sericeus* J. Jiménez Almonte, Anales Univ. Santo Domingo 18: 126. 1953; non (de Candolle) G. Don, 1832.

Shrub or tree, to 12(20) m; branchlets angled or winged. Leaves with the blade elliptic to obovate, 2–10 cm long, 0.5–4(5) cm wide, the apex acute to acuminate, the base cuneate, the margin entire, the upper and lower surfaces glabrous or densely pubescent, the lower surface with pit domatia, the petiole 2–15 cm long, with 2 nectar glands near the blade base. Flowers in heads in a raceme 3–10 cm long (sometimes paniculiform and to 25 cm long), the heads 4–9 mm long; hypanthium funnelform, greenish, ca. 1 mm long; sepals triangular-ovate, subequaling the floral tube; nectary disk on top of the ovary; stamens 5(10), 2–4 mm long. Fruiting heads 6–12 mm long, greenish brown, the fruit scalelike, 3–5 mm long, pubescent distally; seed flat.

Tidal swamps. Frequent; central and southern peninsula. Florida; West Indies, Mexico, Central America, and South America; Africa. All year.

Laguncularia C. F. Gaertn. 1807. WHITE MANGROVE

Shrubs or trees. Leaves opposite, decussate, pinnate-veined, petiolate, estipulate. Flowers in terminal and axillary spikes or panicles, bisexual or unisexual (perfect and staminate on the same plant), bracteate, bracteolate; sepals 5, connate; petals 5, free; stamens 10, free. Fruit a drupe; seeds 2.

A monotypic genus; North America, West Indies, Mexico, Central America, South America, and Africa. [*Laguncularis*, flask- or bottle-shaped, in reference to the shape of the fruit.]

Laguncularia racemosa (L.) C. F. Gaertn. [Flowers/fruits in a raceme.] WHITE MANGROVE.

> *Conocarpus racemosus* Linnaeus, Syst. Nat., ed. 10. 930. 1759. *Laguncularia racemosa* (Linnaeus) C. F. Gaertner, Suppl. Carp. 209. 1807. *Schousboea commutata* Sprengel, Syst. Veg. 2: 332. 1825, nom. illegit. *Rhizaeris alba* Rafinesque, Sylva Tellur. 90. 1838, nom. illegit.
> *Laguncularia glabriflora* C. Presl, Reliq. Haenk. 2: 22. 1831. *Laguncularia racemosa* (Linnaeus) C. F. Gaertner var. *glabriflora* (C. Presl) Stace, in Harling & Sparre, Fl. Ecuador 81: 11. 2007.

Shrub or tree, to 10(20) m. Leaves with the blade ovate to obovate, oblong, or suborbicular, 2–9 cm long, 1.5–5 cm wide, the apex rounded, the base obtuse, rounded, or retuse, the margin entire, the upper and lower surfaces with minute, scattered, sunken, salt-excreting, glandular trichomes, the lower surface with pit domatia, the petiole 6–15 mm long, with 2 distal nectar glands. Flowers in a spike or panicle 2–13 cm long; hypanthium shallowly cupulate, 1–2 mm long; sepals triangular, 1 mm long, green, the outer surface pubescent; petals suborbicular, 1 mm long, greenish white, pubescent; nectary disk on top of the ovary, pubescent; stamens 1–2 mm long. Fruit oblong to obovoid, 13–20 mm long, greenish or gray-green, pubescent, slightly flattened, with 2 major ridges forming spongy wings, these subtended by bracteoles, the hypanthium and calyx persistent.

Tidal swamps. Frequent; peninsula. Florida; West Indies, Mexico, Central America, and South America; Africa. Spring.

Lumnitzera Willd. 1803.

Shrubs or trees. Leaves alternate, spiral, pinnate-veined, petiolate, estipulate. Flowers in axillary spikes, bisexual, bracteate, bracteolate; sepals 5, connate; petals 5, free; stamens 10, free. Fruit a drupe; seeds 2–5.

A genus of 2 species; North America, Africa, Asia, Australia, and Pacific Islands. [Commemorates Istrán Lumnitzer (1750–1806), Hungarian botanist.]

Lumnitzera racemosa Willd. [With racemes.] BLACK MANGROVE.

Lumnitzera racemosa Willdenow, Ges. Naturf. Freunde Berlin Neue Schriffen 4: 187. 1803.

Shrub or tree, to 6 m; branchlets glabrous or sparsely short-pubescent. Leaves with the blade obovate to elliptic, 1.7–9 cm long, 0.6–3 cm wide, fleshy-leathery, the apex rounded to retuse, the base narrowly cuneate, the margin entire, the upper and lower surfaces glabrous, the lower surface with minute pit domatia, the petiole 3–5 mm long. Flowers in an axillary spike 1–3(7) cm long, bracteate, bracteolate; hypanthium cylindric, the free portion 1–3 mm long, glabrous; sepals triangular, ca. 1 mm long, green, the margin ciliate, otherwise glabrous; petals ovate to elliptic, 4–5 mm long, the margin sparsely ciliate, otherwise glabrous; nectary disk on the inner surface of the hypanthium at the base of the free portion; stamens 4–6 mm long. Fruit ovoid, 10–20 mm long, slightly flattened, with 2 major ridges, these subtended by bracteoles, the other ridges only developed distally, green or brown, glabrous.

Tidal swamps. Rare; Miami-Dade County. Escaped from cultivation. Florida; Africa, Asia, Australia, and Pacific Islands. Native to Africa, Asia, Australia, and Pacific Islands. Spring–summer.

Lumnitzera racemosa is listed as a Category I invasive species in Florida by the Florida Exotic Pest Plant Council (FLEPPC, 2015).

Terminalia L., nom. cons. 1767. TROPICAL ALMOND

Trees or shrubs. Leaves alternate, spiral, pinnate-veined, clustered at the ends of erect, short shoots and appearing pseudowhorled, petiolate, stipulate. Flowers in axillary spikes, bracteate, ebracteolate, bisexual or staminate (perfect and staminate on the same plant); sepals 4–5, connate; petals absent; stamens 10, free. Fruit a drupe; seeds 1 or 2.

A genus of about 200 species; North America, West Indies, Mexico, Central America, South America, Africa, Asia, Australia, and Pacific Islands. [*Terminalis*, terminal, in reference to the leaves clustered at the end of the branches.]

Terminalia buceras and *T. molinetii* have usually been placed in *Bucida*. A recent molecular study shows *Bucida* embedded within *Terminalia* (Maurin et al., 2010).

Bucida L., nom. rej. 1759.

1. Fruit 2-winged or -ridged or lacking wings or ridges.
  2. Fruit 4–6 cm long, 2-winged or -ridged .... **T. catappa**
  2. Fruit ca. 2 cm long, lacking wings or ridges .... **T. muelleri**

1. Fruit 5-angled or -winged.
  3. Fruit 2.5–5 cm long...T. arjuna
  3. Fruit to 1 cm long.
    4. Spinescent shrub or small tree; leaves 0.5–2.5(3.5) cm long; flowers few in a short subcapitate spike; fruit 3–4 mm long...T. molinetii
    4. Unarmed tree; leaves 3–9 cm long; flowers numerous in an elongate spike; fruit ca. 8 mm long .... T. buceras

Terminalia arjuna (Roxb. ex DC.) Wight & Arn. [Bright or shining, in Thervada Buddhism, it is said to have been used as the tree for achieved enlightenment by the Tenth Lord Buddha.] ARJUN.

Pentaptera arjuna Roxburgh ex de Candolle, Prodr. 3: 14. 1828. *Terminalia arjuna* (Roxburgh ex de Candolle) Wight & Arnott, Prodr. Fl. Ind. Orient. 314. 1834.

Tree, to 15(25) m; branchlets pale greenish to light gray, smooth. Leaves with the blade oblong-elliptic, 5–18(25) cm long, 4–9 cm wide, the apex obtuse or subacute, the base rounded or cordate, the margin entire or somewhat crenate or serrate in the upper half or throughout, the upper surface glabrous or subglabrous, the lower surface sparsely pubescent, domatia and basal nectar glands absent, the petiole 5–10 mm long, with 2(1) rounded apical glands; stipules minute. Flowers in an axillary or terminal paniculiform spike 3–6 cm long, sessile; hypanthium 4–5 mm long, sericeous at the base; sepal lobes triangular, 15 mm long, glabrous; stamens much exserted; nectar disk barbate. Fruit oblong-ovate, 2.5–5 cm long, 5-winged, the wings striated with 5 ascending veins.

Disturbed sites. Rare; Miami-Dade County. Escaped from cultivation. Florida; Asia and Africa. Native to Asia. Spring.

Terminalia buceras (L.) C. Wright [Ox-horn, in reference to the galls sometimes formed on the fruits, these resembling the horn of an ox.] BLACK OLIVE; OXHORN BUCIDA.

Bucida buceras Linnaeus, Syst. Nat., ed. 10. 1025. 1759. *Terminalia buceras* (Linnaeus) C. Wright, in Sauvalle, Anales Acad. Ci. Méd. Habana 5: 99. 1869. *Myrobalanus buceras* (Linnaeus) Kuntze, Revis. Gen. Pl. 1: 237. 1891. *Buceras buceras* (Linnaeus) Millspaugh, Publ. Field Columb. Mus., Bot. Ser. 1: 515. 1902, nom. inadmiss.

Tree, to 25 m; branches with nodal thorns in juvenile plants, these absent in mature plants, the branchlets rufous-pubescent, becoming glabrous. Leaves with the blade elliptic to obovate, 2–10 cm long, 0.8–4.5 cm wide, the apex obtuse to rounded, sometimes slightly retuse, the base cuneate, the margin entire, the upper and lower surfaces glabrous or sparsely pubescent, domatia and basal nectary glands absent, the petiole 2–15 mm long; stipules minute, appearing as fingerlike multicellular trichomes. Flowers in an axillary pedunculate spike 3–19 cm long, bisexual, greenish to pale yellow; hypanthium ca. 4 mm long, the free portion 1–2 mm long, sericeous-tomentose; sepal lobes triangular, ca 1 mm long; stamens 3–5 mm long; nectary a 5-lobed crateriform disk on top of the ovary, the upper margin sericeous-pilose. Fruit ovoid, (4)5–8 mm long, with 5 poorly developed, rounded lobes, glabrate or pubescent.

Hammocks. Rare; Charlotte and Lee Counties, southern peninsula. Escaped from cultivation. Florida; West Indies, Mexico, and Central America. Native to tropical America. All year.

Terminalia catappa L. [Malayan vernacular name.] WEST INDIAN ALMOND.

Terminalia catappa Linnaeus, Syst. Nat., ed. 12. 674. 1767; Mant. Pl. 128. 1767. *Myrobalanus catappa* (Linnaeus) Kuntze, Revis. Gen. Pl. 1: 237. 1891. *Buceras catappa* (Linnaeus) Hitchcock, Rep. (Annual) Missouri Bot. Gard. 4: 85. 1893.

Tree, to 20(35) m; branchlets glabrous. Leaves with the blade obovate, 6–35 cm long, 2.5–15.5 cm wide, the apex acuminate to obtuse or rounded, the base narrowly cuneate to rounded or narrowly cordate, the margin entire, the upper surface glabrous or glabrate, the midvein densely to sparsely pubescent, at least basally, the lower surface glabrate to moderately pubescent, the midvein and secondary veins sparsely to densely pubescent, with pit domatia at the major vein junctions and nectar glands near the base, the petiole 5–28 mm long; stipules minute, appearing as fingerlike multicellular trichomes. Flowers in an axillary spike 5–25 cm long, with a few bisexual ones proximally and numerous staminate distally; free portion of the hypanthium 1–2 mm long; sepal lobes 5, triangular, 1–3 mm long; nectary a lobed disk on top of the ovary; stamens 3–5 mm long. Fruit ovoid to ellipsoid, 3.5–7 cm long, green or red, slightly flattened, with 2 well-developed ridges or short wings, sparsely pubescent or glabrous.

Disturbed sites. Rare; Broward County, southern peninsula. Escaped from cultivation. Florida; West Indies and Mexico; Africa, Asia, Australia, and Pacific Islands. Native to Africa, Asia, Australia, and Pacific Islands. All year.

Terminalia catappa is listed as a Category II invasive species in Florida by the Florida Exotic Pest Plant Council (FLEPPC, 2015).

Terminalia molinetii M. Gómez [Commemorates Eugenio Molinet Amorós (1861–1959), Brigadier General of the Cuban Army of Liberation.] SPINY BLACK OLIVE.

Terminalia molinetii M. Gómez de la Maza y Jiménez, Anales Soc. Esp. Hist. Nat. 19: 244. 1890. *Bucida molinetii* (M. Gómez de la Maza y Jiménez) Alwan & Stace, Ann. Missouri Bot. Gard. 76: 1127. 1989.
Terminalia spinosa Northrop, Mem. Torrey Bot. Club 12: 54, pl. 13. 1902; non Engler, 1895. *Bucida spinosa* Jennings, Ann. Carnegie Mus. 11: 201. 1917. *Bucida correlliana* Wilbur, Taxon 37: 467. 1988, nom. illegit.

Shrub or tree, to 8 m; branches with divaricate, slender, nodal thorns 3–7 mm long, usually in 3s at the end of the branchlets, the branchlets glabrous. Leaves with the blade oblanceolate to spatulate, 1–2.5(3.5) cm long, 2–8(15) mm wide, the apex obtuse or retuse, the base cuneate, the margin entire, the upper and lower surfaces glabrous or sparsely pubescent, domatia and nectar glands absent, the petiole 1–3 mm long; stipules minute, appearing as fingerlike multicellular trichomes. Flowers in an axillary, short-pedunculate, subcapitate spike 1–4 cm long, bisexual; perianth greenish; hypanthium ca. 3 mm long, the free portion 1–2 mm long, villous within; sepal lobes 5, triangular, ca. 1 mm long, nectary a lobed disk on top of the ovary; stamens 3–4 mm long. Fruit ovoid, 3–6 mm long, green to brown, glabrous or sparsely pubescent.

Hammocks. Rare; Miami-Dade County. Escaped from cultivation. Florida; West Indies, Mexico, and Central America. Native to West Indies, Mexico, and Central America. All year.

Terminalia muelleri Benth. [Commemorates Ferdinand von Mueller (1825–1896), German-born and educated Australian physician and botanist.] AUSTRALIAN ALMOND.

Terminalia muelleri Bentham, Fl. Austral. 2: 500. 1864. *Myrobalanus muelleri* (Bentham) Kuntze, Revis. Gen. Pl. 1: 237. 1891.

Tree, to 10 m; branchlets without thorns, glabrous. Leaves with the blade obovate, 4.2–19.5 cm long, 2–7.5 cm wide, the apex short-acuminate, acute, obtuse, or rounded, the base cuneate, the margin entire, the upper surface glabrous or glabrate to sparsely pubescent, the midvein densely to sparsely pubescent, at least basally, the lower surface glabrate to sparsely pubescent, the midvein and secondary veins densely to sparsely pubescent or glabrous with a few elongate trichomes in the junction of the secondary vein and the midvein and forming pit domatia, usually with nectary glands near the base, the petiole 8–20 mm long; stipules minute, appearing as fingerlike multicellular trichomes. Flowers in an axillary spike 6.5–15 cm long, with several bisexual ones proximally and numerous staminate ones distally; free portion of the hypanthium 1–2 mm long; sepal lobes 4 or 5, triangular, 2–3 mm long; nectary a lobed disk on top of the ovary; stamens 2–4 mm long. Fruit ovoid to ellipsoid, 12–20 mm long, green or red, becoming blue or blue-black, slightly flattened, lacking wings or ridges, sparsely pubescent or glabrous.

Mangrove swamps and disturbed coastal hammocks. Rare; Manatee and Palm Beach Counties southward. Escaped from cultivation. Florida; Australia. Native to Australia. All year.

Terminalia muelleri is listed as a Category II invasive species in Florida by the Florida Exotic Pest Plant Council (FLEPPC, 2015).

LYTHRACEAE J. St.-Hil., nom. cons. 1805. LOOSESTRIFE FAMILY

Trees, shrubs, or herbs. Leaves opposite, alternate, or whorled, sessile or petiolate, estipulate. Flowers in indeterminate, axillary or terminal racemes, spikes, cymes, or thyrses, actinomorphic or zygomorphic, bisexual, the pedicels with paired prophylls, perigynous and producing a hypanthium; sepals 4–6, alternating with epicalyx appendages or the appendages absent; petals 1–7 or absent; anthers dorsifixed, introrse, bilocular, longitudinally dehiscent; ovary 2- to 6-carpellate and -loculate. Fruit a dehiscent capsule, surrounded by the persistent floral tube.

A family of about 30 genera and about 600 species; nearly cosmopolitan.

Selected reference: Graham (1964a).

1. Shrub or small tree.
    2. Flowers in axillary clusters.. **Decodon**
    2. Flowers in terminal panicles...**Lagerstroemia**
1. Herb (sometimes suffruticose).
    3. Hypanthium cylindrical.

    4. Hypanthium symmetrical at the base..**Lythrum**
    4. Hypanthium asymmetrical at the base (enlarged and spurred on 1 side).........................**Cuphea**
  3. Hypanthium campanulate to globose.
    5. Flowers 2 or more in the leaf axils.. **Ammannia**
    5. Flowers solitary in the leaf axils.
      6. Petals present; fruit dehiscent ....**Rotala**
      6. Petals absent; fruit indehiscent ....**Didiplis**

Ammannia L. 1753. REDSTEM

Annual herbs. Leaves opposite, decussate, sessile, estipulate. Flowers in axillary cymes, the paired prophylls subtending the lateral flowers of the cyme, actinomorphic; sepals 4–6, free, alternating with epicalyx appendages; petals 4–6 or absent, free; stamens 4–8; ovary 2- to 4-carpellate, the locules incomplete, the style 1. Fruit an irregularly dehiscent capsule.

    A genus of about 25 species; nearly cosmopolitan. [Commemorates Paul Ammann (1634–1691), German professor of botany.]

1. Style 1.5–3 mm long, filiform... **A. coccinea**
1. Style 0.5 mm long, stout..**A. latifolia**

Ammannia coccinea Rottb. [Red, in reference to the frequent reddish coloration.] VALLEY REDSTEM; SCARLET AMMANNIA.

> *Ammannia coccinea* Rottbøll, Pl. Horti Univ. Rar. Progr. 7. 1773.
> *Ammannia teres* Rafinesque, Autik. Bot. 39. 1840.

Annual herb, to 1 m; stem simple or sparsely branched above the base, glabrous. Leaves with the blade lanceolate to linear-oblong, elliptic, or spatulate, 2–8 cm long, 2–15 mm wide, the apex acute, the base narrowly cuneate, the margin entire, sessile. Flowers (1)3–7(14) in a pedunculate cyme, the peduncle (1)3–9 mm long, the pedicel 1–2 mm long; hypanthium urceolate to campanulate, 3–5 mm long; sepals 4(5), deltate; petals 4(5), ca. 2 mm long, deep rose-purple; stamens 4(7), alternating with thickened epicalyx appendages, the appendages equaling the sepal lobes in length and protruding outward; style 1.5–3 mm long, as long or longer than the ovary, filiform. Fruit subglobose, 3–5 mm long; seeds 200–250, obovoid, concave-convex, covering the surface of the large, globose placenta.

    Marshes, wet flatwoods, and floodplain forests. Occasional; nearly throughout. Pennsylvania and New Jersey south to Florida, West to North Dakota, South Dakota, Nebraska, Kansas, and California; West Indies, Mexico, Central America, and South America; Europe, Africa, Asia, and Pacific Islands. Native to North America and tropical America. Summer–fall.

Ammannia latifolia L. [With broad leaves.] PINK REDSTEM; TOOTHCUPS.

> *Ammannia latifolia* Linnaeus, Sp. Pl. 119. 1753. *Ammannia lythrifolia* Salisbury, Prodr. Stirp. Chap.
>   Allerton 65. 1796, nom. illegit.
> *Ammannia koehnei* Britton, Bull. Torrey Bot. Club 18: 271. 1891.

Ammannia koehnei Britton var. *exauriculata* Fernald, Rhodora 38: 437, t. 449(4–5). 1936. *Ammannia teres* Rafinesque var. *exauriculata* (Fernald) Fernald, Rhodora 46: 50. 1944.

Annual herb, to 1 m; stem simple or sparsely branched from the lower part of the stem. Leaves with the blade linear-lanceolate to oblong, elliptic, or spatulate, 1.5–7(10) cm long, 4–15(21) mm wide, the apex obtuse to rounded, the base cuneate, the margin entire, the upper ones auriculate, sessile. Flowers 3–10 in a sessile or subsessile cyme, the peduncle 0–3 mm long; hypanthium urceolate, 3–4 mm long; sepals 4–6, shallowly deltate, alternating with the epicalyx appendages, the appendages equaling the sepal lobes and protruding outward; petals 4–6 or absent, ca. 1 mm long, pale pink to white; stamens 4(8); style 0.5 mm long, much shorter than the ovary, stout. Fruit subglobose, 4–6 mm long; seeds 200–250, obovoid, concave-convex, covering the surface of the large, globose placenta.

Brackish and freshwater marshes. Frequent; nearly throughout. New Jersey south to Florida, west to Texas; West Indies, Mexico, Central America, and South America. Summer–fall.

Cuphea P. Browne 1756. WAXWEED

Herbs, sometimes suffrutescent. Leaves opposite or whorled, sessile or petiolate, estipulate. Flowers in leafy axillary or terminal, bracteate racemes, zygomorphic; sepals 6, alternating with epicalyx appendages; petals 6; stamens 6 or 11; ovary 2-carpellate and -loculate, with a nectariferous disk at the base. Fruit a dry capsule, dehiscing by a dorsal longitudinal slit.

A genus of about 260 species; North America, West Indies, Mexico, Central America, South America, Australia, and Pacific Islands. [*Cofea* or *cuffia*, headdress, apparently in reference to the gibbous base of the floral tube resembling a lady's voluminous headwear.]

Parsonsia P. Browne, nom. rej. 1756.

1. Leaves (at least some) whorled; flowers with the pedicel 1 cm long or longer .... C. aspera
1. Leaves all opposite; flowers subsessile or with the pedicel up to 3 mm long.
    2. Plant creeping and rooting along the stems, the stems with various types of trichomes or glabrate, but not glandular-hirtellous .... C. strigulosa
    2. Plant upright, the stems with glandular-hirtellous trichomes.
        3. Leaves broadly elliptic to lanceolate, the upper surface evenly scabrid .... C. carthagenensis
        3. Leaves narrowly elliptic, the upper surface scabrid only near the margin .... C. hyssopifolia

Cuphea aspera Chapm. [Rough, in reference to its scabrid surfaces.] TROPICAL WAXWEED.

> *Cuphea aspera* Chapman, Fl. South. U.S. 135. 1860. *Parsonsia chapmanii* A. Heller, Cat. N. Amer. Pl. 5. 1898, nom. illegit. *Parsonsia lythroides* Small, Fl. S.E. U.S. 829, 1335. 1903, nom. illegit. TYPE: FLORIDA: Gulf Co.: St. Josephs, s.d., *Chapman s.n.* (holotype: NY; isotypes: NY).

Perennial herb, to 5 dm; stem minutely white strigose and purple glandular-setose, viscid on the youngest internodes. Leaves opposite below, 3- to 4-whorled at midstem and above, the blade lanceolate to linear, 1–2.5 cm long, 2–5 mm wide, the apex acute, the base cuneate, the upper and lower surfaces glabrate or sparsely strigose, the margin entire, sessile. Flowers 2–4 in a whorl in a terminal, bracted raceme, the prophylls linear, ca. 1 mm long; hypanthium 7–9 mm long, gibbous or with a short spur 1 mm long, deep purple-red dorsally, white strigose with purple glandular setae on the veins, the inner surface villous around the stamens; sepals 6, equal, with an epicalyx thickening between them, this often terminated with a bristle; petals 6, linear or narrowly spatulate, 3–5 mm long, the 2 dorsal ones slightly larger, pale purple or pink; stamens 6. Seeds 3, orbicular, 2–3 mm long.

Flatwoods. Rare; Franklin, Calhoun, and Gulf Counties. Endemic. Spring–summer.

Cuphea aspera is listed as endangered in Florida (Florida Administrative Code, Chapter 5B-40).

Cuphea carthagenensis (Jacq.) J. F. Macbr. [Of Cartagena, Colombia.] COLOMBIAN WAXWEED.

Lythrum carthagenensis Jacquin, Enum. Syst. Pl. 22. 1760. *Cuphea carthagenensis* (Jacquin) J. F. Macbride, Publ. Field Mus. Nat. Hist., Bot. Ser. 8: 124. 1930.

Annual herb, to 6 dm; stem usually much branched, hispid and setose, sometimes also puberulent. Leaves opposite, the blade broadly elliptic to lanceolate, the apex acute to obtuse, the base cuneate, the margin entire, the upper and lower surfaces hispid and setose, sometimes also puberulent, sessile or the petiole to 2 mm long. Flowers in a leafy, indistinct raceme, sometimes 1–3 on an axillary branchlet, the pedicel 1–2 mm long; prophylls linear-lanceolate, ca. 1 mm long; hypanthium 4–6 mm long, the base gibbous, ca. 1 mm long, green or purple dorsally and distally, sparsely coarse setose on the veins; sepals 6, equal, with a short, thickened epicalyx between each, this often terminated with a bristle; petals 6, linear or subspatulate, 2–3 mm long, subequal, deep purple or rose-purple; stamens 6. Seeds (4)6(9), obovate, 1–2 mm long.

Marshes, wet hammocks, and floodplain forests. Frequent; nearly throughout. North Carolina south to Florida, west to Texas; West Indies, Mexico, Central America, and South America; Australia and Pacific Islands. Native to tropical America. Summer–fall.

Cuphea hyssopifolia Kunth [With leaves like *Hyssopus* (Laminaceae).] MEXICAN HEATHER; FALSE HEATHER.

Cuphea hyssopifolia Kunth, in Humboldt et al., Nov. Gen. Sp. 6: 199. 1824. *Parsonsia hyssopifolia* (Kunth) Standley, Contr. U.S. Natl. Herb. 23: 1018. 1924.

Suffrutescent perennial herb, to 70 cm; stem much branched, hispid. Leaves opposite, the blade narrowly lanceolate to linear or oblanceolate, 0.5–3 cm long, 4–8 mm wide, the apex acute, the base cuneate, the margin entire, the upper surface glabrous, the lower surface hispid, at least on the midrib, subsessile. Flowers in a leafy raceme, solitary at a node, the pedicel 3–8 mm long; hypanthium 5–7 mm long, the base only slightly gibbous; sepals 6, alternating with a short, thickened epicalyx appendage, this often terminated with a bristle; petals 6, obovate, 3–5 mm long, subequal, purple, pink, or white; stamens 11; seeds 5–8, orbicular, ca. 1 mm long.

Disturbed sites. Rare; Orange and Brevard Counties. Escaped from cultivation. Florida; West Indies, Mexico, Central America, and South America; Pacific Islands. Native to Mexico, Central America, and South America. Summer–fall.

Cuphea strigulosa Kunth [Having straight bristlelike trichomes.] STIFFHAIR WAXWEED.

Cuphea strigulosa Kunth, in Humboldt et al., Nov. Gen. Sp. 6: 204. 1824. *Cuphea strigulosa* Kunth subsp. *opaca* Koehne, in Martius, Fl. Bras. 13(2): 257. 1877, nom. inadmiss.

Suffrutescent perennial herb, to 1 m; stem sparsely branched, puberulent and often sparsely setose, the trichomes red, glandular-viscous. Leaves opposite, the blade elliptic, 1.5–4.5 cm long, 7–25 mm wide, the apex acute to attenuate, the base broadly cuneate, the margin entire, the upper and lower surfaces puberulent, the petiole 1–2 mm long. Flowers in a simple, leafy raceme, solitary at a node, the pedicel 1–2 mm long; prophylls linear-subulate, ca. 0.5 mm long, scarious; hypanthium 6–8 mm long, the spur ca. 0.5 mm long, purple or green dorsally, green ventrally, puberulent and sparsely glandular-setose; sepals 6, equal, alternating with thickened epicalyx appendages as long as the sepals, this often terminated with a bristle; petals 6, oblong to spatulate, 3–5 mm long, pale rose, subequal; stamens 6, the filaments sparsely villous; nectary thick, erect; seeds 6–13, suborbicular to oblong.

Pond margins. Rare; Palm Beach, Broward, and Miami-Dade Counties. Florida; West Indies, Mexico, Central America, and South America. Native to tropical America. Spring–summer.

EXCLUDED TAXA

Cuphea viscosissima Jacquin—Reported for Florida by Clewell (1985). No Florida specimens seen.

Decodon J. F. Gmel. 1791.

Shrubs. Leaves simple, opposite or whorled, petiolate, estipulate. Flowers in axillary cymose clusters, actinomorphic, bisexual; sepals 5–7, alternating with epicalyx appendages; petals (4)5(7); stamens (8)10; ovary 3(5)-carpellate and -loculate. Fruit a loculicidally dehiscent capsule.

A monotypic genus; North America. [From the Greek *deka*, ten, and *odon*, tooth, in reference to the combination of five sepals and five alternating epicalyx appendages.]

Decodon verticillatus (L.) Elliott [In whorls, in reference to the leaves and flowers.] WILLOW-HERB; SWAMP LOOSESTRIFE.

Lythrum verticillatum Linnaeus, Sp. Pl. 446. 1753. *Decodon verticillatus* (Linnaeus) Elliott, Sketch Bot. S. Carolina 1: 544. 1821. *Nesaea verticillata* (Linnaeus) Kunth, in Humboldt et al., Nov. Gen. Sp. 6: 191. 1824. *Decodon verticillatus* (Linnaeus) Elliott var. *pubescens* Torrey & A. Gray, Fl. N. Amer. 1: 483. 1840, nom. inadmiss.

Shrub, to 3 m; stem glabrous or puberulent. Leaves with the blade lanceolate, 3–20 cm long, 0.5–5 cm wide, the apex acute, the base cuneate, the margin entire, the upper surface glabrous, the lower surface glabrous or sparsely puberulent with pale brown trichomes, the petiole 3–10 mm long. Flowers 12–25 in an axillary cyme, the pedicel 5–13 mm long; hypanthium campanulate, 5–8 mm long; sepals linear-lanceolate, 1–2 mm long, the epicalyx appendages linear,

2–3 mm long, about twice as long as the alternating sepals; petals obovate or elliptic, 8–15 mm long, rose-purple; stamens inserted around or just above the ovary in 2 whorls of 3 lengths, each flower with 2 of the three lengths, the filaments glabrous. Fruit globose, surrounded by the persistent hypanthium; seeds 20–30, obpyramidal, 1–2 mm long.

Swamps. Frequent; northern counties, central peninsula. Quebec south to Florida, west to Ontario, Minnesota, Iowa, Missouri, and Texas. Summer.

Didiplis Raf. 1833.

Herbs. Leaves simple, opposite, subopposite, or whorled, petiolate or epetiolate, estipulate. Flowers solitary, axillary, actinomorphic, bisexual; hypanthium well developed; sepals 4; epicalyx appendages absent; petals absent; stamens 2–4; ovary 2-carpellate and -loculate. Fruit a dehiscent capsule.

A monotypic genus; North America. [From the Greek *dis*, twice, and *diplo*, two, in reference to the 4-merous flowers.]

Didiplis diandra (Nutt. ex DC.) A. W. Wood [From the Greek *di*, two, and *andro*, male, in reference to the two stamens.] WATERPURSLANE.

Peplis diandra Nuttall ex de Candolle, Prodr. 3: 77. 1828. *Didiplis linearis* Rafinesque, Atl. J. 177. 1833, nom. illegit. *Hypobrichia nuttallii* Torrey & A. Gray, Fl. N. Amer. 1: 480. 1840, nom. illegit. *Didiplis diandra* (Nuttall ex de Candolle) A. W. Wood, Amer. Bot. Fl. 124. 1870.

Submerged or emergent annual or short-lived perennial herb, to 4 dm; stem irregularly branched, glabrous. Leaves with the blade of the submerged ones linear, 5–33(40) mm long, 1–5(10) mm wide, the apex acute to obtuse, the base truncate, the margin entire, the upper and lower surfaces glabrous, sessile or subsessile, the aerial ones shorter, narrowly elliptic to lanceolate, the base cuneate. Flowers axillary, solitary, sessile; hypanthium short-campanulate, ca. 2 mm long, green; sepals broadly deltate, ca. 1 mm long, green. Fruit globose, thin-walled; seeds ca. 25, subspatulate, ca. 0.5 mm long, convex-concave.

Ponds and lakes. Rare; Sarasota County. Virginia south to Florida, west to Minnesota, Iowa, Kansas, Oklahoma, and Texas, also Utah. Summer.

Lagerstroemia L. 1759. CRAPEMYRTLE

Shrubs or trees. Leaves simple, alternate or subopposite, petiolate, estipulate. Flowers in terminal or axillary panicles, actinomorphic, bisexual; hypanthium well developed; sepals 6, the epicalyx appendages absent; petals 6; stamens 36–42, dimorphic, single or in clusters of 5 or 6; ovary (3)6-carpellate and -loculate. Fruit a loculicidally dehiscent capsule.

A genus of about 56 species; nearly cosmopolitan. [Commemorates Magnus von Lagerström (1696–1759), Swedish merchant, friend of Linnaeus, and supporter of Uppsala University.]

Lagerstroemia indica L. [Of India.] CRAPEMYRTLE.

Lagerstroemia indica Linnaeus, Syst. Nat., ed. 10. 1076. 1759. *Lagerstroemia pulchra* Salisbury, Prodr. Stirp. Chap. Allerton 365. 1796, nom. illegit. *Lagerstroemia chinensis* Lamarck, Encycl. 3: 375. 1789, nom. illegit. *Murtughas indica* (Linnaeus) Kuntze, Revis. Gen. Pl. 1: 249. 1891.

Shrub or tree, to 7 m; stem with the outer bark smooth, pale, whitish, flaking in thin plates, the inner bark light brown or reddish, the young branchlets glabrous or slightly puberulent. Leaves with the blade obovate or elliptic 2–7 cm long, 1.5–3 cm wide, the apex acute, the base rounded, the margin entire, the upper and lower surfaces glabrous, the petiole ca. 1 mm long. Flowers in a subpyramidal panicle 5–25 cm long, in a 2- to 3-flowered cyme, or solitary; hypanthium campanulate, 7–10 mm long, narrowing to a slender pedicel-like epipodium 2–5 mm long; sepals erect, ½ the length of the floral tube; petals 14–18 mm long, long-clawed, rose, pink, or white; stamens dimorphic, the 6 in front single, the filaments thick and the anthers green, the others in clusters of 5 or 6, the filaments thin, the anthers yellow. Fruit globose to oblong, 9–15 mm long, woody, subtended by the persistent floral tube; seeds numerous, obpyramidal, 7–9 mm long, unilaterally winged from the raphe.

Disturbed sites. Occasional; northern and central peninsula, central and western panhandle. Escaped from cultivation. Maryland south to Florida west to Indiana, Arkansas, and Texas; West Indies, Mexico, Central America, and South America; Europe, Africa, Asia, and Australia. Native to Asia. Summer.

Lythrum L. 1753. LOOSESTRIFE

Herbs. Leaves simple, opposite or alternate, sessile or petiolate or epetiolate, estipulate. Flowers in terminal spikes or racemes or axillary and solitary, paired, or in cymes, actinomorphic, bisexual, distylous; hypanthium well developed; sepals 6, alternating with the epicalyx appendages; petals 6; stamens 6, in 1 or 2 whorls; ovary 2-carpellate and -loculate, with or without a nectariferous ring at the base. Fruit a septicidal, septifragal, or irregularly dehiscent capsule.

A genus of about 35 species; North America, Europe, and Asia. [From the Greek *Lythron*, clotted blood, in reference to the use of *L. salicaria* in stopping hemorrhages.]

1. Leaves opposite throughout.
　2. Midstem leaves linear; flowering stem stiffly erect...**L. lineare**
　2. Midstem leaves elliptic to oblong; flowering stem weak, usually decumbent ....**L. flagellare**
1. Leaves mostly alternate above, opposite below.
　3. Hypanthium 5–6 mm long; calyx lobes ½ the length of the epicalyx appendages; petals 5–6 mm long..**L. alatum**
　3. Hypanthium 3–4 mm long; calyx lobes subequaling the epicalyx appendages; petals ca. 3 mm long...**L. curtissii**

Lythrum alatum Pursh [*Alatus*, winged, in reference to the sometimes winged upper stems.] WINGED LOOSESTRIFE.

Erect perennial herb, to 1.5 m; stem winged, angled, or terete, glabrous. Leaves opposite or subopposite, the blade ovate to oblong or lanceolate to linear or elliptic, 1–7.5 cm long, 2–14 mm wide, the apex acute, the base cuneate or subcordate to rounded, the margin entire, the upper and lower surfaces glabrous, sessile. Flowers in a terminal raceme, alternate, solitary at a node, orientated nearly parallel to the stem, the pedicel 1–3 mm long; prophylls ca. 1 mm long; hypanthium cylindric, 3–7 mm long; sepals deltoid or ovate-acuminate, ca. 1 mm long, the epicalyx appendages ca. twice the sepal length; petals obovate or oblong, 2–7 mm long, purple; nectariferous ring present at the base of the ovary. Seeds numerous, elongate or ovoid, convex on one side.

1. Lower leaves ovate to oblong, the base subcordate to rounded .... var. **alatum**
1. Lower leaves lanceolate to linear or elliptic, the base cuneate .... var. **lanceolatum**

Lythrum alatum Pursh var. **alatum**

> *Lythrum alatum* Pursh, Fl. Amer. Sept. 334. 1814. *Salicaria alata* (Pursh) Lunell, Amer. Midl. Naturalist 4: 480. 1916.
>
> *Lythrum cordifolium* Nieuland, Amer. Midl. Naturalist 3: 265. 1914; non Sessé y Lacasta & Moçiño, 1888. TYPE: FLORIDA/GEORGIA/NORTH CAROLINA: s.d., *Buckley s.n.* (holotype: US; isotypes: GH, NY).

Stem slender, usually less than 1 m, often more than 1 stem arising from an enlarged rootstock. Lower leaves ovate to oblong, the base subcordate to rounded.

Marshes. Rare; Citrus County. Not recently collected. Maine and Ontario south to Florida, west to Montana, Wyoming, Colorado, Oklahoma, and Texas; Mexico. Native to North America. Summer.

Lythrum alatum Pursh var. **lanceolatum** (Elliott) Torr. & A. Gray ex Rothr. [Lance-shaped, in reference to the leaves.]

> *Lythrum lanceolatum* Elliott, Sketch Bot. S. Carolina 1: 544. 1821. *Lythrum alatum* Pursh var. *lanceolatum* (Elliott) Torrey & A. Gray ex Rothrock, Rep. U.S. Geogr. Surv., Wheeler 120. 1879.

Stem robust, often 1 m or more, usually solitary. Lower leaves lanceolate to linear or elliptic, the base cuneate.

Flatwoods, marshes, swamps, and lake margins. Frequent; peninsula west to central panhandle. Virginia south to Florida, west to Oklahoma and Texas; West Indies. Native to North America. Summer.

Lythrum curtissii Fernald [Commemorates Allen Hiram Curtiss (1845–1907), botanist for U.S. Department of Agriculture.] CURTISS' LOOSESTRIFE.

> *Lythrum curtissii* Fernald, Bot. Gaz. 33: 155. 1902. SYNTYPE: FLORIDA: Gadsden Co.: Aspalaga, Oct 1897, *Chapman 6170*.

Erect perennial herb, to 1 m; stem glabrous. Leaves opposite at the lower nodes, alternate at midstem and above, the blade lanceolate or oblong, 2–7.5 cm long, 5–17 mm wide, abruptly reduced on the branches, oblong to narrowly oblong, 3–15 mm long, 2–3 mm wide, the apex acute, the base cuneate, the margin entire, the upper and lower surfaces glabrous, sessile or subsessile. Flowers in a raceme, solitary or paired at a node, alternate, outward orientated, the pedicel 1–2 mm long; prophylls linear, 1 mm long; hypanthium obconic, 3–6 mm long; sepals broadly deltoid, ca. 0.5 mm long, the epicalyx appendages equaling the sepals; petals ca. 2 mm long, deep to pale purple with a dark central vein; nectariferous ring present at the base of the ovary. Seeds numerous, elongate or ovoid, convex on one side.

Swamps, seepage areas, and along rivers. Rare; St. Johns and Putnam Counties, central panhandle. Georgia and Florida. Summer.

Lythrum curtissii is listed as endangered in Florida (Florida Administrative Code, Chapter 5B-40).

Lythrum flagellare Shuttlew. ex Chapm. [Whip-like, in reference to the creeping branches.] FLORIDA LOOSESTRIFE; LOWLAND LOOSESTRIFE.

Lythrum flagellare Shuttleworth ex Chapman, Fl. South. U.S., ed. 2. 620. 1883. TYPE: FLORIDA: Sarasota Co.: Sarasota, s.d., *Garber s.n.*

Decumbent or trailing perennial herb, to 4 dm; stem glabrous. Leaves opposite, the blade oblong, 5–13 mm long, 2–6 mm wide, the apex obtuse to rounded, the base rounded, the margin entire, the upper and lower surfaces glabrous, the petiole to 0.5 mm. Flowers few in a raceme, solitary at a node, often outwardly orientated, the pedicel 1 mm long; prophylls linear, 0.5 mm long; hypanthium slender obconic, 4–5 mm long; sepals deltoid, ca. 0.5 mm long, the epicalyx appendages twice as long as the sepals; petals 3–4 mm long, pale purple to purple; nectariferous ring surrounding the ovary base. Seeds numerous, elongate or ovoid, convex on one side.

Wet prairies. Occasional; central peninsula, Collier County. Endemic. Spring.

Lythrum flagellare is listed as endangered in Florida (Florida Administrative Code, Chapter 5B-40).

Lythrum lineare L. [Linear, in reference to the leaves.] WAND LOOSETRIFE.

Lythrum lineare Linnaeus, Sp. Pl. 447. 1753.

Erect perennial, to 1.5 m; stem glabrous. Leaves opposite or subopposite, the blade narrowly linear to narrowly lanceolate, 5–40 mm long, 1–4 mm wide, those of the upper stems and in the inflorescence 5–10 mm long, the apex acute, the base cuneate, the margin entire, the upper and lower surfaces glabrous, sessile. Flowers in a leafy raceme, solitary at a node, erect to suberect, the pedicel ca. 1 mm long; prophylls to 1 mm long; hypanthium obconic to cylindric, 3–5 mm long; sepals deltoid, ca. 0.5 mm long, the epicalyx appendages equaling or slightly longer than the sepals; petals 3–4 mm long, pale purple or white; nectariferous ring absent. Seeds numerous, elongate or ovoid, convex on one side.

Brackish marshes. Occasional; nearly throughout. New York south to Florida, west to Texas. Summer.

EXCLUDED TAXON

Lythrum vulneraria Aiton ex Schran—Reported by Small (1903), the name of this Mexican species misapplied to *L. flagellare*.

Rotala L. 1771.

Annual or perennial herbs. Leaves opposite or whorled, sessile or subsessile, estipulate. Flowers in racemes, bracteate or subtended by leaves, actinomorphic; prophylls foliose; hypanthium present; sepals 4, alternating with the epicalyx appendages or the epicalyx absent; petals 4 or absent; stamens 4; ovary 2- to 4-carpellate and -loculate. Fruit a septicidally dehiscent capsule.

A genus of about 46 species; nearly cosmopolitan. [*Rota*, wheel, and the diminutive suffix *-al*, in reference to the whorled leaves of the original species, *R. verticillatus.*]

1. Leaves linear to oblanceolate .... **R. ramosior**
1. Leaves broadly elliptic to orbicular .... **R. rotundifolia**

Rotala ramosior (L.) Koehne [Much branched.] LOWLAND ROTALA; TOOTHCUP.

> *Ammannia ramosior* Linnaeus, Sp. Pl. 120. 1753. *Ammannia ramosa* Hill, Veg. Syst. 11: 14. 1769, nom. illegit. *Ammannia purpurea* Lamarck, Encycl. 1: 131. 1783, nom. illegit. *Rotala ramosior* (Linnaeus) Koehne, in Martius, Fl. Bras. 13(2): 194. 1877. *Rotala ramosior* (Linnaeus) Koehne var. *typica* Fernald & Griscom, Rhodora 37: 169. 1935, nom. inadmiss.
>
> *Ammannia humilis* Michaux, Fl. Bor.-Amer. 1: 199. 1803. *Boykinia humilis* (Michaux) Rafinesque, Autik Bot. 9: 1840.
>
> *Ammannia occidentalis* (Sprengel) de Candolle var. *pygmaea* Chapman, Fl. South. U.S. 134. 1860. TYPE: FLORIDA: Monroe Co.: Key West, s.d., *Blodgett s.n.* (holotype: NY?).
>
> *Rotala ramosior* (Linnaeus) Koehne var. *interior* Fernald & Griscom, Rhodora 37: 169, t. 345(1–2). 1935.

Erect annual herb, to 4 dm; stem simple or branched, glabrous. Leaves opposite, the blade oblong-elliptic to oblanceolate, 1–5 cm long, 2–12 mm wide, the apex acute, the base cuneate, the margin entire, membranous, the upper and lower surfaces glabrous, sessile or subsessile. Flowers in a raceme, solitary at a node, sessile or subsessile; prophylls ½ as long as the flower; hypanthium campanulate, 2–4 mm long; sepals 4, less than 1 mm long, the epicalyx segments shorter or longer than the sepals; petals 4 or absent, if present, then less than 1 mm long, white or pink, caducous; stamens inserted near the base of the hypanthium. Fruit globose, ca. 2 mm long, 4-valved; seeds numerous, ovoid, ca. 0.5 mm long.

Marshes and swamps. Frequent; nearly throughout. New Hampshire and Ontario south to Florida, west to British Columbia, Washington, Oregon, and California; West Indies, Mexico, Central America, and South America; Europe and Asia. Native to North America and tropical America. Summer–fall.

Rotala rotundifolia (Buch-Ham. ex Roxb.) Koehne [With rounded leaves.] DWARF ROTALA; ROUNDLEAF TOOTHCUP.

> *Ammannia rotundifolia* Buchanan-Hamilton ex Roxburgh, Fl. Ind. 1: 446. 1820. *Ameletia rotundifolia* (Buchanan-Hamilton ex Roxburgh) Dalzell & A. Gibson, Bombay Fl. 96. 1861. *Rotala rotundifolia* (Buchanan-Hamilton ex Roxburgh) Koehne, Bot. Jahrb. Syst. 1: 175. 1880.

Creeping perennial herb, to 4 dm; stem branched below, simple above, young growth 4-angled, glabrous. Leaves opposite or occasionally 3-whorled, the blade of the aerial ones obovate to orbicular, the apex rounded, the base rounded to subcordate, the margin entire, the upper and lower surfaces glabrous, the blade of the submersed ones linear to orbicular, the lower surface often red or purple, sessile or subsessile. Flowers in a terminal raceme, each flower solitary at a node, bracteate; prophylls scarious, reaching the sepals; hypanthium campanulate, ca. 2 m long; sepals ca. 1 mm long, the epicalyx appendages absent; petals 4, obovate, 1–2 mm long, rose; stamens inserted near the base of the hypanthium. Fruit globose, ca. 2 mm long, 4-valved; seeds numerous, ovoid, ca. 0.5 mm long.

Canals. Rare; Lee and Palm Beach Counties, southern peninsula. Escaped from cultivation. Alabama and Florida; Asia and Australia. Native to Asia. All year.

Rotala rotundifolia is listed as a Category II invasive species in Florida by the Florida Exotic Pest Plant Council (FLEPPC, 2015).

EXCLUDED GENUS

Punica granatum Linnaeus—Reported by Small (1903, 1913a, 1913b, 1913c, 1913d, 1913e, 1933). All Florida material seen is from plants persistent from cultivation.

ONAGRACEAE Juss., nom. cons. 1798. EVENINGPRIMROSE FAMILY

Herbs, sometimes suffruticose. Leaves alternate or opposite, simple, pinnate-veined, petiolate or epetiolate, stipulate or estipulate. Flowers axillary, spicate to subcapitate racemes, actinomorphic or zygomorphic, bisexual; hypanthium well developed; sepals 4–5(6), free; petals 4–6(6), free; stamens 4–10(12); ovary 4- or 5-carpellate and -loculate, surrounded by the adnate hypanthium. Fruit a loculicidally or poricidally dehiscent capsule or indehiscent and nutlike.

A family of about 20 genera and about 650 species; nearly cosmopolitan.

Epilobiaceae Vent. (1799).

Selected reference: Wagner et al. (2007).

1. Hypanthium not prolonged beyond the ovary; sepals persistent after anthesis......................**Ludwigia**
1. Hypanthium prolonged beyond the ovary; sepals deciduous after anthesis..........................**Oenothera**

Ludwigia L. 1753. PRIMROSEWILLOW

Herbs or shrubs. Leaves alternate or opposite, simple, pinnate-veined, stipulate. Flowers solitary, clustered, or in spikes or panicles, bracteolate or ebracteolate; hypanthium not prolonged beyond the ovary; sepals 3–7, persistent after anthesis; petals as many as the sepals or absent; stamens as many or twice as many as the sepals, the anthers versatile or basifixed; nectaries present at the base of the epipetalous stamens or absent; carpels and locules as many as the sepals or rarely more. Fruit a capsule irregularly dehiscent by terminal pores or flaps separating from the valve-like top.

A genus of about 80 species; nearly cosmopolitan. [Commemorates Christian Gottlieb Ludwig (1709–1773), German botanist.]

Natural hybridization is fairly common among members of *Ludwigia* section *Isnardia* (L.) W. L. Wagner & Hoch and section *Microcarpium* Munz (=section *Dantia* (DC.) Munz), which occur primarily in the Atlantic and Gulf Coastal Plain of the United States (see Peng, 1989; Peng et al., 2005; Wagner et al., 2007). Material from Florida seen by us includes hybrids of *L. alata* × *L. pilosa*, *L. alata* × *L. suffruticosa*, *L. arcuata* × *L. repens*, *L. curtissii* × *L. linifolia*, and *L. pilosa* × *L. sphaerocarpa*.

Isnardia L. (1753); *Jussiaea* L. (1753); *Ludwigiantha* (Torr. & A. Gray) Small (1897).

Selected references: Peng (1989); Peng et al. (2005); Ramamoorthy and Zardini (1987); Raven (1963).

1. Leaves opposite.
    2. Flowers or fruits pedunculate.
        3. Pedicel of the fruit as long as or shorter than the subtending leaves; petals 4–5 mm long .... **L. brevipes**
        3. Pedicel of the fruit usually much longer than the subtending leaves; petals 7–11 mm long .... **L. arcuata**
    2. Flowers or fruits sessile.
        4. Plant densely strigulose .... **L. spathulata**
        4. Plant glabrous.
            5. Capsules with 4 green, rugose, callous-like, longitudinal bands .... **L. palustris**
            5. Capsules lacking 4 green, rugose, callous-like, longitudinal bands .... **L. repens**
1. Leaves alternate.
    6. Stamens 8–10, twice as many as the sepals, in 2 series.
        7. Sepals 5, rarely 6 or 7.
            8. Stem erect; peduncles much shorter than the hypanthium.
                9. Capsule narrowly cylindric, 3–5 cm long .... **L. leptocarpa**
                9. Capsule obconic, quadrangular, 1–3 cm long .... **L. peruviana**
            8. Stem decumbent or creeping and rooting at the nodes; peduncles much longer than the hypanthium.
                10. Hypanthium equaling or shorter than the sepals; bracteoles at the hypanthium base lanceolate .... **L. grandiflora**
                10. Hypanthium longer than the sepals; bracteoles at the hypanthium base deltoid or ovate .... **L. peploides**
        7. Sepals 4.
            11. Stem conspicuously 4-winged on the angles.
                12. Leaves lanceolate to ovate-lanceolate or elliptic .... **L. decurrens**
                12. Leaves linear-lanceolate .... **L. longifolia**
            11. Stem not 4-winged or only very faintly winged on the angles.
                13. Stem shaggy-pubescent .... **L. peruviana**
                13. Stem glabrate.
                    14. Fruit more than twice as long as the sepals, with a distinct ridge continuous with the sepal midvein .... **L. octovalvis**

      14. Fruit subequal to slightly longer than the sepals, without a ridge continuous with the sepal midvein .... **L. peruviana**

        15. Calyx lobes 3–4 mm long; petals 4–5 mm long .... **L. erecta**

        15. Calyx lobes (5)10–15 mm long; petals (10)15–30 mm long.

          16. Hypanthium much longer than the sepals; pedicel shorter than the hypanthium ....**L. octovalvis**

          16. Hypanthium subequaling or not much longer than the sepals; pedicel exceeding the hypanthium ....**L. bonariensis**

6. Stamens 4–5, equaling the sepals in number, in 1 series.

  17. Capsule much longer than wide.

    18. Petals absent; capsules longitudinally grooved below each calyx lobe; leaves lanceolate ....**L. glandulosa**

    18. Petals present; capsules not longitudinally grooved; leaves linear.

      19. Calyx lobes 5–6 mm long, ca. ⅔ as long as or longer than the capsule; capsule subcylindric .... **L. linifolia**

      19. Calyx lobes 2–3 mm long, much shorter than the capsule; capsule quadrangular ....**L. linearis**

  17. Capsule little if any longer than wide.

    20. Flowers and capsules distinctly pedicellate.

      21. Leaf base cuneate-attenuate .... **L. alternifolia**

      21. Leaf base rounded or obtuse.

        22. Style 7–10 mm long, longer than the sepals; lower leaves glabrous..............**L. virgata**

        22. Style to 3 mm long, shorter than the sepals; lower leaves pubescent.

          23. Calyx lobes reflexed after anthesis; trichomes short, spreading or curly ....**L. maritima**

          23. Calyx lobes erect or spreading after anthesis; trichomes long, spreading............... ....**L. hirtella**

    20. Flowers and capsules sessile or subsessile.

      24. Flowers in terminal congested spikes ....**L. suffruticosa**

      24. Flowers axillary, in elongate, interrupted spikes or racemes.

        25. Plant hirtellous throughout.

          26. Sepals creamy white within, usually tinged with pink along the midvein and the edge, the apex elongate-acuminate to subcuspidate, reflexed.................... **L. pilosa**

          26. Sepals greenish within, the apex acuminate, ascending........................... **L. ravenii**

        25. Plant glabrous or glabrate.

          27. Capsule 4-angled or narrowly winged.

            28. Sepals subequaling the capsule, creamy white; stem usually distinctly ridged or winged... **L. alata**

            28. Sepals ½ as long as the capsule, greenish; stem nearly smooth or only slightly winged..**L. lanceolata**

          27. Capsules not 4-angled or winged.

            29. Fruit 1–1.5 mm long; seeds reddish brown ....**L. microcarpa**

            29. Fruit 1.5–7 mm long; seeds light tan or yellowish.

              30. Cauline leaves lanceolate or oblong-elliptic........................**L. sphaerocarpa**

              30. Cauline leaves obovate-spatulate to oblanceolate ....**L. curtissii**

Ludwigia alata Elliott [Winged, in reference to the stems.] PRIMROSEWILLOW.

> *Ludwigia alata* Elliott, Sketch Bot. S. Carolina 1: 212. 1817. *Isnardia alata* (Elliott) de Candolle, Prodr. 3: 61. 1828.

Erect or somewhat sprawling stoloniferous herb, to 12(16) m, stem slightly to distinctly winged, glabrous. Leaves alternate, those of the stolons with the blade orbicular, oblanceolate to elliptic, 4–26 mm long, 4–15 mm wide, the apex acute to rounded or retuse, the base tapering into a petiole 2–10 mm long, the margin entire, the upper and lower surfaces glabrous, the cauline leaves with the blade lance-elliptic to elliptic to linear, 1.8–10 cm long, 2–12 mm wide, the apex acute, the base narrowly cuneate, the margin with remote but pronounced hydathodal glands, the upper and lower surfaces glabrous or occasionally minutely papillose-serrulate, sessile or the petiole to 3 mm long; stipules minute. Flowers solitary or several in the leaf axil; bracteoles 4, lance-elliptic, 3–5 mm long; sepals 4, ovate-deltate, 2–4 mm long, the inner surface creamy white, the outer surface pale green, the margin smooth or minutely papillose-serrulate; petals absent; stamens 4, shorter than the sepals; nectary disc 4-lobed. Fruit obpyramidal, 3–5 mm long, winged on the angles, glabrous; seeds ellipsoid, light brown.

Cypress swamps and marshes. Frequent; nearly throughout. Virginia south to Florida, west to Louisiana; West Indies. Spring–fall.

Ludwigia alternifolia L. [With alternate leaves.] SEEDBOX.

> *Ludwigia alternifolia* Linnaeus, Sp. Pl. 118. 1753. *Ludwigia macrocarpa* Michaux, Fl. Bor.-Amer. 1: 89. 1803, nom. illegit. *Isnardia alternifolia* (Linnaeus) de Candolle, Prodr. 3: 60. 1828. *Ludwigia alternifolia* Linnaeus var. *typica* Munz, Bull. Torrey Bot. Club 71: 158. 1944, nom. inadmiss.

Erect perennial herb, to 12 dm; stem branched above the base, subglabrous or strigulose. Leaves alternate, the blade of the midcauline ones elliptic-lanceolate, 4–8(12) cm long, 8–15(24) mm wide, the apex acute to acuminate, the base cuneate, the margin entire, the petiole 3–7(10) mm long; stipules minute. Flowers solitary in the leaf axil; bracteoles 1–3 mm long; sepals 4, ovate, 7–10 mm long; petals 8–10 mm long; stamens 4; nectary disk elevated, with a depressed white-ciliate nectary surrounding the base of each epipetalous stamen. Fruit subglobose-cubical with a rounded base, 5–6 mm long, slightly wing-angled, dehiscent by a terminal pore, the pedicel 3–5 mm long; seeds asymmetric-oblong, pale brown.

Swamps and bogs. Occasional; panhandle. Quebec south to Florida, west to Ontario, Iowa, Nebraska, Colorado, and Texas. Spring–fall.

Ludwigia arcuata Walter [Curved like a bow, in reference to the creeping stems.] PIEDMONT PRIMROSEWILLOW.

> *Ludwigia arcuata* Walter, Fl. Carol. 89. 1788. *Isnardia arcuata* (Walter) Kuntze, Revis. Gen. Pl. 1: 251. 1891. *Ludwigiantha arcuata* (Walter) Small, Bull. Torrey Bot. Club 24: 178. 1897.

Creeping, prostrate perennial herb; stem to 3 dm, rooting at the nodes, glabrous. Leaves opposite, the blade linear to oblanceolate, 7–25 mm long, 1–4 mm wide, the apex obtuse to acute, the base cuneate, the margin entire, the upper and lower surfaces glabrous, sessile or subsessile; stipules minute. Flowers solitary in the leaf axil, the pedicel 1.5–3.5 cm long; longer than the subtending leaves; bracteoles 2, small; hypanthium turbinate, 4–6 mm long; sepals 4, linear-lanceolate, 5–10 mm long; petals 4, obovate, 6–8 mm long, longer than the sepals, yellow; stamens 4, 2–3 mm long. Fruit clavate, 7–10 mm long; seeds ellipsoid, pale brown.

Marshes and lake and pond margins. Frequent; nearly throughout. South Carolina south to Florida, west to Alabama. Summer.

Ludwigia bonariensis (Micheli) H. Hara [Of Buenos Aires, Argentina.] CAROLINA PRIMROSEWILLOW.

> *Jussiaea bonariensis* Micheli, Flora 57: 303. 1874. *Jussiaea suffruticosa* Linnaeus var. *bonariensis* (Micheli) Léveillé, Bull. Soc. Bot. France 54: 427. 1907. *Ludwigia bonariensis* (Micheli) H. Hara, J. Jap. Bot. 28: 291. 1953.
>
> *Jussiaea neglecta* Small, Man. S.E. Fl. 945, 1506. 1933. TYPE: FLORIDA: Escambia Co.: along the Escambia River near Pensacola, Aug 1930, *Small s.n.* (holotype: NY).

Erect perennial herb, to 1 m; stem pubescent. Leaves alternate, the blade elliptic-oblanceolate or -lanceolate to -linear or narrowly linear, 4–15 cm long, 5–10 mm wide, the apex acute to acuminate, the base cuneate, the margin entire, the upper and lower surfaces pubescent, the petiole to 10 cm long. Flowers solitary in the leaf axil; hypanthium obconic, shorter than the calyx in bud, densely pubescent; sepals 4, broadly ovate, ca. 1.5 cm long; petals 4, obovate, 1.5–3 cm long, yellow; stamens 8. Fruit obconic, 2–3.5 cm long, sparsely hirsute; seeds suborbicular, brown.

Pond margins. Rare; Santa Rosa and Escambia Counties. North Carolina, South Carolina, Florida, and Alabama; South America. Native to South America. Summer.

Ludwigia brevipes (B. H. Long) Eames [Short-stalked.] LONG BEACH PRIMROSEWILLOW.

> *Ludwigiantha brevipes* B. H. Long, in Britton & A. Brown, Ill. Fl. N. U.S., ed. 2. 2: 586. 1913. *Ludwigia brevipes* (B. H. Long) Eames, Rhodora 35: 228. 1933.

Creeping herb, to 3 dm; stem rooting at the nodes, glabrous. Leaves opposite, the blade oblong-lanceolate, 1–2.5 cm long, the apex acute, the base cuneate, sessile or tapering to a short, narrow-margined petiole, the margin entire, the upper and lower surfaces glabrous, sessile. Flowers solitary in the leaf axil, on a slender pedicel 5–12 mm long; bracteoles 2, small; hypanthium turbinate, 4–6 mm long, 4-angled; sepals 4, lanceolate or ovate-lanceolate, 4–6 mm long, subequaling the hypanthium; petals 4, broadly elliptic, 4–6 mm long, equaling the sepals, yellow; stamens 4, 2–3 mm long. Fruit clavate, 8–10 mm long; seeds ellipsoid, light brown.

Marshes and lake and pond margins. Rare; Escambia County. New Jersey south to Florida. Spring–fall.

Ludwigia curtissii Chapm. [Commemorates Allen Hiram Curtiss (1845–1907), botanist for U.S. Department of Agriculture.] CURTISS' PRIMROSEWILLOW.

> *Ludwigia curtissii* Chapman, Fl. South. U.S., ed 2, suppl. 681. 1892. TYPE: FLORIDA: Brevard Co.: Cape Malabar, Jul 1879, *Curtiss 922* (holotype: US; isotypes: BM, F, GH, M, MO, NY, PH, US).

Ludwigia simpsonii Chapman, Fl. South. U.S., ed 2, suppl. 682. 1892. *Ludwigia curtissii* Chapman var. *simpsonii* (Chapman) D. B. Ward, Novon 11: 362. 2001. TYPE: FLORIDA: Manatee Co.: Manatee, s.d., *Simpson s.n.* (holotype: US; isotypes: GH, MO, US).

Ludwigia spathulifolia Small, Man. S.E. Fl. 943, 1506. 1933. TYPE: FLORIDA: Miami-Dade Co.: Everglades NW of Perrine, 16 Jan 1909, *Small & Carter 2990* (holotype: NY).

Erect or ascending, rarely creeping perennial herb, to 7.5 dm; stem solitary or branched, glabrous. Leaves alternate, the blade oblanceolate-spatulate or linear, 10–25(30) mm long, 2–8 mm wide, the apex acute or mucronate, the base attenuate into a winged petiole 3–12 mm long, the margin subentire with hydathodal glands, the upper and lower surfaces glabrous; stipules narrowly ovate, reddish purple. Flowers solitary or few in the leaf axil; sepals 4, deltate or triangular, 2–3 mm long, green with a whitish base; petals absent or 1–3 narrowly elliptic, 1–3 mm long, yellow; stamens 4, ca. 1 mm long; nectary disk green, prominently 4-lobed, glabrous. Fruit obconical, (2)2–4(5) mm long, glabrous or occasionally remotely minutely puberulent, the pedicel to 0.5 mm long, subtended by persistent lanceolate bracteoles 2–4 mm long; seeds ellipsoid, light brown.

Marshes, swamps, and pond margins. Frequent; peninsula, west to central panhandle. Florida and Mississippi; West Indies. Spring–fall.

Ludwigia simpsonii is sometimes treated as a distinct species. It frequently occurs in the same locality as *L. curtissii* and overlaps morphologically with it. Peng (1989) recognizes them as distinct and separates them on the basis of chromosome number, which appears to be correlated with mature capsule size ($n=24$, capsules 1.5–2(2.5) mm long in *L. simpsonii* vs. $n=32$, capsules (2)2.5–4(4.7) mm long in *L. curtissii*).

Ludwigia decurrens Walter [Running downward, in reference to the winged petiole.] WINGLEAF PRIMROSEWILLOW.

Ludwigia decurrens Walter, Fl. Carol. 89. 1788. *Jussiaea decurrens* (Walter) de Candolle, Prodr. 3: 56. 1828. *Diplandra decurrens* (Walter) Rafinesque, Autik. Bot. 35. 1840.

Jussiaea tenuifolia Nuttall, Amer. J. Sci. Arts 5: 294. 1822. TYPE: FLORIDA: s.d., *Ware s.n.* (holotype: PH).

Diplandra compressa Rafinesque, Autik. Bot. 36. 1840. TYPE: FLORIDA.

Diplandra ovata Rafinesque, Autik. Bot. 36. 1840. TYPE: "Florida to Louisiana."

Diplandra pumila Rafinesque, Autik. Bot. 36. 1840. TYPE: FLORIDA.

Erect perennial herb, to 2 m; stem much branched, subglabrous, 4-winged from the decurrent leaf bases, the wings 1–2 mm wide. Leaves alternate, the blade lanceolate to elliptical, 2–12 cm long, 0.2–3.5 cm wide, the apex acute to acuminate, the base narrowly cuneate, the margin entire, the upper and lower surfaces subglabrous, subsessile. Flowers solitary in the axil of the upper leaves; bracteoles ca. 1 mm long; sepals 4, lance-ovate, 7–10 mm long, glabrous or minutely puberulent; petals obovate, 8–12 mm long, yellow; stamens 8, the filaments 2–3 mm long, the epipetalous ones shorter; nectary disk not elevated, with a sunken white-ciliate nectary surrounding the base of each epipetalous stamen. Fruit clavate, 1–2 cm long, pale brown with darker ribs, sharply 4-angled, irregularly loculicidal, puberulent or glabrous, subsessile or with a pedicel to 1 cm long; seeds elongate-obovoid, pale brown.

Swamps and marshes. Occasional; nearly throughout. Pennsylvania south to Florida, west to Wisconsin, Kansas, Oklahoma, and Texas; West Indies, Mexico, Central America, and South America; Europe, Africa, and Asia. Native to North America and tropical America. Spring–fall.

Ludwigia erecta (L.) H. Hara [Upright.] YERBA DE JICOTEA.

Jussiaea erecta Linnaeus, Sp. Pl. 388. 1753. *Ludwigia erecta* (Linnaeus) H. Hara, J. Jap. Bot. 28: 292. 1953.

Jussiaea acuminata Swartz, Fl. Ind. Occid. 2: 745. 1800. *Ludwigia acuminata* (Swartz) M. Gómez de la Maza y Jiménez, Anales Soc. Esp. Hist. Nat. 23: 66. 1894.

Erect annual herb, to 3 m; stem branched or unbranched, angled and often winged from decurrent leaf bases, glabrous. Leaves alternate, the blade elliptic or narrowly lanceolate, 2–20 cm long, 0.2–4 cm wide, the apex acute or acuminate, the base cuneate, decurrent, the margin entire, the upper surface glabrous or minutely scabrous along the margin, the lower surface glabrous or sometimes minutely strigulose along the veins, the petiole 2–22 mm long, glabrous or sometimes minutely strigulose; stipules deltoid, ca. 2 mm long. Flowers solitary in the axil of the upper leaves, subsessile; bracteoles scalelike, minute; hypanthium obconic, 4–10 mm long, 4-angled, glabrous or rarely strigulose; sepals 4, ovate or lanceolate, 3–6 cm long, glabrous or strigulose; petals obovate, 3.5–5 mm long, yellow; stamens 8, subequal, the filaments ca. 1.5 mm long. Fruit 1–2.2 cm long, 4-angled, glabrous or rarely strigulose, subsessile; seeds elongate-ovoid, pale brown.

Marshes and swamps. Frequent; nearly throughout. Florida and Mississippi; West Indies, Mexico, Central America, and South America; Africa and Asia. Native to North America and tropical America. Spring–fall.

Ludwigia glandulosa Walter [Glandular.] CYLINDRICFRUIT PRIMROSEWILLOW.

Ludwigia glandulosa Walter, Fl. Carol. 88. 1788. *Ludwigia glandulosa* Walter var. *typica* Munz, Bull. Torrey Bot. Club 71: 164. 1944, nom. inadmiss.

Ludwigia cylindrica Elliott, Sketch Bot. S. Carolina 1: 213. 1817. *Isnardia cylindrica* (Elliott) de Candolle, Prodr. 3: 61. 1828.

Erect, stoloniferous perennial herb, to 8(10) dm; stem usually well branched, glabrous to sparingly strigulose on the ridges formed by the decurrent leaf bases. Leaves alternate, the blade of the main cauline ones elliptic, 3.5–12 cm long, 4–20 mm wide, the apex acute, the base cuneate, the margin densely fringed with minute papillose-strigose trichomes and with the hydathodal glands often visible, the upper surface glabrous, the lower surface minutely strigulose on the veins, the petiole 1–15 mm long, the leaves of the branches smaller (1–4.5 cm long, 3–10 mm wide); stipules ovate-triangular, minute. Flowers numerous and congested in the leaf axil; hypanthium subcylindric, 4–5 mm long, 4-angled, glabrous to minutely papillose-strigulose; sepals 4, ovate-deltoid, ca. 2 mm long, the margin fringed with minute strigulose trichomes; petals absent; anthers ca. 1 mm long; nectary disk raised, 4-lobed, greenish, glabrous. Fruit subcylindric, (4)–7(9) mm long, with 4 shallow grooves, glabrous or minutely papillose-strigulose, sessile or subsessile; seeds kidney-shaped, light brown.

Marshes and swamps. Occasional; panhandle, Nassau County. Maryland south to Florida, west to Kansas, Oklahoma, and Texas. Summer–fall.

Ludwigia grandiflora (Michx.) Greuter & Burdet [Large-flowered.] LARGEFLOWER PRIMROSEWILLOW.

Jussiaea grandiflora Michaux, Fl. Bor.-Amer. 1: 267. 1803. *Jussiaea repens* Linnaeus var. *grandiflora* (Michaux) Micheli, in Martius, Fl. Bras. 13(2): 167. 1875. *Ludwigia clavellina* M. Gómez de la Maza y Jiménez var. *grandiflora* (Michaux) M. Gómez de la Maza y Jiménez, Anales Soc. Esp. Hist. Nat. 23: 66. 1894. *Jussiaea repens* Linnaeus subsp. *grandiflora* (Michaux) P. Fournier, Quatre Fl. France 603. 1937. *Jussiaea michauxiana* Fernald, Rhodora 46: 198. 1944, nom. illegit. *Ludwigia grandiflora* (Michaux) Greuter & Burdet, Willdenowia 16: 448. 1987.

Jussiaea uruguayensis Cambessedes, in A. Saint-Hilaire, Fl. Bras. Merid. 2: 264. 1829. *Jussiaea repens* Linnaeus var. *uruguayensis* (Cambessedes) Hassler, Repert. Spec. Nov. Regni Veg. 12: 276. 1913. *Jussiaea uruguayensis* Cambessedes var. *genuina* Munz, Darwiniana 4: 268. 1942, nom. inadmiss. *Ludwigia uruguayensis* (Cambessedes) H. Hara, J. Jap. Bot. 28: 294. 1953.

Jussiaea hexapetala Hooker & Arnott, in Hooker, Bot. Misc. 3: 312. 1833. *Ludwigia hexapetala* (Hooker & Arnott) Zardoni et al., Ann. Missouri Bot. Gard. 16: 243. 1991. *Ludwigia grandiflora* (Michaux) Greuter & Burdet subsp. *hexapetala* (Hooker & Arnott) G. L. Nesom & Kartesz, Castanea 65: 125. 2000.

Perennial herb, to 1 m; stem decumbent with suberect branches, villous-viscid to glabrous. Leaves alternate, the blade spatulate to oblanceolate, 3–10 cm long, 3–10 mm wide, the apex acute, glandular-mucronate or not, the base narrowly cuneate, the margin entire, the upper and lower surfaces villous or glabrous, the petiole 1–5(25) mm long; stipules ovate-triangular, minute. Flowers solitary in the axil of the upper leaves, the pedicel 1–2(5) cm long; bracteoles lanceolate, 1–2 mm long, at the base of the floral tube; sepals 5(6), 6–14(19) mm long; petals 12–23(30) mm long, yellow; stamens 10(12), the filaments (2)3–7 mm long; nectary disk slightly elevated, with a depressed white-ciliate nectary surrounding the base of each epipetalous stamen. Fruit cylindric, 12–25 mm long, 3–4 mm wide, the pedicel 0.5–5 cm long; seeds obovate, enclosed by the endocarp.

Marshes, swamps, and pond and lake margins. Occasional; peninsula, western panhandle. New York south to Florida, west to Missouri, Oklahoma, and Texas, also Washington, Oregon, and California; Mexico, Central America, and South America. Summer.

Ludwigia hexapetala is considered a distinct taxon by some and is listed as a Category I invasive species in Florida by the Florida Exotic Pest Plant Council (FLEPPC, 2015).

The morphological characters used to distinguish *L. hexapetala* from *L. grandiflora* show considerable overlap (Nesom and Kartesz, 2000) and are here treated as synonymous.

Although the type specimen of *L. grandiflora* was collected in the late 1700s from Georgia, it is unclear if it is truly native to the region.

Ludwigia hirtella Raf. [Hairy.] SPINDLEROOT.

Ludwigia hirtella Rafinesque, Med. Repos., ser. 2. 5: 358. 1808. *Isnardia hirtella* (Rafinesque) Kuntze, Revis. Gen. Pl. 1: 251. 1891.

Erect perennial herb, to 1 m; stem hirsute. Leaves alternate, the blade lanceolate, 1.5–6 cm long, 3–18 mm wide, the apex acute, the base rounded to subcordate, the margin entire, the upper and lower surfaces hirsute, sessile. Flowers solitary in the axil of the upper leaves, the pedicel 6–10(15) mm long; bracteoles ovate to elliptic, 1–2 mm long, above the middle of the pedicel; sepals 4, triangular-ovate, 7–10 mm long; petals 4, 10–15 mm long, yellow; stamens 4; disk elevated, with a depressed white-ciliate nectary surrounding the base of each epipetalous stamen. Capsule subglobose-cubical, 4–6 mm long, dehiscent by a terminal pore, the pedicel 4–8 mm long; seeds ellipsoid, light brown.

Flatwoods and bogs. Occasional; Nassau County, central and western panhandle. New Jersey south to Florida, west to Oklahoma and Texas. Spring–fall.

Ludwigia lanceolata Elliott [Lance-shaped, in reference to the leaves.] LANCELEAF PRIMROSEWILLOW.

> *Ludwigia lanceolata* Elliott, Sketch Bot. S. Carolina 1: 213. 1817. *Isnardia lanceolata* (Elliott) de Candolle, Prodr. 3: 61. 1828.

Erect, stoloniferous perennial herb, to 1 m; stem glabrous, often brownish purple, leafless below, branched above. Leaves alternate, the blade elliptic or oblanceolate to linear, 2–7.5 cm long, 2–7.5(14) mm wide, the apex acute, the base cuneate, the margin entire, with minute hydathodal glands, the petiole winged, to 5 mm long; stipules ovate, minute, dark reddish purple. Flowers solitary in the axil of the upper leaves, sessile; bracteoles ovate to elliptic, 1–2 mm long; sepals 4, triangular or deltate, 2–3 mm long, greenish, glabrous, minutely papillose along the margin; petals absent; stamens 4, the filaments 1–2 mm long; nectary disk raised, 4-lobed, yellowish green, glabrous. Fruit obpyramidal, 4–5 mm long, winged on the corners, glabrous or sometimes minutely strigulose, especially on the wings, the pedicel to 0.5 mm long; seeds oblong, light brown.

Swamps and marshes. Occasional; northern and central peninsula, panhandle. North Carolina south to Florida. Summer–fall.

Ludwigia leptocarpa (Nutt.) H. Hara [From the Greek *lepto*, slender, and *carpos*, fruit, in reference to the slender fruit.] ANGLESTEM PRIMROSEWILLOW.

> *Jussiaea leptocarpa* Nuttall, Gen. N. Amer. Pl. 1: 279. 1818. *Jussiaea leptocarpa* Nuttall var. *genuina* Munz, Darwiniana 4: 255. 1942, nom. inadmiss. *Ludwigia leptocarpa* (Nuttall) H. Hara, J. Jap. Bot. 28: 292. 1953.
>
> *Jussiaea pilosa* Kunth, in Humboldt et al., Nov. Gen. Sp. 6: 101, pl. 532. 1823. *Jussiaea variabilis* G. Meyer var. *pilosa* (Kunth) Kuntze, Revis. Gen. Pl. 1: 251. 1891.

Erect perennial herbs, to 3 m; stems villous, well branched above. Leaves alternate, the blade lanceolate, 3.5–18 cm long, 1–4 cm wide, the apex acuminate, the base narrowly cuneate, the margin entire, the upper and lower surfaces villous, the petiole 0.2–3.5 cm long. Flowers solitary in the leaf axil, the pedicel 1–15 mm long; bracteoles narrowly deltoid or absent; sepals (4)5(6–7), deltoid, 6–11 mm long, villous; petals as many as the sepals, obovate, 5–11 mm long, orange-yellow; stamens twice as many as the sepals, the filaments 2–4 mm long, the epipetalous ones shorter; nectary disk slightly elevated, the base of each epipetalous stamen surrounded by a depressed nectary covered with white matted trichomes. Fruit cylindric, 1.5–5 cm long, 3–4 mm wide, terete, villous, the pedicel 2–20 mm long; seeds obovoid, ca. 1 mm long, pale brown.

Pond and swamp margins. Frequent; nearly throughout. Pennsylvania south to Florida, west to Illinois, Missouri, Oklahoma, and Texas; West Indies, Mexico, Central America, and South America; Africa. All year.

Ludwigia linearis Walter [Linear, in reference to the narrow leaf shape.] NARROWLEAF PRIMROSEWILLOW.

> *Ludwigia linearis* Walter, Fl. Carol. 89. 1788. *Isnardia linearis* (Walter) de Candolle, Prodr. 3: 60. 1828. *Ludwigia linearis* Walter var. *typica* Munz, Bull. Torrey Bot. Club 71: 163. 1944, nom. inadmiss.

Erect perennial herb, to 1 m; stem glabrous, minutely strigulose, or puberulent. Leaves alternate, the blade linear to narrowly elliptical, 2.5–6 cm long, 2–5 mm wide, the apex acute, the base cuneate, the margin entire, the upper and lower surfaces glabrous or puberulent, subsessile. Flowers solitary in the axil of the upper leaves, sessile; sepals 4, 3–4 mm long; petals 4, 4–5 mm long, yellow; stamens 4, the filaments ca. 1 mm long; nectary disk elevated, 4-lobed, glabrous. Fruit elongate-obpyramidal, 6–8 mm long, sessile; seeds oblong-elliptic, pale brown.

Bogs and swamps. Occasional; northern counties, central peninsula. New Jersey south to Florida, west to Oklahoma and Texas. Summer–fall.

Ludwigia linifolia Poir. [With leaves narrowly linear.] SOUTHEASTERN PRIMROSEWILLOW.

> *Ludwigia linifolia* Poiret, in Lamarck, Encycl., Suppl. 5: 513. 1817. *Isnardia linifolia* (Poiret) Kuntze, Revis. Gen. Pl. 1: 251. 1891.

Erect or ascending, stoloniferous perennial herb, to 5.5(6) dm; stem glabrous, usually much branched. Leaves alternate, the blade linear or linear-oblanceolate, 1.5–4 cm long, 1–4(6) mm wide, the apex acute, the base narrowly cuneate, the margin entire, with obscure hydathodal glands, the upper and lower surfaces glabrous, sessile or subsessile; stipules narrowly ovate to narrowly lanceolate. Flowers solitary in the leaf axil, sessile; bracteoles 3–5 mm long; sepals 4, narrowly triangular, (3)4–7 mm long, glabrous or minutely papillose; petals narrowly obovate-elliptic, 4–6 mm long, yellow; stamens 4, the filaments 2–3 mm long. Fruit subcylindric, 5–10(12) mm long, papillose, occasionally also remotely scaberulous, sessile; seeds oblong-elliptic, reddish.

Marshes and wet flatwoods. Occasional; nearly throughout. North Carolina south to Florida, west to Mississippi. Spring–fall.

Ludwigia longifolia (DC.) H. Hara [With long leaves.] LONGLEAF PRIMROSEWILLOW.

> *Jussiaea longifolia* de Candolle, Mém. Soc. Phys. Genève, ser. 2. 2: 141. 1824. *Jussiaea longifolia* de Candolle var. *minor* Micheli, in Martius, Fl. Bras. 13(2): 157. 1875, nom. inadmiss. *Jussiaea octofila* de Candolle forma *longifolia* (de Candolle) H. Léveillé, Bull. Soc. Bot. France, ser. 4. 7: 425. 1907. *Jussiaea peruviana* Linnaeus var. *longifolia* (de Candolle) H. Léveillé ex Bertoni, Descr. Fis. Econ. Paraguay 13: 1910. *Ludwigia longiflora* (de Candolle) H. Hara, J. Jap. Bot. 28: 293. 1953.

Erect annual or perennial herb, to 2.5 m; stem much branched, sharply 4-angled and winged, glabrous. Leaves alternate, the blade oblong-lanceolate or lanceolate, 5–35 cm long, 0.4–2.5

cm wide, the apex acute or short acuminate, the base cuneate, the margin entire, the upper and lower surfaces glabrous, sometimes minutely scabrous along the margin, sessile; stipules setaceous, minute. Flowers solitary in the leaf axil, the pedicel 0.5–4 cm long, sharply 4-angled, narrowly winged; bracteoles setaceous or lanceolate, 5–7 mm long; sepals 4(5), ovate, 1–1.8 cm long, green, sometimes with red or pink within, glabrous, sometimes minutely scabrous along the margin; petals as many as the sepals, suborbicular or obovate, 2–2.5 cm long, yellow; stamens twice as many as the sepals, subequal, the filaments 3–4 mm long; nectary disk flat. Fruit oblong to narrowly obong, 1.1–3.5(4.2) cm long, sharply 4-angled, glabrous or puberulent; seeds oblong, brown.

Swamps and marshes. Rare; Seminole County. Florida; South America and Australia. Native to South America. Spring–fall.

Ludwigia maritima R. M. Harper [Growing by the sea.] SEASIDE PRIMROSEWILLOW.

Ludwigia maritima R. M. Harper, Torreya 4: 163. 1904.

Erect perennial herb, to 6 dm; stem few-branched on the upper half, slightly angled by the decurrent leaf margins, cinereous-puberulent. Leaves alternate, the blade lanceolate to oblong, 3–5 cm long, 0.5–1 cm wide, the apex acute, the base cuneate, the margin entire, the upper and lower surfaces cinereous-pubescent, sessile. Flowers solitary in the axil of the upper leaves, each subtended by a foliaceous bract equaling or exceeding the flower, the pedicel 6–10(15) mm long; bracteoles minute, somewhat above the middle of the pedicel; sepals 4, twice as long as the ovary, reflexed at anthesis, then ascending, finally deciduous; petals absent; stamens 4. Fruit oblong, winged on the angles, light brown.

Flatwoods, bogs, and swamp margins. Frequent; nearly throughout. North Carolina south to Florida, west to Louisiana. Spring–fall.

Ludwigia microcarpa Michx. [Small-fruited.] SMALLFRUIT PRIMROSEWILLOW.

Ludwigia microcarpa Michaux, Fl. Bor.-Amer. 1: 88. 1803. *Isnardia microcarpa* (Michaux) Poiret, in Lamarck, Encycl., Suppl. 3: 188. 1813.

Erect or ascending, stoloniferous perennial herb, to 60 dm; stem unbranched to profusely branched. Leaves alternate, the blade broadly elliptic to suborbicular, 2–17 mm long, 2–5 mm wide, the apex acute to obtuse, the base cuneate, the margin subentire, with prominent hydathodal glands forming minute teeth, the upper and lower surfaces papillose-strigulose, the petiole 2–5 mm long, winged; stipules lanceolate-deltate, ca. 2 mm long, reddish purple. Flowers solitary in the leaf axil, sessile or subsessile; bracteoles linear to oblong, shorter than the ovary; sepals 4, ovate-deltate, 1–2 mm long, green with a whitish thickened base, glabrous; petals absent; stamens 4, to 1 mm long; nectary disk nearly flat, 4-lobed, greenish. Fruit obconical, 1–2 mm long, minutely puberulent; seeds ovate-oblong, reddish brown.

Wet hammocks, marshes, and lake and pond margins. Common; nearly throughout. North Carolina south to Florida, west to Missouri, Arkansas, and Texas. Spring–fall.

Ludwigia octovalvis (Jacq.) P. H. Raven [Eight-valved, in reference to the fruit.] MEXICAN PRIMROSEWILLOW.

> *Oenothera octovalvis* Jacquin, Enum. Syst. Pl. 19. 1760. *Jussiaea octonervia* Lamarck, Encycl. 3: 332. 1789, nom. illegit. *Jussiaea octovalvis* (Jacquin) Swartz, Observ. Bot. 142. 1791. *Jussiaea octofila* de Candolle, Prodr. 3: 57. 1828, nom. illegit. *Jussiaea peruviana* Linnaeus var. *octofila* Bertoni, Descr. Fis. Econ. Paraguay 13: 1910. *Jussiaea suffruticosa* Linnaeus subsp. *octonervia* Hassler, Bull. Soc. Bot. Genève 5: 271. 1913. *Jussiaea suffruticosa* Linnaeus var. *octofila* (Bertoni) Munz, Darwiniana 4: 239. 1942. *Ludwigia octovalvis* (Jacquin) P. H. Raven, Kew Bull. 15: 476. 1962. *Ludwigia octovalvis* (Jacquin) P. H. Raven var. *octofila* (Bertoni) Alain, Bull. Torrey Bot. Club 90: 191. 1963, nom. inadmiss.
>
> *Jussiaea angustifolia* Lamarck, Encycl. 3: 331. 1789. *Jussiaea suffruticosa* Linnaeus var. *angustifolia* (Lamarck) Kuntze, Revis. Gen. Pl. 1: 251. 1891. *Ludwigia angustifolia* (Lamarck) M. Gómez de la Maza y Jiménez, Anales Soc. Esp. Hist. Nat. 23: 66. 1894; non Michaux, 1803. *Jussiaea suffruticosa* Linnaeus forma *angustifolia* (Lamarck) Alston, in Trimen, Handb. Fl. Ceylon 6: 130. 1931.

Erect perennial herb or subshrub, to 4 m; stem much branched, subglabrous to puberulent. Leaves lanceolate or linear, 0.7–14.5 cm long, 0.1–4 cm wide, the base narrowly to broadly cuneate, the apex attenuate, the margin entire, the upper and lower surfaces subglabrous to puberulent, the petiole to 10 cm long. Flowers solitary in the leaf axil, the pedicel to 1 mm long; bracteoles to 1 mm long; sepals 4, ovate or lanceolate, 3–15 mm long; petals 4, broadly obovate, 3–17 mm long, emarginate; stamens 8, the epipetalous ones shorter, the filaments 1–4 mm long; nectary disk slightly raised, with a sunken white-pubescent nectary surrounding the base of each epipetalous stamen. Fruit 1.7–4.5 cm long, pale brown with 8 darker ribs, irregularly loculicidally dehiscent; seeds subglobose, brown.

Marshes. Common; nearly throughout. North Carolina south to Florida, west to Texas; West Indies, Mexico, Central America, and South America; Africa, Asia, Australia, and Pacific Islands. All year.

Ludwigia palustris (L.) Elliott [Of swamps and marshes.] MARSH SEEDBOX.

> *Isnardia palustris* Linnaeus, Sp. Pl. 120. 1753. *Ludwigia palustris* (Linnaeus) Elliott, Sketch Bot. S. Carolina 1: 211. 1817. *Dantia palustris* (Linnaeus) Des Moulins, Actes Soc. Linn. Bordeaux 20: 517. 1859. *Quadricosta palustris* (Linnaeus) Dulac, Fl. Hautes-Pyrénées 329. 1867. *Jussiaea isnardia* E.H.L. Krause, in Sturm, Deutschl. Fl., ed. 2. 9: 185. 1901. *Ludwigia palustris* (Linnaeus) Elliott var. *typica* Fernald & Griscom, Rhodora 37: 176. 1935; nom. inadmiss.
>
> *Ludwigia palustris* (Linnaeus) Elliott var. *nana* Fernald & Griscom, Rhodora 37: 176. 1935.

Creeping or ascending-decumbent perennial herb; stem glabrous. Leaves opposite, the blade broadly elliptic or subovate, 0.7–4.5 cm long, 0.4–2.3 cm wide, the apex subacute, the base broadly cuneate and narrowed to a winged petiole, the margin entire, the upper and lower surfaces glabrous, the petiole to 6 mm long. Flowers usually paired in the leaf axil, sessile or subsessile; bracteoles lacking or to 1 mm long; sepals 4, deltoid, ca. 2 mm long; petals absent; stamens 4, green, the filaments ca. 0.5 mm long; nectary disk elevated, glabrous. Fruit elongate-globose, (2)3–5 mm long, obscurely 4-angled, brown with broadly green band on the angles, irregularly loculicidally dehiscent; seeds elongate-ovoid, ca. 1 mm long, light brown.

Marshes, swamps, and pond margins. Frequent; northern counties, central peninsula. Nearly throughout North America; West Indies, Mexico, Central America, and South America; Europe, Africa, and Asia. Spring–fall.

Ludwigia peploides (Kunth) P. H. Raven subsp. **glabrescens** (Kuntze) P. H. Raven [Resembling *Peplis* (Lythraceae); nearly glabrous.] FLOATING PRIMROSEWILLOW.

Jussiaea repens Linnaeus var. *glabrescens* Kuntze, Revis. Gen. Pl. 1: 251. 1891. *Ludwigia adscendens* (Linnaeus) H. Hara var. *glabrescens* (Kuntze) H. Hara, J. Jap. Bot. 28: 192. 1953. *Ludwigia peploides* (Kunth) P. H. Raven subsp. *glabrescens* (Kuntze) P. H. Raven, Reinwardtia 6: 394. 1963. *Ludwigia peploides* (Kunth) P. H. Raven var. *glabrescens* (Kuntze) Shinners, Sida 1: 386. 1964.

Sprawling perennial herb; stem to 6 dm long, rooting at the nodes, glabrescent. Leaves alternate, the blade elliptical, 1–9.5 cm long, 0.4–3 cm wide, the apex acute to obtuse, the base narrowly cuneate, the margin entire, the upper and lower surfaces glabrescent, the petiole 0.2–3 cm long. Flowers solitary in the axil of the upper leaves, the pedicel 1–6 cm long; bracteoles deltoid, ca. 1 mm long; sepals 5, deltoid, 4–12 mm long, glabrous; petals obovate, 7–17 mm long, yellow with a darker spot at the base; stamens 10, the epipetalous ones slightly shorter, the filaments 3–5 mm long, yellow; nectary disk slightly elevated with a depressed white-pilose nectary surrounding the base of each epipetalous stamen. Fruit linear-oblong, 1–5.8 cm long, glabrate; seeds cuboidal, ca. 1 mm long, pale brown.

Swamps, ponds, and lake margins. Rare; central peninsula, Okaloosa County. New York south to Florida, west to Nebraska, Kansas, Oklahoma, and Texas; West Indies, Mexico, Central America, and South America. Spring–fall.

Ludwigia peruviana (L.) H. Hara [Of Peru.] PERUVIAN PRIMROSEWILLOW.

Jussiaea peruviana Linnaeus, Sp. Pl. 388. 1753. *Jussiaea peruviana* Linnaeus var. *typica* Munz, Darwiniana 4: 232. 1942, nom. inadmiss. *Jussiaea grandiflora* Ruiz López & Pavón, Anales Inst. Bot. Cavanilles 14: 753, nom. illegit; non Michaux, 1803. *Ludwigia peruviana* (Linnaeus) H. Hara, J. Jap. Bot. 28: 293. 1953.

Oenothera hirta Linnaeus, Syst. Nat., ed. 10. 998. 1759. *Jussiaea hirta* (Linnaeus) Swartz, Observ. Bot. 142. 1791; non Lamarck, 1789. *Ludwigia hirta* (Linnaeus) M. Gómez de la Maza y Jiménez, Anales Soc. Esp. Hist. Nat. 23: 66. 1894.

Erect suffrutescent perennial herb or shrub, to 4 m; stem much branched, terete or angled, usually ridged, the bark peeling, glabrous or villous; stipules setaceous, 1–2 mm long, deciduous. Leaves alternate, the blade lanceolate to elliptic or obovate to ovate, 2–45 cm long, 1–10 cm wide, the apex acute to acuminate or rounded and sometimes emarginate, the base cuneate, the margin entire or subentire, gland-toothed, the upper and lower surfaces villous or sometimes glabrous, the petiole 0.5–1.5 cm long. Flowers solitary in the leaf axil, the pedicel 0.5–6.5 cm long, angled or subterete, villous or glabrous; bracteoles ovate, lanceolate, or linear, 5–20 mm long; sepals 4(5), ovate or ovate-lanceolate, 1–2.3 cm long, entire or glandular-serrulate, villous or glabrous; petals 4(5), orbicular or obovate, 1–4 cm long, yellow; stamens 8(10), subequal, the filaments 2–5 mm long, yellow; nectary disk elevated. Fruit obconic, 1–4 cm long, 4(5)-angled, with prominent dark lines on the angles, villous or glabrous; seeds oblong, ca. 1 mm long, brown or reddish brown.

Swamps and lake and pond margins. Common; nearly throughout. North Carolina south to Florida, west to Texas; West Indies, Mexico, Central America, and South America; Asia and Australia. Native to Mexico, Central America, and South America. All year.

Ludwigia peruviana is listed as a Category I invasive species in Florida by the Florida Exotic Pest Plant Council (FLEPPC, 2015).

Ludwigia pilosa Walter [With long ascending trichomes.] HAIRY PRIMROSEWILLOW.

Ludwigia pilosa Walter, Fl. Carol. 89. 1788. *Isnardia pilosa* (Walter) Kuntze, Revis. Gen. Pl. 1: 251. 1891.

Stoloniferous, erect or creeping perennial herb, to 12 dm; stem much branched, hirtellous. Leaves alternate, the blade of the cauline elliptic-lanceolate, 1.5–8(10) cm long, 3–12(14) mm wide, the apex acute, the base cuneate, the margin entire, the hydathodal glands obscure, the upper and lower surfaces hirtellous, sessile or the petiole 2(10) mm long; stipules ovate to lanceolate, minute. Flowers solitary in the axil of the upper leaves, sessile or subsessile; bracteoles linear-lanceolate to elliptic, ca. 5 mm long, hirtellous; sepals 4, ovate-triangular, 3–5 mm long, the inner surface hirtellous; petals absent; stamens 4, the filaments ca. 2 mm long, yellowish; nectary disk indistinctly 4-lobed, hirtellous, yellow. Fruit subglobose or oblong-obovoid, 3–5 mm long, sessile or subsessile; seeds elliptic or oblong-ovoid, brown.

Swamps. Occasional; northern counties, central peninsula. Virginia south to Florida, west to Texas. Summer–fall.

Ludwigia ravenii C.-I Peng [Commemorates Peter Hamilton Raven (b. 1936), botanist and environmentalist, president emeritus of the Missouri Botanical Garden.] RAVEN'S PRIMROSEWILLOW.

Ludwigia ravenii C.-I Peng, Syst. Bot. 9: 129. 1984.

Erect, stoloniferous perennial herb, to 3.5(9) dm; stem densely hirtellous. Leaves alternate, the blade of the cauline lanceolate-elliptic, 1.3–6.5 cm long, 4–15 mm wide, the apex acute, the base cuneate, the margin entire, with minute hydathodal glands, the upper and lower surfaces densely hirtellous, the petiole 1–8 mm long; stipules lanceolate to deltate, minute, reddish purple. Flowers solitary in leaf axil, sessile or subsessile; bracteoles lanceolate to elliptic, 2–4 mm long, hirtellous; sepals ovate-deltate, 2–3 mm long, green, the outer surface hirtellous, the inner surface glabrous, the margin entire; petals absent; stamens 4, the filaments ca. 1 mm long; nectary disk raised, greenish, 4-lobed, glabrous. Fruit oblong-obovoid, 4–5 mm long, hirtellous; seeds elliptic-oblong, light brown.

Pond margins. Rare; Clay County. Virginia, North Carolina, South Carolina, and Florida. Summer.

Ludwigia repens J. R. Forst. [Creeping.] CREEPING PRIMROSEWILLOW.

> *Ludwigia repens* J. R. Forster, Fl. Amer. Sept. 6. 1771, nom. cons. TYPE: FLORIDA: Duval Co.: near Jacksonville, 19 May 1894. *Curtiss 4836* (holotype: MO; isotypes: FLAS, GA, GH, MSC, NA, NY, P, W).
>
> *Ludwigia repens* Swartz, Prodr. 33. 1788; non J. R. Forster, 1771. *Isnardia repens* de Candolle, Prodr. 3: 60. 1828.
>
> *Ludwigia natans* Elliott, Sketch Bot. S. Carolina 1: 581. 1821. *Isnardia natans* (Elliott) Kuntze, Revis. Gen. Pl. 1: 251. 1891. *Ludwigia natans* Elliott var. *typica* Fernald & Griscom, Rhodora 37: 175. 1935.
>
> *Isnardia repens* de Candolle var. *rotundata* Grisebach, Cat. Pl. Cub. 107. 1866. *Ludwigia repens* Swartz var. *rotundata* (Grisebach) M. Gómez de la Maza y Jiménez, Anales Soc. Esp. Hist. Nat. 23: 66. 1894. *Ludwigia natans* Elliott var. *rotundata* (Grisebach) Fernald & Griscom, Rhodora 37: 175. 1935.
>
> *Isnardia intermedia* Small & Alexander ex Small, Man. S.E. Fl. 940, 1506. 1913. TYPE: FLORIDA: Miami-Dade Co.: between Homestead and Cross Key, 21–22 Nov 1906, *Small & Carter 2626* (holotype: NY; isotypes SMU, TENN).

Creeping perennial herb; stem to 5 dm, rooting at the nodes, glabrous or puberulent. Leaves opposite, the blade elliptic to subrotund, 1–4.5 cm long, 0.5–2 cm wide, the apex obtuse to subacute, the base cuneate, the margin denticulate or serrate, the upper and lower surfaces glabrous or puberulent, the petiole 3–25 mm long. Flowers solitary in the leaf axil, sessile or the pedicel to 3 mm long; sepals 4, 3–4 mm long; petals 4, 4–5 mm long, white to pink; stamens 4; nectary disk elevated, 4-lobed, glabrous. Fruit elongate, 3–8 mm long; seeds elliptic, strongly papillose and with a coma, brown.

Marshes and swamps. Common; nearly throughout. Virginia south to Florida, west to California; West Indies, Mexico, Central America, and South America. Spring–fall.

Ludwigia spathulata Torr. & A. Gray [Spoon-shaped, in reference to the leaves.] SPOON PRIMROSEWILLOW.

> *Ludwigia spathulata* Torrey & A. Gray, Fl. N. Amer. 1: 526. 1840. *Isnardia spathulata* (Torrey & A. Gray) Kuntze, Revis. Gen. Pl. 1: 251. 1891. TYPE: FLORIDA: s.d., *Chapman s.n.* (holotype: NY; isotypes: F, GH, GOET, MO, NY, P, UPS, US).

Prostrate and creeping perennial herb; stem rooting at the nodes, hirsute. Leaves opposite, the blade elliptic, 1–2 cm long, 3–5 mm wide, the apex obtuse, the base cuneate, the margin entire, the upper and lower surfaces hirsute, subsessile or the petiole to 1 cm long. Flowers solitary in the leaf axil, sessile; bracteoles absent; sepals 4, broadly ovate, ca. 1 mm long; petals absent; stamens 4. Fruit short obpyramidal, 2–3 mm long, slightly quadrangular; seed obliquely ovate, dark reddish brown with longitudinal stripes.

Pond margins. Rare; central panhandle, Walton County. South Carolina, Georgia, Florida, and Alabama. Summer.

Ludwigia sphaerocarpa Elliott [From the Greek *sphaero*, globose, and *carpos*, fruit.] GLOBEFRUIT PRIMROSEWILLOW.

> *Ludwigia sphaerocarpa* Elliott, Sketch Bot. S. Carolina 1: 213. 1817. *Isnardia sphaerocarpa* (Elliott) de Candolle, Prodr. 3: 61. 1828. *Ludwigia sphaerocarpa* Elliott var. *typica* Fernald & Griscom, Rhodora 37: 174. 1935, nom. inadmiss.

Erect, stoloniferous perennial herb, to 11 dm; stem glabrous to densely strigulose. Leaves alternate, narrowly elliptic or lanceolate to linear, (3)6–10 cm long, 5–11(16) mm wide, the apex acute, the base cunete, the margin entire, the hydathodal glands often visible on the main cauline ones, the upper and lower surfaces glabrous to strigulose, the petiole 1–4(10) mm long; stipules lanceolate or deltoid, ca. 0.5 mm long, reddish-purple. Flowers solitary in axil of the upper leaves, sessile or subsessile; bracteoles lanceolate, ca. 1 mm long; sepals 4, ovate-deltoid, 2–4 mm long, green on the outer surface, yellow on the inner, glabrous or densely strigulose on both surfaces, the margin entire; petals absent; stamens 4, the filaments 1–2 mm long, yellow; nectary disk raised, 4-lobed, yellow, glabrous or short hirtellous between the lobes. Fruit subglobose, 2–4 mm long, pinkish, glabrous or strigulose; seeds elliptic, brown.

Marshes and cypress swamps. Occasional; northern peninsula west to central panhandle, Volusia County. Massachusetts south to Florida, west to Michigan, Illinois, and Texas. Summer–fall.

Ludwigia suffruticosa Walter [Somewhat woody.] SHRUBBY PRIMROSEWILLOW.

> *Ludwigia suffruticosa* Walter, Fl. Carol. 90. 1788. *Ludwigia capitata* Michaux, Fl. Bor.-Amer. 1: 90. 1803, nom. illegit. *Isnardia capitata* de Candolle, Prodr. 3: 60. 1828, nom. illegit. *Isnardia suffruticosa* (Walter) Kuntze, Revis. Gen. Pl. 1: 251. 1891.

Erect, stoloniferous perennial herb, to 9 dm; stem unbranched or slightly branched, puberulous to hirtellous in the inflorescence and at the base, otherwise usually glabrous. Leaves alternate, the blade of the cauline ones lance-elliptic to linear, the lower ones oblong-oblanceolate to oblong or oblong-lanceolate, 2.5–9.5 cm long, (1)3–9 mm wide, the apex acuminate or acute, the base rounded or obtuse, the margin entire, the hydathodal glands usually obscure, the upper and lower surfaces glabrous or somewhat pilose, sessile; stipules deltoid, ca. 0.5 mm long. Flowers in a dense terminal raceme 1–5(12) cm long, the subtending leaves reduced and bract-like, lanceolate or elliptic-lanceolate, 4–12(18) mm long glabrous or hirtellous, sessile or subsessile; bracteoles lanceolate, 3–5 mm long; sepals 4, ovate or deltoid, 2–3(4) mm long, glabrous, whitish, the margin entire; petals absent; stamens 4, the filaments 1–2 mm long, yellowish; nectary disk raised, 4-lobed, yellow, glabrous. Fruit obpyramidal or subspherical, 3–4 mm long, glabrous or minutely strigulose, the pedicel 0.5–2 mm long; seeds elliptic-oblong, brown.

Marshes, wet flatwoods, and pond margins. Common; peninsula west to central panhandle. North Carolina south to Florida, west to Alabama. Spring–fall.

Ludwigia virgata Michx. [*Virgula*, long and slender.] SAVANNAH PRIMROSEWILLOW.

> *Ludwigia virgata* Michaux, Fl. Bor.-Amer. 1: 89. 1803. *Isnardia virgata* (Michaux) de Candolle, Prodr. 3: 60. 1828.

Erect perennial herb, to 1 m; stem simple to much-branched below, pubescent or glabrate. Leaves alternate, the blade oblong to linear-oblong, 2–5 cm long, 2–4 mm wide, reduced upward to bracts, the apex rounded to obtuse, the base cuneate, the margin entire, the upper and lower surfaces pubescent or glabrate, sessile. Flowers solitary in the leaf axil, forming wand-like racemes, the pedicels 6–10(15) mm long, with 2 opposite or subopposite bracteoles above the middle; sepals 4, ovate to oblong-triangular, 5–10 mm long, strongly reflexed at anthesis, the outer surface short-pubescent, the inner surface short-pubescent or glabrous; petals 4, oblong-obovate, 10–15 mm long, yellow; stamens 4; style 7–9 mm long; nectary disk raised, short-pubescent. Fruit short-oblong to subglobose, 5–7 mm long, 4-lobed, dehiscent by a central pore at the style base; seeds oblong, ca. 1 mm long, buff-colored to brown, smooth.

Wet flatwoods. Frequent; northern and central peninsula, central and western panhandle. Virginia south to Florida, west to Mississippi. Spring–fall.

HYBRIDS

Ludwigia ×simulata Small (*L. lanceolata* × *L. pilosa*) [To resemble another species of *Ludwigia*.]

> *Ludwigia simulata* Small, Fl. S.E. U.S. 816, 1335. 1903, pro. sp. TYPE: FLORIDA: "West Florida," (holotype: NY).

Swamps and marshes. Rare; Gulf, Franklin, and Highlands Counties. Florida and North Carolina. Summer–fall.

EXCLUDED TAXA

> *Ludwigia glandulosa* Walter subsp. *brachycarpa* (Torrey & A. Gray) C.-I Peng—Reported for Florida by Munz (1944, as *L. glandulosa* var. *torreyi* Munz), but this taxon, as typified and delineated by Peng (1986, 1989) does not include the Florida material, which is of the typical subspecies.
> *Ludwigia octovalvis* subsp. *sessiliflora* (Micheli) P. H. Raven—Reported by Small (1903, 1913a, 1933, all as *Jussiaea scabra* Willd.), who misapplied this name to material of *L. octovalvis* subsp. *octovalvis*.
> *Ludwigia stolonifera* (Guillemin & Perrott) P. H. Raven—Misapplied to Florida material of *L. peploides* subsp. *glabrescens* by Small (1903, 1913a, 1933, all as *Jussiaea diffusa* Forsskål).
> *Ludwigia suffruticosa* (Linnaeus) M. Gómez de la Maza y Jiménez—Reported by Small (1903), who misapplied the name to material of *L. bonariensis*. *Jussiaea suffruticosa* Linnaeus is referred to as *L. octovalvis* subsp. *sessiliflora* by Raven (1963), a taxon not present in Florida.

Oenothera L. 1753. EVENINGPRIMROSE

Annual, biennial, or perennial herbs. Leaves alternate or opposite, simple, petiolate or epetiolate, estipulate. Flowers solitary in the leaf axil, forming a racemiform, spiciform, or corymbiform inflorescence; hypanthium well developed, prolonged beyond the ovary; sepals (3)4; petals (3)4; stamens (6)8; ovary (3)4-carpellate and -loculate. Fruit a woody, (3)4-angled or winged, dehiscent capsule with numerous seeds or indehiscent and nutlike with 1–4 seeds.

A genus of about 145 species; North America, Mexico, Central America, and South America. [From the Greek *oinos*, wine, and *ther*, wild animal, in reference to the ancient belief that the roots of which, soaked in wine, would tame wild animals.]

Gaura L. (1753); *Hartmannia* Spach (1835); *Kneiffia* Spach (1835); *Peniophyllum* Pennell (1919); *Raimannia* Rose ex Britton & A. Br. (1905).

Selected reference: Munz (1938).

1. Fruit indehiscent, nutlike; seeds 1–4.
  2. Fruiting pedicel slender, 2–8 mm long.
    3. Petals 4–6 mm long..**O. filipes**
    3. Petals 8–10 mm long...**O. sinuosa**
  2. Fruiting pedicel thick, less than 1 mm long.
    4. Sepals 2–3 mm long; petals 2–3 mm long...**O. curtiflora**
    4. Sepals 4 mm long or longer; petals 4 mm long or longer.
      5. Stem sparsely loose-uncinate; sepals 4–10(12) mm long; petals 4–10 mm long ....
      ....**O. simulans**
      5. Stem short-hirsute; sepals 12–15 mm long; petals 12–20 mm long.................**O. lindheimeri**
1. Fruit a dehiscent capsule; seeds numerous.
  6. Leaves linear-filiform, less than 1 mm wide ....**O. linifolia**
  6. Leaves more than 1 mm wide.
    7. Petals white or pink..**O. speciosa**
    7. Petals yellow.
      8. Capsules obovoid to broadly clavate ....**O. fruticosa**
      8. Capsules subcylindric.
        9. Seeds prismatic.
          10. Petals 4–6 cm long; calyx lobes 2.5–5 cm long; styles 2–5 mm long, longer than the
          anthers..**O. grandiflora**
          10. Petals 1–2.5 cm long; calyx lobes 1–2.5 cm long; styles 3–17 mm long, subequaling
          the anthers.
            11. Hypanthium villous or hirsute..**O. biennis**
            11. Hypanthium glabrate ....**O. nutans**
        9. Seeds round.
          12. Petal apex acute to rounded...**O. curtissii**
          12. Petal apex truncate or emarginate.
            13. Plant densely strigulose; leaf margins subentire to remotely shallowly dentate;
            bracts flat.
              14. Sepals 1.3–3.3 cm long; petals 2–4.5 cm long............................**O. drummondii**
              14. Sepals 0.3–1.1 cm long; petals 0.5–1.6 cm long................................**O. humifusa**
            13. Plant sparsely to moderately strigulose, usually villous; leaf margins deeply lobed
            to dentate or rarely subentire; bracts revolute.
              15. Petals 2.5–4 cm long.. **O. grandis**
              15. Petals 0.5–2.5 cm long..**O. laciniata**

Oenothera biennis L. [Biennial, completing its life cycle in two years.] COMMON EVENINGPRIMROSE.

Oenothera biennis Linnaeus, Sp. Pl. 346. 1753. *Onagra biennis* (Linnaeus) Scopoli, Fl. Carniol., ed. 2. 1: 269. 1771. *Onagra europea* Spach, Hist. Nat. Vég. 4: 359. 1835, nom. illegit. *Onagra vulgaris* Spach, Nouv. Ann. Mus. Hist. Nat. 4: 353. 1836 ("1835"), nom. illegit. *Oenothera biennis* Linnaeus var. *vulgaris* Torrey & A. Gray, Fl. N. Amer. 1: 492. 1840, nom. indamiss. *Pseudo-oenothera virginiana*

Ruprecht, Fl. Ingr. 1: 365. 1860, nom. illegit. *Brunyera biennis* (Linnaeus) Burbani, Fl. Pyren. 2: 649. 1899. *Oenothera communis* K. Léveillé, Bull. Acad. Int. Geogr. Bot. 19: 328. 1909, nom. illegit. *Oenothera biennis* Linnaeus subsp. *centralis* Munz, in Britton, N. Amer. Fl., ser. 2. 5: 134. 1965.

Erect biennial herb, to 2 m; stem simple or branched, glandular-pubescent. Basal rosette leaves with the blade lanceolate, 0.6–3 dm long, 1–7 cm wide, the apex acute to obtuse, the base gradually narrowed to a winged petiole, the margin liniate-lobed, denticulate, to subentire, the petiole to 12 cm long, cauline leaves with the blade narrowly lanceolate to elliptic or oblong, 2.5–15 cm long, 1–2.5 cm wide, the apex acute, the base cuneate, the margin denticulate, the upper and lower surfaces pubescent, sessile or the petiole to 8 mm long. Flowers in a terminal spike, opening in the evening; hypanthium 2–4 cm long, glabrate; sepals 1.2–2(2.8) cm long, green or yellowish, rarely reddish, reflexed at anthesis, the free tips 1–3 mm long; petals obovate, 1–2.5(3) cm long, yellow; stamens subequaling the petals; styles 2–5 mm long, the stigma lobes 4–7 mm long, longer than the anthers. Fruit a dehiscent capsule, subcylindric, 1.4–2.5 cm long, strigose to subglabrous, sessile or subsessile; seeds numerous, prismatic.

Disturbed sites. Occasional; nearly throughout. Nearly throughout North America; South America; Europe, Africa, Asia, Australia, and Pacific Islands. Native to North America. Summer–fall.

Oenothera curtiflora W. L. Wagner & Hoch [*Curtus*, short, in reference to the short inflorescence.] VELVETWEED.

Gaura parviflora Douglas ex Lehmann, Nov. Stirp. Pug. 2: 15. 1830. *Gaura parviflora* Douglas ex Lehmann var. *typica* Munz, Bull. Torrey Bot. Club 65: 109. 1938, nom. inadmiss. *Oenothera curtiflora* W. L. Wagner & Hoch, in W. L. Wagner et al., Syst. Bot. Monogr. 83: 211. 2007.

Annual or biennial herb, to 2(3) m; stem erect, simple below, simple or freely branching above, villous and with shorter glandular trichomes. Basal rosette leaves with the blade broadly oblanceolate, 5–15 cm long, 1–2 cm wide, the apex acute, the base gradually narrowed to a winged petiole, the margin remotely sinuate-denticulate, the upper and lower surfaces pubescent, the cauline leaves lance-ovate to lanceolate, 3–10 cm long, 1–2.5(3) cm wide, the apex acuminate to acute, the base cuneate, the margin entire or remotely sinuate-denticulate, the upper and lower surfaces pubescent, sessile or subsessile. Flowers in a terminal spike 1–3 dm long; floral bracts linear-lanceolate, 2–5 mm long, caducous; hypanthium 2–3 mm long, the outer surface glabrous to short pubescent, the inner surface pubescent; sepals oblong-lanceolate, 2–3 mm long, greenish to reddish, separate and reflexed at anthesis, glabrous or puberulent; petals spatulate, 2–3 mm long, pink to rose; stamens slightly shorter than the sepals, the episepalous ones slightly longer than the epipetalous, the filaments reddish, flattened; style subequaling the sepals, reddish, usually pubescent basally. Fruit indehiscent, nutlike, subfusiform, 6–10 mm long, ca. 2 mm wide, 4-nerved and obtusely 4-angled above, the pedicel thick, less than 1 mm long; seeds 1–4.

Disturbed sites. Rare; Alachua, Leon, and Wakulla Counties. Virginia south to Florida, west to Washington, Oregon, and California. Spring–summer.

Oenothera curtissii Small [Commemorates Allen Hiram Curtiss (1845–1907), botanist for U.S. Department of Agriculture.] CURTISS' EVENINGPRIMROSE.

Oenothera curtissii Small, Fl. S.E. U.S., ed 2. 1353, 1375. 1913. *Raimannia curtissii* (Small) Rose ex Small, Man. S.E. Fl. 947. 1933. *Oenothera heterophylla* Spach var. *curtissii* (Small) Fosberg, Amer. Midl. Naturalist 27: 763. 1942.

Erect perennial herb, to 1 m; stem sparsely appressed pubescent. Leaves with the blade lanceolate to linear, the apex acute, the base cuneate, the margin entire or finely to coarsely toothed or pinnatifid. Flowers solitary in the leaf axil or aggregated in a terminal spike; sepals 4, lanceolate, 9–11 mm long, with short free tips, glabrous or sparsely pilose; petals obovate, ca. 1.5 cm long, the apex acute to rounded, yellow. Fruit a dehiscent capsule, subcylindric, 12–16 mm long, appressed pubescent; seeds numerous, round.

Dry, disturbed sites. Occasional; northern counties, Marion County. South Carolina south to Florida, west to Mississippi. Summer–fall.

Oenothera drummondii Hook. [Commemorates Thomas Drummond (1793–1835), Scottish-born botanist who collected in the southwestern United States.] BEACH EVENINGPRIMROSE.

Oenothera drummondii Hooker, Bot. Mag. 61: t. 3361. 1834. *Oenothera sinuata* Linnaeus var. *drummondii* (Hooker) H. Léveillé, Monogr. Oenothera 351. 1909. *Raimannia drummondii* (Hook.) Rose ex Sprague & Riley, Bull. Misc. Inform. Kew 1921: 200. 1921.

Erect or ascending perennial herb, to 6 dm; stem branched, gray-strigulose and sparsely appressed to erect long-villous. Basal rosette leaves with the blade narrowly oblanceolate, 6–18 cm long, 1–1.5 cm wide, the apex acute, the base gradually narrowed to the petiole, the margin distantly sinuate-toothed, the cauline narrowly elliptic to oblanceolate, 1.5–8 cm long, 0.5–2 cm wide, the apex acute, the base narrowly cuneate, the margin remotely shallow-dentate to subentire, sessile or short petiolate. Flowers solitary in the leaf axil; bracts narrowly elliptic, the margin remotely toothed to entire, flat, sessile; hypanthium 2–4 cm long; sepals linear-lanceolate, 1–2 mm long; petals broadly ovate, 2.5–5 cm long, the apex truncate or emarginated, yellow; filaments 12–21 mm long, the anthers 7–10 mm long; style 5–8 cm long, the stigma lobes 4–8 mm long. Fruit a dehiscent capsule, subcylindric, 2.5–4 cm long; seeds numerous, round.

Coastal dunes. Rare; St. Johns, Volusia, Brevard, and Miami-Dade Counties. North Carolina, South Carolina, Florida, Louisiana, and Texas; Mexico and South America. Native to Texas and Mexico. Spring–summer.

Oenothera filipes (Spach) W. L. Wagner & Hoch [Threadlike, apparently in reference to the slender pedicel.] SLENDER BEEBLOSSOM.

Gaura filipes Spach, Nouv. Ann. Mus. Hist. Nat. 4: 379. 1835. *Gaura filipes* Spach var. *typica* Munz, Bull. Torrey Bot. Club 65: 216. 1938, nom. inadmiss. *Oenothera filipes* (Spach) W. L. Wagner & Hoch, in W. L. Wagner et al., Syst. Bot. Monogr. 83: 212. 2007.
Gaura michauxii Spach, Nouv. Ann. Mus. Hist. Nat. 4: 379. 1835.
Gaura filipes Spach var. *major* Torrey & A. Gray, Fl. N. Amer. 1: 517. 1840.

Ascending to erect perennial herb, to 18 dm; stem, puberulent, simple below, paniculate-branched above. Cauline leaves with the blade linear to oblanceolate or lanceolate, 3–6 cm long, 3–15 mm wide, the apex acute to obtuse, the base cuneate, the margin sinuate-denticulate, the upper surface subglabrous to strigulose, the lower surface subglabrous to strigulose, sometimes villous on the veins, sessile or the petiole to 2 mm long, the upper leaves linear-lanceolate, 0.5–2 cm long, 2–4 cm wide, the apex acuminate, the base cuneate, the margin entire. Flowers in an openly branched paniculate-spike 2–5 dm long; floral bracts linear-lanceolate, 3–5 mm long, caducous, the spikes slender and lax, peduncle to 10 cm long; hypanthium narrow-funnelform, 3–5 mm long, the outer gray-strigulose, the inner surface white-puberulent; sepals linear-lanceolate, 5–10 mm long, separate and reflexed at anthesis; petals broadly lanceolate, 4–6 mm long, white, turning rose; filaments 4–5 mm long, subequal, slightly enlarged upward; style pubescent at the base, slightly exceeding the stamens. Fruit indehiscent, nutlike, obovoid-clavate, 4–5 mm long, sharply 4-angled, narrowed into a slender pedicel-like base 2–8 mm long; seeds 1–4.

Sandhills and flatwoods. Occasional; northern peninsula, central and western panhandle. Indiana and Illinois south to Florida and Louisiana. Summer–fall.

Oenothera fruticosa L. [Shrubby, bushy.] SUNDROPS; NARROWLEAF EVENINGPRIMROSE.

Oenothera fruticosa Linnaeus, Sp. Pl. 346. 1753. *Oenothera florida* Salisbury, Prodr. Stirp. Chap. Allerton 278. 1796, nom. illegit. *Kneiffia suffruticosa* Spach, Hist. Nat. Vég. 4: 374. 1835, nom. illegit. *Oenothera fruticosa* Linnaeus var. *vera* Hooker, Bot. Mag. 64: t. 3545. 1837, nom. inadmiss. *Kneiffia fruticosa* (Linnaeus) Raimann, in Engler & Prantl, Nat. Pflanzenfam. 3(7): 214. 1893.

Oenothera linearis Michaux, Fl. Bor.-Amer. 1: 255. 1803. *Kneiffia linearis* (Michaux) Spach, Hist. Nat. Vég. 4: 376. 1835. *Kneiffia angustifolia* Spach, Nouv. Ann. Mus. Hist. Nat. 4: 367. 1835, nom. illegit. *Oenothera fruticosa* Linnaeus var. *linearis* (Michaux) S. Watson, Proc. Amer. Acad. Arts 8: 584. 1873. *Oenothera fruticosa* Linnaeus var. *angustifolia* Spach ex H. Léveillé, Monogr. Oenothera 108. 1902, nom. illegit. *Oenothera fruticosa* Linnaeus forma *angustifolia* H. Léveillé, Monogr. Oenothera 108. 1902.

Oenothera riparia Nuttall, Gen. N. Amer. Pl. 1: 247. 1818. *Oenothera tetragona* Roth var. *riparia* (Nuttall) Munz, Bull. Torrey Bot. Club 64: 302. 1937.

Kneiffia longipedicellata Small, Bull. Torrey Bot. Club 23: 178. 1896. *Oenothera longipedicellata* (Small) B. L. Robinson, Rhodora 10: 34. 1908.

Kneiffia brevistipata Pennell, Bull. Torrey Bot. Club 46: 369. 1919. *Oenothera tetragona* Roth var. *brevistipata* (Pennell) Munz, Bull. Torrey Bot. Club 64: 301. 1937.

Kneiffia semiglandulosa Pennell, Bull. Torrey Bot. Club 46: 369. 1919.

Erect or decumbent perennial herb, to 8(12) dm; stem simple or few branched from the base or many above, strigose or rarely glandular-pubescent. Basal rosette leaves with the blade oblanceolate to obovate, 3–10 cm long, 0.5–2 cm wide, the petiole 1–4 cm long, the cauline leaves with the blade 2–6(8) cm long, 0.2–1.5(2), narrowly elliptic to narrowly ovate, the margin subentire, the upper and lower surfaces densely strigose or velutinous to subglabrous, the petiole 0.2–2(4) cm long. Flowers solitary in the axil of the upper leaves; bracts linear to lanceolate, 0.5–4 cm long, ¼ to subequaling the flower length; hypanthium 0.5–2 cm long; sepals lanceolate, 0.5–2 cm long, the free tips 1(6) mm long; petals ovate, (0.8)1.5–2.5(3) cm long, yellow, the apex truncate to cleft; filaments 5–15 mm long, the anthers 4–7 mm long; style 12–20 mm long, the stigmatic lobes 3–5 mm long, divergent. Fruit a dehiscent capsule, obovate to clavate, (5)10–17(20) mm long, 4-angled, the stipe 3–10 mm long; seeds numerous.

Flatwoods and hammocks. Occasional; northern counties. Quebec south to Florida, west to Manitoba, Illinois, Missouri, Oklahoma, and Louisiana. Spring–summer.

Oenothera grandiflora L'Hér. ex Aiton [Large-flowered.] LARGEFLOWER EVENINGPRIMROSE.

> *Oenothera grandiflora* L'Héritier de Brutelle ex Aiton, Hort. Kew. 2: 2. 1789. *Oenothera grandiflora* L'Héritier de Brutelle ex Aiton var. *glabra* Seringe, in de Candolle, Prodr. 3: 46. 1828, nom. inadmiss. *Oenothera biennis* Linnaeus var. *grandiflora* (L'Héritier de Brutelle ex Aiton) Torrey & A. Gray, Fl. N. Amer. 1: 492. 1840. *Oenothera biennis* Linneus forma *grandiflora* (L'Héritier de Brutelle ex Aiton) D. S. Carpenter, in Dole, Fl. Vermont, ed. 3. 198. 1937.

Erect perennial herb, to 3 m; stem simple or few-branched from the base or above, sparsely strigose or pubescent. Leaves with the blade elliptic to lanceolate, the apex acute or acuminate, the base cuneate, the margin denticulate to subentire, the upper and lower surfaces sparsely pubescent to subglabrous, the petiole to 2 mm long. Flowers solitary in the axil of the upper leaves; bracts linear to lanceolate, 1–3 cm long, shorter than the flower length; sepals 2.5–5 cm long, the free tips 1–4 mm long; petals obovate, 4–6 cm long, yellow; styles 2–5 mm long, longer than the anthers. Fruit a dehiscent capsule, subcylindric, 3–4 cm long, 4-angled; seeds numerous, prismatic.

Disturbed sites. Occasional; northern and central peninsula, central and western panhandle. Peninsula collections are probably escapes from cultivation. Quebec south to Florida, west to Kentucky, Tennessee, and Mississippi. Summer–fall.

Oenothera grandis Smyth [Large, in reference to the showy flowers.] SHOWY EVENINGPRIMROSE.

> *Oenothera sinuata* Linnaeus var. *grandiflora* S. Watson, Prodr. Amer. Acad. Arts 8: 581. 1873. *Oenothera sinuata* Linnaeus var. *grandis* Britton, Mem. Torrey Bot. Club 5: 358. 1894, nom. illegit. *Oenothera laciniata* Hill var. *occidentalis* Small, Bull. Torrey Bot. Club 23: 173. 1896, nom. illegit. *Oenothera laciniata* Hill var. *grandis* Britton, in Britton & A. Brown, Ill. Fl. N. U.S. 2: 487. 1897, nom. illegit. *Oenothera grandis* Smyth, Trans. Kansas Acad. Sci. 16. 160. 1899. *Oenothera laciniata* Hill var. *grandiflora* (S. Watson) B. L. Robinson, Rhodora 10: 34. 1908. *Raimannia grandis* (Smyth) Rose ex Britton & A. Brown, Ill. Fl. N. U.S., ed. 2. 2: 597. 1913.

Erect annual herb, to 0.6(1) m; stem simple or much branched from the base and above, the lower branches usually decumbent, strigulose and with long spreading trichomes. Basal rosette leaves with the blade oblanceolate or broadly elliptic, 1.5–6 cm long, the margin entire or sinuate-dentate, often pinnatifid below, the upper and lower surfaces glabrate or strigulose, the petiole usually longer than the blade, the cauline leaves alternate, the blade lanceolate to oblanceolate or elliptic, 2–6 cm long, 5–17 mm wide, the margin sinuate-denticulate to sinuate-pinnatifid, rarely entire, sometimes deeply lobed, the upper and lower surfaces glabrate to strigulose or villous-hirsute, sessile or the petiole to 6 mm long. Flowers solitary in the axil of the upper leaves; buds erect, but the stem tip nodding; hypanthium 2.5–5 cm long, sparsely to densely villous; sepals lanceolate, (1.5)2–3 cm long, remaining coherent at anthesis and reflexed to 1 side, the free tips 2–5 mm long; petals obovate, (2)2.5–3.5(4) cm long, the apex truncate or emarginated, yellow, often fading pinkish; stamens ⅔ as long as the petals, the anthers 7–9 mm long; styles longer than the stamens, the stigmas 7–8 mm long. Fruit a dehiscent capsule, subcylindric, (1)2.5–3.5(4) cm long, strigulose and with spreading trichomes; seeds numerous, round.

Disturbed sites. Rare; Polk County. New York south to Florida, west to Colorado and New Mexico, Mexico. Native to the Great Plains. Summer–fall.

Oenothera humifusa Nutt. [Spread out over the ground, procumbent.] SEABEACH EVENINGPRIMROSE.

Oenothera humifusa Nuttall, Gen. N. Amer. Pl. 1: 245. 1818. *Oenothera sinuata* Linnaeus var. *humifusa* (Nuttall) Torrey & A. Gray, Fl. N. Amer. 1: 494. 1840. *Raimannia humifusa* (Nuttall) Rose ex Britton & A. Brown, Ill. Fl. N. U.S., ed. 2. 2: 597. 1913. TYPE: FLORIDA: Nassau Co.: sea beach near Cumberland Island [GA], s.d., *Baldwin 883* (holotype: PH).

Oenothera niveifolia Gandoger, Bull. Soc. Bot. France 65: 27. 1918. TYPE: FLORIDA: Escambia Co.: Perdido, 25 May 1903, *Tracy 8719* (holotype: P; isotypes: BM, CU, E, F, G, GH, MIN, MO, MSC, NCU, NY, PENN, TAES, UMO, US, WIS).

Ascending or decumbent perennial herb, to 5 dm; stems much branched at the base, silky-canescent. Leaves with the blade oblong-spatulate to oblanceolate or lanceolate, 2–3.5 cm long, 5–10 mm wide, the apex acute, the base cuneate to rounded, sessile, the margin undulate-repand or shallowly dentate to subentire. Flowers solitary in the leaf axil; bracts flat, hypanthium 2–3 cm long; sepals linear-lanceolate, 3–11 mm long; petals obovate, 2–4.5 cm long, the apex truncate to emarginated, yellow. Fruit a dehiscent capsule, subcylindric, 2–3 cm long, striate; seeds numerous, round.

Coastal dunes and beaches. Frequent; peninsula, central and western panhandle. Pennsylvania south to Florida, west to Louisiana. All year.

Oenothera laciniata Hill [Slashed into narrow divisions, in reference to the leaf margin.] CUTLEAF EVENINGPRIMROSE.

Oenothera laciniata Hill, Syst. Veg. 12(App.): 64, pl. 10. 1767. *Raimannia laciniata* (Hill) Rose ex Britton & A. Brown, Ill. Fl. N. U.S., ed. 2. 2: 597. 1913.

Decumbent or ascending annual or short-lived perennial herb, to 5(8) dm; stem much branched at the base, glabrous or strigulose and sometimes sparsely villous. Basal rosette leaves with the blade oblanceolate, (1)3–7(9) cm long, (0.5)1–2(3) cm wide, the margin entire, sinuate-dentate, or pinnatifid, the upper and lower surfaces glabrate to strigulose or spreading villous, the petiole usually as long as the blade, the cauline leaves alternate, the blade lanceolate, elliptic, or oblanceolate to spatulate, 2.5–5 cm long, 1–3 cm wide, the apex acute or obtuse, the base tapered to a winged petiole or cuneate and sessile, the margin sinuate-dentate to pinnate or subentire, the upper and lower surfaces glabrate to strigulose or sparsely villous, sessile or the petiole to 3 mm long. Flowers solitary in the leaf axil; buds and stem tips erect or nodding; hypanthium 3–3.5 cm long; sepals ca. 1 cm long, reflexed; petals obovate, the apex truncate or emarginate, yellow. Fruit a dehiscent capsule, subcylindric, 2.5–3.5 cm long; seeds numerous, rounded, strongly pitted.

Open pinelands, open woodlands, and dry, open disturbed sites. Common; nearly throughout. Maine south to Florida, west to Ontario, Colorado, New Mexico, and California; Mexico. All year.

Oenothera lindheimeri (Engelm. & A. Gray) W. L. Wagner & Hoch [Commemorates Ferdinand Jacob Lindheimer (1801–1879), German-born and educated Texas botanist.] LINDHEIMER'S BEEBLOSSOM.

Gaura lindheimeri Engelmann & A. Gray, Boston J. Nat. Hist. 5: 217. 1845. *Oenothera lindheimeri* (Engelmann & A. Gray) W. L. Wagner & Hoch, in W. L. Wagner et al., Syst. Bot. Monogr. 83: 213. 2007.

Erect perennial herb, to 9 dm; stem usually freely branched above, short-hirsute. Lower cauline leaves with the blade spatulate to oblong-oblanceolate, 3–9 cm long, 7–15 mm wide, the apex acute to obtuse, the base cuneate to a short winged petiole or subsessile, the margin sinuate-dentate or -denticulate, the upper leaves linear-elliptic, 5–20 mm long, the margin subentire. Flowers in a simple spike 2–6 dm long, the peduncle 5 cm long or longer, the axis glabrous or pilose and glandular pubescent; hypanthium slender, 5–8 mm long, reddish, the outer surface with glandular-tipped trichomes and longer eglandular ones, the inner surface white-pubescent; sepals linear-lanceolate, 12–15 mm long, pilose and glandular- pubescent, reddish, separate and reflexed at anthesis; petals spatulate-rhomboid, 12–20 mm long, white, turning rose; stamens subequal, the filaments 10–15 mm long, slightly flattened upward, the anthers red; style slightly exceeding the stamens, pubescent at the base. Fruit indehiscent, nutlike, elliptic-oblong, 7–9 mm long, sharply 4-angled, subglabrous or pilose, the pedicel thick, less than 1 mm long; seeds 1–4.

Dry, disturbed sites. Rare; Liberty County. Florida, Louisiana, and Texas. Native to Texas and Louisiana. Spring–summer.

Oenothera linifolia Nutt. [With very narrow leaves.] THREADLEAF EVENINGPRIMROSE.

Oenothera linifolia Nuttall, J. Acad. Nat. Sci. Philadelphia 2: 120. 1821. *Kneiffia linifolia* (Nutt.) Spach, Nouv. Ann. Mus. Hist. Nat. 4: 368. 1835. *Kneiffia linearifolia* Spach, Ann. Sci. Nat., Bot., ser. 2. 4: 167, nom. illegit. *Peniophyllum linifolium* (Nuttall) Pennell, Bull. Torrey Bot. Club 46: 373. 1919. *Oenothera linifolia* Nuttall var. *typica* Munz, Bull. Torrey Bot. Club 64: 289. 1937, nom. inadmiss.

Erect annual herb, to 5 dm; stem simple or few to many branched from the base or above, sparsely pilose at the base, strigulose or glandular-puberulent above. Basal rosette leaves with the blade ovate to obovate or narrowly elliptic, 1–2(4) cm long, 2–6 mm wide, the margin entire to remotely dentate, the upper and lower surfaces glabrous to sparsely strigulose or glandular-puberulent, especially along the petiole, the winged petiole 2–10(15) mm long, the lower cauline leaves becoming abruptly sessile, linear or filiform, 1–4 cm long, less than 1 mm wide. Flowers in a terminal, unbranched spike (1)3–6(12) cm long, glandular-puberulent to strigulose; bracts ovate to deltoid-ovate, 0.5–2 mm long; hypanthium 1–2 mm long; sepals ca. 2 mm long, without free tips; petals obovate, 3–5(7) mm long, yellow; filaments 1–2 mm long, the anthers 0.5–1 mm long; style 1–2 mm long, stigma shallowly 4-lobed. Fruit a dehiscent capsule, subcylindric, 4–6(10) mm long, 4-angled, sessile or the stipe 1–4 mm long; seeds numerous, oblong, pale reddish brown, minutely verrucose.

Dry, disturbed sites. Rare; Washington County. Virginia south to Florida, west to Kansas, Oklahoma, and Texas. Spring.

Oenothera nutans G. F. Atk. & Bartlett [Nodding]. NODDING EVENINGPRIMROSE.

Oenothera nutans G. F. Atkinson & Bartlett, Science, ser. 2. 37: 717. 1913; Rhodora 15: 83. 1913. *Oenothera biennis* Linnaeus var. *nutans* (G. F. Atkinson & Bartlett) Wiegand, Rhodora 26: 3. 1924.

Erect biennial herb, to 1.5 m; stem sparsely pubescent. Leaves with the blade elliptic-lanceolate, 15–24 cm long, 3.5–5 cm wide, the apex acute, the base cuneate, the margin sinuate-dentate or entire. Flowers solitary in the leaf axil, hypanthium glabrate; sepal lobes 1–2.5 cm long; petals obovate, 1–2.5 cm long, yellow; styles 5–17 mm long, subequaling the anthers; fruit a dehiscent capsule, subcylindric; seeds numerous, prismatic.

Dry, open sites. Rare; Jackson County. Maine south to Florida, west to Ontario, Missouri, and Arkansas. Summer–fall.

Oenothera simulans (Small) W. L. Wagner & Hoch [In reference to its resemblance to *Gaura angustifolia*.] SOUTHERN BEEBLOSSOM.

Gaura angustifolia Michaux, Fl. Bor.-Amer. 1: 226. 1803. *Gaura angustifolia* Michaux var. *typica* Munz, Bull. Torrey Bot. Club 65: 117. 1938, nom. inadmiss.

Gaura simulans Small, Bull. New York Bot. Gard. 3: 432. 1905. *Gaura angustifolia* Michaux var. *simulans* (Small) Munz, Bull. Torrey Bot. Club 65: 117. 1938. *Oenothera simulans* (Small) W. L. Wagner & Hoch, in W. L. Wagner et al., Syst. Bot. Monogr. 83: 213. 2007. TYPE: FLORIDA: Miami-Dade Co.: between Coconut Grove and Cutler, Nov 1913, *Small & Carter 766* (holotype: NY; isotypes: F, GH, MIN).

Gaura eatonii Small, Fl. S.E. U.S., ed. 2. 1353, 1375. 1913. *Gaura angustifolia* Michaux var. *eatonii* (Small) Munz, Bull. Torrey Bot. Club 65: 118. 1938. TYPE: FLORIDA: Lee Co.: Punta Rassa, 4 Mar 1905, *Eaton 1289* (holotype: NY).

Gaura angustifolia Michaux var. *strigosa* Munz, Bull. Torrey Bot. Club 65: 118. 1938. TYPE: FLORIDA: Lake Co.: Eustis, 16–31 Jul 1894, *Nash 1464* (holotype: NY; isotypes: GH, US).

Erect annual or biennial herb, to 18 dm; stem simple below, with slender divaricate branches above, sparsely loose-uncinate. Basal rosette leaves with the blade narrowly oblanceolate, 5–18 cm long, 7–15 mm wide, the apex acute, the base gradually narrowed into a winged petiole 1–7 cm long, the margin irregularly sinuate-dentate, the main cauline leaves narrowly oblanceolate to narrowly lanceolate, 3–6 cm long, 5–8 mm wide, the apex acute, the margin subentire to sinuate-dentate, sessile or short-petiolate, the upper cauline leaves with the blade lance-linear,

1–3 cm long, 1–3 mm wide, gradually reduced upward, the margin subentire, the apex acute, sessile. Flowers in a terminal simple or openly branched spike 1–3 dm long; hypanthium slender, 3–5 mm long, pubescent, reddish; sepals lanceolate, 4–10(12) mm long, separate and reflexed or somewhat adherent in anthesis, pubescent to glabrous; petals spatulate to spatulate-oblong, 4–10 mm long, white to pink; stamens slightly unequal, 4–5 mm long, the filaments flattened, glabrous, each with a minute fleshy, basal scale, the anthers white to pinkish; style slightly exceeding the stamens, pubescent in the lower part. Fruit indehiscent, nutlike, elliptic to narrowly ovoid, 5–10 mm long, sharply 3- to 4-angled, narrowed into a thick pedicel-like base less than 1 mm long, pubescent or glabrous; seeds 2.

Sandhills and dry, disturbed sites. Common; nearly throughout. North Carolina south to Florida, west to Mississippi. Spring–summer.

Oenothera sinuosa W. L. Wagner & Hoch [Wavy, in reference to the leaf margin.] WAVYLEAF BEEBLOSSOM.

Gaura sinuata Nuttall ex Seringe, in de Candolle, Prodr. 3: 44. 1828. Oenothera sinusoa W. L. Wagner & Hoch, in W. L. Wagner et al., Syst. Bot. Monogr. 83: 214. 2007.

Perennial herb, to 8 dm; stem simple or with several ascending branches from above the base, glabrous or subglabrate. Basal rosette leaves with the blade oblanceolate to oblong-lanceolate, 3–8 cm long, 1–2 cm wide, the apex obtuse to acute, the base tapering to a short, winged petiole, the margin sinuate-dentate, the upper and lower surfaces subglabrous or puberulent, especially on the margin, the cauline leaves with the blade spatulate to lanceolate or linear, 1–5 cm long, the margin sinuate-dentate to subentire, often wavy, the upper and lower surfaces subglabrous to strigulose. Flowers in a simple or branched spike 2–5 dm long, the peduncle 1 dm long or longer; floral bracts lanceolate or ovate, 1–3 mm long, subglabrous or with a few short trichomes on the margins, caducous; hypanthium narrow-funnelform, 2.5–3 mm long, the outer surface subglabrous or sparsely strigulose, the inner surface white-pubescent; sepals linear-lanceolate, 7–10 mm long, grayish-strigulose, separate and reflexed at anthesis; petals broadly elliptic, 8–10 mm long, white, becoming red; filaments unequal, 8–10 mm long, slightly enlarged upward, the anthers reddish; style longer than the stamens, puberulent at the base. Fruit indehiscent, nutlike, fusiform, 5–9 mm long, obtusely 4-angled, gradually tapering to a rather thick pedicel-like base 2–8 mm long; seeds 1–4.

Disturbed sites. Rare; Alachua, Marion, and Calhoun Counties. Native to Texas, Oklahoma, and Mexico. Summer–fall.

Oenothera speciosa Nutt. [Showy, splendid.] PINKLADIES.

Oenothera speciosa Nuttall, J. Acad. Nat. Sci. Philadelphia 2: 119. 1821. Xylopleurum nuttallii Spach, Nouv. Ann. Mus. Hist. Nat. 4: 371. 1835, nom. illegit. Xylopleurum speciosum (Nuttall) Raimann, in Engler & Prantl, Nat. Pflanzenfam. 3(7): 214. 1893. Hartmannia speciosa (Nuttall) Small, Bull. Torrey Bot. Club 23: 181. 1896. Oenothera speciosa Nuttall var. typica Munz, Amer. J. Bot. 19: 764. 1932, nom. inadmiss.

Erect or ascending, rhizomatous perennial herb, to 5 dm; stem simple or branched, strigulose. Basal rosette leaves with the blade oblanceolate to obovate, 2–9 cm long, 4–25 mm wide, the apex obtuse, the base cuneate, margin sinuate-pinnatifid with lanceolate or ovate lateral lobes, the petiole to 3 cm long, the cauline leaves with the blade oblong-lanceolate, oblanceolate, or elliptic, 2–10 cm long, the apex acute, the base cuneate, the margin subentire, sinuate-dentate, to pinnatifid, the upper and lower surfaces strigose, reduced upward to lance-linear bracts in the inflorescence. Flowers solitary in the axil of the upper leaves, the inflorescence sharply nodding, the flowers opening in the evening or morning; bracts lance-linear to elliptic, 1–2 cm long; hypanthium 1–2 cm long, strigose, the free tips 1–4(5) mm long; sepals lanceolate, 1.5–3 cm long, remaining coherent and reflexed to one side at anthesis, the free tips 1–4 mm long; petals obcordate, 2.5–4 cm long, white or fading reddish or rose-purple; stamens ⅔ as long to nearly as long as the petals, the anthers 10–12 mm long; style as long as the petals, the stigmas 3–6 mm long. Fruit a dehiscent capsule, subcylindric, 1–1.5 cm long, strigose, ribbed distally, sessile; seeds numerous.

Open, disturbed sites. Frequent; nearly throughout. Connecticut south to Florida, west to California; Mexico. Native to the lower Midwestern United States. Spring–fall.

EXCLUDED TAXA

Oenothera heterophylla Spach—Reported by Small (1903), who misapplied the name to material of *O. curtissii.*

Oenothera mollissima Linnaeus—Reported for Florida by Small (1933, as *Raimannia mollissima* (Linnaeus) Sprague & L. Riley), apparently misapplied to material of *O. humifusa.*

Oenothera rhombipetala Nuttall ex Torrey & A. Gray—Reported for Florida by Radford et al. (1964, 1968) and Wilhelm (1984), both misapplications of the name to material of *O. curtissii.*

MYRTACEAE Juss., nom. cons. 1789. MYRTLE FAMILY

Trees or shrubs. Leaves opposite or alternate, simple, pinnate-veined or the secondary veins basal and parallel, petiolate, estipulate. Flowers in axillary racemes, panicles, or spikes, bracteate or ebracteate, bracteolate or ebracteolate, actinomorphic, bisexual; calyx 4–5(7) or absent, the lobes free or fused into a calyptra; petals usually equaling the sepals, free or connate into a calyptra; ovary inferior and partly so; nectary disk present, glandular or eglandular; stamens 10–many, the anther basifixed or dorsifixed, dehiscent by slits; ovary 1- to 6-carpellate and -loculate. Fruit a berry, capsule with apical dehiscence, or nutlike; seeds 1–many.

A family of about 100 genera and about 3,500 species; nearly cosmopolitan.

1. Leaves alternate; fruit a woody capsule.
   2. Leaves pinnate-veined; petals fused into a calyptra (lid), this circumscissile and dehiscent at anthesis..**Eucalyptus**
   2. Leaves with the secondary veins basal and parallel; petals free.. **Melaleuca**
1. Leaves opposite; fruit a fleshy berry.
   3. Corolla rose-pink; buds and the lower leaf surface whitish-tomentose ....**Rhodomyrtus**
   3. Corolla white, yellowish white, or absent; buds and the lower leaf surface glabrous or pubescent, but not whitish-tomentose.

4. Flowers borne in a compound panicle.
    5. Calyx circumscissile, the terminal part a calyptra (lid); petals absent.............. **Calyptranthes**
    5. Calyx not circumscissile; petals present.
        6. Leaf apex acute to acuminate; fruit 1-seeded .... **Syzygium**
        6. Leaf apex obtuse; fruit many-seeded ....**Pimenta**
4. Flowers solitary or borne in a dichasium or a raceme.
    7. Calyx closed or nearly so in bud, splitting to the disk into 4 or 5 lobes at anthesis; fruit many-seeded.
        8. Fruit ca. 1 cm long; flowering pedicel slender, 2–3.5 cm long; leaves 1–5 cm long; perianth 4-merous .... **Mosiera**
        8. Fruit 2–4 cm long; flowering pedicel stout, 1–2 cm long; leaves 4–14 cm long; perianth 5(4)-merous ....**Psidium**
    7. Calyx open in bud; fruit few-seeded.
        9. Inflorescence cymose .... **Myrcianthes**
        9. Inflorescence racemose (sometimes this short, congested, and appearing fasciculate) or the flowers solitary.
            10. Leaves to ca. 6 cm long; flowers to ca. 1 cm long.. **Eugenia**
            10. Leaves 10 cm long or longer; flowers 3–4 cm long .... **Syzygium**

Calyptranthes Sw., nom. cons. 1788. LIDFLOWER

Shrubs or trees. Leaves opposite, pinnate-veined, petiolate, estipulate. Flowers in subterminal panicles; sepals 4, connate, the calyx forming a circumscissile calyptra, but usually remaining attached at one side; petals absent; stamens ca. 200, inserted on the margin of the hypanthium, which is projected above the ovary, longitudinally dehiscent; ovary 2-carpellate and -loculate. Fruit a berry.

A genus of about 200 species; Florida, West Indies, Mexico, Central America, and South America. [From the Greek *kalyptra*, cap or covering, and *anthos*, flower, in reference to the cap-like calyx.]

1. Inflorescence tomentulose, the flowers in sessile clusters at the ends of the panicle branches; calyptra obtuse to rounded; young stems narrowly 2-winged ....**C. pallens**
1. Inflorescence glabrous, the flowers pedicellate; calyptra apiculate; young stems not winged.................
    .... **C. zuzygium**

Calyptranthes pallens Griseb. [Pale, in reference to the leaf undersurface.] PALE LIDFLOWER; SPICEWOOD.

Calyptranthes chytraculia (Linnaeus) Swartz var. *pauciflora* O. Berg, Linnaea 27: 27. 1855. *Calyptranthes pallens* Grisebach, Abh. Königl. Ges. Wiss, Göttingen 7: 215. 1857. *Chytraculia pauciflora* (O. Berg) Kuntze, Revis. Gen. Pl. 1: 238. 1891, nom. illegit. *Chytraculia chytraculia* (Linnaeus) Sudworth var. *pauciflora* (O. Berg) Sudworth, U.S.D.A. Div. Forest. Bull. 14: 305. 1897. *Chytraculia pallens* (Grisebach) Britton, in Shattuck, Bahama Isl. 260. 1905.

Shrub or tree, to 10 m; branchlets narrowly 2-winged, tomentulose with light brown to yellowish white trichomes, the bark pale gray, smooth or scaly. Leaves with the blade, elliptic to ovate, 3–6(10) cm long, 2–4(6) cm wide, the apex acute to acuminate, the base obtuse to cuneate, lateral veins 10–15 pairs, with a single marginal vein 1–2 mm from the margin, the margin entire, the upper surface appressed-puberulous or glabrous, lustrous, the midvein sulcate, the lower surface appressed pubescent to glabrescent, the petiole 4–8 mm long, puberulent. Flowers (20)30–50+ in a panicle 7–15 cm long, in sessile clusters at the ends of the panicle branches, tomentulose, the peduncle slender, 2–4 cm long; bracts ovate, deciduous, bracteoles minute, early deciduous, the pedicel 1–2 mm long; hypanthium narrow crateriform; calyptra obtuse to rounded; stamens ca. 5 mm long. Fruit subglobose, 5–8 mm long, purplish black, lustrous, crowned by a cylindric hypanthium, glabrous or sparsely pubescent; seed 1, subglobose, 4–5 mm long.

Hammocks. Occasional; Miami-Dade County, Monroe County keys. Florida; West Indies, Mexico, and Central America. Spring–fall.

Calyptranthes pallens is listed as threatened in Florida (Florida Administrative Code, Chapter 5B-40).

Calyptranthes zuzygium (L.) Sw. [Native vernacular name.]
MYRTLE-OF-THE-RIVER.

Myrtus zuzygium Linnaeus, Syst. Nat., ed. 10. 1056. 1759. *Calyptranthes zuzygium* (Linnaeus) Swartz, Prodr. 79. 1788. *Calyptranthes chytraculia* (Linnaeus) Swartz var. *zuzygium* (Linnaeus) O. Berg, Linnaea 27: 28. 1855. *Chytraculia zuzygium* (Linnaeus) Kuntze, Revis. Gen. Pl. 1: 238. 1891. *Chytraculia chytraculia* (Linnaeus) Sudworth var. *zuzygium* (Linnaeus) Sudworth, U.S.D.A. Div. Forest. Bull. 14: 305. 1897.

Shrub or tree, to 12 m; branchlets terete to slightly compressed, not winged, glabrous, the bark pale gray, smooth. Leaves with the blade elliptic to obovate or ovate, 4–6(7) cm long, 2–4 cm wide, the base cuneate, the apex obtuse to abruptly blunt-tipped, the lateral veins ca. 20 pairs, the marginal vein 1, 1–2 mm from the margin, the margin entire, the upper surface glabrous, lustrous, the midvein convex, the lower surface glabrous, the petiole 2–4 mm long, glabrous. Flowers 9–20 in a solitary or paired axillary panicle, 1–3 on pedicels to 5 mm long at the ends of the panicle branches, the peduncle 2.5–5 cm long; bracts and bracteoles early deciduous; hypanthium crateriform; calyptra apiculate; stamens ca. 4 mm long. Fruit spheroid to oblate, 8–10 mm long, bluish black, glaucous; seed 1, subglobose, 7–10 mm long.

Hammocks. Rare; Miami-Dade County, Monroe County keys. Florida; West Indies. Spring–summer.

Calyptranthes zuzygium is listed as endangered in Florida (Florida Administrative Code, Chapter 5B-40).

EXCLUDED TAXON

Calyptranthes chytraculia (Linnaeus) Swartz—Reported for Florida by Chapman (1860, 1883, 1897) and Small (1903, as *Chytraculia chytraculia* (Linnaeus) Sudworth), the name of this West Indian species misapplied to material of *C. pallens*.

Eucalyptus L'Hér. 1789.

Trees. Leaves alternate, pinnate-veined, petiolate, estipulate. Flowers in terminal or subterminal axillary umbels or panicles of umbels; perianth parts 4–5, fused into a calyptra; stamens numerous; ovary 3- to 6-carpellate and -loculate. Fruit a capsule; seeds numerous.

A genus of about 700 species; nearly cosmopolitan. [From the Greek *eu*, true, and *kalyptos*, covered, in reference to the calyptra covering the stamens in the flower.]

Corymbia Hill & L.A.S. Johnson (1995).

1. Young leaves and branchlets reddish hirsute; leaves ovate-lanceolate, the apex obtuse......**E. torelliana**
1. Young leaves and branchlets glabrous; leaves linear-lanceolate, the apex long-acuminate.
   2. Adult leaves 0.7–2 cm wide; fruit hemispheric or ovoid .... **E. camaldulensis**
   2. Adult leaves 2–4 cm wide; fruit subpyriform or cylindric.
      3. Fruit subpyriform, 5–8 mm long, valves exserted; bark shedding, smooth on the upper trunk and branches..**E. grandis**
      3. Fruit cylindric, 10–18 mm long, valves included; bark rough throughout....................**E. robusta**

Eucalyptus camaldulensis Dehnh. subsp. **acuta** Brooker & M. W. McDonald
[Named for a private estate garden near the Camaldoli Monastery near Naples, Italy, where the specimen used to describe the plant came from; with acute leaves.] RIVER REDGUM.

> *Eucalyptus camaldulensis* Dehnhardt subsp. *acuta* Brooker & M. W. McDonald, Austral. Syst. Bot. 22: 270, f. 7. 2009.

Tree, to 25 m; bark gray or tan, smooth or nearly so, the branchlets glabrous. Leaves with the blade narrowly lanceolate, 6–20 cm long, 1.5–2.5 cm long, often falcate, the apex acuminate, the base obtuse, the margin entire, the upper and lower surfaces glabrous, the petiole 0.3–2 cm long. Flowers usually 7 in an umbel; hypanthium hemispheric, 2–3 mm long; calyptra subhemispheric, conic; stamens white, the anthers dorsifixed. Fruit hemispheric, 5–9 mm long, glabrous, the valves 3–5, exserted.

Disturbed sites. Rare; Charlotte County. Escaped from cultivation. Florida; Australia. Native to Australia. Summer.

Eucalyptus grandis W. Hill ex Maiden [Large.] GRAND EUCALYPTUS.

> *Eucalyptus grandis* W. Hill ex Maiden, J. Proc. Roy Soc. New South Wales 52: 501. 1919. *Eucalyptus saligna* Smith var. *pallidivalvis* R. T. Baker & H. G. Smith, Res. Eucalypts 32. 1902.

Tree, to 50 m; bark white, grayish white, or bluish gray, smooth, the branchlets glabrous. Leaves with the blade lanceolate to elliptic, 9.5–16 cm long, 2–5 cm wide, often falcate, the apex acuminate, the base rounded to obtuse, the margin entire, the upper and lower surfaces glabrous, grayish green or yellow-green, the petiole 1–2.2 cm long. Flowers 7–11 in an umbel; hypanthium obconic or campanulate, 3–4 mm long; calyptra conic, 3–4 mm long; stamens white, the anthers dorsifixed. Fruit subpyriform, 5–8 mm long, glabrous, the valves 4 or 5, exserted.

Disturbed sites. Rare; Pinellas, Glades, Hendry, and Palm Beach Counties. Escaped from cultivation. Florida and California; South America; Europe, Africa, Asia, Australia, and Pacific Islands. Native to Australia. Summer.

Eucalyptus robusta Sm. [Large size.] SWAMPMAHOGANY.

> *Eucalyptus robusta* Smith, Spec. Bot. New Holland 39, t. 13. 1795.

Tree, to 30 m; bark reddish brown, rough, deeply furrowed. Leaves with the blade broadly lanceolate, 8.5–17 cm long, 2.5–7 cm wide, the apex acuminate, the base cuneate, the margin entire, the upper and lower surfaces glabrous, the petiole 1.5–3 cm long. Flowers 9–15 in an umbel, sessile, the peduncle 1.5–3 cm long; hypanthium obconic to pyriform, 6–7 mm long; calyptra conic to rostrate, 10–12 mm long; stamens white, the anthers dorsifixed. Fruit cylindric, 10–18 mm long, glabrous, the valves 3–4, exserted.

Disturbed sites. Occasional; Volusia, Brevard, St. Lucie, Martin, Pinellas, Charlotte, and Lee Counties. Escaped from cultivation. Florida and California; Asia, Australia, and Pacific Islands. Native to Australia. Summer.

Eucalyptus torelliana F. Muell. [Commemorates Count I. de Torelli (1810–1887), a member of the Italian Senate who promoted the use of *Eucalyptus* to dry up the malarial marshes near Rome.] TORELL'S EUCALYPTUS; CADAGA.

> *Eucalyptus torelliana* F. Mueller, Fragm. 10: 106. 1877. *Corymbia torelliana* (F. Mueller) K. D. Hill & L.A.S. Johnson, Telopea 6: 385. 1995.

Tree, to 30 m; bark slate-green, gray, or black, glabrous, smooth, the branchlets reddish, hirsute or setose. Leaves with the blade cordate to ovate or elliptic, 8–15 cm long, 5–11 cm wide, the apex obtuse or rounded, the base peltate, the margin undulate, the upper surface glabrous, the lower surface reddish hirsute, at least on the veins, the petiole 2–2.5 cm long, reddish hirsute. Flowers 3–7 in a panicle of umbels, the peduncle ca. 4 cm long, the pedicel 2–5 mm long; hypanthium ovoid, 6–8 mm long; calyptra rounded to conic to slightly rostrate; stamens white, the anthers versatile. Fruit urn-shaped or truncate-globose, 9–15 mm long, glabrous, the valves 3, deeply included.

Disturbed sites. Rare; Palm Beach and Lee Counties. Escaped from cultivation. Florida; Africa, Asia, and Australia. Native to Australia. Summer.

Eucalyptus torelliana is sometimes placed in *Corymbia*, a genus of about 113 species. However, there is still considerable uncertainty regarding eucalypt classification, especially regarding whether *Corymbia* is monophyletic or paraphyletic with another *Eucalyptus* segregate *Angophora*, and it seems best at this time to retain the species in *Eucalyptus* s.s.

Eugenia L. 1753. STOPPER

Shrubs or trees. Leaves opposite, pinnate-veined, petiolate, estipulate. Flowers in axillary racemes or solitary; bracteoles 2, free or connate, forming an involucre below the hypanthium; hypanthium not prolonged beyond the ovary; sepal lobes 4, in 2 opposing equal or markedly unequal pairs; petals 4; stamens 25–70; nectary disk present; ovary 2-carpellate and -loculate. Fruit a berry; seeds 1(2).

A genus of about 1,000 species; Florida, West Indies, Mexico, Central America, South America, Africa, and Asia. [Commemorates Prince Eugene of Savoy (1663–1736).]

1. Pedicels less than 5 mm long, stout.
    2. Leaf blades oblanceolate (rarely elliptic), the apex rounded or obtuse .... E. foetida
    2. Leaf blades ovate or lanceolate (rarely elliptic), the apex acute .... E. axillaris
1. Pedicels more than 5 mm long, slender.
    3. Calyx lobes 4–8 mm long, lanceolate, equal .... E. uniflora
    3. Calyx lobes less than 4 mm long, reniform, in 2 equal series.
        4. Leaf blades with a blunt acuminate apex, the upper surface dull .... E. rhombea
        4. Leaf blades with a slender acuminate apex, the upper surface shiny .... E. confusa

Eugenia axillaris (Sw.) Willd. [In the leaf axil, in reference to the flowers.] WHITE STOPPER.

Myrtus axillaris Swartz, Prodr. 78. 1788. *Eugenia axillaris* (Swartz) Willdenow, Sp. Pl. 2: 960. 1799. *Eugenia anthera* Small, Man. S.E. Fl. 935, 1506. 1933. TYPE: FLORIDA: Indian River Co.: near Roseland, Aug 1928, *Mosier s.n.* (holotype: NY).

Tree, to 10 m; bark gray or brown, the branchlets terete or compressed at the nodes, glabrous. Leaves with the blade ovate or elliptic, 4–8 cm long, 2–4 cm wide, leathery, the apex acute to rounded, the base cuneate, decurrent on the distal edge of the petiole, the margin entire, the upper and lower surfaces with scattered glands, these obscure on the upper surface, the petiole splayed or flattened, 3–8 mm long. Flowers 4–8 in an axillary raceme 3–6 mm long, these solitary or 2 superposed, the pedicel 1–3 mm long; bracteoles ovate, ca. 0.5 mm long, the base connate and involucrate or free; hypanthium campanulate, ca. 1 mm long; sepals elliptic, in 2 unequal pairs, the larger pair ca. 1 mm long, the margin ciliate; petals elliptic, ca. 3 mm long; stamens 30–50, 2–3 mm long. Fruit subglobose, 6–9 mm long, purplish black, the calyx persistent.

Coastal hammocks, rarely inland. Frequent; central and southern peninsula. Florida; West Indies, Mexico, and Central America. All year.

Eugenia confusa DC. [Confused as to its identity.] REDBERRY STOPPER; REDBERRY EUGENIA.

Eugenia confusa de Candolle, Prodr. 3: 279. 1828.
Eugenia garberi Sargent, Gard. & Forest 2: 28, t. 87. 1889. TYPE: FLORIDA.

Tree or shrub, to 6 m; trunk bark reddish brown, tan, or gray, the branchlets glabrous or glabrate, terete or weakly compressed. Leaves with the blade ovate to elliptic-ovate, 2.5–6 cm long, 1–3 cm wide, leathery, the apex caudate-acuminate, the base rounded to cuneate, the margin entire, the upper and lower surfaces with evident or obscure glands, glabrous, the petiole 3–9 mm long, channeled. Flowers 2–8 in an axillary raceme, the axis 1–4 mm long, or solitary; bracteoles ovate, ca. 1 mm long, free, caducous at anthesis, the pedicel 6–15 mm long;

hypanthium obconic, ca. 2 mm long; sepals in 2 unequal pairs, the larger 1–2 mm long; petals obovate, 3–4 mm long; stamens ca. 40, 2–4 mm long. Fruit globose or obovoid, 6–9 mm long, bright red, the calyx persistent.

Tropical hammocks. Rare; Martin and Miami-Dade Counties, Monroe County keys. Florida; West Indies. All year.

Eugenia confusa is listed as endangered in Florida (Florida Administrative Code, Chapter 5B-40).

Eugenia foetida Pers. [Stinking.] SPANISH STOPPER; BOXLEAF STOPPER.

> *Eugenia foetida* Persoon, Syn. Pl. 2: 29. 1806. *Eugenia foetida* Persoon var. *genuina* O. Berg, Linnaea 27: 211. 1856, nom. inadmiss.
>
> *Myrtus buxifolia* Swartz, Prodr. 78. 1788. *Eugenia buxifolia* (Swartz) Willdenow, Sp. Pl. 2: 960. 1799; non Lamarck, 1789. *Eugenia myrtoides* Poiret, in Lamarck, Encycl., Suppl. 3: 125. 1813. *Eugenia triplinervia* O. Berg var. *buxifolia* (Swartz) O. Berg, Linnaea 27: 191. 1856.

Shrubs or trees, to 10 m; bark tan or gray, the branchlets puberulent. Leaves with the blade elliptic or obovate, 2.5–8 cm long, 0.8–3.5 cm wide, papery, the apex rounded or bluntly acute, sometimes retuse, the base cuneate, decurrent along the petiole, the margin entire, revolute, the upper and lower surfaces with minute glands, pubescent along the midvein and margin or glabrate, the petiole 2–5 mm long, terete or sulcate, puberulent. Flowers 4–8 in an axillary raceme, the axis 1–4 mm long, solitary or 2–3 superposed, the pedicel 1–3 mm long; bracteoles ovate, ca. 1 mm long, free, the margin scarious, persistent; hypanthium campanulate, ca. 1 mm long, pubescent; sepals in subequal pairs, ca. 1 mm long, the margin ciliate, the outer surface glabrous or with scattered trichomes; petals elliptic to widely ovate, 2–4 mm long, the margin ciliate; stamens 25–30, 3–5 mm long. Fruit globose, 4–6 mm long, black, the calyx persistent.

Coastal hammocks and dunes. Frequent; Brevard and Manatee Counties southward. Florida; West Indies, Mexico, and Central America. All year.

Eugenia rhombea (O. Berg) Krug & Urb. [Rhombic in shape, in reference to the leaves.] RED STOPPER.

> *Eugenia foetida* Persoon var. *rhombea* O. Berg, Linnaea 27: 212. 1856. *Eugenia rhombea* (O. Berg) Krug & Urban, in Urban, Bot. Jahrb. Syst. 19: 644. 1895. SYNTYPE: FLORIDA.

Shrub or tree, to 5(10) m; bark gray, the branchlets glabrous, compressed at the nodes. Leaves with the blade narrowly ovate to elliptic, 2–8 cm long, 1–3.5 cm wide, leathery, the apex bluntly acute or acuminate, the base rounded or cuneate, the margin entire, the upper and lower surfaces glabrous, glandular, the petiole 3–5 mm long, sulcate. Flowers solitary or 2–8, in a superficially fasciculate raceme ca. 2 mm long, the pedicel 8–20(30) mm long; bracteoles ovate to lanceolate, ca. 0.5 mm long, free, the margin minutely scarious; hypanthium globose or campanulate, ca. 2 mm long; sepals in unequal pairs, the larger pair 2–4 mm long, the margin ciliolate; petals elliptic, ca. 3 mm long, the margin ciliolate; stamens ca. 60, 2–4 mm long. Fruit globose or oblong, 4–7 mm long, dark red or purple, the calyx persistent.

Tropical hammocks. Rare; Miami-Dade County, Monroe County keys. Florida; West Indies, Mexico, and Central America. All year.

Eugenia rhombea is listed as endangered in Florida (Florida Administrative Code, Chapter 5B-40).

Eugenia uniflora L. [One-flowered.] SURINAM CHERRY.

Eugenia uniflora Linnaeus, Sp. Pl. 470. 1753. *Eugenia michelii* Lamarck, Encycl. 3: 203. 1789, nom. illegit. *Stenocalyx michelii* O. Berg, Linnaea 27: 310. 1856, nom. illegit. *Stenocalyx uniflorus* (Linnaeus) Kausel, Lilloa 32: 331. 1967.

Shrub or tree, to 10 m; bark reddish, the branchlets glabrous, compressed distally. Leaves with the blade ovate, 3–6 cm long, 1.5–3 cm wide, papery, the apex acute to acuminate, the base rounded, the margin entire, the upper and lower surfaces with small, numerous, raised glands, glabrous, the petiole 1–3 mm long, channeled. Flowers 2–6 in a short raceme 1–2 mm long, and appearing fasciculate, rarely solitary, the pedicel 15–25 mm long; bracteoles oblong-lanceolate, ca. 1 mm long, free, the margin ciliate; hypanthium campanulate, 1–2 mm long, 8-ribbed; sepals subequal, oblong, 2–4 mm long, the margin ciliate; petals obovate, 4–6 mm long, the margin ciliate; stamens 40–70, 4–6 mm long. Fruit subglobose, 12–15 mm long, 8-ribbed, red, the calyx persistent.

Disturbed hammocks. Occasional; central and southern peninsula. Escaped from cultivation. Florida; South America. Native to South America. All year.

Eugenia uniflora is listed as a Category I invasive species in Florida by the Florida Exotic Pest Plant Council (FLEPPC, 2015).

EXCLUDED TAXA

Eugenia monticola (Swartz) de Candolle—Reported for Florida by Chapman (1860, 1883, 1897) and Small (1903), the name misapplied to material of *E. axillaris*.

Eugenia procera (Swartz) Poiret—Reported for Florida by Chapman (1860, 1883, 1897) and Small (1903, 1913c), based on a misapplication of the name to material of *E. rhombea*.

Melaleuca L., nom. cons. 1767.

Shrubs or trees. Leaves alternate or opposite, pinnate-veined or the secondary veins basal and parallel, petiolate, estipulate. Flowers in psuedoterminal or axillary spikes, clusters, or sometimes solitary; hypanthium adnate to the ovary proximally to ¾ the ovary length; sepal lobes 5; petals 5; stamens numerous, the filaments connate proximally into 5 bundles; ovary 3-carpellate and -loculate. Fruit a capsule partially enclosed by a woody or subwoody hypanthium; seeds numerous, obovoid to oblong.

A genus of about 300 species; North America, Asia, Australia, and Pacific Islands. [From the Greek *melas*, black, and *leukos*, white, apparently in reference to the bark of the tree of some species, which may be black and white.]

Callistemon R. Br. (1814).

1. Leaves opposite..**M. linariifolia**
1. Leaves alternate.
  2. Flowers joined to the inflorescence axis in clusters of 3; filaments white, cream, greenish, or yellow..**M. quinquenervia**
  2. Flowers joined to the inflorescence axis singly; filaments red or crimson.....................**M. viminalis**

Melaleuca linariifolia Sm. [With narrow leaves.] CAJEPUT TREE.

Melaleuca linariifolia Smith, Trans. Linn. Soc. London 3: 278. 1797. *Myrtoleucodendron linariifolium* (Smith) Kuntze, Revis. Gen. Pl. 1: 241. 1891. *Melaleuca linariifolia* Smith var. *typica* Domin, Biblioth. Bot. 89: 456. 1928, nom. inadmiss.

Shrub or tree, to 10 m; bark papery. Leaves opposite, the blade linear, 1.7–4.5 cm long, 1–4 mm wide, the main veins 3, longitudinal, the apex acute, the base narrowly cuneate, the margin entire, the upper and lower surfaces glabrescent, the petiole to 2 mm long. Flowers 4–20 in a pseudoterminal raceme, in monads, sometimes also axillary distally; sepals setaceous, the outer surface glabrous, the margin scarious; petals ca. 3 mm long, deciduous; stamens in 5 bundles of 32–73 per bundle, white or cream, 8–24 mm long, the bundle claw (6)8–16 mm long. Fruit subglobose, 3–4 mm long, glabrous.

Disturbed sites. Rare; Osceola County. Escaped from cultivation. Florida; Australia. Native to Australia. Spring–fall.

Melaleuca quinquenervia (Cav.) S. T. Blake [Five-nerved, in reference to the leaves, which frequently have 5 nearly parallel nerves.] PUNKTREE.

Metrosideros quinquenervia Cavanilles, Icon. 4: 19, t. 333. 1797. *Melaleuca quinquenervia* (Cavanilles) S. T. Blake, Proc. Roy. Soc. Queensland 69: 76. 1958.

Tree, to 18 m; bark papery. Leaves alternate, the blade elliptic, sometimes somewhat falcate, 5.5–12 cm long, 1–3.1 cm wide, the main veins 5–7, longitudinal, the apex acute to obtuse, the base cuneate, the margin entire, the upper and lower surfaces glabrescent, the petiole 2–5 mm long. Flowers 15–54 in a pseudoterminal raceme, in triads, sometimes also axillary distally; sepals lanceolate, the outer surface glabrous, the margin scarious; petals 2–4 mm long, deciduous; stamens in 5 bundles of 5–10 per bundle, white, cream, green, or yellow, 10–20 mm long, the bundle claw 1–2 mm long. Fruit subglobose, 3–4 mm long, glabrous.

Wet flatwoods, pinelands, mangrove swamp margins, hydric hammocks, wet prairies, and wet disturbed sites. Frequent; central and southern peninsula. Escaped from cultivation. Florida and Louisiana; Asia, Australia, and Pacific Islands. Native to Asia, Australia, and Pacific Islands. All year.

Melaleuca quinquenervia is listed as a Category I invasive species in Florida by the Florida Exotic Pest Plant Council (FLEPPC, 2015).

Melaleuca viminalis (Sol. ex Gaertn.) Byrnes [Having long, flexible shoots.] BOTTLEBRUSH.

Metrosideros viminalis Solander ex Gaertner, Fruct. Sem. Pl. 1: 171, t. 34. 1788. *Callistemon viminalis* (Solander ex Gaertner) G. Don ex Loudon, Hort. Brit. 197. 1830. *Melaleuca viminalis* (Solander ex Gaertner) Byrnes, Austrobaileya 2: 75. 1984.

Shrub or tree, to 35 m; bark fibrous. Leaves alternate, the blade narrowly elliptic to elliptic, 2.5–6(13) cm long, 0.3–2 cm wide, the veins pinnate, the apex acute, the base cuneate, the margin entire, the upper and lower surfaces glabrescent, the petiole to 2 mm long. Flowers 15–20 in a pseudoterminal raceme, in monads, sometimes also axillary distally; sepals lanceolate, the outer surface pubescent or glabrescent, the margin herbaceous; petals 3–6 mm long, deciduous; stamens 9–14 per bundle, red or crimson, the bundle claw to 2 m long. Fruit subglobose, 4–5 mm long, glabrous.

Disturbed sites. Occasional; Highlands, Martin, and Collier Counties, southern peninsula. Escaped from cultivation. Florida and California; Australia. Native to Australia. Spring–summer.

Melaleuca viminalis is listed as a Category II invasive species in Florida by the Florida Exotic Pest Plant Council (FLEPPC, 2015).

EXCLUDED TAXON

> *Melaleuca leucadendron* (Linnaeus) Linnaeus—Reported for Florida by Small (1933), who misapplied the name to material of *M. quinquenervia*.

Mosiera Small 1933.

Trees or shrubs. Leaves opposite or whorled, pinnate-veined, petiolate, estipulate. Flowers axillary, solitary or in dichasia or racemes; bracteoles 2; hypanthium well developed; sepal lobes 4; petals 4; stamens 76–120, free; ovary 2- to 4-carpellate and -loculate. Fruit a berry.

A genus of about 20 species; North America, West Indies, Mexico, and Central America. [Commemorates Charles A. Mosier (1871–1936), Superintendent of Royal Palm State Park, now part of the Everglades National Park.]

Selected reference: Salywon (2003).

Mosiera longipes (O. Berg) Small [Long-stalked, in reference to the long peduncle.] MANGROVEBERRY.

> *Eugenia longipes* O. Berg, Linnaea 27: 150. 1856. *Anamomis longipes* (O. Berg) Britton ex Small, Fl. Miami 132, 200. 1913. *Mosiera longipes* (O. Berg) Small, Man. S.E. Fl. 937, 1506. 1933. *Psidium littorale* Raddi var. *longipes* (O. Berg) Fosberg, Proc. Biol. Soc. Wash. 54: 180. 1941. *Psidium longipes* (O. Berg) McVaugh, J. Arnold Arbor. 54: 312. 1973. TYPE: FLORIDA: s.d., *Leitner s.n.* (holotype: B?).
> *Eugenia bahamensis* Kiaerskov, Bot. Tiddsskr. 17: 266, pl. 8A. 1890. *Anamomis bahamensis* (Kiaerskov) Britton ex Small, Fl. Florida Keys 104, 155. 1913. *Myrtus bahamensis* (Kiaerskov) Urban, Ark. Bot. 12A(5): 18. 1927. *Mosiera bahamensis* (Kiaerskov) Small, Man. S.E. Fl. 937, 1506. 1933.

Tree or shrub, to 4 m; bark smooth or peeling in small flakes, the branchlets reddish brown, gray, or yellowish green, terete, smooth, glabrous or sometimes sparsely to densely puberulent, glandular, flattened near the nodes, the older ones gray, bronze, or reddish brown. Leaves with the blade elliptic or ovate, (1.1)1.8–4(5.2) cm long, (0.2)0.8–3(3.8) cm wide, subcoriaceous, the apex acute to rounded, mucronate or emarginated, the base cuneate to rounded, the margin entire, the upper and lower surfaces glabrous or rarely sparsely pubescent. Flowers 1–3(5) in the leaf axil or in a leafless node on young branchlets, the peduncle (0.5)1–4(5) cm long; bracteoles elliptic to orbicular, 2–5 mm long; hypanthium campanulate, 2–3 mm long; sepals ovate to hemiorbiculate, 3–4 mm long, glabrous, the margin ciliate; petals obovate or suborbicular, 4–6 mm long, white, the margin ciliate. Fruit subglobose to elliptic, 7–10 mm long, dark purple, red, or black, densely glandular, glabrous or rarely sparsely puberulent; seeds several, ca. 2 mm long, yellow, lustrous.

Tropical hammocks. Rare; Miami-Dade County, Monroe County keys. Florida; West Indies. All year.

Mosiera longipes is listed as threatened in Florida (Florida Administrative Code, Chapter 5B-40).

Myrcianthes O. Berg 1856.

Trees or shrubs. Leaves opposite, pinnate-veined, petiolate, estipulate. Flowers axillary, solitary or in dichasia; bracteate, bracteolate; hypanthium not prolonged beyond the summit of the ovary; sepal lobes 4; petals 4; stamens numerous, free; ovary 2-carpellate and loculate. Fruit a berry.

A genus of about 40 species; Florida; West Indies, Mexico, Central America, and South America. [From the Greek *myrci*, and *anthos*, in reference to the flowers like *Myrica*.] *Anamomis* Griseb. (1860).

Myrcianthes fragrans (Sw.) McVaugh [Fragrant.] TWINBERRY; SIMPSON'S STOPPER.

Myrtus fragrans Swartz, Prodr. 79. 1788, nom. cons. *Eugenia fragrans* (Swartz) Willdenow, Sp. Pl. 2: 964. 1799. *Anamomis fragrans* (Swartz) Grisebach, Fl. Brit. W.I. 240. 1860. *Myrcianthes fragrans* (Swartz) McVaugh, Fieldiana, Bot. 29: 485. 1963.

Myrtus dichotoma Poiret, in Lamarck, Encycl., Suppl. 4: 53. 1816; non Salisbury, 1796. *Eugenia dichotoma* de Candolle, Prodr. 3: 278. 1828. *Anamomis dichotoma* (de Candolle) Sargent, Gard. & Forest 6: 130. 1893.

Eugenia dicrana O. Berg, Linnaea 27: 259. 1856. *Anamomis dicrana* (O. Berg) Britton, in Britton & Schafer, N. Amer. Trees 728. 1908. *Myrcianthes dicrana* (O. Berg) K. A. Wilson, J. Arnold Arbor. 41: 276. 1960. TYPE: FLORIDA.

Eugenia fragrans (Swartz) Willdenow var. *brachyrrhiza* Krug & Urban, in Urban, Bot. Jahrb. Syst. 19: 665. 1895. *Anamomis fragrans* (Swartz) Grisebach var. *brachyrrhiza* (Krug & Urban) Stehlé et al., Fl. Guadeloupe 3: 76. 1949.

Anamomis simpsonii Small, Torreya 17: 222. 1917. *Eugenia simpsonii* (Small) Sargent, Man. Trees, ed. 2. 775. 1922. *Myrcianthes simpsonii* (Small) K. A. Wilson, J. Arnold Arbor. 41: 276. 1960. *Myrcianthes fragrans* (Swartz) McVaugh var. *simpsonii* (Small) R. W. Long, Rhodora 72: 23. 1970. *Myrcianthes fragrans* (Swartz) McVaugh subsp. *simpsonii* (Small) E. Murray, Kalmia 13: 9. 1983. TYPE: FLORIDA: Miami-Dade Co.: Arch Creek Hammock, 12 May 1917, *Simpson & Small s.n.* (holotype: NY).

Tree or shrub, to 20 m; bark reddish brown, smooth, exfoliating, the branchlets terete or compressed, sparsely to densely appressed-pubescent with white or cinereous trichomes, soon glabrescent. Leaves with the blade elliptic to obovate, 2–9, 1.7–3 cm wide, leathery, the apex acuminate to bluntly acute to rounded, retuse, the base cuneate, decurrent onto the petiole, the margin entire, flat or revolute basally, the upper and lower surfaces with numerous small glands, glabrate or with a few scattered appressed trichomes along the midrib, the petiole 2.5–10 mm long, sericeous or glabrate. Flowers axillary, solitary or few in a dichasium, the axis sericeus or glabrate, the pedicel of the lateral flowers 3–10 mm long, compressed, the peduncle 2–6 cm long; bracts and bracteoles linear, 2–4 mm long, caducous; hypanthium 2–3 mm long, coarsely sericeous; sepals deltate or broadly ovate, ca. 2 mm long, sericeous, the outer surface persistently so; petals oblong or obovate, 4–5 mm long, convex, the margin ciliate; nectary disk round or quadrate, 3–4 mm in diameter, the staminal ring usually pubescent; stamens 3–9 mm long. Fruit globose or ovoid, 6–15 mm long, purplish black, the calyx persistent; seeds 1–4, reniform.

Coastal hammocks, rarely inland. Occasional; peninsula. Florida; West Indies, Mexico, Central America, and South America. All year.

Myrcianthes fragrans is listed as threatened in Florida (Florida Administrative Code, Chapter 5B-40).

Pimenta Lindl. 1821.

Trees. Leaves opposite, pinnate-veined, petiolate, estipulate. Flowers in axillary panicles, bracteolate, bisexual or sometimes unisexual; sepal lobes 4; petals 4; stamens numerous, free; ovary 2-carpellate and -loculate. Fruit a berry.

A genus of 15 species; Florida; West Indies, Central America, and South America. [Greek name, in reference to seeds rich in oil.]

Selected reference: Landrum (1986).

Pimenta dioica (L.) Merr. [Dioecious, apparently in error.] ALLSPICE.

Myrtus dioica Linnaeus, Syst. Nat., ed. 10. 1056. 1759. *Pimenta dioica* (Linnaeus) Merrill, Contr. Gray Herb. 165: 37. 1947. *Evanesca crassifolia* Rafinesque, Sylva. Tellur. 105. 1838, nom. illegit.

Tree, to 15 m; bark pale brown, nearly smooth, the branchlets flattened, 4-angled. Leaves with the blade oblong to elliptic, (7)9–12(20) cm long, 3–9 cm wide, thin-coriaceous, the apex acute to obtuse, the base acute to rounded, the margin entire, the upper and lower surfaces pellucid-glandular, glabrate, the veins 9–12 pairs, the midvein deeply sulcate on the upper surface, the petiole 1.5–2 cm long. Flowers 50–100 in a panicle 6–12 cm long, 3- to 4-times compound, the axis puberulent; bracts and bracteoles oblong-ovate, 1–2 mm long, pubescent, caducous; hypanthium obconic, 1–2 mm long, puberulent, sepals subequal, broadly suborbicular 1–2 mm long, puberulent; petals ca. 2 mm long, white; stamens ca. 150; nectary disk 2–3 mm wide, sunken around the base of the style; ovules 1(2) per locule. Fruit subglobose or oblong to pyriform, (4)6–8 mm long, black, verrucose; seeds (1)2(3), suborbicular, ca. 4 mm long, compressed, brown.

Disturbed sites. Rare; Miami-Dade County. Escaped from cultivation. Florida; West Indies, Mexico, and Central America. Native to West Indies, Mexico, and Central America. Summer.

Psidium L. 1753. GUAVA

Shrubs or trees. Leaves opposite, pinnate-veined, petiolate, estipulate. Flowers 1 or 3 forming axillary cymes; hypanthium projecting beyond the ovary; sepal lobes 5, basally connate; petals 5; stamens numerous, free; ovary 3- to 6-carpellate and -loculate. Fruit a berry; seeds numerous.

A genus of about 100 species; nearly cosmopolitan. [Greek vernacular name for *Punica*, which it resembles because of its edible fruit.]

1. Leaves glabrous on the lower surface, the veins not conspicuously impressed on the upper surface or raised on the lower surface .... **P. cattleianum**
1. Leaves pubescent on the lower surface, the veins conspicuously impressed on the upper surface and raised on the lower surface.
   2. Leaves with the lateral veins 9–22 on each side of the midrib .... **P. guajava**
   2. Leaves with the lateral veins 6–10 on each side of the midrib .... **P. guineense**

Psidium cattleianum Sabine [Commemorates Sir William Cattley (1788–1835), English horticulturist.] STRAWBERRY GUAVA.

Psidium cattleianum Sabine, Trans. Hort. Soc. London 4: 317, t. 11. 1821. *Psidium variabile* O. Berg, in Martius Fl. Bras. 14(1): 400. 1857, nom. illegit. *Guajava cattleiana* (Sabine) Kuntze, Revis. Gen. Pl. 1: 239. 1891.

Psidium littorale Raddi, Alc. Sp. Pero. 6, t. 1(2). 1821. *Psidium cattleianum* Sabine var. *littorale* (Raddi) Mattos, Loefgrenia 85: 1. 1984.

Psidium cattleianum Sabine forma *lucidum* O. Degener, Fl. Hawaii Fam. 273. 1939. *Psidium littorale* Raddi var. *lucidum* (O. Degener) Fosberg, Proc. Biol. Soc. Wash. 54: 180. 1941.

Shrub or tree, to 8 m; bark reddish brown, smooth or scaly, the branchlets light reddish brown to light gray, flattened, becoming subterete, glabrous. Leaves with the blade obovate, oblanceolate, or elliptic, 5–10 cm long, 2–5.8 cm wide, leathery, the apex acute or acuminate to broadly rounded, the base cuneate, the margin entire, the upper and lower surfaces glabrous, the midvein flat to shallowly impressed on the upper surface, prominent on the lower surface, the lateral veins 8–13 pairs, weak to obscure, the petiole 2–14 mm long, channeled, glabrous. Flowers solitary in the leaf axil, the pedicel ca. 7 mm long; bract ovate, lanceolate, or oblong, 1–2 mm long, caducous at anthesis; hypanthium tube and basally connate calyx extending beyond the ovary, irregularly tearing at anthesis; petals suborbicular to elliptic, 3–6 mm long, whitish; nectary disk 4–6 mm; stamens 300–400, 3–8 mm long. Fruit pyriform to subglobose, 1.5–3 cm long, red or yellow; seeds to 100, round or reniform, ca. 5 mm long.

Disturbed sites. Occasional; central and southern peninsula. Escaped from cultivation. Florida; West Indies, Mexico, and South America; Europe, Africa, Australia, and Pacific Islands. Native to South America. All year.

Psidium cattleianum is listed as a Category I invasive species in Florida by the Florida Exotic Pest Plant Council (FLEPPC, 2015).

Psidium guajava L. [A pre-Linnaean vernacular name for the plant in India.] GUAVA.

> *Psidium guajava* Linnaeus, Sp. Pl. 470. 1753. *Psidium pyriferum* Linnaeus, Sp. Pl., ed. 2. 672. 1762, nom. illegit. *Guajava pyriformis* Gaertner, Fruct. Sem. Pl. 1: 185. 1788. *Guajava pyrifera* Kuntze, Revis. Gen. Pl. 1: 239. 1891, nom. illegit. *Myrtus guajava* (Linnaeus) Kuntze, Revis. Gen. Pl. 3(2): 91. 1898. *Myrtus guajava* (Linnaeus) Kuntze var. *pyrifera* Grisebach ex Kuntze, Revis. Gen. Pl. 3(2): 91. 1898, nom. inadmiss.

Tree or shrub, to 8 m; bark light brown, reddish brown or light grayish green, with large flaky scales, the branchlets quadrangular, slightly to strongly winged, glabrate to densely appressed-pubescent. Leaves with the blade elliptic, oblong, or lanceolate, 4.5–14 cm long, 2.4–7.5 cm wide, leathery to submembranous, the apex acute, acuminate, or rounded, the base obtuse, the margin entire, the upper surface glabrate, the midvein impressed, the lower surface sparsely to densely appressed-pubescent, the midvein prominent, the lateral veins 9–22 pairs, the petiole 2–5 mm long, channeled, densely pubescent to glabrate. Flowers 1 or 3 in the leaf axil, the pedicel 1–3.5 cm long; bract linear to narrowly triangular, 2–5 mm long, sparsely pubescent; hypanthium tube and basally connate calyx extending beyond the ovary, tearing irregularly at anthesis; petals obovate to elliptic, 13–22 mm long, white; nectary disk 4–6 mm in diameter; stamens 300–700, 7–15 mm long. Fruit globose or pyriform, 2–6(8) cm long, green or yellow; seeds ca. 50+, subreniform, 3–4 mm long.

Hammocks and disturbed sites. Frequent; central and southern peninsula. Escaped from cultivation. Florida and Louisiana; West Indies, Mexico, Central America, and South America; Europe, Africa, Asia, Australia, and Pacific Islands. Native to tropical America. All year.

Psidium guajava is listed as a Category I invasive species in Florida by the Florida Exotic Pest Plant Council (FLEPPC, 2015).

Psidium guineense Sw. [Of Guinea.] GUINEA GUAVA.

> *Psidium guineense* Swartz, Prodr. 77. 1788. *Guajava guineensis* (Swartz) Kuntze, Revis. Gen. Pl. 1: 239. 1891. *Myrtus guineensis* (Swartz) Kuntze, Revis. Gen. Pl. 3(2): 91. 1898. *Mosiera guineensis* (Swartz) Bisse, Revista Jard. Bot. Nat. Univ. Habana 6(3): 4. 1986 ("1985").

Shrub or small tree, to 2.5(6) m; bark gray, smooth or scaly, the branchlets pale or coppery red pubescent, terete or compressed, rarely angled below the leaf base. Leaves with the blade broadly elliptic or obovate, 6–10 cm long, the apex rounded or obtuse, the base rounded, obtuse, or cuneate, the margin entire, the upper surface glabrous or glabrate, the lower surface sparsely to densely pubescent, the midvein slightly impressed on the upper surface, raised on the lower surface, the lateral veins convex strongly arching-ascending, 6–10 on each side of the midrib, the petiole 5–10 mm long. Flowers solitary or 2 or 3 in the leaf axil, the peduncle (1.2)1.7–3.2 cm long, the lateral flowers with the pedicel 8–12 mm long; bract filiform to subulate, 2–3 mm long; hypanthium tube and basally connate calyx 10–12 mm long, the calyx splitting at anthesis into 4 or 5 irregular lobes; petals 10–14 mm long; stamens 150–200; style 7–11 mm long. Fruit globose or pyriform, 1–2 cm long, yellow or yellowish green; seeds numerous, ca. 3 mm long.

Disturbed wooded sites. Rare; Manatee County. Escaped from cultivation. Florida; West Indies, Mexico, Central America, and South America; Africa, Asia, and Australia. Native to tropical America. All year.

Rhodomyrtus (DC.) Rchb. 1814.

Shrubs or trees. Leaves opposite, pinnate-veined and with a distinct submarginal nerve (3-veined), petiolate, estipulate. Flowers 1 or 3, axillary, solitary or in dichasia, bracteate; hypanthium enclosing the ovary to the summit; sepal lobes free, 4 or 5; petals 4 or 5; stamens ca. 150; ovary 3-carpellate and -loculate, each locule with a false septa. Fruit a berry.

A genus of about 18 species; North America, Asia, Australia, and Pacific Islands. [From the Greek *Rhodo*, rose, and *Myrtus*.]

Rhodomyrtus tomentosa (Aiton) Hassk. [With densely matted trichomes.] ROSE MYRTLE.

> *Myrtus tomentosa* Aiton, Hort. Kew. 2: 159. 1789. *Rhodomyrtus tomentosa* (Aiton) Hasskarl, Flora 25(2, Beibl.): 35. 1842.

Shrub or small tree, to 2 m; branchlets densely white or yellowish tomentose. Leaves with the blade elliptic to oblong-elliptic, 3.5–7 cm long, 2–4 cm wide, the apex rounded to obtuse, the base cuneate, with 2 arching veins arising from the midvein slightly above the base and uniting near the apex, the margin entire, the upper surface glabrous, the lower surface densely white or yellowish tomentose, the petiole 0.5–1 cm long, white or yellowish tomentose. Flowers 1 or 3 in the leaf axil, the axis white or yellow tomentose; bracteoles ovate, 1–3 mm long, white or yellow tomentose; hypanthium campanulate, white or yellow tomentose; sepal lobes hemiorbicular to ovate, unequal, ca. 5 mm long, white or yellowish tomentose; petals elliptic-oblong, ca. 2 cm long, pink or red; stamens pink or red. Fruit subglobose, elliptical, or elongate-cylindric, 1–1.5 cm long, purplish black, whitish pubescent; seeds 20+, reniform, laterally compressed.

Flatwoods. Occasional; central peninsula, Collier County. Escaped from cultivation. Florida; Asia and Pacific Islands. Native to Asia. Spring.

Rhodomyrtus tomentosa is listed as a Category I invasive species in Florida by the Florida Exotic Pest Plant Council (FLEPPC, 2015).

Syzygium P. Br. ex Gaertn., nom. cons. 1788.

Trees or shrubs. Leaves opposite, pinnate-veined, petiolate, estipulate. Flowers in terminal or axillary panicles or racemes; bractate and bracteolate; hypanthium forming a tube well beyond the ovary summit; sepal lobes 4; petals 4, free or connate and forming a calyptra; stamens numerous; ovary 2(4)-carpellate and -loculate. Fruit a berry; seed 1.

A genus of ca. 1,200 species; North America, West Indies, Central America, South America, Africa, Asia, Australia, and Pacific Islands. [From the Greek *syzgos*, joined, in reference to the paired leaves and bracts.]

1. Inflorescence paniculate; flower buds 5–6 mm long; calyx truncate, the lobes inconspicuous; petals connate, calyptrate at anthesis...**S. cumini**
1. Inflorescence racemose; flower buds 25–30 mm long; calyx 4-lobed; petals free....................**S. jambos**

Syzygium cumini (L.) Skeels [An Indo-Malayan vernacular name.] JAVA PLUM.

Myrtus cumini Linnaeus, Sp. Pl. 471. 1753. *Syzygium cumini* (Linnaeus) Skeels, Bull. Bur. Pl. Industr., U.S.D.A. 248: 25. 1912. *Eugenia cumini* (Linnaeus) Druce, Bot. Exch. Club Soc. Brit. Isles 3: 48. 1914.

Tree or shrub, to 20 m; bark white, smooth, the branchlets weakly compressed, glabrous. Leaves with the blade ovate, elliptic, or oblong, 8–17 cm long, 3.5–7 cm wide, leathery, the apex acuminate or obtuse, the tip bluntly acute, the base cuneate, obtuse, or rounded, the margin entire, the upper and lower surfaces glabrous, glandular-punctate. Flowers 15–100 in an axillary panicle (dichasia 1–3 times compound), short-pedicellate or sessile, the buds pyriform, 5–6 mm long; bracts and bracteoles caducous; hypanthium obconic to narrowly campanulate, 3–5 mm long; calyx truncate, the lobes inconspicuous; petals connate, forming a calyptra, this falling as a unit at anthesis; stamens 50–100, 3–5 mm; style 6–7 mm long. Fruit ellipsoid, 1.5–2 cm long, purple-black.

Disturbed hammocks. Occasional; central and southern peninsula. Escaped from cultivation. Florida; West Indies and South America; Africa, Asia, Australia, and Pacific Islands. Native to Africa and Asia. All year.

Syzygium cumini is listed as a Category I invasive species in Florida by the Florida Exotic Pest Plant Council (FLEPPC, 2015).

Syzygium jambos (L.) Alston [An Indo-Malayan vernacular name.] MALABAR PLUM; ROSE APPLE.

Eugenia jambos Linnaeus, Sp. Pl. 470. 1753. *Myrtus jambos* (Linnaeus) Kunth, in Humboldt et al., Nov. Gen. Sp. 6: 144. 1823. *Jambosa vulgaris* de Candolle, Prodr. 3: 286. 1828, nom. illegit. *Jambosa jambos* (Linnaeus) Millspaugh, Publ. Field Columb. Mus., Bot. Ser. 2: 80. 1900. *Syzygium jambos* (Linnaeus) Alston, in Trimen, Handb. Fl. Ceylon 6: 116. 1931.

Tree or shrub to 10 m; bark reddish brown, flaky, the branchlets terete or quadrangular, glabrous. Leaves with the blade narrowly elliptic or lanceolate, 12–24 cm long, 3–5 cm wide, leathery, the apex acuminate, the base cuneate, the margin entire, the upper and lower surfaces glabrous, glandular-punctate, the petiole 0.5–1 cm long. Flowers 2–8 in a terminal raceme, the pedicel 7–15 mm long, the buds pyriform, 25–30 mm long; bracts and bracteoles caducous; hypanthium infundibular or obconic, 12–17 mm long; sepal lobes in unequal pairs, 4–8 mm long, the margin scarious; petals free, orbicular, 10–15 mm long, the margin scarious; stamens ca. 300, 2–4 cm long; style 4–6 cm long. Fruit subglobose, 3–4 cm long, the calyx lobes persistent, erect.

Disturbed hammocks. Occasional; central and southern peninsula. Escaped from cultivation. Florida; West Indies, Mexico, Central America, and South America; Africa, Asia, and Pacific Islands. Native to Asia. All year.

Syzygium jambos is listed as a Category II invasive species in Florida by the Florida Exotic Pest Plant Council (FLEPPC, 2015).

MELASTOMATACEAE Juss., nom. cons. 1789. MELASTOME FAMILY

Annual or perennial herbs, shrubs, or trees. Leaves opposite or verticillate, simple, palmately and parallel-veined, the veins diverging at the base and converging at the apex, strongly cross-veined, petiolate or epetiolate, estipulate. Flowers in terminal or paniculate axillary cymes or solitary in the leaf axil, bisexual, zygomorphic, bracteate, bracteolate; hypanthium urceolate or campanulate; sepal lobes 4–5; petals 4–5, free; stamens 8 or 10 (twice the number of the petals), the filaments free, unequal, the anthers basifixed, introrse, 2- or 4-locular, dehiscent by 1 or 2 apical pores or short longitudinal slits, the connective sometimes appendaged; ovary superior or inferior, 3- or 4-carpellate and -loculate, the style and stigma 1. Fruit a berry or a loculicidally or an irregularly transverse-dehiscent capsule; seeds numerous.

A family of 150–70 genera and about 5,000 species; North America, West Indies, Mexico, Central America, South America, Africa, Asia, Australia, and Pacific Islands.

Selected reference: Wurdack and Kral (1982).

1. Herb; perianth 4-merous; ovary superior; fruit a loculicidal capsule.
    2. Anther connective with ventral appendages .... **Acisanthera**
    2. Anther connective with dorsal appendages.. **Rhexia**
1. Shrub; perianth 5-merous; ovary inferior; fruit a berry or an irregulary transverse-dehiscent capsule.
    3. Leaf upper surface glabrous (sometimes with a few white-lepidote scales), smooth to the touch, the lower surface densely white-lepidote; fruit a berry ....**Miconia**
    3. Leaf upper surface with coarse, appressed trichomes, rough to the touch, the lower surface strigose-pilose to sericeous and not white-lepidote; petals purple; fruit an irregularly transverse-dehiscent capsule ....**Melastoma**

Acisanthera P. Browne 1756.

Herbs. Stems 4-angled. Leaves with 3 primary longitudinal veins, petiolate, estipulate. Flowers solitary in the leaf axis; sepal lobes 4; petals 4; stamens 8, dimorphic, in 2 whorls, the anthers 4-locular, upcurved, apically 1-poricidal, the connective with ventral appendages; ovary free from the hypanthium (superior), 2- to 4-carpellate and -loculate. Fruit a loculicidal capsule; seeds numerous.

A genus of about 20 species; North America, West Indies, Mexico, Central America, and South America. [From the Greek *acis*, pointed, and *anthera*, anther.]

Selected reference: Kriebel (2008).

Acisanthera erecta J. St.-Hil. [Erect, in reference to the habit.] DUSTSEED.

Rhexia acisanthera Linnaeus, Syst. Nat., ed. 10. 988. 1759. *Acisanthera erecta* J. Saint-Hilaire, Expos. Fam. Nat. 2: 178. Feb–Apr 1805. *Acisanthera quadrata* Persoon, Syn. Pl. 1: 477. Apr–Jun 1805, nom. illegit. *Tibouchina quadrata* M. Gómez de la Maza y Jiménez, Anales Soc. Esp. Hist. Nat. 23: 67. 1894, nom. illegit. *Acisanthera acisanthera* (Linnaeus) Britton, in Britton & P. Wilson, Bot. Porto Rico 6: 2. 1925, nom. inadmiss.

Annual or perennial herb, to 4 dm; stem much branched, 4-angled and narrowly winged, glandular-pubescent or glabrate, the nodes setose-pilose. Leaves with the blade ovate, ovate-oblong, or ovate-lanceolate, 1–2 cm long, 5–10 mm wide, the apex acute to obtuse, the base broadly cuneate, the margin denticulate, the upper and lower surfaces glandular-pubescent or glabrous, the petiole 2–5 mm long. Flowers solitary in the leaf axil, the pedicel 1–2 mm long; hypanthium broadly campanulate, 2–3 mm long; sparsely glandular-pubescent; sepal lobes 4, triangular, ca. 2 mm long, sparsely glandular-pubescent; petals 4, obovate, 5–6 mm long, the apex subretuse, pink or rose; stamens 8, with the filaments crimson upwardly, the anthers yellow the connective of the larger anthers with ventral spur-like appendages, the smaller anthers with short-lobed appendages. Fruit globose, ca. 4 mm long, sparsely glandular-pubescent or glabrous; seeds subellipsoid, with minute pits in longitudinal rows.

Wet, disturbed sites. Rare; Collier County. Florida; West Indies, Mexico, Central America, and South America. Native to tropical America. Summer–fall.

Melastoma L. 1753.

Shrubs. Leaves with 1 or 2 pairs of lateral primary veins, petiolate, estipulate. Flowers in terminal or subterminal cymes, pedicellate, bracteate; calyx lobes 5; petals 5; stamens 10, unequal, in 2 whorls, the anthers 2-locular, dehiscent by apical pores; ovary 5-carpellate and -loculate. Fruit an irregularly dehiscent capsule.

A genus of 22 species; North America, West Indies, Mexico, Asia, Australia, and Pacific islands. [From the Greek *melas*, black, and *stomos*, mouth, in reference to the black stain left in the mouth coming from the dark blue or black, fleshy placenta when the fruits of some species are eaten.]

Selected reference: Meyer (2001).

Melastoma malabathricum L. [Of Malabar, India.] STRAITS RHODODENDRON; MALABAR MELASTOME.

Melastoma malabathricum Linnaeus, Sp. Pl. 390. 1753. *Malabathris nigra* Rafinesque, Sylva Tellur. 97. 1838, nom. illegit.

Shrub, to 2(3) m; bark scaly, the branchlets 4-angled, with dense, short-appressed scales, pilose. Leaves with the blade lanceolate to elliptic-lanceolate or oblong, 4–15 cm long, 2–5 cm wide, the apex acuminate, the base rounded to subcordate, with a prominent midvein and 1 or 2 pairs of longitudinal veins, the marginal pairs sometimes inconspicuous, the secondary veins numerous, the margin entire, the upper surface with coarse, appressed trichomes, rough to the touch, the lower surface strigose-pilose to -sericeous, the petiole 0.5–1.9 cm long. Flowers (1)2–7, in a terminal or subterminal cyme, the pedicel 2–8(10) mm long, strigose; bracts 2, foliate, basal; hypanthium campanulate to globose-urceolate, 5–9 mm long, densely appressed-scaly; sepal lobes triangular-lanceolate, 5–13 mm long, densely appressed-scaly; petals obovate, 2.5–3.5(4) cm long, pink to purple or lavender; stamens dimorphic, the episepalous ones with purple upcurved anthers and long connectives, the epipetalous with yellow straight anthers and short connectives. Fruit urceolate, 5–12 mm long, pinkish-tan, succulent, appressed scaly, irregularly transverse-dehiscent.

Disturbed flatwoods. Rare; Martin County. Escaped from cultivation. Florida; West Indies and Mexico; Asia, Australia, and Pacific Islands. Native to Asia, Australia, and Pacific Islands. Spring.

Miconia Ruiz & Pavon, nom. cons. 1794. CLOVER ASH

Shrubs or trees. Leaves with 3 prominent primary veins, petiolate, estipulate. Flowers in paniculate cymes, bracteate, pedicellate; hypanthium well developed; sepal lobes 5; petals 5; stamens 10, subequal, in 1 whorl, apically porocidal, the connective bases unappendaged; ovary 4- or 5-carpellate and -loculate, the style curved. Fruit a berry; seeds numerous.

A genus of about 1,800 species; North America, West Indies, Mexico, Central America, South America, Africa, Asia, Australia, and Pacific Islands. [Commemorates Francisco Micó (1528–1592), Spanish physician and botanist.]

Tetrazygia Rich. ex DC. (1828).

Selected reference: Judd et al. (2014).

Miconia bicolor (Mill.) Triana. [Two-colored, in reference to the difference in the upper and lower leaf surfaces.] FLORIDA CLOVER ASH.

Melastoma bicolor Miller, Gard. Dict., ed. 8. 1768. *Miconia bicolor* (Miller) Triana, Trans. Linn. Soc. London 28: 103. 1871. *Tetrazygia bicolor* (Miller) Cogniaux, in A. de Candolle, Monogr. Phan. 7: 724. 1891.

Shrub or tree, to 10 m; stem bark thin, gray-brown, the branchlets tawny to gray-brown lepidote. Leaves with the blade, narrowly ovate-lanceolate to lanceolate or oblong-lanceolate, 5.5–12(20) cm long, 2–3.5 cm wide, the apex acuminate, base rounded to obtuse, the margin entire, the upper surface green, glabrous, the lower surface tawny-lepidote, the petiole 1–2.5 cm long, tawny-lepidote. Flowers in a paniculate-cyme 4–10 cm long, the pedicel 5–10; hypanthium urceolate, 4–5 mm long, tawny-lepidote; calyx subtruncate; petals obovate to obtriangular, 5–7 mm long, white to pink; anthers 4–5 mm long, downcurved, yellow. Fruit subglobose, 7–9 mm long, purple to purplish black, the sepal lobes persistent.

Tropical rockland hammocks and pine rocklands. Rare; Miami-Dade County. Florida; West Indies; Pacific Islands. Native to North America and West Indies. Summer.

Miconia bicolor is listed as threatened in Florida (Florida Administrative Code, Chapter 5B-40).

Rhexia L. 1753. MEADOWBEAUTY

Herbs. Stems 4-angled. Leaves with 3 primary longitudinal veins, petiolate or epetiolate, estipulate. Flowers in cymes, bracteate; sepal lobes 4; petals 4, asymmetric; stamens 8, subequal, in 2 whorls, the anthers 4-locular, usually downcurved, apically or subapically poricidal; ovary 4-carpellate and loculate. Fruit a loculicidal capsule; seeds numerous.

A genus of 13 species; North America and West Indies. [From the Greek *rhexis*, a breaking or busting forth, apparently in reference to the capsule dehiscence.]

Selected references: James (1956); Kral and Bostick (1969).

1. Anthers ca. 2 mm long, straight or nearly so.
  2. Petals yellow; leaves oblong or spatulate to linear; stem internodes with at least a few trichomes ... . **R. lutea**
  2. Petals pink; leaves ovate; stem internodes glabrous.
    3. Hypanthium stiptate-glandular; leaf margins with blunt teeth or the teeth with short trichomes.. **R. nuttallii**
    3. Hypanthium glabrous except for a few trichomes between the calyx lobes; leaf margins with distinctly ciliate teeth..**R. petiolata**
1. Anthers 5–11 mm long, curved.
  4. Stem and leaves glabrous; seeds wedge-shaped .... **R. alifanus**
  4. Stem and leaves with at least some trichomes; seeds cochleate.
    5. All 4 stem faces subequal and flat at midstem.
      6. Leaves linear to narrowly oblong or elliptic, the blade twisted vertically .... **R. salicifolia**
      6. Leaves lanceolate to ovate, the blade horizontal.
        7. Petals white; leaves with a short petiole to 5 mm long ....**R. parviflora**
        7. Petals lavender; leaves subsessile..**R. virginica**
    5. Stem faces unequal, 1 opposing pair wide and rounded to convex, the other pair narrow and flat or concave.
      8. Petals white.
        9. Bracts foliaceous, nearly as wide as the hypanthium..**R. parviflora**
        9. Bracts not foliaceous, much narrower than the hypanthium ....**R. mariana**
      8. Petals lavender.
        10. Hypanthium to 1 cm long ....**R. mariana**
        10. Hypanthium more than 1 cm long.
          11. Hypanthium stiptate-glandular.. **R. cubensis**
          11. Hypanthium glabrous..**R. nashii**

Rhexia alifanus Walter [*Alifana*, earthenware drinking vessel, in reference to the resemblance of the hypanthium to a vessel or urn.] SAVANNAH MEADOWBEAUTY.

Rhexia alifanus Walter, Fl. Carol. 130. 1788. *Rhexia glabella* Michaux var. *alifanus* (Walter) Pursh, Fl. Amer. Sept. 258. 1814.
Rhexia glabella Michaux, Fl. Bor.-Amer. 1: 222. 1803.

Erect perennial herb, to 1(2) dm; stem usually branched, or few-branched distally, the faces equal, longitudinally striate, glabrous. Leaves with the blade lanceolate-ovate to lanceolate or elliptic, 3.5–7.5 cm long, 4–15 mm wide, the apex acute to cuminate, the base narrowly cuneate, the margin entire or with low, remote, blunt-tipped teeth, the upper and lower surfaces glabrous, subsessile. Flowers few in a diffuse inflorescence, not obscured by bracts; hypanthium subglobose to ovoid, longer than the constricted neck, 8–10 mm long, stipitate-glandular; sepal lobes deltate to lanceolate, 1–2 mm long; petals broadly obovate, 2–2.5 cm long, spreading, lavender-rose; anther 7–8 mm long, curved. Fruit broadly pyriform, 5–7 mm long, glabrous or with a few stipitate-glandular trichomes; seeds cuneate-prismatic, 1–2 mm long, the surfaces nearly smooth.

Wet prairies, wet flatwoods, and bogs. Occasional; northern counties south to Orange County. North Carolina south to Florida, west to Texas. Spring–summer.

Rhexia cubensis Griseb. [Of Cuba.] WEST INDIAN MEADOWBEAUTY.

Rhexia cubensis Grisebach, Cat. Pl. Cub. 104. 1866.
Rhexia floridana Nash, Bull. Torrey Bot. Club 22: 150. 1895. TYPE: FLORIDA: Lake Co.: Hick's Prairies, vicinity of Eustis, 1–15 Jul 1894, *Nash 1218* (holotype: NY: isotypes: GH, MO, PH, US).

Erect perennial herb, to 6 dm; stem unbranched or few branched, the faces strongly unequal, 1 opposing pair wide and rounded to convex, the other pair narrow and flat or concave, sparsely stipitate-glandular. Leaves with the blade linear or linear-elliptic to oblong or narrowly spatulate, 2–4 cm long, 1–4(8) mm wide, the apex acute to obtuse, the base cuneate, the margin serrate, the upper and lower surfaces stipitate-glandular, sessile. Flower in a diffuse inflorescence, not obscured by bracts; hypanthium ovoid to subglobose, (10)14–15(16) mm long, about as long as the constricted neck, sparsely stipitate-glandular; sepal lobes triangular; petals obovate, 1.5–2 cm long, lavender-rose, spreading; anthers 7–10 mm long, curved. Fruit globose, 5–8 mm long, glabrous; seeds cochleate, the surface smooth.

Wet flatwoods, pond margins, and marshes. Frequent; nearly throughout. North Carolina south to Florida, west to Louisiana; West Indies. Spring–fall.

Rhexia lutea Walter [Yellow, in reference to the petal color.] YELLOW MEADOWBEAUTY.

Rhexia lutea Walter, Fl. Carol. 130. 1788.

Erect perennial herb, to 4 dm; stem branched proximally, the faces subequal, flat to convex, hirsute. Leaves with the blade spatulate to oblanceolate or elliptic, 2–3 cm long, 2–8 mm wide, the apex acute, the base cuneate, the margin subentire to shallowly serrate, the upper and lower surfaces loosely yellowish strigose, subsessile. Flowers in a diffuse inflorescence, not obscured by bracts; hypanthium globose, 6–7 mm long, much longer than the constricted neck, hirsute to villous; sepal lobes triangular, 1–3 mm long; petals obovate, 7–13 mm long, 1–1.5 cm long, yellow, erect; anthers ca. 2 mm long, straight. Fruit subglobose, glabrous; seeds cochleate, the surface papillate to nearly smooth.

Wet prairies, wet flatwoods, and bogs. Occasional; northern peninsula south to Volusia County, central and western panhandle. Spring–summer.

Rhexia mariana L. [Of Maryland]. PALE MEADOWBEAUTY; MARYLAND MEADOWBEAUTY.

Rhexia mariana Linnaeus, Sp. Pl. 346. 1753. *Rhexia mariana* Linnaeus var. *typica* Fernald, Rhodora 37: 171. 1935, nom. inadmiss.
Rhexia lanceolata Walter, Fl. Carol. 129. 1788. *Rhexia mariana* Linnaeus var. *exalbida* Michaux, Fl. Bor.-Amer. 1: 221. 1803. *Rhexia trichotoma* Rottbøll var. *exalbida* (Michaux) Persoon, Syn. Pl. 1:

406. 1805. *Rhexia angustifolia* Nuttall, Gen. N. Amer. Pl. 1: 244. 1818, nom. illegit. *Rhexia mariana* Linnaeus var. *lanceolata* (Walter) T. F. Wood & G. McCarthy, J. Elisha Mitchell Sci. Soc. 3: 97. 1866, nom. illegit.
Rhexia filiformis Small, Bull. Torrey Bot. Club 25: 468. 1898.

Erect perennial herb, to 8 dm; stem few to several branched, the faces strongly unequal, 1 pair of opposing faces round to convex, the other narrower, flat or concave, stipitate-glandular. Leaves with the blade linear, lanceolate, elliptic, or narrowly ovate, 2–4 cm long, (5)8–15(20) mm wide, the apex acute, the base cuneate, the margin serrate, the upper and lower surfaces loosely strigose to strigose-hirsute or villous, sessile or the petioles to 2 mm long. Flowers in a diffuse inflorescence, not obscured by bracts; hypanthium ovoid to subglobose, 6–10 mm long, about as long as the constricted neck, stipitate-glandular or glabrate; sepal lobes triangular, 1–3 mm long; petals obovate, 12–15 mm long, white to pale lavender, spreading; anthers 5–8 mm long, curved. Fruit globose, 5–7 mm long, glabrous; seeds cochleate, the surface tuberculate, papillate or dome-shaped, laterally flattened processes.

Wet prairies, wet flatwoods, bogs, and marshes. Common; nearly throughout. Massachusetts south to Florida, west to Michigan, Kansas, Oklahoma, and Texas. Spring–fall.

Rhexia nashii Small [Commemorates George Valentine Nash (1864–1921), agrostologist and head gardener at the New York Botanical Garden and collector of Florida plants.] MAID MARIAN.

Rhexia nashii Small, Fl. S.E. U.S. 824, 1335. 1903. TYPE: FLORIDA: Lake Co.: vicinity of Eustis, 1–15 Jun 1894, *Nash 863* (holotype: NY; isotypes: BH, GH, MO, PH, US).
Rhexia mariana Linnaeus var. *purpurea* Michaux, Fl. Bor.-Amer. 1: 221. 1803.

Erect perennial herb, to 1.5 m; stem usually unbranched, faces strongly unequal, 1 opposing pair rounded to convex, the other pair narrower, flat or concave, stipitate-glandular. Leaves with the blade ovate to ovate-lanceolate or elliptic, 3–7 cm long, (5)7–12(15) mm wide, the apex acute, the base cuneate, the margin finely to coarsely serrate, the upper and lower surfaces hirsute, the petiole 1–2 mm long. Flowers in a diffuse inflorescence, not obscured by bracts; hypanthium ovoid to subglobose, 10–15(20) mm long, about as long as the constricted neck, glabrous or glabrate, the rim stipitate-glandular; sepal lobes triangular, 1–3 mm long, stipitate-glandular; petals obovate, 2–2.5 cm long, pale lavender, spreading; anther curved, 8–11 mm long. Fruit pyriform, 5–7 mm long, glabrous or sometimes with a few stipitate-glands; seeds cochleate, the surface with dome-shaped, laterally flattened processes.

Bogs and wet flatwoods. Frequent; northern counties, central peninsula. Maryland south to Florida, west to Louisiana. Summer.

Rhexia nuttallii C. W. James [Commemorates Thomas Nuttall, (1786–1859), English-born American botanist who traveled and collected extensively in North America, including west Florida.] NUTTALL'S MEADOWBEAUTY.

Rhexia serrulata Nuttall, Gen. N. Amer. Pl. 1: 243. 1818; non Richard, 1813. *Rhexia nuttallii* C. W. James, Brittonia 8: 214. 1956.

Erect perennial herb, to 3(4) dm; stem branched near the base, the faces subequal, flat to convex, sparsely hirsute. Leaves with the blade ovate to suborbicular, 1–1.5 cm long, 3–10 mm wide, the apex acute to obtuse, the base rounded to cordate or broadly obtuse, the margin serrate, the upper surface sparsely villous, the lower surface glabrous, the petiole ca. 1 mm long. Flowers in a condensed inflorescence, mostly obscured by foliaceous bracts; hypanthium globose, 5–7 mm long, much longer than the constricted neck, stipitate-glandular; sepal lobes deltate, 1–2 mm long, stipitate-glandular; petals obovate, 1–1.2 cm long, lavender-rose, ascending; anthers ca. 2 mm long, straight. Fruit subglobose, 3–4 mm long, glabrous; seeds cochleate, the surface with short, irregularly spaced ridges.

Wet flatwoods, bogs, and wet prairies. Frequent; nearly throughout. Florida and Georgia. Spring–summer.

Rhexia parviflora Chapm. [Small-flowered]. WHITE MEADOWBEAUTY; APALACHICOLA MEADOWBEAUTY.

Rhexia parviflora Chapman, Fl. South. U.S., ed. 3. 156. 1897. TYPE: FLORIDA: Franklin Co.: Apalachicola, s.d., *Chapman s.n.* (lectotype: MO). Lectotypified by James (1956: 218).

Erect perennial herb, to 4 dm; stem branched or unbranched, the faces subequal, glabrous or sparsely stipitate-glandular, the nodes hirsute. Leaves with the blade ovate to elliptic, 1.5–3 cm long, 4–8 mm wide, the apex rounded to obtuse, the base cuneate, the margin finely serrate, the upper and lower surfaces strigose, the petiole 1–5 mm long. Flowers in a diffuse inflorescence, not obscured by bracts; hypanthium subglobose, 5–7 mm long, much longer than the constricted neck, sparsely stipitate-glandular apically; sepal lobes triangular, ca. 2 mm long; petals obovate, 0.8–1(1.3) cm long, white to pale lavender, spreading; anthers slightly curved, 3–3.5 cm long. Fruit subglobose, 3–4 mm long, glabrous; seeds cochleate, the surface with grooved tubercles.

Cypress swamp margins, bogs, and pond margins. Occasional; central and western panhandle. Florida, Georgia, and Alabama. Summer.

Rhexia parviflora is listed as endangered in Florida (Florida Administrative Code, Chapter 5B-40).

Rhexia petiolata Walter [Petiolate.] FRINGED MEADOWBEAUTY.

Rhexia petiolata Walter, Fl. Carol. 130. 1788.
Rhexia ciliosa Michaux, Fl. Bor.-Amer. 1: 221. 1803.

Erect perennial herb, to 5 dm; stem unbranched or few-branched, the faces subequal, flat to convex, the nodes sparsely stipitate-glandular, otherwise glabrous. Leaves with the blade ovate to short-elliptic or suborbicular, 1–2 cm long, 4–14 mm wide, the apex acute to obtuse, the base rounded to truncate, the margin serrate, the upper surface sparsely villous, the lower surface glabrous, petiole 1–2 mm long. Flowers in a dense cluster, usually obscured by the foliaceous bracts; hypanthium globose, 5–7(9) mm long, much longer than the constricted neck, glabrous; sepal lobes oblong-lanceolate, 2–4 mm long, ciliate; petals obovate, 1–2 mm long, lavender-rose, spreading; anthers straight, 1–2 mm long. Fruit subglobose, 4–6 mm long, glabrous; seeds cochleate, ca. 0.6 mm long, the surface pebbled or with ridges or dome-like processes.

Wet prairies, wet flatwoods, bogs, and pond and lake margins. Frequent; nearly throughout. Maryland south to Florida, west to Texas. Spring–summer.

Rhexia salicifolia Kral & Bostick [With leaves like *Salix* (Salicaceae).] PANHANDLE MEADOWBEAUTY.

> *Rhexia salicifolia* Kral & Bostick, Sida 3: 402. 1969. TYPE: FLORIDA: Bay Co.: Merial Lake, along FL 77 WNW of Vicksburg, 6 Oct 1961, *Godfrey & Houk 61554* (holotype: FSU; isotype: FSU).

Erect perennial herb, to 5.5 dm; stem usually several-branched distally, the faces subequal, the angles winged, stipitate-glandular. Leaves with the blade narrowly elliptic or oblong to narrowly oblanceolate or linear, 1.5–4 cm long, 1–5 mm wide, the apex acute to obtuse, the base cuneate, the margin entire or minutely crenulate, ciliate with gland-tipped trichomes, the upper and lower surfaces sparsely to moderately stipitate-glandular, sessile. Flowers in a diffuse inflorescence, not obscured by bracts; hypanthium globose, (4)5–7(8) mm long, longer than the constricted neck, sparsely stipitate-glandular; sepal lobes narrowly triangular, ca. 2 mm long; petals obovate, 1.1–1.2 mm long, pink to lavender-rose, spreading; anthers curved, 4–5 mm long. Fruit subglobose or ovoid, 4–5 mm long, glabrous; seeds cochleate, the surface with 3–5 prominent, broad longitudinal ridges or contiguous dome-like tubercles in lines.

Pond margins and seepage bogs. Occasional; central and western panhandle. Florida and Alabama. Summer–fall.

Rhexia salicifolia is listed as threatened in Florida (Florida Administrative Code, Chapter 5B-40).

Rhexia virginica L. [Of Virginia.] HANDSOME HARRY.

> *Rhexia virginica* Linnaeus, Sp. Pl. 346. 1753.
> *Rhexia stricta* Pursh, Fl. Amer. Sept. 258. 1814; non Bonpland, 1804. *Rhexia purshii* Sprengel, Syst. Veg. 5: 590. 1828. *Rhexia virginica* Linnaeus var. *purshii* (Sprengel) C. W. James, Brittonia 8: 227. 1956.

Erect perennial herb, to 1 m; stem unbranched or few branched, the faces subequal, the angles narrowly winged, sparsely stipitate-glandular or glabrous. Leaves with the blade lanceolate, ovate, or elliptic, 3–5(7) cm long, (0.7)1–2(3.5) cm wide, the apex acute, the base rounded to broadly cuneate, the margin finely serrate, the upper surface villous, the lower surface glabrate, sessile. Flowers in a diffuse inflorescence, not obscured by bracts; hypanthium globose, (6)7–10 mm long, longer than the constricted neck, glabrous or sparsely stipitate-glandular; sepal lobes narrowly triangular, 1–3 mm long; petals obovate, 1.5–2 cm long, lavender-rose, spreading; anthers curved, 5–6 mm long. Fruit pyriform, 5–7 mm long, glabrous or sparsely stipitate-glandular at the apex; seeds cochleate, the surface papillose or tuberculate in lines.

Wet flatwoods, bogs, and lake and cypress pond margins. Frequent; northern counties. Nova Scotia and Maine south to Florida, west to Ontario, Iowa, Missouri, Oklahoma, and Texas. Summer–fall.

STAPHYLEACEAE Martinov, nom. cons. 1820. BLADDERNUT FAMILY

Shrubs or trees. Leaves opposite, odd-pinnately compound, petiolate, stipulate. Flowers in terminal panicles, actinomorphic, bisexual, bracteate; sepals 5, connate; petals 5, connate; stamens 5, the anthers 2-loculate, longitudinally dehiscent; nectary disk present; ovary superior, 3-carpellate and -loculate, the styles 3. Fruit a capsule.

A family of 2 genera and about 45 species; North America, West Indies, Mexico, Central America, South America, Europe, and Asia.

Selected reference: Spongberg (1971).

Staphylea L. 1753. BLADDERNUT

Shrubs or trees. Leaves opposite, odd-pinnately compound, petiolate, stipulate. Flowers in terminal panicles, bisexual, bracteate; sepals 5, basally connate; petals 5, basally connate; stamens 5, free, the anthers 2-loculate, longitudinally dehiscent, basally 2-lobed; hypogynous nectar disk present; ovary superior, 3-carpellate, the styles 3, free below, connate apically. Fruit an inflated, indehiscent capsule, the carpels separating at maturity; seeds 1–2 per carpel.

A genus of 23 species; North America, West Indies, Mexico, Central America, South America, Europe, and Asia. [From the Greek *staphyle*, a cluster, in reference to the infructescence.]

Staphylea trifolia L. [With 3 leaves, in reference to the leaves having 3 leaflets.] AMERICAN BLADDERNUT.

> *Staphylea trifolia* Linnaeus, Sp. Pl. 270. 1753. *Staphylodendron trifoliatum* Crantz, Inst. Rei. Herb. 2: 438. 1766. *Staphylea trifolia* Linnaeus var. *typica* C. K. Schneider, Ill. Handb. Laubholzk. 2: 190. 1907, nom. inadmiss.

Shrub or small tree, to 4(5) m; bark gray to nearly black, somewhat mottled, the branchlets brown, with pale, elongate to rounded lenticels. Leaves opposite, 3(5)-foliolate, the lateral veins adnate to the midrib for a short distance and then abruptly divergent, the blade elliptic to ovate-elliptic, 3–10 cm long, 2–5 cm wide, the apex short-acuminate, the base rounded to broadly cuneate, sometimes oblique, the margin finely serrate, the upper surface glabrous or short-pubescent when young, glabrous in age, the lower surface sparsely pubescent, the lower pair of leaflets sessile or subsessile, the terminal one usually with a petiolule 1.3–3 cm long, the petiole 4–10 cm long, the rachis with 2 small, stipule-like glands, the stipules linear-subulate, caducous. Flowers few in a loose, usually pendent panicle in the axil of the uppermost leaf of a short, lateral branchlet, pedicels 1–2 cm long, slender, flexuous; bracts 2, linear-subulate, pubescent, membranous; sepals 5, basally connate, the lobes linear-oblong, 6–8 mm long, erect, greenish white; petals 5, basally connate, the corolla campanulate, slightly longer than the sepals, the lobes erect, cream with green stripes; stamens 5, 6–8 mm long, subequaling the petals in length, inserted outside and below a fleshy nectary disk, pubescent below; ovary 3-loculate, pubescent, the styles 3, free below and connate at the apex, glabrous. Fruit elliptic to obovate,

3–5 cm long, ca. 3 cm wide, bladderlike, the carpels 3-lobed and weakly 3-beaked with the persistent styles, separating at the apex as the fruit matures and becoming inflated, ellipsoid to obovate, membranaceous; seeds subglobose, 5–7 mm long, brown, somewhat flattened, smooth.

Floodplain woods, wooded bluffs, and ravine slopes. Rare; Gadsden and Liberty Counties. Quebec south to Florida, west to Ontario, Minnesota, Nebraska, Kansas, Oklahoma, and Louisiana. Spring.

Staphylea trifolia is listed as endangered in Florida (Florida Administrative Code, Chapter 5B-40).

PICRAMNIACEAE Fernando & Quinn 1995. BITTERBUSH FAMILY

Trees or shrubs. Leaves alternate, odd-pinnately compound, petiolate, estipulate. Flowers in terminal or axillary racemes or panicles, actinomorphic, unisexual (plant dioecius); sepals 5, free or basally connate; petals 5, free, sometimes absent in carpellate flowers; stamens 5, reduced to staminodia or absent in carpellate flowers; nectar disk present; ovary 2- to 3-carpellate, sessile on the infrastaminal disk, rudimentary or absent in the staminate flowers, the ovule 1- to 3-carpellate and -loculate. Fruit samaroid or a berry; seeds few.

A family of 3 genera and about 50 species; North America, West Indies, Mexico, Central America, and South America.

Alvaradoa and *Picramnia* were previously included in the Simaroubaceae, but recent molecular data have shown that these genera are best placed in the Picramniaceae (Fernando and Quinn, 1995).

Selected references: Brizicky (1962b).

1. Leaflets 1–2.5 cm long; fruit a samara, the margin ciliate ....**Alvaradoa**
1. Leaflets 3–12 cm long; fruit a few-seeded berry, glabrous ....**Picramnia**

Alvaradoa Liebm. 1854.

Shrubs and trees. Leaves alternate, odd-pinnately compound, the leaflets alternate, petiolate, estipulate. Flowers in terminal or axillary racemes, actinomorphic, unisexual (plants dioecious); sepals 5, free in carpellate flowers, basally connate in the staminate; petals 5 (absent in the carpellate flowers), free; stamens 5, inserted below and between the lobes of the nectary disk (absent in the carpellate flowers), the anthers basifixed, with a swollen connective, 4-locular, introrse; intrastaminal nectar disk 5-lobed (scarcely so in the carpellate flowers); ovary 3-carpellate and -loculate, but only 1 carpel fertile (absent in the staminate flowers), the styles 3, free. Fruit a samaroid capsule; seed 1.

A genus of 5 species; North America, West Indies, Mexico, Central America, and South America. [Commemorates Pedro de Alvarado y Contreras (ca. 1485–1541), Spanish conquistador and Governor of Guatemala.]

Alvaradoa amorphoides Liebm. [Resembling the genus *Amorpha* (Fabaceae).] MEXICAN ALVARADOA.

Alvaradoa amorphoides Liebmann, Vidensk. Meddel. Dansk Naturhist. Foren. Kjøbenhavn 1853: 100. 1854. *Alvaradoa amorphoides* Liebmann subsp. *typica* Cronquist, Brittonia 5: 135. 1944, nom. inadmiss.

Alvaradoa psilophylla Urban, Repert. Spec. Nov. Regni Veg. 20: 304. 1924. *Alvaradoa amorphoides* Liebmann subsp. *psilophylla* (Urban) Cronquist, Brittonia 5: 135. 1944.

Shrub or tree, to 10 m; branchlets pubescent. Leaves odd-pinnately compound, the leaflets 19 or more, alternate, the blade ovate to oblong, 1–2.5 cm long, 5–1.5 cm wide, the apex rounded, the base rounded, the margin entire, the upper surface glabrous, the lower surface pale and sericeous, the petiolule 1–2 mm long, the petiole to ca. 3 cm long. Flowers green or yellowish; staminate flowers in a raceme ca. 2 dm long, the carpellate ones in a dense, plume-like raceme to ca. 1.3 dm long; sepals ovate, sericeous; petals ovate-lanceolate, the margin ciliate; stamens exserted, basally pilose. Fruit narrowly oblong-lanceolate, 1–1.5 cm long, densely pilose, the margin densely ciliate with long spreading trichomes.

Hammocks. Rare; Miami-Dade County. Florida; West Indies, Mexico, Central America, and South America. Winter–spring.

Alvaradoa amorphoides is listed as endangered in Florida (Florida Administrative Code, Chapter 5B-40).

Picramnia Sw., nom. cons. 1788. BITTERBUSH

Shrubs or trees. Leaves alternate, odd-pinnately compound, the leaflets alternate, petiolate, estipulate. Flowers in terminal, paniculate racemes, actinomorphic, unisexual (plants dioecious); sepals 5, basally connate; petals 5, free; stamens 5, inserted below and between the lobes of the intrastaminal disk (reduced to staminodia in the carpellate flowers), the anthers basifixed, 4-locular, longitudinally dehiscent, introrse; ovary 2(3)-carpellate and -loculate (rudimentary in the staminate flowers). Fruit a berry; seeds 1–2.

A genus of about 40 species; North America, West Indies, Mexico, Central America, and South America. [From the Greek *pikeos*, bitter, and *thamnos*, shrub, in reference to the bitterness of the plants.]

Picramnia pentandra Sw. [From the Greek *penta*, five, and *andrus*, male, in reference to the 5 stamens.] FLORIDA BITTERBUSH.

Picramnia pentandra Swartz, Fl. Ind. Occid. 1: 220. 1797.

Shrub or small tree, to 10 m; branchlets finely appressed-pubescent. Leaves odd-pinnately compound, the leaflets 5–9, alternate, the blade ovate to oblong or lanceolate, 3–12 cm long, 1–5 cm wide, the apex acuminate to obtuse, the base cuneate, the margin entire, the upper and lower surfaces finely appressed pubescent when young, glabrous at maturity, the petiolule 2–5 mm long, the petiole 2–3 cm long. Flowers in a slender, loosely branched panicle 7–15 cm; sepals ovate, ca. 2 mm long, appressed pubescent; petals linear-lanceolate, 2–3 mm long; stamens 5, 4–5 mm long. Fruit globose to oblong or obovoid, 9–15 mm long, red, glabrous.

Hammocks. Rare; Miami-Dade County. Florida; West Indies and South America. All year.

Picramnia pentandra is listed as endangered in Florida (Florida Administrative Code, Chapter 5B-40).

BURSERACEAE Kunth, nom. cons. 1824. GUMBO LIMBO FAMILY

Trees. Leaves alternate, odd-pinnately compound, petiolate, estipulate. Flowers in axillary, racemiform panicles, actinomorphic, bisexual or unisexual (plants polygamodioecious); sepals 3–5, basally connate; petals 3–5, free; stamens 6–10, free (staminodal in the carpellate flowers), the anthers dorsifixed, longitudinally dehiscent; intrastaminal nectar disk present; ovary superior, 3-carpellate, 1-loculate (rudimentary in the staminate flowers). Fruit a several-seeded drupe.

A family of 19 genera and about 755 species; North America, West Indies, Mexico, Central America, South America, Africa, Asia, and Pacific Islands.

Selected reference: Brizicky (1962b).

Bursera Jacq. ex L., nom. cons. 1824.

Trees. Leaves alternate, odd-pinnately compound, the leaflets opposite, petiolate, estipulate. Flowers in axillary racemiform panicles, actinomorphic, bisexual or unisexual (plants polygamodioecious), actinomorphic; sepals 3–5, basally connate; petals 3–5, free; stamens 6–10, free (staminodal in the carpellate flowers), the anthers dorsifixed, longitudinally dehiscent; intrastaminal nectar disk present; ovary superior, 3-carpellate, 1-loculate (rudimentary in the staminate flowers), the stigma capitate, 3-lobed, the style short. Fruit a several-seeded drupe.

A genus of about 100 species; North America, West Indies, Mexico, Central America, and South America. [Commemorates Joachim Burser (1583–1639), German physician and botanist.]

Elaphrium Jacq., nom. rej. (1760).

Bursera simaruba (L.) Sarg. [Resembling a species of *Simarouba* (Simaroubaceae).] GUMBO LIMBO.

Pistacia simaruba Linnaeus, Sp. Pl. 1026. 1753. *Terebinthus brownii* Jacquin, Enum. Syst. Pl. 18. 1760, nom. illegit. *Bursera gummifera* Linnaeus, Sp. Pl., ed. 2. 471. 1762, nom. illegit. *Bursera simaruba* (Linnaeus) Sargent, Gard. & Forest 3: 260. 1890. *Icicariba simaruba* (Linnaeus) M. Gómez de la Maza y Jiménez, Fl. Haban. 214. 1897. *Terebinthus simaruba* (Linnaeus) W. Wight, Contr. U.S. Natl. Herb. 10: 122. 1906. *Elaphrium simaruba* (Linnaeus) Rose, in Britton, N. Amer. Fl. 25: 246. 1911.

Tree, to 15 m; bark greenish brown, becoming light-red to dark reddish brown and peeling off in thin, gland-dotted, paper-like sheets, the branchlets stout, light greenish red, with numerous lenticels, glabrous. Leaves odd-pinnately compound, 9–35 cm long, the leaflets (3)5–9(11), opposite, the blade broadly ovate to ovate-oblong, or elliptic, 3–12 cm long, 1.5–5 cm wide, the apex acute to acuminate or abruptly acuminate-cuspidate, the base obliquely rounded, the margin entire, the upper surface glabrous, the lower surface sparsely strigose at the base of the midrib, otherwise glabrous, the petiolule 1–5 mm long, the petiole 6–11 cm long or longer. Flowers in an axillary racemiform panicle 2.5–10 cm long, the staminate inflorescence larger than the carpellate, glabrous, white to cream-colored or yellowish green. Staminate flowers with the sepals 5, the calyx ca. 1 mm long, the lobes rounded; petals 5, elliptic, 2–3 mm long, spreading reflexed; stamens 10; ovary rudimentary; pedicel 2–4 mm long. Carpellate flowers with the sepals 3–4, ca. 1 mm long, the lobes rounded; petals 3–4, 2–3 mm long, ovate-elliptic, spreading-reflexed; staminodes 6–8; pedicel 3–4 mm long. Fruit obliquely ovoid, 1–1.5 cm long, 3-angled, 3-valvate, dark red, the pyrenes 1(2).

Coastal hammocks and shell middens. Occasional; central and southern peninsula. Florida; West Indies, Mexico, Central America, and South America. Winter–spring.

ANACARDIACEAE R. Br., nom. cons. 1818. CASHEW FAMILY

Trees, shrubs, or woody vines. Leaves alternate, simple, 3-foliolate, or odd-pinnately compound, the venation pinnate, petiolate, estipulate. Flowers in cauliflorous, terminal, and/or axillary, paniculate, racemiform, or spiciform thyrses, bracteate, actinomorphic, bisexual or unisexual (plants dioecious, monoecious, or polygamodioecious); sepals 4–5(6), basally connate; petals 4–5(6), free; nectary disk intrastaminal or extrastaminal; stamens (1)5–10(20), free, the anthers dorsifixed or basifixed; ovary 3- to 5-carpellate, 1- to 5-loculate, the styles 1–5, the stigmas 1–5. Fruit a fleshy or dry drupe.

A family of about 80 genera and about 800 species; nearly cosmopolitan.

Spondiadaceae Martinov (1820).

Selected reference: Brizicky (1962c).

1. Leaves simple ....**Mangifera**
1. Leaves 3-foliolate or pinnately compound.
    2. Leaf rachis winged.
        3. Fruit glabrous; stamens 10 ....**Schinus**
        3. Fruit glandular-pubescent; stamens 5 ....**Rhus**
    2. Leaf rachis not winged.
        4. Leaves all 3-foliolate.
            5. Fruit glandular-pubescent, red; inflorescences both terminal and axillary ....**Rhus**
            5. Fruit glabrous or with a few remote trichomes, white or greenish white; inflorescences all axillary.. **Toxicodendron**
        4. Leaves 3- to 25-foliolate.
            6. Leaflet margin crenate or serrate (rarely with some entire).
                7. Fruit glandular-pubescent ....**Rhus**
                7. Fruit glabrous...**Schinus**
            6. Leaflet margin entire.

8. Leaflets sessile or with a petiolule rarely more than 1 mm long; fruit 2.5–4 cm long .... **Spondias**
8. Leaflets with a petiolule 3–30 mm long; fruit less than 2 cm long.
9. Leaflet margin conspicuously thickened ....**Metopium**
9. Leaflet margin not conspicuously thickened.
10. Leaflet apex acuminate, the lower surface evidently paler than the upper ....**Toxicodendron**
10. Leaflet apex rounded to emarginate, often with a short acumen, the lower surface not evidently paler than the upper ....**Sorindeia**

Mangifera L. 1753. MANGO

Trees. Leaves simple, petiolate, estipulate. Flowers in terminal and/or axillary thyrses, bisexual or unisexual (plants polygamodioecious); sepals 5, basally connate; petals 5, free; nectary disk extrastaminal; stamens 8, 1–2 fertile, the others reduced to staminodia, the anthers dorsifixed; style 1. Fruit a fleshy 1-loculate drupe; seed 1.

A genus of about 70 species; North America, West Indies, Mexico, Central America, South America, Africa, Asia, Australia, and Pacific Islands. [From the Malayalam and Tamil *manga*, a vernacular name for the mango, and the Latin *fero*, to bear.]

Mangifera indica L. [Of India.] MANGO.

Mangifera indica Linnaeus, Sp. Pl. 200. 1753. *Mangifera domestica* Gaertner, Fruct. Sem. Pl. 2: 95. 1791, nom. illegit.

Tree, to 40 m; bark gray or dark brown, longitudinally fissured, the branchlets glabrous. Leaves with the blade oblong-lanceolate, 8–40 cm long, 2–10 cm wide, subcoriaceous, the young leaves reddish, the mature leaves dark green, the apex acute to acuminate, the base cuneate, the margin entire, the upper and lower surfaces glabrous, the petiole 0.8–6 cm long. Flowers in a thyrse 10–40(60) cm long, the axis pubescent; sepal lobes lanceolate to ovate, 2–3 mm long, the outer surface pubescent; petals elliptic to oblanceolate, 3–5 mm long, white with yellow or pink glandular ridges on the inner surface, reflexed, sparsely pubescent apically. Fruit subreniform, 8–30 cm long, the exocarp green, yellow, or red, glabrous, the endocarp woody, fibrous, ridged; seed oblong-ovoid, compressed.

Disturbed sites. Occasional; central and southern peninsula. Escaped from cultivation. Florida; West Indies, Mexico, Central America, South America; Asia, Africa, Australia, and Pacific Islands. Native to Asia. Spring.

Metopium P. Browne 1756. FLORIDA POISON TREE

Trees or shrubs. Leaves odd-pinnately compound, the leaflets opposite, petiolulate, petiolate, estipulate. Flowers in axillary panicles, unisexual (plants dioecious); sepals 5, basally connate; petals 5, free; stamens 5, the anthers basifixed; nectary disk intrastaminal; carpels 3, the style 1, the stigma 3-lobed. Fruit a 1-loculate drupe; seed 1.

A genus of 3 species; North America, West Indies, Mexico, and Central America. [*Metopion*, the name for a gum tree, perhaps in reference to the exudate.]

Metopium toxiferum (L.) Krug & Urb. [*Toxicum*, poison, and *fero*, to bear, in reference to its poisonous property.] FLORIDA POISON TREE; POISONWOOD.

Amyris toxifera Linnaeus, Syst. Nat., ed. 10. 1000. 1759. *Metopium toxiferum* (Linnaeus) Krug & Urban, in Urban, Bot. Jahrb. Syst. 21: 612. 1869.

Tree or shrub, to 14 m; bark thin, scaly, peeling off in plates revealing a yellow to orange underbark, the branchlets glabrous. Leaves 3- to 7-foliolate, the leaflets deltoid, ovate, obovate, or cordate, 2.5–10 cm long, 1.5–8 cm wide, coriaceous, the apex acute to obtuse, the base rounded to cordate, the margin entire or undulate, the upper and lower surfaces glabrous, the petiolule 0.5–3 cm long, the petiole 2–10 cm long. Flowers in a panicle 14.5–34 cm long; sepal lobes ca. 1 mm long; petals 2–2.5 mm long, ca. 1.5 mm wide, yellow-green with dark veins. Fruit ellipsoid to obovate, 8–14 mm long, the exocarp orange to brown, glabrous, the mesocarp thin, resinous, the endocarp brittle, smooth; seeds subquadrangular, compressed.

Hammocks and pinelands. Occasional; Martin and Palm Beach Counties, southern peninsula. Florida; West Indies. Spring.

EXCLUDED TAXON

Metopium linnaei Engler—Reported for Florida by Chapman (1860, 1883, 1897, all as *Rhus metopium* Linnaeus) and Small (1903, as *M. metopium* (Linnaeus) Small), all a misapplication of the name to material of *M. toxiferum*.

Rhus L. 1753. SUMAC

Shrubs or trees. Leaves odd-pinnately compound, the leaflets opposite or subopposite, petiolulate, petiolate, estipulate. Flowers in terminal and/or axillary thyrses, panicles, racemes, or spikes, actinomorphic, bisexual or unisexual (plants dioecious or polygamodioecious), bracteate; sepals 5, basally connate; petals 5, free; nectary disk intrastaminal; stamens 5, the anthers dorsifixed; ovary 3-carpellate, 1-loculate, the style 1. Fruit a drupe; seed 1.

A genus of about 35 species; North America, West Indies, Mexico, Central America, Europe, Asia, Africa, and Pacific Islands. [From the greek *rhodo*, red, in reference to fruit; also from the Greek *rhous* or *rhoys*, an ancient name for the Sicilian sumac *R. coriaria*.]

1. Leaves 3-foliolate...**R. aromatica**
1. Leaves 9- to 25-foliolate.
  2. Stem glabrous ....**R. glabra**
  2. Stem pubescent.
    3. Leaflet margin entire or nearly so, rarely crenate or serrate, the rachis always winged................
    ....**R. copallinum**
    3. Leaflet margin evenly serrate, the rachis not winged or only narrowly winged ......**R. michauxii**

Rhus aromatica Aiton [Fragrant.] FRAGRANT SUMAC.

Rhus aromatica Aiton, Hort. Kew. 1: 367. 1789. *Lobadium amentaceum* Rafinesque, Amer. Monthly
Mag. & Crit. Rev. 4: 358. 1819, nom. illegit. *Schmaltzia aromatica* (Aiton) Desvaux ex Small, Fl.
S.E. U.S. 728. 1903.

Shrub, to 3 m; stem grayish brown, glabrate. Leaves 3-foliolate, the junction of the leaflets
glabrous or pilose, the leaflet blade ovate to rhomboid, 1.2–7.5 cm long, 1–5.5 cm wide, subcoriaceous, the apex acute, the base cuneate to rounded, the margin crenate in the upper half, the
upper surface glabrous to puberulent, the lower surface puberulent to pilose, sessile or subsessile, the petiole 0.3–4.3 cm long. Flowers in a dense panicle, spike, or raceme, the branches
densely pubescent and with eglandular and glandular trichomes; sepal lobes oblong, 0.5–1 mm
long, puberulent; petals broadly lanceolate, 1–3 mm long, yellow to pinkish, pubescent along
the central veins on the inner surface. Fruit subglobose, 4–6 mm long, red, the surface with
whitish eglandular and shorter reddish glandular trichomes.

Rocky, open hammocks. Rare; Jackson and Escambia Counties. Quebec south to Florida,
west to Ontario, Minnesota, North Dakota, Nebraska, Kansas, Oklahoma, and Texas. Spring.

Rhus copallinum L. [Ancient name for some plant.] WINGED SUMAC.

Rhus copallinum Linnaeus Sp. Pl. 266. 1753. *Rhus pistachiifolia* Salisbury, Prodr. Stirp. Chap. Allerton
169. 1796, nom. illegit. *Rhus lentiscifolia* Stokes, Bot. Mat. Med. 2: 164. 1812, nom. illegit. *Toxicodendron copallinum* (Linnaeus) Kuntze, Revis. Gen. Pl. 1: 153. 1891. *Schmaltzia copallinum* (Linnaeus)
Small, Fl. S.E. U.S. 728, 1334. 1903.

Rhus leucantha Jacquin, Pl. Hort. Schoenbr. 3: 50, t. 342. 1798. *Rhus copallinum* Linnaeus var. *leu-
cantha* (Jacquin) de Candolle, Prodr. 2: 68. 1825. *Rhus copallinum* Linnaeus forma *angustialata*
Engler, in A. de Candolle, Monogr. Phan. 4: 384. 1883. *Rhus copallinum* Linnaeus subsp. *leucantha*
(Jacquin) E. Murray, Kalmia 13: 28. 1983.

Rhus copallinum Linnaeus forma *integrifolia* Engler, in A. de Candolle, Monogr. Phan. 4: 384. 1883.
SYNTYPE: FLORIDA.

Schmaltzia obtusifolia Small, Fl. S.E. U.S. 729, 1334. 1903. *Rhus obtusifolia* (Small) Small, Fl. Miami
112. 1913. *Rhus copallinum* Linnaeus var. *obtusifolia* (Small) Fernald & Griscom, Rhodora 37: 168.
1935. TYPE: FLORIDA: Lake Co.: vicinity of Eustis, 16–25 Aug 1894, *Nash 1659* (holotype: NY).

Shrub or small tree, to 7(10) m; stem light brown, pubescent, glabrate in age. Leaves odd-pinnately compound, 10–30(35) cm long, the rachis winged, pubescent, the leaflets 9–25, the blade
linear to ovate, 0.7–11.5 cm long, 0.5–4 cm wide, straight or somewhat falcate, chartaceous, the
apex acuminate to obtuse, the base cuneate to rounded, asymmetrical, the margin entire or
sparsely serrate or crenate, the upper surface sparsely pubescent to glabrous, pubescent on the
primary veins, the lower surface densely pubescent to glabrate, sessile or subsessile, the petiole
2.5–7.5 cm long. Flowers in a panicle (5)10–30(35) cm long, pubescent, the pedicel 1–2 mm
long; sepal lobes ovate, 0.5–1 mm long, the outer surface pubescent, the margin ciliate; petals
lanceolate, 1–2.5 mm long, greenish, yellowish, or pinkish, pubescent at the base and along the
veins on the outer surface. Fruit subglobose, 4–5 mm long, dull red, the surface with translucent eglandular and short-stipitate or sessile, reddish, glandular trichomes.

Sandhills, flatwoods, and dry hammocks. Frequent; nearly throughout. Maine south to
Florida, west to Ontario, Kansas. Oklahoma, and Texas. Spring.

Rhus glabra L. [Smooth.] SMOOTH SUMAC.

Rhus glabra Linnaeus, Sp. Pl. 265. 1753. *Toxicodendron glabrum* (Linnaeus) Kuntze, Revis. Gen. Pl. 1: 154. 1891; non Miller, 1768. *Schmaltzia glabra* (Linnaeus) Small, Fl. S.E. U.S. 729, 1334. 1903.

Shrub or small tree, to 6(7) m; stem gray, glabrous, glaucous. Leaves odd-pinnately compound, 21–45 cm long, the rachis not winged, glabrous, the leaflets 9–25, the blade lanceolate to narrowly elliptic, 5–12 cm long, 1.5–3(4) cm wide, straight or somewhat falcate, chartaceous, the apex acuminate, the base rounded, the margin serrate, rarely subentire, the upper surface green, sparsely pubescent on the veins, the lower surface glabrous, glaucous, sessile or subsessile, the petiole 4–11 cm long. Flowers in a panicle (5)10–25(45) cm long, pubescent, the pedicel 1–2 mm long, densely pubescent; sepal lobes lanceolate, the outer surface glabrous to pubescent, the margin ciliate; petals lanceolate, 2–5 mm long, greenish yellow to greenish white, pubescent, the margin ciliate. Fruit subglobose, 3–5 mm long, red, the surface with whitish, simple, eglandular and reddish, glandular trichomes.

Open hammocks and thickets sites. Occasional; central panhandle, Escambia County. Nearly throughout North America; Mexico. Summer.

Rhus michauxii Sarg. [Commemorates André Michaux (1746–1802), French botanist who collected in North America.] FALSE POISON SUMAC; MICHAUX'S SUMAC.

Rhus pumila Michaux, Fl. Bor.-Amer. 1: 182. 1803; non Meerburgh, 1798. *Toxicodendron pumilum* Kuntze, Revis. Gen. Pl. 1: 154. 1891. *Rhus michauxii* Sargent, Gard. & Forest 8: 404. 1895. *Schmaltzia michauxii* (Sargent) Small, Fl. S.E. U.S. 729, 1334. 1903, nom. illegit.

Shrub, to 2 m; stem reddish brown-gray, pubescent. Leaves odd-pinnately compound, (16)21–34 cm long, the rachis pubescent, not winged, the leaflets 9–15, opposite or subopposite, the blade lanceolate to ovate, 3.5–11 cm long, 1.5–6 cm wide, chartaceous, the apex acute to obtuse, the base cuneate to cordate, the margin serrate or dentate, the upper and lower surfaces pubescent, the petiole 4.5–10 cm long. Flowers in a panicle 8–18 cm long, pubescent, the pedicel 1–2 mm long; sepal lobes ovate, 1–2 mm long, the outer surface pubescent, the margin ciliate; petals lanceolate, 2–4 mm long, the outer surface sparsely pubescent. Fruit subglobose, 3–5 mm long, red, the surface with clavate or glandular trichomes.

Dry hammocks. Rare; Alachua County. Virginia south to Florida. Spring.

Rhus michauxii is listed as endangered in Florida (Florida Administrative Code, Chapter 5B-40) and in the United States (U.S. Fish and Wildlife Service, 50 CFR 23).

Schinus L. 1753. PEPPERTREE

Shrubs or trees. Leaves alternate, odd-pinnately compound, petiolate, estipulate. Flowers in axillary paniculate thyrses, actinomorphic, bisexual or unisexual (plants dioecious), bracteate; sepals 5, basally connate; petals 5, free; stamens 10, in 2 series of unequal length, the anthers dorsifixed; nectar disk intrastaminal; ovary 3-carpellate, 1-loculate, the styles 3, basally connate. Fruit a drupe; seed 1.

A genus of about 30 species; nearly cosmopolitan. [From the Greek *schinos*, ancient vernacular name for the mastic tree, *Pistacia lentiscus* (Anacardiaceae), in reference to the similarity of the gummy exudate.]

Schinus terebinthifolia Raddi [With leaves like *Pistacia terebinthus* (Anacardiaceae).] BRAZILIAN PEPPER.

Schinus terebinthifolia Raddi, Quar. Piant. Nouv. Bras. 20. 1820.

Shrub or small tree, to 7 m; branchlets short-pilose to glabrate. Leaves odd-pinnately compound, 7–22 cm long, the rachis sometimes winged, the leaflets (3)5–15, opposite or subopposite, the blade ovate to elliptic or obovate, 1.5–7.5 cm long, 0.8–3 cm wide, chartaceous to subcoriaceous, the apex acute to rounded, the base cuneate to rounded, the margin entire or serrulate, the upper and lower surfaces pilose to glabrate, sessile or subsessile, the petiole 2–4 cm long. Flowers in an axillary paniculate thyrse 2–15 cm long, the pedicel 1–3 cm long; sepals lobes deltate, ca. 1 mm long, the margin ciliate; petals lanceolate, to obovate, 2–3 mm long, white, yellow, or green. Fruit subglobose, 4–6 mm long, red, glabrous or sparsely pubescent.

Disturbed moist to mesic sites, often invading coastal communities. Common; peninsula, Franklin County. Escaped from cultivation. Florida; Georgia, Alabama, Texas, and California; West Indies, Mexico, Central America, and South America; Europe, Africa, Asia, Australia, and Pacific Islands. Native to South America. Spring–summer.

Schinus terebinthifolia is listed as a Category I invasive species in Florida by the Florida Exotic Pest Plant Council (FLEPPC, 2015).

EXCLUDED TAXON

Schinus molle Linnaeus—Reported for Florida by the USDA/PLANTS database (USDA, NRCS, 2015), apparently based on an old (probably McFarlin) collection from Polk County housed at NCU. There are also specimens at FLAS, one a McFarlin gathering from near Lakeland, Polk County, the other a Haime specimen from Merritt Island, Brevard County. Until better evidence is found that this occasionally cultivated plant has ever escaped in Florida, it will not be treated as part of the flora.

Sorindeia Thouars 1806.

Trees or shrubs. Leaves alternate, odd-pinnately compound, the leaflets opposite or subopposite, petiolate, estipulate. Flowers in cauliflorus, paniculiform thyrses, actinomorphic, unisexual (plants dioecious or monoecious); sepals 5, basally connate; petals 5, free; stamens 10–20, the anthers dorsifixed; ovary 3-carpellate, 1-loculate, the style 1, the stigma 3 lobed. Fruit a drupe; seed 1.

A genus of 9 species; North America and Africa. [*Sondriry* and *tsirondrano*, the Madagascan vernacular names for the species, perhaps from the Greek *soros*, heap, in reference to the large cluster of fruits produced by the type species (*S. madagascariensis*).]

Selected reference: Breteler (2003).

Sorindeia madagascariensis DC. [Of Madagascar.] MTIKIZA.

Sorindeia madagascariensis de Candolle, Prodr. 2: 80. 1825.

Shrub or tree, to 30 m; bark dark brown. Leaves odd-pinnately compound, 2.5–5.5 cm long, the leaflets (3)7–13, the blade ovate-elliptic to oblanceolate, ovate, oblong, or elliptic, 2.5–20(30) cm long, 1.5–6(10) cm wide, chartaceous, the apex acute to obtuse or rounded to emarginate, the base cuneate, often asymmetric, the margin entire, the upper surface glabrous, the lower surface minutely puberulent on the midrib, the petiolules 0.5–5 cm long, the petiole 5.5–18.5 cm long. Flowers in a cauliflorous, paniculate thyrse 3–10 dm long, pedicellate; calyx cupulate, the sepal lobes deltoid, 1–2 mm long, glabrous; petals lanceolate, 2–4 mm long, pinkish to red, glabrous; stamens inserted outside and on the nectar disk, the anthers pink to red. Fruit subellipsoid, 2–3 mm long, sometimes apiculate, the exocarp yellow, smooth, glabrous, the mesocarp thin, fleshy, the endocarp woody; seed ellipsoid.

Disturbed sites. Rare; Miami-Dade County. Escaped from cultivation. Florida; Africa. Native to Africa. Summer.

Spondias L. 1753. MOMBIN

Trees. Leaves alternate, odd-pinnately compound, petiolate, estipulate. Flowers in axillary or cauliflorus racemiform or paniculate thyrses, bisexual or unisexual (plants polygamodioecious); sepals (4)5, basally connate; petals (4)5, free; stamens (8)10, the anthers dorsifixed; ovary (4)5-carpellate and -loculate, the styles and stigmas (4)5. Fruit a drupe; pyrenes 4(5).

A genus of about 18 species; North America, West Indies, Mexico, Central America, South America, Africa, and Asia. [From the Greek *spondias*, in reference to the similarity in fruits.]

Spondias purpurea L. [Purple, in reference to the fruit.] PURPLE MOMBIN.

Spondias myrobalanus Linnaeus, Fl. Jamaic. 16. Dec 1759; non Linnaeus, May–Jun 1759. *Spondias purpurea* Linnaeus, Sp. Pl., ed. 2. 613. 1762.

Tree, to 8 m; bark gray, smooth or rough in age, the branchlets glabrous. Leaves odd-pinnately compound, 6–30 cm long, the leaflets (5)11–25, alternate to subopposite, the blade elliptic to ovate or obovate or lanceolate, 2.5–6 cm long, 2–3 cm wide, the apex acute to obtuse or rounded, the base cuneate to rounded, usually asymmetrical, the margin shallowly crenate or serrulate near the apex, the upper surface glabrous, the lower surface pubescent on the midrib, soon glabrous, the petiole to 7 cm long. Flowers in a cauliflorous racemiform or paniculate thyrse 1–10 cm long; sepal lobes deltate, ca. 0.5 mm long; petals lanceolate, 3–4 mm long, pink; stamens inserted below the nectar disk; nectar disk copular, crenate. Fruit ovoid or oblong, 2–3.5 cm long, red, orange, purple, or yellow, glabrous, the mesocarp fleshy, the endocarp boney, with a fibrous outer layer.

Disturbed hammocks. Rare; Collier County. Escaped from cultivation. Florida; West Indies, Mexico, Central America, and South America; Asia and Africa. Native to Mexico, Central America, and South America. Spring–summer.

Toxicodendron Mill. 1754. POISON OAK

Shrubs, woody vines, or trees. Leaves alternate, 3-foliolate or odd-pinnately compound, petiolate, estipulate. Flowers in axillary paniculate thyrses, bisexual or unisexual (dioecious or polygamodioecious); sepals 5, basally connate; petals 5, free; stamens 5, free, the anthers dorsifixed; ovary 3-carpellate, 1-loculate, the styles 3. Fruit a drupe; seed 1.

A genus of about 25 species. North America, Central America, Asia, and Pacific Islands. [From the Greek *taxikon*, poison, and *dendron*, tree, in reference to a tree with toxic properties.] Selected reference: Gillis (1971).

1. Leaves pinnately compound .... **T. vernix**
1. Leaves 3-foliolate.
    2. Leaflets with the apex usually acute to acuminate, with tufts of trichomes in the vein axils on the lower surface ....**T. radicans**
    2. Leaflets with the apex usually obtuse to rounded, lacking tufts of trichomes in the vein axils on the lower surface ....**T. pubescens**

Toxicodendron pubescens Mill. [Hairy.] ATLANTIC POISON OAK; EASTERN POISON OAK.

Toxicodendron pubescens Miller, Gard. Dict., ed. 8. 1768. *Rhus pubescens* (Miller) Farwell, Rep. (Annual) Commiss. Parks & Boulev. Detroit 11: 73. 1900; non Thunberg, 1794; nec Ecklon & Zeyher, 1836.

Rhus toxicodendron Linnaeus, Sp. Pl. 266. 1753. *Rhus toxicarium* Salisbury, Prodr. Stirp. Chap. Allerton 170. 1796, nom. illegit. *Rhus radicans* Linnaeus var. *toxicodendron* (Linnaeus) Persoon, Syn. Pl. 1: 325. 1805. *Toxicodendron toxicodendron* (Linnaeus) Britton, in Britton & A. Brown, Ill. Fl. N. U.S., ed. 2. 2: 484. 1913, nom. inadmiss. *Toxicodendron toxicarium* Gillis, Rhodora 73: 402. 1971.

Rhus toxicodendron Linnaeus, var. *quercifolia* Michaux, Fl. Bor.-Amer. 1: 183. 1803. *Toxicodendron quercifolium* (Michaux) Greene, Leafl. Bot. Observ. Crit. 1: 127. 1905. *Rhus quercifolia* (Michaux) Steudel ex B. L. Robinson & Fernald, in A. Gray, Manual, ed. 7. 553. 1908; non Goeppert, 1855.

Shrub, to 1 m; stem reddish gray to brown, pubescent. Leaves 3-foliolate, the leaflet blade ovate to elliptic, 6–11 cm long, 3–8 cm wide, the apex acute to rounded, the base cuneate to rounded, the margin dentate to lobate, the upper and lower surfaces pubescent, the petiolule 1–3 mm long, the petiole 5–15 cm, densely to sparsely pubescent. Flowers in a paniculate thyrse 1–7 cm long, the pedicel 1–3 mm long, pubescent but glabrate apically; sepal lobes deltate, 1–2 mm long, glabrous; petals ovate-lanceolate, 2–3 mm long, greenish to yellowish white, glabrous. Fruit subglobose, 2–7 mm long, greenish white, glabrous or glabrate to densely pubescent, smooth or longitudinally, irregularly ridged.

Sandhills and dry hammocks. Rare; northern counties, Levy and Marion Counties. New Jersey south to Florida, west to Kansas, Oklahoma, and Texas. Spring.

Toxicodendron radicans (L.) Kuntze [Spreading outward from a common point.] EASTERN POISON IVY.

Rhus radicans Linnaeus, Sp. Pl. 266. 1753. *Rhus radicans* Linnaeus var. *opaca* Aiton, Hort. Kew. 1: 367. 1789, nom. inadmiss. *Rhus scandens* Salisbury, Prodr. Stirp. Chap. Allerton 170. 1796, nom.

illegit. *Rhus toxicodendron* Linnaeus var. *radicans* (Linnaeus) Eaton, Man. Bot., ed. 2. 400. 1818. *Toxicodendron radicans* (Linnaeus) Kuntze, Revis. Gen. Pl. 1: 153. 1891. *Rhus toxicodendron* Linnaeus forma *radicans* (Linnaeus) McNair, Publ. Field Mus. Nat. Hist., Bot. Ser. 4: 68. 1925. *Rhus toxicodendron* Linnaeus subsp. *radicans* (Linnaeus) R. T. Clausen, Cornell Univ. Agric. Exp. Sta. Mem. 291: 8. 1949.

Rhus blodgettii Kearney, Bull. Torrey Bot. Club 21: 486. 1894. *Toxicodendron blodgettii* (Kearney) Greene, Leafl. Bot. Observ. Crit. 1: 126. 1905. TYPE: FLORIDA: Monroe County: Pine Key, near Key West, s.d., *Blodgett s.n.* (holotype: NY).

Rhus floridana Mearns, Proc. Biol. Soc. Wash. 15: 149. 1902. TYPE: FLORIDA: Lee Co.: Alva, Jul–Aug 1900, *Hitchcock 39* (holotype: US; isotypes: GH, MIN, NY).

Rhus littoralis Mearns, Proc. Biol. Soc. Wash. 15: 148. 1902. *Rhus radicans* Linnaeus var. *littoralis* (Mearns) Deam, Fl. Indiana 651. 1924. *Toxicodendron radicans* (Linnaeus) Kuntze var. *littoralis* (Mearns) Barkley, Ann. Missouri Bot. Gard. 24: 434. 1937.

Toxicodendron goniocarpum Greene, Leafl. Bot. Observ. Crit. 1: 125. 1905. TYPE: FLORIDA: Columbia Co.: Lake City, 25 Jun 1901, *McCulloch 45* (holotype: US; isotypes: FLAS, US).

Shrub or woody vine, to 30 m; stem reddish brown to black, pubescent to glabrate, often with numerous adventitious roots. Leaves 3-foliolate, the leaflet blade lanceolate to ovate or orbicular, 3–17(20) cm long, (1.3)5–13 cm wide, asymmetrical, the apex acute to acuminate or rounded, the base cuneate to rounded or truncate, the margin entire, serrate, dentate, or lobate, the upper surface glabrous, the lower surface pubescent, sometimes only along the veins, the petiolules 1–5 mm long, the petiole 3–9(12.5) cm long. Flowers in a paniculate thyrse 2–9 cm long, the pedicel 1–2 mm long; sepal lobes deltate, 1–2 mm long; petals oblong, 2–4 mm long, white to yellowish green with purplish venation, glabrous. Fruit subglobose, 3–5 mm long, glabrous or pubescent, longitudinally shallow-grooved.

Wet hammocks, floodplain forests, swamps, and wet, disturbed sites. Common; nearly throughout. Quebec south to Florida, west to Ontario, Missouri, Oklahoma, and Texas; West Indies, Mexico, and Central America. Spring–summer.

Toxicodendron vernix (L.) Kuntze [White waxy skin covering of a fetus, in reference to its glaucous nature.] POISON SUMAC.

Rhus vernix Linnaeus, Sp. Pl. 265. 1753. *Toxicodendron pinnatum* Miller, Gard. Dict., ed. 8. 1768, nom. illegit. *Rhus vernicifera* Salisbury, Prodr. Stirp. Chap. Allerton 169. 1796, nom. illegit. *Rhus venenata* de Candolle, Prodr. 2: 68. 1825, nom. illegit. *Toxicodendron vernix* (Linnaeus) Kuntze, Revis. Gen. Pl. 1: 153. 1891.

Shrub or small tree, to 7.5 m; stem gray, smooth, glaucous. Leaves odd-pinnately compound, (10)15–35 cm long, the leaflets 7–11(15), the blade ovate or elliptic to oblanceolate, 3.5–8(15) cm long, 2–4(6) cm wide, the apex acuminate, the base cuneate to rounded, the margin entire, the upper surface green, glabrous, the lower surface whitish, glabrate, the petiolule 0.5–3.5 cm long, the petiole 3.5–11(20) cm long. Flowers in a paniculate thyrse 7–20(24) cm long, the pedicel 1–2 mm long; sepal lobes deltate, ca. 1 mm long, glabrous; petals ovate-lanceolate, 1–2 mm long, greenish white, glabrous. Fruit globose, 4–7 mm long, slightly laterally compressed, longitudinally shallow-grooved, glabrous.

Swamps and wet hammocks. Occasional; northern counties, central peninsula. Quebec south to Florida, west to Ontario, Minnesota, Illinois, and Texas. Spring–summer.

EXCLUDED TAXON

Toxicodendron crenatum Miller—Reported for Florida by Small (1913e, 1933, all as *Schmaltzia crenata* (Miller) Greene), based on a misapplication of the name to material of *Rhus aromatica*. *Toxicodendron crenatum* has been typified as conspecific with the western *T. rydbergii* (Small ex Rydberg) Greene, and at the same time proposed for rejection, a course recommended by the proper committee.

SAPINDACEAE Juss., nom. cons. 1789. SOAPBERRY FAMILY

Trees, shrubs, or woody vines. Leaves alternate or opposite, simple, pinnately or palmately compound, petiolate or epetiolate, stipulate or estipulate. Flowers in terminal and/or axillary thyrses, racemes, corymbs, or umbel-like fascicles on short branchlets, actinomorphic or zygomorphic, unisexual or bisexual; sepals 4–5, free or basally connate; petals 4–5, free; nectariferous disk intrastaminal or extrastaminal; stamens 6–10, free or basally connate, the anthers versatile, introrse, longitudinally dehiscent; ovary superior, 2- to 4-carpellate and -loculate, style 1, the stigmas 2–3. Fruit a drupe, berry, capsule, or schizocarp.

A genus of about 140 genera and about 1,630 species; nearly cosmopolitan.

Aceraceae Juss., nom. cons. (1789); *Aesculaceae* Burnett (1835); *Dodonaeaceae* Kunth ex Small, nom. cons. (1903); *Hippocastanaceae* A. Rich., nom. cons. (1823).

Selected reference: Brizicky (1963).

1. Vine with axillary tendrils ....**Cardiospermum**
1. Tree or shrub.
    2. Leaves simple.
        3. Leaves lobed; fruit a samaroid schizocarp ....**Acer**
        3. Leaves unlobed; fruit a winged capsule .... **Dodonaea**
    2. Leaves compound.
        4. Leaves palmately compound, with 5–7 leaflets..**Aesculus**
        4. Leaves 3-foliolate or pinnately or bipinnately compound.
            5. Leaves bipinnately compound; fruit inflated.. **Koelreuteria**
            5. Leaves 3-foliolate or pinnately compound; fruit not inflated.
                6. Fruit a samaroid schizocarp..**Acer**
                6. Fruit other than a samaroid schizocarp.
                    7. Leaves 3-foliolate..**Hypelate**
                    7. Leaves pinnately compound.
                        8. Fruit a dehiscent capsule.
                            9. Leaflet margin crenate...**Cupania**
                            9. Leaflet margin entire.
                                10. Locules 2- to several-seeded... **Harpullia**
                                10. Locules 1-seeded... **Cupaniopsis**
                        8. Fruit indehiscent.
                            11. Leaves with 3 or more pairs of leaflets.
                                12. Petals clawed, with 2 basal scales; fruit smooth or nearly so.......... **Sapindus**

12. Petals not clawed, lacking basal scales; fruit rugose or slightly tuberculate
.... **Dimocarpus**
  11. Leaves with 1–2(3) pairs of leaflets.
     13. Flowers 4-merous; inflorescence branches racemose ....**Melicoccus**
     13. Flowers 5-merous; inflorescence branches corymbiform cymes..... **Exothea**

Acer L. 1753. MAPLE

Trees or shrubs. Leaves opposite, simple and palmate-veined or odd-pinnately compound and pinnate-veined, petiolate, stipulate or estipulate. Flowers in terminal racemiform or corymbiform thyrses, racemes, corymbs, or terminal or lateral umbelliform fascicles on short, few-leaved or leafless branchlets, actinomorphic, unisexual or sometimes also bisexual (plants dioecious or polygamodioecious); sepals 4–5, free or basally connate; petals the same number as the calyx parts or sometimes absent, free; nectariferous disk extrastaminal or intrastaminal, or wanting; stamens 4–8(12), inserted on or within the nectary disk; ovary 2-carpellate and -loculate, the stigmas 2. Fruit a samaroid schizocarp composed of 2 1-seeded mericarps.

A genus of about 125 species; nearly cosmopolitan. [Classical Latin name of the maple.]

Acer was previously placed in the family Aceraceae.

Argentacer Small (1933); *Negundo* Boehm. (1760); *Rufacer* Small (1933); *Rulac* Adans. (1763); *Saccharodendron* (Raf.) Nieuwl. (1914).

1. Leaves odd-pinnately compound, the leaflets 3–7(9), pinnate-veined ....**A. negundo**
1. Leaves simple, palmate-veined.
  2. Leaf margin entire to remotely dentate, the apex abruptly rounded; flowers in terminal clusters, appearing with the leaves..**A. saccharum**
  2. Leaf margin evenly serrate to doubly serrate, the apex acute; flowers in axillary clusters, appearing before the leaves.
    3. Leaves cut less than halfway to the main vein at the base, typically 3-lobed; petals present (similar to the sepals) .... **A. rubrum**
    3. Leaves cut more than halfway to the main vein at the base; petals absent .... **A. saccharinum**

Acer negundo L. [Aboriginal name.] BOXELDER.

Acer negundo Linnaeus, Sp. Pl. 1056. 1753. *Negundo aceroides* Moench, Methodus 334. 1794. *Negundo virginianum* Medikus, Beytr. Pfl.-Anat. 439. 1801, nom illegit. *Negundium fraxinifolium* Rafinesque, Med. Repos., ser 2. 5: 352. 1808, nom. illegit. *Acer fraxinifolium* Nuttall, Gen. N. Amer. Pl. 1: 253. 1818, nom. illegit. *Negundo negundo* (Linnaeus) H. Karsten, Deut. Fl. 596. 1882, nom. inadmiss. *Acer negundo* Linnaeus var. *vulgare* Pax, Bot. Jahrb. Syst. 7: 211. 1886, nom. inadmiss. *Acer negundo* Linnaeus subsp. *typicum* Wesmael, Bull. Soc. Roy. Bot. Belgique 29: 42. 1890, nom. inadmiss. *Acer negundo* Linnaeus var. *normalis* Kuntze, Revis. Gen. Pl. 1: 146. 1891, nom. inadmiss. *Acer negundo* Linnaeus subsp. *vulgare* Schwerin, Gartenflora 42: 200. 1893, nom. inadmiss. *Rulac negundo* (Linnaeus) Hitchcock, Key Spring Fl. Manhattan 25. 1894. *Rulac nuttallii* Nieuwland, Amer. Midl. Naturalist 2: 137. 1911, nom. illegit. *Negundo nuttallii* Rydberg, Bull. Torrey Bot. Club 40: 55. 1913, nom. illegit.

Acer negundo Linnaeus var. *latifolium* Pax, Bot. Jahrb. Syst. 11: 75. 1889. *Acer negundo* Linnaeus subsp. *latifolium* (Pax) Schwerin, Gartenflora 42: 205. 1893. *Acer negundo* Linnaeus forma *latifolium* (Pax) Sargent, Bot. Gaz. 67: 239. 1919.

Tree, to 25 m; branchlets green, glabrous, sometimes glaucous, the older ones with scattered brown or buff-colored, circular lenticels. Leaves odd-pinnately compound, the leaflets 3–7(9), the blade pinnate-veined, ovate to elliptic, 5–10 cm long, 5–7.5 cm wide, the apex acuminate, the base cuneate to rounded, that of the lateral ones usually asymmetrical, the margin coarsely and irregularly serrate, the terminal leaflet sometimes palmately 3-lobed and -veined, the upper surface pubescent, becoming glabrous, the lower surface pubescent, sometimes glabrate or pubescent primarily on the veins, the petiolule 1–15 mm long, the petiole 1–3 cm long, the base dilated and covering the axillary bud. Staminate flowers fasciculate, 5–10 per fascicle, the axis lax, filiform, and pendulous, pubescent, greenish yellow or red; sepals 4 or 5, oblong-lanceolate to obovate or linear, unequal; petals absent; stamens (4)5(6). Carpellate flowers in a lax, stalked, pendulous raceme, yellowish green, the pedicel 8–12 mm long, at anthesis, glabrous. Fruit with each winged half 2.5–3.5 cm long, yellow.

Floodplain forests, wet hammocks, and stream banks. Occasional; northern and central peninsula, west to central panhandle. Nearly throughout North America; Mexico, Central America, and South America; Europe, Asia, Australia, and Pacific Islands. Native to North America, Mexico, and Central America. Spring.

Acer rubrum L. [Red, in reference to the color of the flowers and fruits.] RED MAPLE.

Acer rubrum Linnaeus, Sp. Pl. 1055. 1753. *Acer rubrum* Linnaeus var. *coccineum* Aiton, Hort. Kew. 3:434. 1789, nom. inadmiss. *Acer rubrum* Linnaeus var. *eurubrum* Pax, Bot. Jahrb. Syst. 7: 181. 1886, nom. inadmiss. *Acer rubrum* Linnaeus subsp. *normale* Wesmael, Bull. Soc. Roy. Bot. Belgique 29: 28. 1890, nom. inadmiss. *Acer rubrum* Linnaeus var. *normale* Schwerin, Gartenflora 42: 166. 1893, nom. inadmiss. *Rufacer rubrum* (Linnaeus) Small, Man. S.E. Fl. 826, 1505. 1933.

Acer carolinianum Walter, Fl. Carol. 251. 1788. *Acer rubrum* Linnaeus subsp. *carolinianum* (Walter) W. Stone, Pl. S. New Jersey 544. 1912. *Rufacer carolinianum* (Walter) Small, Man. S.E. Fl. 826, 1505. 1933.

Acer drummondii Hooker & Arnott ex Nuttall, N. Amer. Sylva 2: 83, pl. 70. 1846. *Acer rubrum* Linnaeus var. *drummondii* (Hooker & Arnott ex Nuttall) Sargent, Rep. For. N. Amer. 50. 1884. *Acer rubrum* Linnaeus forma *drummondii* (Hooker & Arnott ex Nuttall) Schwerin, Gartenflora 42: 167. 1893. *Rufacer drummondii* (Hooker & Arnott ex Nuttall) Small, Man. S.E. Fl. 826. 1933. *Acer rubrum* Linnaeus subsp. *drummondii* (Hooker & Arnott ex Nuttall) E. Murray, Kalmia 1: 29. 1969.

Acer rubrum Linnaeus var. *trilobum* Torrey & A. Gray ex K. Koch, Hort. Dendrol. 80. 1853.

Acer rubrum Linnaeus var. *tridens* A. W. Wood, Class-book Bot., ed. 1861. 286. 1861. *Acer rubrum* Linnaeus forma *tridens* (A. W. Wood) B. Boivin, Naturaliste Canad. 93: 432. 1966.

Tree, to 15 m; bark furrowed, ultimately divided into long, narrow ridges covered with scaly plates, the branchlets purplish red or gray, smooth. Leaves simple, the blade ovate to orbicular or obovate, usually palmately 3-lobed, sometimes also with a pair of small basal lobes, 5–12 cm long, 4–10 cm wide, the apex acuminate, the base truncate, rounded, or subcordate, the margin serrate-dentate, cut less than halfway to the main vein at the base, the upper surface green, glabrous or sparsely short-pubescent along the principal veins, the lower surface grayish or silvery, glabrous or pubescent, the petiole 3–10 cm long. Flowers in a fascicle, deep red to pinkish, brownish, or yellowish, the pedicel ca. 1 cm long at anthesis, glabrous; sepals 5, oblong to linear, ca. 2 mm long; petals similar to the sepals; stamens 5–8. Fruit with each winged half 2.5–3.5 cm long, reddish.

Floodplain forests and swamps. Frequent; nearly throughout. Newfoundland and Quebec south to Florida, west to Ontario, Minnesota, Iowa, Missouri, Oklahoma, and Texas. Winter–spring.

Acer saccharinum L. [Sugary.] SILVER MAPLE.

Acer saccharinum Linnaeus, Sp. Pl. 1055. 1753. *Acer saccharinum* Linnaeus var. *normale* Pax, in Engler, Pflanzenr. 4(Heft 8): 39. 1902, nom. inadmiss. *Saccharodendron saccharinum* (Linnaeus) Nieuwland, Amer. Midl. Naturalist 3: 183. 1914. *Sacchrosphendamnus saccharinus* (Linnaeus) Nieuwland, Amer. Midl. Naturalist 3: 183, 379. 1914. *Argentacer saccharinum* (Linnaeus) Small, Man. S.E. Fl. 825, 1505. 1933.

Acer dasycarpum Ehrhart, Gartenkalender 4: 202. 1784. *Acer dasycarpum* Ehrhart subvar. *normale* Pax, Bot. Jahrb. Syst. 7: 180. 1886, nom. inadmiss. *Acer dasycarpum* Ehrhart var. *typicum* Schwerin, Gartenflora 42: 162. 1893, nom. inadmiss. *Acer dasycarpum* Ehrhart forma *genuinum* Schwerin, Gartenflora 42: 162. 1893, nom. inadmiss.

Tree, to 25 m; bark breaking into long, loose, scaly plates, the branchlets green, becoming brown or reddish brown, with scattered buff-colored, vertically lenticular lenticels, glabrous. Leaves simple, the blade ovate to orbicular, 6–12 cm long and wide, palmate-veined and palmately 3- to 5-lobed, the terminal lobe more than half the blade length, the sinus between the terminal lobe and the major laterals narrowly acute to broadly V-shaped, the apex long acuminate, the base truncate to cordate, the margin irregularly prominently and sharply toothed, the upper surface green and glabrous, the lower surface at first shaggy-pubescent, maturing to a short, appressed-pubescent visible with magnification, the petiole 5–10 cm long, usually red. Flowers in compact fascicles, the pedicel very short and the flower appearing subsessile; calyx funnel-shaped, glabrous; petals absent; stamens 3–7. Fruit with each winged half 3–6 cm long.

Wet hammocks and riverbanks. Occasional; central and western panhandle, Citrus County. New Brunswick and Quebec south to Florida, west to Ontario, North Dakota, South Dakota, Nebraska, Kansas, and Oklahoma, also Manitoba, Washington, New Mexico, and California. Spring–summer.

Acer saccharum Marshall [Sugar.] SUGAR MAPLE.

Tree, to 30 m; bark grayish, shallowly furrowed and scaly ridged, the branchlets green at first, becoming reddish brown, with scattered vertically lenticular lenticels. Leaves simple, the blade ovate to orbicular, 3–9 cm long and wide, palmate-veined and 3- to 5-lobed, the margin of the 3 major lobes usually with 3 smaller lobes distally, the apex acuminate, the base truncate to cordate, the upper surface glabrous, the lower surface pubescent, green or glaucous, the petiole 4–6 cm long. Flowers borne in a fascicle terminating a short shoot, the pedicel 1–2 cm long, sparsely pubescent; calyx cuplike or saucerlike, the lobes broadly triangular; petals absent; stamens 7–8. Fruit with each winged half 2.5–3 cm long.

1. Leaves with the lower surface glaucous, the terminal lobe usually with the sides tapering in toward the base .... subsp. **floridanum**

1. Leaves with the lower surface green, the terminal lobe widest at the base or with the sides parallel......
....subsp. **leucoderme**

Acer saccharum subsp. **floridanum** (Chapm.) Desmarais [Of Florida.] FLORIDA MAPLE.

> *Acer saccharinum* Linnaeus var. *floridanum* Chapman, Fl. South. U.S. 81. 1860. *Acer floridanum* (Chapman) Pax, Bot. Jahrb. Syst. 7: 243. 1886. *Acer saccharinum* Linnaeus subsp. *floridanum* (Chapman) Wesmael, Bull. Soc. Roy. Bot. Belgique 29: 61. 1890. *Acer barbatum* Michaux var. *floridanum* (Chapman) Sargent, Gard. & Forest 4: 148. 1891. *Acer saccharum* Marshall var. *floridanum* (Chapman) Small & A. Heller, Mem. Torrey Bot. Club 3(1): 24. 1892. *Saccharodendron floridanum* (Chapman) Nieuwland, Amer. Midl. Naturalist 3: 182. 1914. *Acer barbatum* Michaux forma *floridanum* (Chapman) Fernald, Rhodora 47: 160. 1945. *Acer saccharum* Marshall subsp. *floridanum* (Chapman) Desmarais, Brittonia 7: 382. 1952. *Acer nigrum* F. Michaux var. *floridanum* (Chapman) Fosberg, Castanea 19: 27. 1954. TYPE: FLORIDA: "Middle Florida," s.d., *Chapman s.n.* (holotype: NY).

Leaves with the lower surface glaucous, the terminal lobe usually with the sides tapering in toward the base.

Mesic hammocks. Occasional; northern counties south to Polk County. Virginia south to Florida, west to Illinois, Missouri, Oklahoma, and Texas. Spring.

Acer saccharum subsp. **leucoderme** (Small) Desmarais [From the Greek *leuco*, white, and *derma*, skin, in reference to the white bark.] CHALK MAPLE.

> *Acer floridanum* (Chapman) Pax var. *acuminatum* Trelease, Rep. (Annual) Missouri Bot. Gard. 5: 99. 1894. *Acer leucoderme* Small, Bull. Torrey Bot. Club 22: 367. 1895. *Acer saccharum* Marshall var. *leucoderme* (Small) Sudworth, U.S.D.A. Div. Forest. Bull. 14: 285. 1897, nom. illegit. *Saccharodendron leucoderme* (Small) Nieuwland, Amer. Midl. Naturalist 3: 182. 1914. *Acer saccharum* Marshall subsp. *leucoderme* (Small) Desmarais, Brittonia 7: 384. 1952. *Acer nigrum* F. Michaux var. *leucoderme* (Small) Fosberg, Castanea 19: 27. 1954. *Acer saccharum* Marshall var. *acuminatum* (Trelease) E. Murray, Kalmia 13: 3. 1983.

Leaves with the lower surface green, the terminal lobe widest at the base or with the sides parallel.

Bluffs and ravine forests. Occasional; central panhandle. North Carolina south to Florida, west to Oklahoma and Missouri. Spring.

EXCLUDED TAXON

> *Acer barbatum* Michaux—This synonym of the typical subspecies of *A. saccharum* (fide Ward, 2004) was reported for Florida by Kurz and Godfrey (1962) and Correll and Johnston (1970), who misapplied the name to our material of *A. saccharum* subsp. *floridanum*.

Aesculus L. 1753. BUCKEYE.

Shrubs or trees. Leaves opposite, palmately compound, petiolate, estipulate. Flowers in terminal panicles, unisexual and staminate or bisexual, usually both in the same inflorescence; sepals 5, connate; petals 4, free, unequal; stamens 6–8, dorsifixed; nectariferous disk extrastaminal; ovary 3-carpellate and -loculate, the style 1. Fruit a loculicidal capsule.

A genus of about 13 species; North America, Mexico, Europe, and Asia. [Ancient name of some oak].

Aesculus was previously placed in the Hippocastanaceae.

Selected reference: Hardin (1957).

Aesculus pavia L. [Commemorates Peter Paaw (Latin: Pavius) (1573–1617), Dutch physician and botanist.] RED BUCKEYE.

Aesculus pavia Linnaeus, Sp. Pl. 344. 1753. *Pavia octandria* Miller, Gard. Dict., ed. 8. 1768.
Pavia glauca Rafinesque, Alsogr. Amer. 71. 1838. TYPE: FLORIDA.
Pavia parviflora Rafinesque, Alsogr. Amer. 73. 1838. TYPE: "Carol. and Florida."

Shrub or small tree, to 4(12) m; bark brown-gray to light gray, smooth, the branchlets reddish brown. Leaves with 5 or 7 leaflets, the leaflet blade oblong, obovate, narrowly elliptic, or oblanceolate, membranous to subcoriaceous, 6–17 cm long, 3–6 cm wide, the apex acuminate, the base cuneate, the margin irregularly serrate or crenate-serrate, the upper surface with the veins deeply impressed, glabrous except for a few scattered trichomes on the main veins, the lower surface glabrous to densely tomentose, the petiolule 1–19 mm long, glabrous or pubescent, the petiole 3–17 cm long, glabrous or pubescent. Flowers in an elongate panicle 10–25 cm long, the pedicel 5–12 mm long, short tomentose; calyx tubular-campanulate or tubular with a gibbous base, 8–18 mm long, villous with scattered stipitate glands, the lobes 5, red, rounded, glandular-pubescent, unequal; petals 4, red, unequal, the upper 2.5–4 cm long, the claw ca. 2 cm long, villous, the blade oblong-obovate to suborbicular, the lateral petals 2–3 cm long, the claw 1–1.5 cm long, villous, the blade oblong-obovate; stamens 2.5–3.5 cm long, the filaments white, villous on the lower half, the anthers yellow, with a few trichomes at the top and bottom; ovary villous, the style equal to the stamens or longer, glabrous. Fruit subglobose to obovoid, 3.5–6 cm long, the pericarp slightly pitted, light brown; seeds 1–3(6), subglobose, 2–3 cm long, dark chestnut-brown.

Calcareous hammocks and sandhills. Frequent; northern counties south to Sumter, Lake, and Orange Counties. Ontario south to Florida and Texas. Spring.

EXCLUDED TAXA

Aesculus sylvatica W. Bartram—Reported for Escambia County by Sargent (1926, as *A. georgiana* Sargent). Also reported for Florida by Wilhelm (1984) and Clewell (1985), based on Sargent's report. No Florida specimens known. Excluded from Florida by Hardin (1957).

Aesculus ×*mutabilis* (Spach) Schelle (*A. pavia* × *A. sylvatica* W. Bartram)—The citation of two specimens of this hybrid by Hardin (1957) is dubious at best, since one of the putative parents (*A. sylvatica*) is found some 200 miles north of Florida at its nearest approach. All Florida material seen is *A. pavia*.

Cardiospermum L. 1753. HEARTSEED

Vines, with axillary tendrils. Leaves biternately or ternately compound, petiolate, stipulate. Flowers in axillary corymb-like reduced thyrses with two opposing tendrils below the top of the peduncle, unisexual, sometimes also bisexual; sepals 5, free, the two outer shorter than the two inner ones; petals 4, free, appendaged, the petaloid scales of the two upper petals equilateral, cucullate crested, bearing a tonguelike appendage below the apex, this pointing downward, those of the two lower petals inequilateral, with a dorsal winglike crest; nectariferous disk with a gland opposite each of the two upper petals; stamens 8, unequal, deflexed, connate at the base; ovary 3-carpellate and -loculate, the stigmas 3. Fruit an inflated, 3-angled, septifragal capsule.

A genus of 12 species; North America, West Indies, Mexico, Central America, South America, Africa, Asia, Australia, and Pacific Islands. [From the Greek *cardia*, heart, and *sperma*, seed, in reference to the heart-shaped hilum on the seeds.]

1. Fruit ca. 1 cm long, obpyramidal to subglobose..**C. microcarpum**
1. Fruit 2–4 cm long, fusiform to subglobose or obovoid.
    2. Seed hilum semicircular to emarginate, less than ½ the width of the seed....................**C. corindum**
    2. Seed hilum broadly cordate, 2-lobed to about the middle or more, nearly as wide as the seed
    ....**C. halicacabum**

Cardiospermum corindum L. [*Corindum*, a hard mineral, in reference to the hard seed.] FAUX PERSIL.

> *Cardiospermum corindum* Linnaeus, Sp. Pl., ed. 2. 526. 1762. *Cardiospermum corindum* Linnaeus forma *villosum* Radlkofer, in Martius, Fl. Bras. 13(3): 447. 1897, nom. inadmiss.
> *Cardiospermum corindum* Linnaeus forma *subglabratum* Radlkofer, in Martius, Fl. Bras. 13(3): 447. 1897. *Cardiospermum corindum* Linnaeus var. *subglabratum* (Radlkofer) Barkley, Lilloa 28: 152. 1957.
> *Cardiospermum keyense* Small, Man. S.E. Fl. 827, 1505. 1933. TYPE: FLORIDA: Monroe Co.: Key Largo, 9 May 1923, *Small et al. 10952* (holotype: NY).

Subherbaceous perennial vine, to 2 m; stem finely pubescent. Leaves biternately compound, the leaflet blade ovate to broadly elliptic, 1.7–5.7 cm long, 1.1–3.3 cm wide, the median one the largest, the apex acute to obtuse or mucronulate, the base decurrent on the petiolule, the margin crenate, the petiole 3–26 cm long. Flowers in an axillary corymbiform reduced thyrse 4–8.7 cm long, the flowers 4–6 mm long, whitish yellow; sepals obovate, hispid; petals obovate, gland-dotted within; nectar disk appendages short and rounded; stamen filaments pilose basally; ovary pilose. Fruit fusiform or subglobose, 2–3 cm long, chartaceous, reddish tan to stramineous, puberulent; seeds globose, 3–4 mm long, black, the hilum semicircular to emarginate, less than ½ the seed width, whitish.

Tropical hammocks. Rare; Miami-Dade and Monroe Counties. Florida, Texas, and Arizona; West Indies, Mexico, Central America, and South America; Africa and Asia. Native to West Indies, South America, Africa, and Asia. All year.

Cardiospermum halicacabum L. [From the Greek, meaning "salt barrel," in reference to the barrel-like fruit.] LOVE-IN-A-PUFF.

> *Cardiospermum halicacabum* Linnaeus, Sp. Pl. 366. 1753. *Corindum halicacabum* (Linnaeus) Medikus, Malvenfam. 110. 1787. *Cardiospermum inflatum* Salisbury, Prodr. Stirp. Chap. Allerton 379. 1796, nom. illegit.

Herbaceous perennial vine, to 5 m; stem sparsely pilose to subglabrous. Leaves biternately or ternately compound, the leaflet blade rhombic, elliptic, or lanceolate, 3.8–4.7 cm long, 1.8–2.8 cm wide, the median one the largest, the apex acute to acuminate, the base abruptly narrowed, cuneate, the margin serrate or incised, the petiole 1.3–6 cm long. Flowers in an axillary corymbiform thyrse 1.2–6.4 cm long, the flowers 3–4 mm long, white; sepals obovate, pilose on the margin or glabrous; petals obovate; nectar disk appendages short and rounded; stamens with the filaments pilose; ovary pilose. Fruit subglobose to obovoid or 3-angled, 2.3–3.1 cm long, chartaceous, stramineous to reddish tan, puberulous; seeds globose, ca. 5 mm long, black, the hilum cordate, 2-lobed to the middle or more, nearly as wide as the seed, white.

Disturbed sites. Occasional; central and southern peninsula. Escaped from cultivation. Massachusetts and New York south to Florida, west to Michigan, Illinois, Kansas, Oklahoma, and Texas; West Indies, Mexico, Central America, and South America; Africa, Asia, Australia, and Pacific Islands. Native to tropical America, Africa, Asia, and Pacific Islands. All year.

Cardiospermum microcarpum Kunth [With small fruit.] HEARTSEED.

Cardiospermum microcarpum Kunth, in Humboldt et al. Nov. Gen. Sp. 5: 104. 1821. *Cardiospermum halicacabum* Linnaeus var. *microcarpum* (Kunth) Blume, Rumphia 3: 185. 1847.

Herbaceous annual vine, to 2 m; stem finely pubescent. Leaves biternately compound, the leaflet blade ovate, elliptic, or lanceolate, 2.8–4.4 cm long, 1.5–2.3 cm wide, the apex acute to acuminate, the base truncate to attenuate, the margin coarsely dentate to incised, the median one the largest, the petiole 1–5.3 cm long. Flowers in an axillary corymbiform thyrse 3.7–6.6 cm long, the flower ca. 4 mm long, white to stramineous; sepals obovate, glabrous; petals obovate, 2–3 mm long; nectar disk appendages short and rounded; stamens with the filaments pilose; ovary pilose. Fruit obpyramidal to subglobose, 0.9–1.3 cm long, strongly 2-angled or -winged, the apex truncate, chartaceous, stramineous, puberulent; seeds globose, ca. 3 mm long, blue-black, the hilum bilobed, white.

Hammocks and pinelands. Occasional; peninsula. Florida; West Indies, Mexico, Central America, and South America; Africa, Asia, and Australia. Native to tropical America, Africa, and Asia. All year.

Cupania L. 1753.

Trees or shrubs. Leaves even-pinnately compound, petiolate, estipulate. Flowers in axillary thyrses, actinomorphic, unisexual or sometimes also bisexual, bracteate, bracteolate; sepals 5, free; petals 5, free; nectariferous disk present; stamens 8(10), inserted within the nectar disk; ovary 3-carpellate and -loculate, style 1, the stigmas 3. Fruit a loculicidally dehiscent capsule; seeds arillate.

A genus of about 45 species; Florida, West Indies, Mexico, and Central America. [Commemorates Francis Cupani (1657–1710), Sicilian monk, physician, and botanist.]

Cupania glabra Sw. [Smooth.] AMERICAN TOADWOOD.

Cupania glabra Swartz, Prodr. 61. 1788.

Shrub or tree, to 15(25) m; branchlets yellow-tomentose, glabrate in age. Leaves with 6–10 leaf-lets, the leaflet blade obovate to oblong-elliptic, 6–20 cm long, 2–7 cm wide, subcoriaceous, the apex obtuse, the margin subentire crenate-denticulate, the upper and lower surfaces glabrate to pubescent. Flowers in an axillary panicle 11–20 cm long, subequal or shorter than the leaves, the upper axis puberulent; sepals suborbicular to elliptic, ca. 2 mm long, white, puberulent on the inner surface; petals ca. 2 mm long, white. Capsule turbinate-globose, 1.5–2 cm long, 3-lobed, glabrous.

Tropical hammocks. Rare; Monroe County keys. Florida; West Indies, Mexico, and Central America. Spring.

Cupania glabra is listed as endangered in Florida (Florida Administrative Code, Chapter 5B-40).

Cupaniopsis Radlk. 1879. CARROTWOOD

Trees. Leaves even-pinnately compound, the leaflets alternate or subopposite, petiolate, estipulate. Flowers in axillary paniculiform thyrses, actinomorphic, unisexual; sepals 5, free or basally connate, in 2 unequal series; petals 5, free; nectariferous disk present; stamens 6–10; ovary 3-carpellate and -loculate. Fruit a loculicidally dehiscent, fleshy capsule; seed arillate.

A genus of about 60 species; North America, Asia, and Australia. [To resemble *Cupania*, in reference to its close relationship to that genus.]

Selected reference: Adema (1991).

Cupaniopsis anacardioides (A. Rich.) Radlk. [To resemble a species of *Anacardia* (Anacardiaceae).] CARROTWOOD.

Cupania anacardioides A. Richard, in Dumont d'Urville, Voy. Astrolabe 2: 33, t. 13. 1834. *Cupaniopsis anacardioides* (A. Richard) Radlkofer, Sitzungsber. Math.-Phys. Cl. Königl. Bayer. Akad. Wiss. München 9: 530. 1879. *Cupaniopsis anacardioides* (A. Richard) Radlkofer forma *genuina* Radlkofer, in Engler, Pflanzenr. 4(Heft 98): 1187. 1934, nom. inadmiss.

Tree, to 15 m; branchlets appressed-pubescent to glabrous, lenticulate. Leaves with the rachis 1.5–13.5 cm long, the leaflets 4–8(12), the blade obovate or elliptic, 4.5–19 cm long, 1.5–7.5 cm wide, the apex obtuse to retuse, the base broadly cuneate, the margin entire, the upper surface glabrous, the lower surface glabrous or puberulent, the petiolules 2–7 mm long, the petiole 3–7.5 cm long. Flowers in a panicle 8–35 cm long, the axis pubescent, the pedicel 3–7 mm long; sepals 3–4 mm long, puberulent or glabrous; petals ovate, 2–3 mm long. Fruit subglobose, 1.5–3 cm long, apiculate, slightly carinate at the sutures, yellow, tinged with red, puberulent; seeds ellipsoid, the aril cupular, nearly enclosing the seed, yellow to orange-red.

Disturbed sites. Occasional; central and southern peninsula. Escaped from cultivation. Florida; Asia, Australia, and Pacific Islands. Native to Asia and Australia. Spring–summer.

Cupaniopsis anacardioides is listed as a Category I invasive species in Florida by the Florida Exotic Pest Plant Council (FLEPPC, 2015).

Dimocarpus Lour. 1790. LONGAN

Trees or shrubs. Leaves even-pinnately compound, petiolate, estipulate. Flowers in terminal and axillary thyrses, bracteate, unisexual; sepals 5, basally connate; petals 5, free; stamens (6)8(10), equal; ovary 2-carpellate and -loculate. Fruit with only 1 carpel developed, forming a 1-seeded berry; seed 1, arillate.

A genus of 5 species; North America, Asia, Australia, and Pacific Islands. [*Diminutus*, made small, and the Greek *carpo*, fruit, in reference to a single lobe developing on the fruit.]

Selected reference: Leenhouts (1971).

Dimocarpus longan Lour. [Vernacular name in southeast Asia.] LONGAN.

Dimocarpus longan Loureiro, Fl. Cochinch. 233. 1790. *Euphoria verruculosa* Salisbury, Prodr. Stirp. Chap. Allerton 280. 1796, nom. illegit. *Euphoria longan* (Loureiro) Steudel, Nomencl. Bot. 328. 1821. *Scytalia longan* (Loureiro) Roxburgh, Fl. Ind., ed. 1832. 2: 270. 1832. *Nephelium longan* (Loureiro) Hooker, Bot. Mag. 70: t. 4096. 1844.

Tree or shrub, 40 m; branchlets with 5 faint grooves, whitish to dark brown, ferruginous-tomentose, glabrescent in age. Leaves with 2–4(6) leaflets, the leaflet blade lanceolate or elliptic, 3–45 cm long, 1.5–20 cm wide, the apex acute to acuminate to rounded or emarginate, the base equal-sided to oblique, cuneate to rounded, the margin entire, the upper surface glabrous or tomentose on the midrib near the base, the lower surface tufted-tomentose, mainly on the midrib and nerves, rarely subglabrous, the petiolules 1–35 mm long, the petiole 1–20 cm long. Flowers in a terminal or axillary thyrse 8–40 cm long, the axis densely tufted-tomentose, the cymules (1)3- to 5-flowered, the pedicel 1–4 mm long; bracts oblong-ovate to narrowly lanceolate, 2–5 mm long; sepal lobes 2–5 mm long, yellow-brown, the inner surface, short-pubescent; petals oblong-lanceolate, 2–6 mm long, yellow-brown, both sides densely woolly pubescent to subglabrous; nectar disk pubescent; stamens 1–6 mm long. Fruit broadly ellipsoid to globose, 1–3 cm long, orange-yellow, smooth, warty, or sometimes aculeate, glabrescent.

Disturbed sites. Rare; Palm Beach County. Escaped from cultivation. Florida; Asia, Australia, and Pacific Islands. Native to Asia. Spring–summer.

Dodonaea Mill. 1754. VARNISHLEAF

Shrubs or trees. Leaves alternate, simple, petiolate, estipulate. Flowers in terminal and axillary thyrses, actinomorphic, unisexual and bisexual (plants polygamous and polygamodioecious); sepals 3–5, free, subequal; petals absent; nectariferous disk stipe-like in perfect and carpellate flowers, absent in staminate; stamens 5–9, free; ovary 2- to 3-carpellate and -loculate, the styles 3, connate. Fruit a winged capsule; seeds exarillate.

A genus of about 70 species; North America, West Indies, Mexico, Central America, South America, Africa, Asia, Australia, and Pacific Islands. [Commemorates Rembert Dodoens (1518–1585), Dutch herbalist.]

Selected reference: Leenhouts (1983).

1. Fruit mostly 3-winged, the wings adnate to the style; leaves with the primary veins inconspicuous or scarcely more evident than the tertiary ones.. **D. elaeagnoides**
1. Fruit mostly 2-winged, the wings free from the style; leaves with the primary veins more prominent than the tertiary ones..**D. viscosa**

Dodonaea elaeagnoides Rudolphi ex Ledeb. & Alderstam [Resembling *Elaeagnus* (Elaeagnaceae), in reference to the leaves.] SMALLFRUIT VARNISHLEAF; KEYS HOPBUSH.

Dodonaea elaeagnoides Rudolphi ex Ledebour & Alderstam, Diss. Bot. Pl. Doming. 18. 1805. *Dodonaea viscosa* Jacquin forma *elaeagnoides* (Rudolphi ex Ledebour & Alderstam) Radlkofer, in Engler, Pflanzenr. 4(Heft 98): 1371. 1934. *Dodonaea viscosa* Jacquin subsp. *elaeagnoides* (Rudolphi ex Ledebour & Alderstam) Acevedo-Rodríguez, in Acevedo-Rodríguez & M. T. Strong, Smithsonian Contr. Bot. 98: 876. 2012.

Dodonaea microcarya Small, Torreya 25: 39. 1925. TYPE: FLORIDA: Monroe Co.: northern part of Big Pine Key, 8 May 1919, *Small et al. 9105* (holotype: NY; isotype: NY).

Shrub, to 3 m; branches grayish, scurfy glandular. Leaves with the blade obovate, 2.5–8(10) cm long, 1–2.5(4) cm wide, chartaceous, the apex rounded to nearly truncate, sometimes apiculate or emarginate, the base cuneate, the primary veins inconspicuous or scarcely more evident than the tertiary ones, the margin entire, the upper and lower surfaces glabrous, usually not very glandular and accordingly not appearing varnished, subsessile or short-petiolate. Flowers in a short terminal or axillary thyrse, glabrous, variably glandular, unisexual (plants dioecious, rarely monoecious); sepals 4; stamens 6–7; ovary 2(4)-carpellate and -loculate. Fruit suborbicular, laterally compressed, 4–7 mm long, the wings 2–4 mm wide, adnate to the style base for 1–3 mm, thin-pergamentaceous, glabrous, usually not evidently glandular, brownish to reddish.

Coastal hammocks and strands. Rare; Monroe County keys. Florida; West Indies. Summer–fall.

Dodonaea elaeagnoides is listed as endangered in Florida (Florida Administrative Code, Chapter 5B-40).

Dodonaea viscosa Jacq. [Sticky.] VARNISHLEAF; FLORIDA HOPBUSH.

Ptelea viscosa Linnaeus, Sp. Pl. 118. 1753.

Dodonaea viscosa Jacquin, Enum. Syst. Pl. 19. 1760. *Dodonaea viscosa* Jacquin var. *vulgaris* Bentham, Fl. Austral. 1: 476. 1863, nom. inadmiss. *Dodonaea viscosa* Jacquin forma *typica* Herter, Revista Sudamer. Bot. 5: 35. 1937, nom. inadmiss.

Dodonaea angustifolia Linnaeus f., Suppl. Pl. 218. 1782 ("1781"). *Dodonaea viscosa* Jacquin var. *angustifolia* (Linnaeus f.) Bentham, Fl. Austral. 1: 476. 1863. *Dodonaea viscosa* Jacquin forma *angustifolia* (Linnaeus f.) Sherff, Amer. J. Bot. 32: 214. 1945. *Dodonaea viscosa* Jacquin subsp. *angustifolia* (Linnaeus f.) J. G. West, Brunonia 7: 39. 1984.

Dodonaea bialata Kunth, in Humboldt et al., Nov. Gen. Sp. 5: 134, t. 422. 1822.

Dodonaea jamaicensis de Candolle, Prodr. 1: 616. 1824.

Dodonaea asplenifolia Rudge var. *arborescens* Hooker f., J. Bot. (Hooker) 2: 415. 1840. *Dodonaea viscosa* Jacquin var. *arborescens* (Hooker f.) Sherff, Amer. J. Bot. 32: 214. 1945. *Dodonaea viscosa* Jacquin forma *arborescens* (Hooker f.) Sherff, Amer. J. Bot. 32: 214. 1945. SYNTYPE: FLORIDA.

Shrub, to 3 m; branches blackish, glabrous or rarely scurfy glandular. Leaves with the blade obovate, 5–12.5 cm long, 1.5–4.5 cm wide, thin-pergamentaceous, the apex rounded and apiculate, the base cuneate, with the primary veins more prominent than the tertiary ones, the margin entire, the upper and lower surfaces glabrous, inconspicuously glandular, not appearing varnished, subsessile or short-petiolate. Flowers in a short panicle, glabrous or slightly glandular, bisexual; sepals 4 or 5; stamens (5)8(9); ovary 2(4)-carpellate and -loculate. Fruit suborbicular, laterally compressed, 9–13 mm long, the wings 4–6 mm wide, free from the style base, membranaceous, glabrous, usually not evidently glandular, stramineous or reddish.

Dunes, coastal pinelands, and hammocks. Occasional; peninsula. Florida, Arizona, and California; West Indies, Mexico, Central America, and South America; Africa, Asia, Australia, and Pacific Islands. Summer–fall.

Exothea Macfad. 1837.

Shrubs or trees. Leaves even-pinnately compound, petiolate, estipulate. Flowers in axillary, subterminal corymbiform thyrses, bracteate, unisexual and also bisexual (plants polygamo-dioecious); sepals 5, basally connate; petals 5, free; nectariferous disk lobed; stamens (7)8(10), inserted on the disk in depressions near the margin; ovary 2-carpellate and -loculate. Fruit a berry.

A genus of 3 species; Florida, West Indies, Mexico, and Central America. [From the Greek *exo*, outside, and *theke*, case, to remove, in reference to the genus having been segregated from the Amyrideae (Ruraceae).]

Exothea paniculata (Juss.) Radlk. [Flowers in panicles.] INKWOOD; BUTTERBOUGH.

Melicocca paniculata Jussieu, Mém. Mus. Hist. Nat. 3: 187, t. 5. 1817. *Hypelate paniculata* (Jussieu) Cambessedes, Mém. Mus. Hist. Nat. 18: 32. 1829. *Exothea paniculata* (Jussieu) Radlkofer, in T. Durand, Index Gen. Phan. 81. 1888.

Shrub or tree, to 20 m; bark scaly reddish-brown, the branchlets glabrous with evident lenticels. Leaves with 2–6 leaflets, the leaflet blade elliptic-obovate, 5–13 cm long, 1.5–4 cm wide, the apex acute or emarginate, the base cuneate, the margin entire, the upper and lower surfaces glabrous, sessile or the petiolule to 2 mm long, the petiole 0.5–2.5 cm long. Flowers in an axillary, subterminal corymbiform thyrse to 14 cm long, the axis pubescent, the pedicel 2–3 mm long; sepals ovate, 2–4 mm long, pubescent; petals oblong-ovate, 2–4 long, white, pubescent; nectar disk annular or lobed, pubescent; stamens borne inside the disk, 2–4 mm long, the staminodes 1–2 mm long; ovary slightly pubescent. Fruit subglobose, 10–13 mm long, red or blackish purple, the pericarp thin, the mesocarp orange.

Coastal hammocks and shell middens. Occasional; Volusia County southward along the east coast, southern peninsula. Florida; West Indies and Central America. Winter–spring.

Harpullia Roxb. 1824.

Trees. Leaves even-pinnately compound, petiolate, estipulate. Flowers in terminal and axillary racemiform thyrses, bracteate, bracteolate, actinomorphic, unisexual (plants dioecious); sepals 5, free; petals 5, free; nectariferous disk present; stamens 5; ovary 2(4)-carpellate and -loculate. Fruit a capsule; seed arillate or exarillate.

A genus of about 26 species; Florida, Asia, and Australia. [Philippine vernacular name for the species.]

Harpullia arborea (Blanco) Radlk. [Tree.] UAS.

Ptelea arborea Blanco, Fl. Filip. 63: 1837. *Blancoa arborea* (Blanco) Blume, Rumphia 3: 181. 1847. *Harpullia arborea* (Blanco) Radlkofer, Sitzungsber. Math.-Phy. Cl. Königl. Bayer. Akad. Wiss. München 16: 404. 1886.

Tree, to 30 m; branchlets brown tomentose. Leaves with 6–10 leaflets, the leaflet blade elliptic-oblong or obovate, 10–19 cm long, 4–7.5 cm wide, thin coriaceous, the apex acute to obtuse, the base cuneate, the margin entire, the upper surface glabrous, the lower surface pubescent on the nerves, the petiolule 4–6 mm long, the petiole 4–9 cm long. Flowers in a terminal or axillary racemiform thyrse 4–22 cm long, the pedicel 1–5 cm long; sepals obovate, 6–7 mm long, tomentose; petals obovate, 12–14 mm long, the blade glabrous, the claw pubescent; nectar disk pubescent; ovary pubescent, the style 1–1.5 cm long, twisted. Fruit broadly obcordate, 1.5–2.3 cm long, orange-yellow to red, reticulate, puberulent, the valves drying somewhat woody; seed 1 per locule, black, shiny, the aril minute or absent.

Disturbed sites. Rare; Miami-Dade County. Escaped from cultivation. Florida; Asia and Australia. Native to Asia and Australia. Spring–summer.

Hypelate P. Browne 1756.

Shrubs or trees. Leaves alternate, 3-foliolate, petiolate, estipulate. Flowers in axillary, subterminal paniculate thyrses, bracteate, bracteolate, slightly zygomorphic, unisexual or bisexual (plants monoecious or polygamodioecious); sepals (4)5, free, unequal; petals (4)5, free; nectariferous disk present; stamens 8–10, inserted on the disk lobes; ovary 3-carpellate and -loculate; style 1, the stigma 3-lobed. Fruit a drupe; seed exarillate.

A monotypic genus; Florida; West Indies. [Ancient Greek name for *Ruscus* (Ruscaceae).]

Hypelate trifoliata Sw. [Leaves with three leaflets.] INKWOOD; WHITE IRONWOOD.

Hypelate trifoliata Swartz, Prodr. 61. 1788.

Shrub or tree, to 10 m; branchlets glabrous. Leaves 3-foliolate, the leaflet blade obovate or oblanceolate, 2–5 cm long, 1–2.5 cm wide, the apex rounded to obtuse, sometimes emarginate or acute, the base cuneate, the margin entire, the upper and lower surfaces glabrous, sessile, the petiole 1–3.5 cm long. Flowers in an axillary, subterminal, paniculate thyrse 4–9 cm long;

sepals ovate, 2–3 mm long, unequal, the margin ciliate; petals suborbicular, ca. 3 mm long, white, the margin ciliate; nectar disk lobed; stamens ca. 3 mm long. Fruit ovoid, 5–7 mm long, red turning black.

Tropical pinelands and hammocks. Rare; Miami-Dade County, Monroe County keys. Florida; West Indies. Spring–summer.

Hypelate trifoliata is listed as endangered in Florida (Florida Administrative Code, Chapter 5B-40).

Koelreuteria Laxm. 1772.

Trees. Leaves alternate, odd-bipinnately compound, petiolate, estipulate. Flowers in terminal panicles, slightly zygomorphic, unisexual (plants monoecious, bracteate; sepals 5, basally connate, the lobes 3 long and 2 short; petals 4–5, inserted at the base of the androgynophore; nectariferous disk present; stamens 8(9), on the nectar disk, the anthers antrorse; ovary 3-locular. Fruit a capsule.

A genus of 3 species; North America, Mexico, Europe, Africa, Asia, Australia, and Pacific Islands. [Commemorates Joseph Gottlieb Koelreuter (1733–1806), German professor of natural history.]

Selected reference: Meyer (1976).

Koelreuteria elegans (Seemann) A. C. Sm. subsp. **formosana** (Hayata) F. G. Mey. [Elegant; of Formosa (Taiwan).] FLAMEGOLD.

> *Koelreuteria formosana* Hayata, Icon. Pl. Formos. 3: 64, pl. 13. 1913. *Koelreuteria elegans* (Seemann) A. C. Smith subsp. *formosana* (Hayata) F. G. Meyer, J. Arnold Arbor. 57: 162. 1976.

Tree, to 25 m; bark peeling in square plates, rough, somewhat corky, furrowed lengthwise, the branchlets with pustulate, corky, cinnamon-brown lenticels. Leaves bipinnately compound, 25–60 cm long, 15–44 cm wide, the odd terminal leaflet much reduced or sometimes absent, the axis glabrous or short-pubescent on the somewhat grooved upper side, the leaflets 8–17 on the major divisions, lanceolate to narrowly ovate to elliptic, (4.5)6–10 cm long, 1.8–3(4.2) cm wide, the apex long-acuminate to caudate, the base strongly oblique, margin usually coarsely serrate, the upper surface glabrous or with scattered trichomes and glands on the veins, the lower surface sparsely pubescent with tufts of trichomes in the vein axils, the petiolule 4–5(10) mm long, the petiole 4–8 cm long. Flowers in a terminal panicle 30–50 cm long, densely pubescent and glandular; sepal lobes ovate, 4–5 mm long, ciliate-glandular; petals lanceolate, 6–7 mm long, the blade appendages lobulate-undulate, the claw ca. 2 mm long, sparsely villous; stamens 5–6 mm long, the filaments densely villous near the base; ovary sparsely villous, the styles 3–4(5) mm long. Fruit ellipsoidal, the valves rounded to suborbicular, 3.4–5(6) cm long, reticulate-veined, rose-purple turning brownish, glabrous; seeds pyriform to subglobose, ca. 5 mm long, black, slightly rugose.

Disturbed sites. Occasional; peninsula. Escaped from cultivation. Florida, Mississippi, Louisiana, and Texas, also California; West Indies and Mexico; Europe, Africa, Asia, Australia, and Pacific Islands. Native to Asia. Fall.

Koelreuteria elegans subsp. *formosana* is listed as a Category II invasive species in Florida by the Florida Exotic Pest Plant Council (FLEPPC, 2015).

EXCLUDED TAXON

Koelreuteria paniculata Laxmann—Reported for Florida by Long and Lakela (1971), the name misapplied to material of *K. elegans* subsp. *formosana*.

Melicoccus P. Browne 1756.

Trees. Leaves even-pinnately compound, petiolate, estipulate. Flowers in terminal and axillary panicles, bracteate, subactinomorphic, unisexual (plants polygamodioecious); sepals 4, free, subequal; petals 4, free; nectariferous disk extrastaminal; stamens 8, the anthers dorsifixed, extrorse, 2-locular; ovary 2(3)-carpellate and -loculate, the style 1, the stigmas 2(3)-lobed. Fruit a berry; seed 1(2), exarillate.

A genus of 2 species; North America, West Indies, Mexico, Central America, South America, Africa, and Asia. [From the Greek *meli*, honey, and *coccos*, grain, seed, or berry, apparently in reference to the sweet taste of the seed coat.]

Melicoccus bijugatus Jacq. [Leaves with two pairs of leaflets.] SPANISH LIME.

Melicoccus bijugatus Jacquin, Enum. Syst. Pl. 19. 1760. *Melicocca bijuga* Linnaeus, Sp. Pl., ed. 2. 495. 1762, nom. illegit. *Melicocca carpopodea* Jussieu, Mém. Mus. Hist. Nat. 3: 187, t. 4. 1817. *Stadmannia bijuga* D. Dietrich, Syn. Pl. 2: 1304. 1840, nom. illegit.

Tree, to 12 m; bark grayish white, the branchlets glabrous. Leaves with 2(3) pairs of leaflets, the leaflet blade elliptic, ovate-elliptic, or obovate, 5–12 cm long, 2–6 cm wide, the apex acute or acuminate, the base cuneate, somewhat unequal, the margin entire, undulate, the upper and lower surfaces glabrous, sessile or the petiolules to 2 mm long, the petiole 1.5–6 cm long. Flowers in a terminal or axillary panicle 6–20 cm long, the branches racemose, the pedicel 4–6 mm long; sepals obovate, ca. 2 mm long, subequal, the margin ciliate; petals obovate, 3–4 mm long, greenish white, the margin ciliate; nectar disk flat, 4- or 5-lobed; stamens glabrous; ovary glabrous. Fruit subglobose, 2–3 cm long, yellowish green, the pericarp glabrous, the pulp gelatinous; seed 1.5–2 cm long, cream-colored, the aril white.

Disturbed sites. Rare; Palm Beach County, Monroe County keys. Escaped from cultivation. Florida; West Indies, Mexico, Central America, and South America; Africa and Asia. Native to South America. Spring–summer.

Sapindus L. 1753. SOAPBERRY

Trees. Leaves alternate, even-pinnately compound, petiolate, estipulate. Flowers in terminal thyrses, zygomorphic, unisexual or bisexual (plants polygamodioecious), bracteate, bracteolate; sepals 5, basally connate, unequal; petals 5, free, equal, clawed, with a single bifid scale and 2 scales above the claw on the inner side; nectariferous disc present; stamens 8(10), inserted within the nectar disk; ovary (2)3(4)-carpellate, (2)3(4)-loculate, the style 1, the stigma 3-lobed. Fruit a schizocarp of 1–2(3) mericarps, the mericarps drupe-like; seed exarillate.

A genus of about 12 species; North America, West Indies, Mexico, Central America, South America, Africa, Asia, Australia, and Pacific Islands. [*Sapo*, soap, and *indicus*, of India, in reference to its soapy properties and use of the fruits for washing.]

Sapindus saponaria L. [Soapy.] SOAPBERRY.

Sapindus saponaria Linnaeus, Sp. Pl. 367. 1753. *Sapindus alatus* Salisbury, Prodr. Stirp. Chap. Allerton 280. 1796, nom. illegit. *Sapindus saponaria* Linnaeus forma *genuinus* Radlkofer, in Martius, Fl. Bras. 13(3): 517. 1900, nom. inadmiss.

Sapindus marginatus Willdenow, Enum. Pl. 432. 1809.

Sapindus falcatus Rafinesque, Med. Fl. 2: 261. 1830. TYPE: FLORIDA.

Sapindus acuminatus Rafinesque, New Fl. 3: 22. 1838 ("1836").

Sapindus manatensis Shuttleworth ex Radlkofer, Sitzungsber. Math.-Phys. Cl. Königl. Bayer. Akad. Wiss. München 8: 318. 1878. TYPE: FLORIDA: Manatee Co.

Tree, to 20 m; bark pale brown or grayish brown, sloughing in small, loose plates, the branchlets glabrous, with lenticular, buff-colored lenticels. Leaves 15–32 cm long, the rachis winged or wingless, the leaflets 6–12, opposite or alternate, the blade oblong to lanceolate, 3–15 cm long, 1.5–5 cm wide, occasionally falcate, chartaceous, the apex acute, acuminate, or rarely obtuse, the base cuneate, the margin entire, the upper and lower surfaces glabrous, sessile or short-petiolulate, the petiole 5–12 cm long. Flowers in a thyrsoid panicle to 25 cm long, the axis puberulent; bracts and bracteoles small; sepals triangular, ca. 2 mm long, the outer 2 smaller than the inner 3, white; petals obovate, ca. 2 mm long, the apex infolded, white, each with a pubescent, bifid scale on the inner side, the margin ciliate; nectar disk glabrous; stamens with the filaments 3–4 mm long, the filaments pilose. Fruit subglobose, 1–2 cm long, brown or orange; seed globose, ca. 1.2 cm long.

Hammocks and shell middens. Occasional; nearly throughout. South Carolina south to Florida, west to Colorado, and Arizona; West Indies, Mexico, Central America, and South America; Africa, Asia, and Pacific Islands. Native to North America and tropical America. All year.

EXCLUDED GENERA

Talisia pedicellaris Radlkofer—Reported for Florida by Small 1913a, 1913c, 1913d, 1933) and by Long and Lakela (1971). No Florida specimens seen.

RUTACEAE Juss., nom. cons. 1789. CITRUS FAMILY

Trees or shrubs. Leaves alternate or opposite, simple or compound, estipulate. Flowers solitary and axillary or several in terminal and/or axillary inflorescences, actinomorphic, bisexual and/or unisexual (plants monoecious, dioecious, or polygamous); sepals 4–5, free or connate; petals 4–5, free; stamens 5–many, the filaments free or connate, the anthers 2-locular, versatile, introrse, longitudinally dehiscent; nectariferous disk intrastaminal or absent; ovary (1)2- to

5(many)-carpellate and -loculate, the styles basal, lateral, or terminal, free or connate. Fruit a drupe, follicle, hesperidium, or berry.

A family of about 160 genera and about 2,070 species; nearly cosmopolitan.

Selected reference: Brizicky (1962a).

1. Leaves all 1-foliolate, appearing simple.
    2. Stamens 20 or more; fruit yellow, orange, or green, pulpy; petiole often winged .... **Citrus**
    2. Stamens 10; fruit black, not pulpy; petiole never winged .... **Atalantia**
1. Leaves 3-foliolate or pinnately compound (1-foliolate leaves sometimes also present).
    3. Branches armed with axillary thorns or prickles.
        4. Branches with prickles; leaflets 5–9; fruit of 1–5 follicles .... **Zanthoxylum**
        4. Branches with axillary thorns; leaflets (1–2)3; fruit a berry.
            5. Petiole winged; stamens 20 or more; fruit resembling a small orange, 4–5 cm in diameter
            ....**Poncirus**
            5. Petiole not winged; stamens 6–10; fruit berrylike, to 1.5 cm in diameter ....**Triphasia**
    3. Branches unarmed.
        6. Fruit a samara ....**Ptelea**
        6. Fruit a drupe, follicle, or a few-seeded berry.
            7. Fruit of 1–5 follicles...**Zanthoxylum**
            7. Fruit a drupe or a few-seeded berry.
                8. Fruit a blue or black drupe...**Amyris**
                8. Fruit a white, pink, or red berry.
                    9. Leaflets 3–9, obovate, 1–3.5 cm long; fruit red .... **Murraya**
                    9. Leaflets 1–3, lanceolate, 10–20 cm long; fruit white or pink.........................**Glycosmis**

Amyris P. Browne 1756. TORCHWOOD

Shrubs or trees. Leaves opposite, odd-pinnately compound, petiolate, estipulate. Flowers in terminal or axillary paniculate cymes, actinomorphic, bisexual; sepals 4, basally connate; petals 4, free; stamens 8, in 2 series, inserted at the base of an intrastaminal nectariferous disk; ovary 1-carpellate and -loculate. Fruit a drupe.

A genus of about 40 species; North America, West Indies, Mexico, Central America, South America. [From the Greek *a*, with, and *myron*, balsamic resin, in reference to the balsamic properties of the genus.]

1. Inflorescence puberulent; ovary puberulent .... **A. balsamifera**
1. Inflorescence glabrous; ovary glabrous ....**A. elemifera**

Amyris balsamifera L. [Balsam bearing.] BALSAM TORCHWOOD.

Amyris balsamifera Linnaeus, Syst. Nat., ed. 10. 1000. 1759. *Elemifera balsamifera* (Linnaeus) Kuntze, Revis. Gen. Pl. 1: 100. 1891.

Shrub or tree, to 13 m; branchlets puberulent, glabrous in age. Leaves 3- to 7-foliolate, the leaflet blade lanceolate to ovate or rhombic-ovate, 4–8 cm long, 2–4 cm wide, the apex acute to long-acuminate, the base cuneate to somewhat rounded, the margin crenulate to entire, the upper and lower surfaces glabrous. Flowers in an axillary or terminal paniculate cyme, the rachis puberulent; calyx 4-lobed, puberulent; petals 4, elliptic, oval, or obovate, 3–4 mm long; stigma sessile, the ovary puberulent. Fruit oblong-ovoid to elliptical or obovoid, 9–14 mm long, black; seed 1.

Hammocks. Rare; Miami-Dade County, Monroe County keys. Florida; West Indies, Central America, and South America. All year.

Amyris elemifera L. [*Elemi*, gum, and *ferens*, bearing, in reference to its resin bearing nature.] SEA TORCHWOOD.

> *Amyris elemifera* Linnaeus, Syst, Nat., ed. 10. 1000. 1759. *Amyris sylvatica* Jacquin, Select. Stirp. Amer. Hist. 107. 1763, nom. illegit. *Amyris maritima* Jacquin var. *angustifolia* A. Gray, Proc. Amer. Acad. Arts 23: 226. 1888. *Elemifera floridana* M. Gómez de la Maza y Jiménez & Roig y Mesa, Fl. Cuba 73. 1914.
> *Amyris maritima* Jacquin, Enum. Syst. Pl. 19. 1760. *Elemifera maritima* (Jacquin) Kuntze, Revis. Gen. Pl. 1: 100. 1891.
> *Amyris floridana* Nuttall, Amer. J. Sci. Arts 5: 294. 1822. TYPE: FLORIDA: s.d., *Ware s.n.* (holotype: PH?).

Shrub or tree, to ca. 5 m; branchlets glabrous or sparsely short pilose. Leaves 3- or 5-foliolate, the leaflet blade ovate-lanceolate, to broadly ovate or rhombic-ovate, 2–7 cm long, 1–4.5 cm wide, the apex rounded or acute to acuminate, the base cuneate to subtruncate, the margin crenulate to entire, the upper and lower surfaces glabrous, the petiolules to 1 cm long; the petiole to 3 cm long. Flowers in a terminal or axillary panicle to 7 cm long, the rachis glabrous; calyx 4-lobed, the lobes ovate to triangular, glabrous; petals narrowly obovate to oval, 2–4 mm long; ovary glabrous. Fruit globose to obovoid, 5–8 mm long, black, glaucous; seed 1.

Hammocks. Occasional; Flagler County, central and southern peninsula along the east coast. Florida; West Indies, Central America, and South America. All year.

Atalantia Corrêa, nom. cons. 1805. BOXORANGE

Shrubs. Leaves alternate, simple, petiolate, estipulate. Flowers axillary, solitary or in fascicles; sepals 5, basally connate; petals 5, free; stamens 10, free or sometimes with a few basally connate; ovary 5-carpellate and -loculate. Fruit a berry.

A genus of about 17 species; North America and Asia. [Atalanta, the swift-footed huntress of Greek mythology.]

Severinia Ten. ex Endl. 1842.

Selected reference: Bayer et al. (2009).

Atalantia buxifolia (Poir.) Oliver [With Leaves like *Buxus* (Buxaceae).] CHINESE BOXORANGE.

> *Citrus buxifolia* Poiret, in Lamarck, Encycl. 4: 580. 1797. *Severinia buxifolia* (Poiret) Tenore, Index Sem. Hort. Neap. App. 1840: [9]. 1840. *Sclerostylis buxifolia* (Poiret) Bentham, Hooker's J. Bot. Kew Gard. Misc. 3: 326, 1851. *Atalantia buxifolia* (Poiret) Oliver ex Bentham, Fl. Hongk. 51. 1861.
> *Limonia monophylla* Loureiro, Fl. Cochinch. 271. 1790; non Linnaeus, 1767. *Severinia monophylla* Tanaka, J. Bot. (Morot) 68: 232. 1930.

Shrub, to 2.5 m; branchlets green, with axillary spines to ca. 4 cm long, or unarmed, the branches grayish-brown. Leaves with the blade ovate, obovate, elliptic, or suborbicular, 2–6 cm long, 1–5 cm wide, leathery, the apex rounded to obtuse, retuse or emarginate, the base broadly cuneate, the margin entire, the upper and lower surfaces glabrous, the petiole 1–7 mm long. Flowers solitary or several in a fascicle, subsessile; calyx 4-lobed, the lobes triangular, 1–2 mm long, glabrous; petals 3–4 mm long, with oil glands; stamens with the filaments white. Fruit globose or subellipsoid, 0.8–1.2 cm long, bluish black, smooth, glabrous; seeds 1–2.

Disturbed hammocks and shell middens. Occasional; peninsula. Escaped from cultivation. Florida; Asia. Native to Asia. Spring.

EXCLUDED TAXON

> *Atalantia monophylla* (Linnaeus) de Candolle—This Asia species was reported for Florida by Wunderlin (1982, 1998, both as *Severinia monophylla* (Linnaeus) Tanaka) and Wunderlin and Hansen (2003, 2011, both as *Severinia monophylla* (Linnaeus) Tanaka), the name misapplied to our material of *A. buxifolia*.

Citrus L. 1753.

Shrubs or trees, often armed with spines. Leaves alternate, 1-foliolate, pellucid-glandular, petiolate, estipulate. Flowers in axillary corymbiform cymes or fascicles or solitary, actinomorphic, bisexual or unisexual (staminate); sepals (4)5, basally connate; petals (4)5, free; stamens 15–60, the filaments free or variously connate, inserted around a nectariferous disk supporting the gynoecium; ovary 10- to 14-carpellate and -loculate. Fruit a hesperidium.

A genus of about 30 species; nearly cosmopolitan. [Classical Latin name, originally used for the wood of *Tetraclinis articulata* (Cupressaceae), the African sandarac tree, but transferred to the citron in the first century.]

Selected references: Mabberley (1997, 1998); Swingle (1944); Zhang and Mabberley (2008).

1. Leaves without an apparent articulation at the junction of the blade with the petiole, the petiole not winged or margined; fruit 15–25 cm long, oval or oblong, with a thick rind........................**C. medica**
1. Leaves with an evident articulation at the junction of the blade with the petiole, the petiole winged or margined; fruit less than 10 cm long, with a thin rind...**C. reticulata**

Citrus medica L. [Vernacular name.] CITRON.

> *Citrus medica* Linnaeus, Sp. Pl. 782. 1753. *Citrus fragrans* Salisbury, Prodr. Stirp. Chap. Allerton 378. 1797, nom. illegit. *Citrus aurantium* Linnaeus var. *medica* (Linnaeus) Wight & Arnott, Prodr. Fl. Ind. Orient. 1: 98. 1834. *Citrus medica* Linnaeus subsp. *genuina* Engler, in Engler & Prantl, Nat. Pflanzenfam. 3(4): 200. 1896, nom. inadmiss. *Aurantium medica* (Linnaeus) M. Gómez de la Maza y Jiménez, Fl. Haban. 205. 1897.

Tree, to 3 m; branchlets with short, stout, axillary spines. Leaves with the leaflet blade elliptic-ovate or ovate-lanceolate, 8–20 cm long, 3–9 cm wide, the apex rounded to obtuse, the base cuneate to rounded, the margin serrulate, the petiole not winged or margined, without an apparent articulation at the junction with the blade. Flowers in a terminal panicle or fasciculate in the leaf axils; calyx 5-lobed; petals 5, large, whitish above, pinkish below; stamens 30–40, or more. Fruit oval or oblong, 15–25 cm long, somewhat mammillate at the apex, yellow, the rind very thick, the pulp scant, acidic.

Disturbed hammocks. Rare; Franklin and Miami-Dade Counties. Escaped from cultivation. Florida; West Indies and South America; Europe and Asia. Widely cultivated in the tropics of the Old and New World and probably naturalized elsewhere. Native to Asia. Spring.

Citrus reticulata Blanco [Net-veined, in reference to the fruit.] TANGERINE.

> *Citrus reticulata* Blanco, Fl. Filip. 610. 1837. *Citrus nobilis* Loureiro subforma *reticulata* (Blanco) M. Hiroe, Forest Pl. Hist. Jap. Islands 1: 212. 1974.

Tree, to 8 m; branchlets with few spines, these mainly on the inner branches. Leaves with the leaflet blade broadly or narrowly lanceolate to ovate or obovate, 4–10 cm long, the apex attenuate-acuminate, retuse, the base cuneate, the margin crenulate, the petiole 6–14 mm long, slightly winged distally. Flowers solitary or in small clusters in the axis, the pedicel 3–4 mm long; calyx 5-lobed, glandular-punctate; petals 5, oblanceolate to oblong, 7–8 mm long; stamens 15–20. Fruit depressed-globose, 5–8 cm long, orange-yellow, the rind easily separating from the pulp, the pulp sweet.

Hammocks and disturbed sites. Rare; Hillsborough, Pinellas, Manatee, Highlands, and Glades Counties. Escaped from cultivation. Florida; West Indies; Asia and Australia. Widely cultivated in the tropics of the Old and New World and probably naturalized elsewhere. Native to Asia. Spring.

HYBRID TAXA

Feral *Citrus* encountered in Florida are hybrids involving *C. medica*, *C. reticulata*, and *C. maxima* (Burm.) Merr. (POMELO).

Citrus ×aurantiifolia (Christm.) Swingle (*C. maxima* × ?) [With leaves like *Citrus aurantium*.] KEY LIME.

> *Limonia acidissima* Houttuyn, Nat. Hist. 2: 444. 1774; non Linnaeus, 1762. *Limonia aurantiifolia* Christmann, in Christmann & Panzer, Vollst. Pflanzensyst. 1: 618. 1777. *Citrus ×aurantiifolia* (Christmann) Swingle, J. Wash. Acad. Sci. 3: 465. 1913, pro sp. *Citrus medica* Linnaeus forma *aurantiifolia* (Christmann) M. Hiroe, Forest Pl. Hist. Jap. Islands 1: 219. 1974.
> *Citrus limetta* Risso, Ann. Mus. Natl. Hist. Nat. 20: 195, t. 2(1). 1813. *Citrus aurantium* Linnaeus var. *limetta* (Risso) Wight & Arnott, Prodr. Fl. Ind. Orient. 1: 98. 1834. *Citrus medica* Linnaeus var. *limetta* (Risso) Hooker f., Fl. Brit. India 1: 515. 1872.
> *Citrus lima* Lunan, Hort. Jamaic. 1: 451. 1814.

Disturbed sites. Rare; Pinellas and Lee Counties, southern peninsula. Spring.

Citrus ×aurantium L. (*C. maxima* × *C. reticulata*) [Orange, in reference to the fruit color.] SOUR ORANGE, SWEET ORANGE, GRAPEFRUIT.

Citrus ×*aurantium* Linnaeus, Sp. Pl. 782. 1753, pro sp. *Citrus florida* Salisbury, Prodr. Stirp. Chap. Allerton 378. 1796, nom. illegit. *Aurantium citrum* M. Gómez de la Maza y Jiménez, Fl. Haban. 206. 1897.

Citrus aurantium Linnaeus var. *sinensis* Linnaeus, Sp. Pl. 783. 1753. *Citrus sinensis* (Linnaeus) Osbeck, Reise Ostindien 250. 1765. *Aurantium sinense* (Linnaeus) Miller, Gard. Dict., ed. 8. 1768. *Citrus aurantium* Linnaeus subsp. *sinensis* (Linnaeus) Engler, in Engler & Prantl, Nat. Pflanzenfam. 3(4): 198. 1896. *Citrus aurantium* Linnaeus subforma *sinensis* (Linnaeus) M. Hiroe, Forest Pl. Hist. Jap. Islands 1: 222. 1974.

Citrus vulgaris Risso, Ann. Mus. Natl. Hist. Nat. 10: 190. 1813. *Citrus amara* Link, Handbuch 2: 346. 1829, nom. illegit. *Citrus aurantium* Linnaeus var. *vulgaris* (Risso) Wight & Arnott, Prodr. Fl. Ind. Orient. 1: 97. 1834. *Aurantium vulgare* (Risso) M. Gómez de la Maza y Jiménez, Fl. Haban. 205. 1897.

Citrus paradisi Macfadyen, Bot. Misc. 1: 304. 1830. *Citrus decumana* (Linnaeus) Linnaeus var. *paradisi* (Macfadyen) H. A. Nicholls, Bull. Misc. Inform. Kew 21: 205. 1888.

Frequent; central panhandle, peninsula. Spring.

Citrus ×jambhiri Lush (*C. medica* × *C. reticulata*) [Jámbhir, Sanskrit vernacular name for *Citrus medica.*] MANDARIN LIME; ROUGH LIME.

Citrus ×*jambhiri* Lush, Indian Forester 36: 342. 1910, pro sp.

Disturbed sites. Rare; Monroe County keys. Spring.

Citrus ×limon (L.) Burm. f. (*C. medica* × *C. ×aurantium*) [Vernacular name.] LEMON.

Citrus medica Linnaeus var. *limon* Linnaeus, Sp. Pl. 782. 1753. *Citrus* ×*limon* (Linnaeus) Osbeck, Reise Ostindien 250. 1765, pro sp. *Limon vulgaris* Miller, Gard. Dict., ed. 8. 1768. *Citrus medica* Linnaeus forma *limon* (Linnaeus) M. Hiroe, Forest Pl. Hist. Jap. Islands 1: 218. 1974.

Citrus limonum Risso, Ann. Mus. Natl. Hist. Nat. 20: 201. 1813. *Citrus aurantium* Linnaeus var. *limonum* (Risso) Wight & Arnott, Prodr. Fl. Ind. Orient. 1: 98. 1834. *Citrus medica* Linnaeus var. *limonum* (Risso) Hooker f., Fl. Brit. India 1: 515. 1872. *Citrus medica* Linnaeus subsp. *limonum* (Risso) Hooker f. ex Engler, in Engler & Prantl, Nat. Pflanzenfam. 3(4): 200. 1896.

Occasional; central and southern peninsula. Spring.

Citrus ×microcarpa Bunge (*C. reticulata* × *C. japonica* Thunberg) [From the Greek *micro*, small, and *carpo*, fruit.] KUMQUAT; CALAMONDIN.

Citrus ×*microcarpa* Bunge, Enum. Pl. China Bor. 10. 1833, pro sp. ×*Citrofortunella microcarpa* (Bunge) Wijnands, Baileya 22: 135. 1984.

Disturbed sites. Rare; Leon, Hillsborough, and Glades Counties. Spring.

Glycosmis Corrêa, nom. cons. 1805.

Shrubs or trees. Leaves alternate, odd-pinnately compound, sometimes 1-foliolate and appearing simple, petiolate, estipulate. Flowers in spiciform cymes arranged in axillary panicles, actinomorphic, bisexual; sepals (4)5, free; petals (4)5, free; stamens (8)10; ovary 2- to 5-carpellate and -loculate. Fruit a berry.

A genus of about 50 species; North America, West Indies, Africa, Asia, and Australia. [From the Greek *glycos*, sweet, and *osme*, scent, in reference to the fragrant flowers.]

Selected reference: Stone (1985).

Glycosmis parviflora (Sims) Little [Small-flowered.] FLOWER AXISTREE.

> *Limonia parviflora* Sims, Bot. Mag. 50: t. 2416. 1823. *Glycosmis parviflora* (Sims) Little, Phytologia 2: 463. 1948.

Shrub or tree, to 3 m. Leaves (1)2–4(5)-foliolate, the leaflet blade elliptic, oblong, or lanceolate, 5–19 cm long, 2.5–8 cm wide, the apex acuminate or obtuse, the base cuneate, the margin entire, the upper and lower surfaces glabrous, the petiolules 1–5 mm long, the petiole 2–4 cm long (to 4 mm if 1-foliolate). Flowers in an axillary or terminal panicle 3–5(14) cm long; sepals ovate, ca. 1 mm long; petals oblong, ca. 4 mm long, white; stamens (8)10. Fruit globose to ellipsoid, 1–1.5 cm long, pale yellowish white, turning reddish to dark vermilion; seeds (1)2–3.

Disturbed hammocks. Rare; southern peninsula. Escaped from cultivation. Florida; West Indies; Africa. Widely cultivated in the tropics of the Old and New World and possibly naturalized elsewhere. Native to Asia. Spring–summer.

EXCLUDED TAXON

> *Glycosmis citrifolia* (Willdenow) Lindley—Reported for Florida by Small (1933) based on a misapplication of the name to material of *G. parviflora*.

Murraya J. König ex L., nom. cons. 1771.

Shrubs or trees. Leaves alternate, odd-pinnately compound, the leaflets alternate or subopposite, petiolate, estipulate. Flowers in terminal or terminal and axillary cymes, actinomorphic, bisexual; sepals 5, basally connate; petals 5, free; stamens 10, free; ovary 2- to 5-carpellate and -loculate. Fruit a berry.

A genus of about 12 species; North America, West Indies, Mexico, Africa, Asia, Australia, and Pacific Islands. [Commemorates Johan Andreas Murray (1740–1791), Swedish physician and botanist.]

Murraya paniculata (L.) Jack [Flowers in panicles.] ORANGE JESSAMINE.

> *Chalcas paniculata* Linnaeus, Mant. Pl. 68. 1767. *Murraya paniculata* (Linnaeus) Jack, Malayan Misc. 1: 31. 1820. *Murraya exotica* Linnaeus var. *paniculata* (Linnaeus) Thwaites, Enum. Pl. Zeyl. 45. 1858. *Murraya scandens* Hasskarl, Abh. Naturf. Ges. Halle 9: 233. 1866, nom. illegit.

Shrubs or trees, to 10 m; branchlets grayish white to pale yellowish gray, puberulous to glabrate. Leaves 3- to 9-foliolate, 4–11 cm long, the leaflet blade ovate to rhombic-ovate or elliptic, 1–5 cm long, 7–24 mm wide, the apex obtuse to rounded or acuminate, the base cuneate, the margin entire or crenulate, the upper and lower surfaces glabrous, the petiolules less than 1 cm long, the petioles 1.5–2 cm long. Flowers in a terminal or terminal and axillary raceme to 5 cm long, the rachis puberulent; sepal lobes 5, triangular, 1–2 mm long; petals narrowly elliptic to oblanceolate, 1.2–2.5 cm long, glandular-punctate, the apex tomentulose. Fruit ovoid, 1–2 cm long, red, glabrous, glandular-punctate, verrucose; seeds 1, villous.

Disturbed sites. Rare; southern peninsula. Escaped from cultivation. Florida; West Indies and Mexico; Africa, Asia, Australia, and Pacific Islands. Native to Asia. All year.

Murraya paniculata is listed as a Category II invasive species in Florida by the Florida Exotic Pest Plant Council (FLEPPC, 2015).

Poncirus Raf. 1838.

Shrubs or trees. Leaves alternate, 1- to 3-foliolate, petiolate, estipulate. Flowers axillary, solitary or in pairs, actinomorphic, bisexual; sepals (4)5(7), basally connate; petals (4)5(7), free; stamens ca. 20, unequal in length; ovary (6)7(8)-carpellate and -loculate. Fruit a hesperidium.

A monotypic genus; North America and Asia. [From the French *poncire*, a name applied to a variety of the citron (*Citrus medica*).]

Poncirus trifoliata (L.) Raf. [With 3 leaflets.] HARDY ORANGE.

Citrus trifoliata Linnaeus, Sp. Pl., ed. 2. 1101. 1763. *Citrus trifolia* Thunberg, Fl. Japon. 294. 1784, nom. illegit. *Aegle sepiaria* de Candolle, Prodr. 1: 538. 1824, nom. illegit. *Poncirus trifoliata* (Linnaeus) Rafinesque, Sylva Tellur. 143. 1838. *Pseudaegle sepiaria* Miquel, Ann. Mus. Bot. Lugduno-Batavi 2: 83. 1865, nom. illegit. *Pseudaegle trifoliata* (Linnaeus) Makino, Bot. Mag. (Tokyo) 16: 15. 1902.

Shrub or tree, to 5 m; bark in long, irregular green and buff-colored stripes, the branchlets green, glabrous or glabrate, with stout, basally flattened axillary thorns to 4 cm long. Leaves usually 3-foliolate, the leaflet blade obovate or ovate or elliptic, 2–5 cm long, 1–3 cm wide, the apex rounded to obtuse, the base cuneate, the margin crenulate distally from about the middle, the upper surface minutely pubescent along the midvein, sometimes also on the lateral veins, the lower surface glabrous, the petiole 0.5–3 cm long, narrowly winged the length or on the distal half. Flowers axillary, solitary or in pairs; sepal lobes (4)5(7), obovate, ca. 5 mm long, the margin somewhat wrinkled, membranous; petals (4)5(7), free, spatulate to obovate, 1.5–3 cm long, white; the ovary densely short-pubescent. Fruit subglobose to pyriform, 4–5 cm long, dull yellow, densely short-pubescent; seeds ca. 20, ovoid, ca. 1 cm long.

Disturbed hammocks. Occasional; northern counties, Marion and Hernando Counties. Escaped from cultivation. Pennsylvania south to Florida, west to Oklahoma and Texas; Asia. Native to Asia. Spring.

Ptelea L. 1753. HOPTREE

Shrubs or trees. Leaves alternate, 3(5)-foliolate, petiolate, estipulate. Flowers in terminal corymbiform, cymose panicles, actinomorphic, bisexual and/or unisexual (plants polygamous); sepals 4 or 5, free; petals 4 or 5, free; stamens 4 or 5, inserted at the base of the nectariferous disk; ovary usually 2-carpellate. Fruit a samara.

A genus of about 3 species; North America and Mexico. [Classical name of *Ulmus* (Ulmaceae), used for this genus because of the similar fruit.]

Selected reference: Bailey (1962).

Ptelea trifoliata L. [With 3 leaflets.] COMMON HOPTREE; WAFER ASH.

> *Ptelea trifoliata* Linnaeus, Sp. Pl. 118. 1753. *Ptelea viticifolia* Salisbury, Prodr. Stirp. Chap. Allerton 68. 1796, nom. illegit. *Ptelea trifoliata* Linnaeus var. *typica* Koehne, Deut. Dendrol. 348. 1893, nom. inadmiss. *Ptelea trifoliata* Linnaeus forma *typica* Schelle, in Beissner et al., Handb. Laubholzben. 278. 1903, nom. inadmiss.
>
> *Ptelea baldwinii* Torrey & A. Gray, Fl. N. Amer. 1: 215. 1838. *Ptelea trifoliata* Linnaeus var. *baldwinii* (Torrey & A. Gray) D. B. Ward, Novon 11: 363. 2001. TYPE: FLORIDA: Duval Co.: St. John's [Fort George Island], s.d., *Baldwin s.n.* (holotype: PH).
>
> *Ptelea trifoliata* Linnaeus var. *mollis* Torrey & A. Gray, Fl. N. Amer. 1: 680. 1840. *Ptelea mollis* (Torrey & A. Gray) M. A. Curtis, Amer. J. Sci. Arts 57: 406. 1849.
>
> *Ptelea obcordata* Greene, Torreya 5: 99. 1905. TYPE: FLORIDA: Lake Co.: vicinity of Eustis, Jun 1894, *Nash 976* (holotype: US).

Shrub or small tree; bark light brown to dark reddish brown, the branchlets terete, pustulardotted, glabrous or pubescent. Leaves 3-foliolate, the leaflet blade elliptic, oval, or obovate, the apex acute to obtuse or rounded, the base cuneate, the margin entire, the upper surface glabrous or sparsely pubescent, sometimes only along the midrib or principal veins, the lower surface glabrous or sparsely to evenly pubescent, sessile. Flowers in a panicle subtended by the uppermost leaf, the axis glabrous or pubescent; sepals 4–5, free, minute, glabrous or pubescent; petals 4–5, free, narrowly oblong, elliptic-oblong, or oblanceolate, 4–6 mm long, greenish white, glabrous or pubescent. Fruit a flat, suborbicular samara, 1.5–2 cm long, glandularpunctate, the wings united around the seed body.

Mesic hammocks. Occasional; northern counties south to Polk County. Quebec south to Florida, west to Ontario, Minnesota, Nebraska, Utah, and Arizona; Mexico. Spring.

Triphasia Lour. 1790.

Shrubs. Leaves alternate, (1)3-foliolate, petiolate, estipulate. Flowers solitary or 2–3 in axillary cymes, bracteolate, actinomorphic, bisexual; sepals 3, basally connate; petals 3, free; stamens 6, free; nectariferous disk present; ovary 3-carpellate, 3-locular. Fruit a berry.

A genus of 3 species; North America, West Indies, Mexico, Central America, South America, Africa, Asia, Australia, and Pacific Islands. [From the Greek *triphasios*, three-fold, in reference to the trifoliolate leaves and trimerous flowers of the type species.]

Triphasia trifolia (Burm. f.) P. Wilson [Three-leaved, in reference to the 3-foliolate leaves.] LIMEBERRY.

> *Limonia trifolia* Burman f., Fl. Indica 103, t. 35(1). 1768. *Limonia trifoliata* Linnaeus, Mant. Pl. 237. 1771, nom. illegit. *Triphasia trifoliata* de Candolle, Prodr. 1: 536. 1824, nom. illegit. *Triphasia trifolia* (Burman f.) P. Wilson, Torreya 9: 33. 1909.
> *Triphasia aurantiola* Loureiro, Fl. Cochinch. 153. 1790.

Shrub, to 4 m; branchlets with mostly paired spines, glabrous. Leaves (1) 3-foliolate, the leaflet blade leathery, pellucid-dotted, the margin sinuate or entire, the upper and lower surfaces glabrous, the terminal leaflet rhombic, 4–5 cm long, 2–3 cm wide, the base cuneate, the apex obtuse or rounded and emarginate, the lateral ones oval or oblong, 1–2 cm long, 0.4–1 cm wide, the apex rounded, emarginate, the base cuneate to rounded, the petioles to 1 cm long. Flowers solitary or 2–3 in an axillary cyme, short pedicellate; calyx 3-lobed, 1–2 mm long, green, glabrous; petals 3, linear-oblong, ca. 1.5 cm long, white, glabrous; stamens 6, the filament 9–11 mm long, the anther ca. 2 mm long, glabrous. Fruit globose or oblong, 1–2 cm long, red, conspicuously glandular-dotted and pitted.

Disturbed hammocks. Rare; southern peninsula. Escaped from cultivation. Florida and Texas; West Indies, Mexico, Central America, and South America; Africa, Asia, Australia, and Pacific Islands. Native to Asia. Winter–spring.

Zanthoxylum L. 1753. PRICKLYASH

Shrubs or trees. Leaves alternate, odd- or even-pinnately compound, petiolate, estipulate. Flowers in axillary spikes or cymose fascicles or in terminal corymbiform panicles, actinomorphic, unisexual (plants monoecious or dioecious); sepals 3–5, free, or absent; petals 3–5, free; stamens 3–5, free; intrastaminal disk small or obscure; ovary of (1)2–5 free carpels. Fruit a 1-seeded follicle.

A genus of about 225 species; North America, West Indies, Central America, South America, Africa, Asia, Australia, and Pacific Islands. [From the Greek *xanthos*, yellow, and *xylon*, wood.]

Fagara L., nom. cons. (1759).

Selected reference: Porter (1976).

1. Flowers in small axillary clusters.
    2. Petiole and rachis winged ....**Z. fagara**
    2. Petiole and rachis not winged ....**Z. americanum**
1. Flowers in terminal cymose panicles.
    3. Leaves even-pinnate..**Z. coriaceum**
    3. Leaves odd-pinnate.
        4. Plant with prickles..**Z. clava-herculis**
        4. Plant lacking prickles..**Z. flavum**

Zanthoxylum americanum Mill. [Of America.] COMMON PRICKLYASH.

Zanthoxylum americanum Miller, Gard. Dict., ed. 8. 1768. *Zanthoxylum clava-herculis* Linnaeus var. *americanum* (Miller) Du Roi, Observ. Bot. 57. 1771.
Zanthoxylum fraxineum Willdenow, Berlin. Baumz. 413. 1796. *Thylax fraxineum* (Willdenow) Rafinesque, Med. Fl. 2: 114. 1830.

Shrub or small tree, to 3(6) m; young stem armed with paired short, broadly flat-based axillary prickles. Leaves odd-pinnate, the leaf axis sparsely pubescent, the leaflets 5–11, the leaflet blade lanceolate to ovate, 1.5–6 cm long, slightly inequilateral, the apex acute to acuminate, the base cuneate to rounded, the margin crenate or entire, with pale, sessile glands, the upper and lower surfaces pubescent, also somewhat stipitate-glandular, the lateral leaflets subsessile, the terminal one usually petiolulate, the petiole 3–6 cm long. Flowers in an axillary cluster on the wood of the previous season; sepals absent; petals 4–5, oblong, apically fringed, green; stamens 4–5; carpels 2–5. Fruit subglobose, ca. 5 mm long, the surface glandular-punctate; seed obovoid to subglobose, ca. 5 mm long, the surface wrinkled-reticulate, black, lustrous.

Rocky hammocks. Rare; Gadsden, Jackson, and Levy Counties. Quebec south to Florida, west to North Dakota, South Dakota, Kansas, Nebraska, Oklahoma, and Louisiana. Summer.

Zanthoxylum americanum is listed as endangered in Florida (Florida Administrative Code, Chapter 5B-40).

Zanthoxylum clava-herculis L. [The club of Hercules.] HERCULES-CLUB.

Zanthoxylum clava-herculis Linnaeus, Sp. Pl. 270. 1753. *Zanthoxylum clavatum* Saint-Lager, Ann. Soc. Bot. Lyon 7: 70. 1880, nom. illegit. *Fagara clava-herculis* (Linnaeus) Small, Fl. S.E. U.S. 675, 1333. 1903.
Zanthoxylum carolinianum Lamarck, Encycl. 2: 39. 1786. *Fagara caroliniana* (Lamarck) Engler, in Engler & Prantl, Nat. Pflanzenfam. 3(4): 117. 1896.
Zanthoxylum fraxinifolium Walter, Fl. Carol. 243. 1788; non Marshall, 1785. *Fagara fraxinifolia* Lamarck, Tabl. Encycl. 1: 334. 1792. *Zanthoxylum tricarpum* Michaux, Fl. Bor.-Amer. 2: 233. 1803. *Kampmania fraxinifolia* (Lamarck) Rafinesque, Med. Respos., ser. 2. 5: 352. 1808.

Shrub or tree, to 10 m; bark irregularly longitudinally splitting, the branchlets minutely pustular-punctate, glabrous, the trunk and branches armed with short, broadly flat-based prickles. Leaves odd-pinnate, the leaflets 5–19, the leaflet blade lanceolate to ovate, 2.5–7 cm long, 2–3 cm wide, inequilateral and falcate, the apex acute to somewhat obtuse, the base rounded, strongly oblique, the upper surface glabrous, glandular-punctate, lustrous, the lower surface dull, glabrous, glandular-punctate, the margin crenate, with buttonlike glands in the notches, the petiole usually 4–5 cm long, glandular-punctate. Flowers in a terminal diffuse-paniculate cyme on the branches of the season; sepals 4–5, minute, or absent; petals 4–5, oblong-ovate or ovate, 3–4 mm long, greenish yellow; stamens 4–5; carpels 1–5. Fruit subglobose, 5–6 mm long, the surface glandular-punctate; seed obovoid to subglobose, 5–6 mm long, the surface wrinkled-reticulate, black, lustrous.

Hammocks. Frequent; nearly throughout. Virginia south to Florida, west to Oklahoma and Texas. Summer.

Zanthoxylum coriaceum A. Rich. [Leathery, in reference to the leaflets.] BISCAYNE PRICKLYASH; LEATHERY PRICKLYASH.

Zanthoxylum coriaceum A. Richard, in Sagra, Hist. Phys. Cuba, Bot. Pl. Vasc. 326, t. 34. 1841. *Fagara coriacea* (A. Richard) Krug & Urban, Bot. Jahrb. Syst. 21: 591. 1896.

Shrub or small tree, to 7 m; the branchlets with short, broadly flat-based prickles, glabrous, the older branches with conic corky protuberances. Leaves even-pinnate, the leaflets 4–12, the leaflet blade obovate to oblong-obovate, 2–5.5 cm long, 1–2 cm wide, the apex obtuse to emarginate or bluntly acuminate, the base cuneate, the margin entire, the upper surface dark green, glabrous, lustrous, the lower surface pale green, often with prickles, the petiole ca. 6 cm long. Flowers in a terminal or rarely axillary cyme; sepals 3, ovate, ca. 1 mm long; petals 3, oblong-ovate, ca. 4 mm long, yellowish white; stamens 3; carpels 3. Fruit subglobose to ellipsoid, 5–6 mm long, brown, rough; seed subglobose, 3–4 mm long, black, lustrous.

Coastal hammocks. Rare; Palm Beach, Broward, and Miami-Dade Counties. Florida; West Indies. All year.

Zanthoxylum coriaceum is listed as endangered in Florida (Florida Administrative Code, Chapter 5B-40).

Zanthoxylum fagara (L.) Sarg. [Generic name of another Rutaceae used because of its resemblance.] WILD LIME; LIME PRICKLYASH.

Schinus fagara Linnaeus, Sp. Pl. 389. 1753. *Zanthoxylum fagara* (Linnaeus) Sargent, Gard. & Forest 3: 186. 1890. *Fagaras fagara* (Linnaeus) Kuntze, Revis. Gen. Pl. 3(2): 34. 1898. *Fagara fagara* (Linnaeus) Small, Fl. S.E. U.S. 675, 1333. 1903, nom. inadmiss.

Fagara pterota Linnaeus, Syst. Nat., ed. 10. 897. 1759. *Zanthoxylon pterota* (Linnaeus) Kunth, in Humboldt et al., Nov. Gen. Sp. 6: 3. 1823.

Shrub or small tree, to 10 m; branchlets puberulent. Leaves odd-pinnate, the rachis winged, the leaflets 5–13, the leaflet blade obovate to elliptic or suborbicular, 1–2 mm long, 7–15 mm wide, the apex rounded or emarginate, the base cuneate, the margin crenulate, the upper and lower surfaces glabrous, the petiole to 2 cm long. Flowers in an axillary, short spike; sepals 4, subobovate, ca. 0.5 mm long; petals 4, obovate, 2–3 mm long, yellowish green; stamens 4; carpels 4. Fruit globose, 3–4 mm long, red to yellowish, slightly rough; seed globose, 2–3 mm long, black, smooth, lustrous.

Hammocks. Frequent; central and southern peninsula. Florida and Texas; West Indies, Mexico, Central America, and South America. All year.

Zanthoxylum flavum Vahl [Yellow, in reference to the yellowish wood.] WEST INDIAN SATINWOOD; YELLOWWOOD; YELLOWHEART.

Zanthoxylum flavum Vahl, Eclog. Amer. 3: 48. 1807. *Fagara flava* (Vahl) Krug & Urban, in Urban, Bot. Jahrb. Syst. 21: 571. 1896.

Zanthoxylum floridanum Nuttall, N. Amer. Sylv. 3: 14, pl. 85. 1849. *Zanthoxylum caribaeum* Lamarck var. *floridanum* (Nuttall) A. Gray, Proc. Amer. Acad. Arts 23: 225. 1888. TYPE: FLORIDA: Monroe Co.: Key West, s.d., *Blodgett s.n.* (holotype: PH?).

Tree, to 12 m; bark light gray, nearly smooth, the branchlets finely pubescent, soon glabrous. Leaves odd-pinnate, the leaflets (3)5–7(11), the leaflet blade ovate to lanceolate or elliptic, 6–8(12) cm long, 2–3(4.5) cm wide, the apex obtuse to subacute, the base rounded to subtruncate or cuneate, the margin crenulate, the upper and lower surfaces finely stellate-pubescent when young, glabrous in age, the petiole 4–10 cm long. Flowers in a terminal panicle 5–15 cm long, the rachis stellate-pubescent, the pedicel 1–4 mm long, stellate-pubescent; sepals 4–5, ovate, ca. 1 mm long; petals ovate-oblong to elliptic-lanceolate, 3–5 mm long, greenish white; stamens (4)5; carpels 1–3. Fruit obovoid, 5–9 mm long; seed subglobose, 4–5 mm long, black, lustrous.

Hammocks. Rare; Monroe County keys. Florida; West Indies. All year.

Zanthoxylum flavum is listed as endangered in Florida (Florida Administrative Code, Chapter 5B-40).

EXCLUDED TAXA

Zanthoxylum caribaeum Lamarck—Reported for Florida by Chapman (1897), the name misapplied to material of *Z. flavum*.

Zanthoxylum clava-herculis Linnaeus var. *fruticosum* (A. Gray) S. Watson—Reported for Florida by Chapman (1860, 1883, both as *Z. carolinianum* Lamarck var. *fruticosum* A. Gray; 1897) and Wilson (1911). The name of this western taxon was misapplied to Florida material of *Z. clava-herculis*.

SIMAROUBACEAE DC. 1811. QUASSIA FAMILY

Trees or shrubs. Leaves alternate, simple or pinnately compound, petiolate, estipulate. Flowers in axillary subterminal panicles or catkins, actinomorphic, unisexual and bisexual (plants polygamodioecious or dioecious); bract and bracteoles present or absent; sepals 3–8, basally connate, or absent; petals (4)5(6), free, or absent; stamens 3–10(15), the anthers versatile, 2-locular; ovary superior, 5-carpellate, the carpels free or connate. Fruit samaroid or a drupe.

Leitneriaceae Benth. & Hook. f. (1880).

Selected reference: Brizicky (1962b).

1. Leaves simple .... **Leitneria**
1. Leaves compound.
    2. Leaves odd-pinnate, the leaflets with 1–5 basal teeth, each prominently gland-tipped; fruit samaroid .... **Ailanthus**
    2. Leaves even-pinnate, the leaflets lacking basal teeth; fruit a drupe .... **Simarouba**

Ailanthus Desf., nom. cons. 1788.

Trees. Leaves odd-pinnately compound, petiolate, estipulate. Flowers in axillary, subterminal panicles, actinomorphic, unisexual and bisexual (plants polygamodioecious), bracteate and bractiolate; sepals 5(6), basally connate; petals 5, free; stamens 10, 5 in each of 2-series, inserted at the base of the annular, intrastaminal, nectariferous disk; carpels 5, free. Fruit samaroid.

A genus of about 15 species; North America, Mexico, South America, Europe, Africa, Asia, Australia, and Pacific Islands. [From the Moluccan name *ailanto*, in reference to the tree height.]

Ailanthus altissima (Mill.) Swingle [Tall]. TREE-OF-HEAVEN.

Toxicodendron altissimum Miller, Gard. Dict., ed. 8. 1768. *Ailanthus altissima* (Miller) Swingle, J. Wash. Acad. Sci. 6: 495. 1916.

Ailanthus glandulosa Desfontaines, Mém. Acad. Sci. (Paris) 1786: 265, pl. 8. 1788. *Ailanthus pongelion* J. F. Gmelin, Syst. Nat. 2: 726. 1791, nom. illegit. *Pongelion glandulosum* (Desfontanes) Pierre, Fl. Forest. Cochinch. 3: sub pl. 294. 1893.

Tree, to 30 m; bark pale gray, roughened and irregularly furrowed, the branchlets chestnut-brown, pubescent, soon glabrous, smooth, with raised lenticels. Leaves odd-pinnately compound, 1.5–9 cm long, the rachis pubescent, glabrous in age, the leaflets opposite, subopposite, or alternate, the leaflet blade 2–15 cm long, 1–2 cm wide, lanceolate-ovate or -oblong, sometimes falcate, the apex acute to acuminate, the base rounded or truncate, oblique, the margin with 1–2(3) blunt teeth on each side near the base, each bearing a gland, the upper surface sparsely pubescent, the lower surface densely pubescent, the petiolule 1–3 mm long, the petiole 2–15 cm long. Flowers in an axillary, subterminal panicle; bracts and bractioles minute; sepal lobes minute; petals lanceolate, 2–3 mm long, the apex infolded (boatlike), greenish yellow, the inner surface pubescent proximally; outer 5 stamens opposite the petals and spreading, the inner 5 suberect; styles 5, connivent at anthesis and appear as 1, the sigmas umbrella-like. Fruit samaroid, 4–5 cm long, the seed central, elongate tapering to the apex and the base.

Disturbed sites. Rare; Alachua, Hillsborough, and Jefferson Counties, central panhandle. Escaped from cultivation. Quebec south to Florida, west to Washington, Oregon, and California; Mexico and South America; Europe, Africa, Asia, Australia, and Pacific Islands. Native to Asia. Spring.

Leitneria Chapm. 1860. CORKWOOD

Shrubs or trees. Leaves simple, petiolate, estipulate. Flowers in axillary catkins, actinomorphic, unisexual; staminate compound, the cymules 3-flowered, sessile; stamens clustered, (3)10–12(15) (apparently representing about 3 flowers), free; bracts 40–50; bracteoles and perianth absent; carpellate simple, spiciform, the flowers solitary; sepals (3)4(8); bracts 10–12, subtending 2 small bracteoles; staminodes absent; ovary superior, 1-carpellate and -loculate. Fruit a dry drupe.

A genus of 2 species; Kentucky, Georgia, Florida, Mississippi, Missouri, Arkansas, and Texas. [Commemorates Edward Frederick Leitner (1812–1838), German-born physician, naturalist, and explorer of Florida.]

Long considered in the monotypic family Leitneriaceae until placed in the Simaroubaceae by various workers and divided into two species based on molecular data (Schrader and Graves, 2011).

Selected references: Bogle (1997); Schrader and Graves (2011).

Leitneria floridana Chapm. [Of Florida.] CORKWOOD.

Leitneria floridana Chapman, Fl. South. U.S. 428. 1860. *Myrica floridana* (Chapman) A. W. Wood,
Amer. Bot. Fl. 309. 1870. TYPE: FLORIDA: Franklin Co.: Apalachicola, Apr–May, *Chapman 2754a*
(holotype: MO; isotypes (GH, MO).

Shrub or small tree, to 4(8) m; bark gray or dark reddish brown, with numerous gray lenticels, the branchlets pubescent, becoming glabrous, the branches reddish brown. Leaves with the blade lanceolate to elliptic-lanceolate 5–20 cm long, 3.5–6 cm wide, leathery, the apex acute to acuminate, the base cuneate, the margin entire, the upper and lower surfaces subglabrous, the trichomes restricted to the midrib, secondary veins, and the margin, the petiole (3)3.5–4.2(5) cm long, pubescent. Staminate catkins lax, 2–6 cm long, bracteate. Carpellate catkins stiffly erect, 1–3 cm long; sepals irregularly inserted, unequal; ovary ca. 2 mm long, finely pubescent. Fruit elliptic to oblong-elliptic, 1–2.5 cm long, erect, often compressed in drying, green, becoming chestnut-brown, glabrous, the endocarp brown, bony, the surface rough-reticulate, subtended by the persistent bracts, bracteoles, and sepals; seed 1, compressed.

Freshwater and brackish swamps and sloughs. Occasional; Levy County, west to central panhandle. Georgia and Florida. Spring.

Leitneria floridana is listed as threatened in Florida (Florida Administrative Code, Chapter 5B-40).

Simarouba Aubl., nom. cons. 1775.

Trees. Leaves even-pinnately compound, petiolate, stipulate. Flowers in terminal and axillary panicles, actinomorphic, unisexual or uni- and bisexual (plants polygamodioecious or dioecious); sepals (4)5(6), basally connate; petals (4)5(6), free; stamens (8)10(12); ovary (4)5(6)-carpellate, partly syncarpous, the carpels connate. Fruit a drupe.

A genus of about 6 species; North America, West Indies, Mexico, Central America, and South America. [Guianan vernacular name.]

Simarouba glauca DC. [Glaucous, with a whitish bloom.] PARADISETREE.

Simarouba glauca de Candolle, Ann. Mus. Natl. Hist. Nat. 17: 424. 1811. *Quassia glauca* (de Candolle)
Sprengel, Syst. Veg. 2: 319. 1825. *Simarouba glauca* de Candolle var. *typica* Cronquist, Bull. Torrey
Bot. Club 71: 231. 1944, nom. inadmiss.
Simarouba officinalis de Candolle forma *glabra* Krug & Urban, Bot. Jahrb. Syst. 15: 305. 1892. SYN-
TYPE: FLORIDA.
Simarouba glauca de Candolle var. *latifolia* Cronquist, Bull. Torrey Bot. Club 71: 231. 1944. TYPE:
FLORIDA: Monroe Co.: Key West, 18 Apr 1896, *Curtiss 5625* (holotype: NY; isotypes: MO, US).

Tree, to 10 m; bark gray rough, the branchlets light brown to brownish gray, smooth, glabrous. Leaves even-pinnately compound, to ca. 4 dm long, the leaflets 10–20, the leaflet blade narrowly oblong to oblong-ovate, 5–10 cm long, 3–5 cm wide, the apex rounded, the base cuneate, slightly asymmetrical, the margin entire, the upper surface glabrous, the lower surface glabrous, pale or glaucescent, the petiole to ca. 7 cm long. Flowers in an open, lax panicle; sepal lobes ovate to triangular, 3–3.5 mm long, whitish, ciliolate; petals ovate to oblong, 4–6 mm long, whitish. Fruit oval to oblong-oval, 1.5–2 cm long, slightly oblique, glabrous, bright red, turning black, the pulp white.

Coastal hammocks. Rare; Brevard County southward along the east coast, southern peninsula. Florida; West Indies, Mexico, Central America, and South America. Spring.

MELIACEAE Juss., nom. cons. 1789. MAHOGANY FAMILY

Trees or shrubs. Leaves alternate, pinnately compound, petiolate, estipulate. Flowers in axillary panicles, actinomorphic, bisexual or unisexual (plants monoecious, dioecious, or polygamous); sepals 4–6, free; petals 4–6, free; stamens 8–12, the filaments connate into a tube; ovary superior, 5- to 6-carpellate and -loculate, the stigma discoid. Fruit a capsule or drupe.

A family of about 50 genera and about 650 species; nearly cosmopolitan.

Selected reference: Miller (1990).

1. Leaves 2- or 3-pinnately compound; leaflet margins serrate or dentate ....**Melia**
1. Leaves pinnately compound; leaflet margin entire.
    2. Leaflets strongly oblique at the base .... **Swietenia**
    2. Leaflets subequal at the base.. **Khaya**

Khaya A. Juss. 1830. AFRICAN MAHOGANY

Trees. Leaves alternate, even-pinnately compound, the leaflets alternate or subopposite, pinnate-veined, petiolate, stipulate. Flowers in axillary panicles, actinomorphic, unisexual (plants monoecious); sepals 4 or 5, basally connate; petals 4 or 5, free; stamens 8–10, connate into a tube; nectariferous disk present; ovary 4- to 5-carpellate and -loculate, the style 1, the head discoid with 4–5 radiating stigmatic ridges. Fruit a septicidal capsule, opening from the apex.

A genus of 7 species; North America and Africa. [From the indigenous African Xhosa language, swelling, apparently in reference to the buttressed tree base.]

Selected reference: Styles and White (1991).

Khaya senegalensis (Desr.) A. Juss. [Of Senegal.] AFRICAN MAHOGANY.

    Swietenia senegalensis Desrousseaux, in Lamarck, Encycl. 3: 679. 1791. *Khaya senegalensis* (Desrousseaux) A. Jussieu, in Guillemin, Bull. Sci. Nat. Géol. 23: 238. 1930.

Tree, to 20 m; bark smooth but exfoliating in small circular scales ca. 3 cm in diameter and leaving a pock-marked mottled gray and brown surface. Leaves even-pinnately compound, to 25 cm long, the leaflets 4–10, alternate or subopposite, coriaceous, the blade leaflet ovate-elliptic or elliptic, 8–12 cm long, 4–5 cm wide, the apex acuminate, the base cuneate, slightly asymmetric, the margin entire, the upper surface bright green, glabrous, lustrous, the lower surface pale grayish green, glabrous, the petiolules 5–10 mm long, the petiole 7–8 cm long. Flowers in a panicle to 20 cm long; calyx 1–2 mm long, 4- to 5-lobed nearly to the base, the lobes suborbicular, ciliate; petals 4–5, elliptic, ca. 6 mm long, white, somewhat hooded; staminal tube ca. 8 mm long, with 8–10 suborbicular, the somewhat overlapping appendages ca. 4 mm long alternating with the anthers or antherodes; nectar disk of the staminate flowers cushion-shaped, reddish, fused to the base of the pistillode, but free from the base of the staminal tube, less conspicuous in the carpellate flowers. Fruit subglobose, ca. 5 cm long, woody, opening by 4–5 valves from the apex, the valves remaining joined at the base, the columella not extending to the apex, with 4–5, sharp, hard woody ridges; seeds 6–12 per locule, ellipsoid or suborbicular, ca. 1 cm long, brown, narrowly winged.

Disturbed rockland hammocks. Rare; Collier and Broward Counties. Escaped from cultivation. Florida; Africa. Native to Africa. Summer.

Melia L. 1753.

Trees. Leaves alternate, odd-bipinnately compound, petiolate, stipulate. Flowers in axillary panicles, actinomorphic, bisexual (plants monoecious); sepals 5–6, basally connate; petals 5–6, free; stamens 10–12; ovary 5- to 6-carpellate and loculate, the stigma 5- to 6-lobed. Fruit a drupe.

A genus of 2–3 species; nearly cosmopolitan. [From the Greek name for the ash tree.]

Melia azedarach L. [Aboriginal name.] CHINABERRY TREE.

Melia azedarach Linnaeus, Sp. Pl. 384. 1753. *Azedarach deleteria* Medikus, Malvenfam. 116. 1787. *Melia florida* Salisbury, Prodr. Stirp. Chap. Allerton 317. 1796, nom. illegit. *Azedarach speciosa* Rafinesque, Fl. Ludov. 135. 1817, nom. illegit. *Azedarach amena* Rafinesque, Med. Fl. 2: 199. 1830, nom. illegit. *Melia azedarach* Linnaeus var. *glabrior* C. de Candolle, in A. de Candolle, Monogr. Phan. 1: 452. 1878, nom. inadmiss. *Azedarach sempervirens* Kuntze, Revis. Gen. Pl. 1: 109. 1891, nom. illegit. *Azedarach sempervirens* Kuntze var. *glabrior* Kuntze, Revis. Gen. Pl. 109. 1891, nom. inadmiss. *Azedarach vulgaris* M. Gómez de la Maza y Jiménez, Repert. Med.-Pharm. Ci. Auxilliares 5: 296. 1894, nom. illegit.

Melia azedarach Linnaeus var. *umbraculifera* G. Knox, Gard. Monthly & Hort. 27: 260. 1885. *Melia azedarach* Linnaeus forma *umbraculifera* (G. Knox) Rehder, Bibl. Cult. Trees 387. 1949.

Tree, to 15 m; bark brown, with narrow furrows, the branchlets with sessile stellate and sinuous, barbellate trichomes. Leaves odd-bipinnately compound, 30–80 cm long, the leaflets 15–35 pairs, alternate or subopposite, the leaflet blade lanceolate, 3–7 cm long, 2–4 cm wide, the apex long-acuminate, the base cuneate to rounded, the margin incised, serrate, or lobed, the upper and lower surfaces glabrous at maturity. Flowers in a panicle 8–25 cm long; sepals lanceolate to elliptic, 1.5–3 mm long, erect, very short-connate at the base, pubescent, stipitate-glandular; petals oblanceolate, 8–12 mm long, pink, blue, or white, the outer surface pubescent; staminal tube 7–8 mm long, dilated at the apex, 10- to 12-toothed, dark purple; ovary glabrous, the stigma 5- to 6-lobed. Fruit subglobose, 12–18 mm long, yellow or orange; seeds oblong, laterally compressed.

Disturbed sites. Frequent; nearly throughout. Escaped from cultivation. New York south to Florida, west to California; West Indies, Mexico, Central America, and South America; Europe, Africa, Asia, Australia, and Pacific Islands. Native to Asia, Australia, and Pacific Islands. Spring.

Melia azedarach is listed as a Category II invasive species in Florida by the Florida Exotic Pest Plant Council (FLEPPC, 2015).

Swietenia Jacq. 1760. MAHOGANY

Trees. Leaves alternate, even-pinnately compound, petiolate, stipulate. Flowers actinomorphic, unisexual (plant monoecious); sepals 5, basally connate; petals 5, free; stamens 10, connate into a tube; ovary 5-carpellate and -loculate; stigma discoid. Fruit a capsule; seeds apically winged.

A genus of 3 species; Florida, West Indies, Mexico, Central America, and South America. [Commemorates Gerard van Swieten (1700–1772), Dutch-born Austrian botanist and physician.]

Swietenia mahagoni (L.) Jacq. [Native vernacular name in the Antilles.] WEST INDIAN MAHOGANY.

> *Cedrela mahagoni* Linnaeus, Syst. Nat., ed. 10. 940. 1759. *Swietenia mahagoni* (Linnaeus) Jacquin, Enum. Syst. Pl. 20. 1760. *Cedrus mahagoni* (Linnaeus) Miller, Gard. Dict., ed. 8. 1768. *Swietenia fabrilis* Salisbury, Prodr. Stirp. Chap. Allerton 317. 1796, nom. illegit. *Swietenia acutifolia* Stokes, Bot. Mat. Med. 2: 479. 1812, nom. illegit.

Tree, to 10 m; bark brown, thick and flaking, the branchlets glabrous. Leaves 10–30 cm long, the leaflets 2–5 pairs, the leaflet blade ovate to lanceolate, 3–8 cm long, 0.5–2 cm wide, coriaceous, the apex acuminate, the base rounded to cuneate, strongly asymmetric, the margin entire, the upper and lower surfaces glabrous. Flowers in a panicle 5–10 cm long, the pedicel 1.5–3 mm long; sepal lobes 5, rounded; petals 5, oblong, 3–5 mm long, pale yellow-white; staminal tube urceolate to cylindric, 3–4 mm long; ovary glabrous. Fruit ovoid, 6–10 cm long, 3–6 cm wide, woody, erect, brown, opening from the base; seeds 4–5 cm long (including the apical wing), chestnut-brown.

Tropical hammocks and disturbed sites. Rare; Lee County, southern peninsula. Florida; West Indies. Spring–summer.

MUNTINGIACEAE C. Bayer et al. 1998. MUNTINGIA FAMILY

Trees or shrubs. Leaves alternate, simple, petiolate, stipulate. Flowers axillary, solitary or in clusters, actinomorphic, unisexual (plants monoecious); sepals 5, basally connate; petals 5, free; nectariferous disk present; stamens numerous, the filaments free or basally connate, the anthers versatile, longitudinally dehiscent; ovary superior, 5- to 7-carpellate and -loculate, the style and stigma 1. Fruit a berry.

A family of 3 monotypic genera; North America, West Indies, Mexico, Central America, South America, Africa, Asia, Australia, and Pacific Islands.

Our sole species, *Muntingia calabura*, was previously placed in the Elaeocarpaceae, Tiliaceae, or Flacourtiaceae by various workers until the Muntingiaceae was erected by Bayer et al. (1998).

Selected references: Brizicky (1965); Strother (2015a).

Muntingia L. 1753.

Trees or shrubs. Leaves alternate, simple, palmate-veined, petiolate, stipulate. Flowers axillary, solitary or few in clusters, pedicellate, actinomorphic, unisexual (plants monoecious); sepals 5, basally connate; petals 5, free; stamens 10–75+, inserted on a nectar disk; ovary 5- to 7-carpellate and -loculate, the style 1, the stigma capitate. Fruit a berry; seeds numerous.

A monotypic genus; North America, West Indies, Mexico, central America, South America, Africa, Asia, Australia, and Pacific Islands. [Commemorates Abraham Munting (1626–1683), Dutch botanist.]

Muntingia calabura L. [Vernacular name.] STRAWBERRY TREE.

Muntingia calabura Linnaeus, Sp. Pl. 509. 1753.

Shrub or tree, to 12 m; bark brown, roughened and slightly furrowed, the branchlets with simple and stellate-pubescent and glandular trichomes. Leaves 2-ranked, the blade lanceolate to oblong-lanceolate, 6–15 cm long, 2–5 cm wide, membranous, the apex long-acuminate, the base obliquely subcordate, asymmetrical, the margin coarsely and irregularly serrate, the upper surface glabrate or sparsely pubescent, the lower surface densely stellate-tomentose, the petiole 2–5 mm long; stipules subulate or filiform, one 2–5 mm long, the other rudimentary or wanting. Flowers solitary or 2–3 in a sessile cluster with 3 filiform bracts at the base, the pedicel 2–3 cm long; sepals 5, basally connate, the lobes lanceolate, 8–12 mm long, pubescent on both surfaces; petals 5, free, obovate or spatulate, 1.2–2 cm long; stamens 10–75+, inserted on a nectar disk; ovary subtended by rings of trichomes. Fruit subglobose, 1–1.5 cm long, 5- to 7-carpellate and -loculate, red; seeds numerous, less than 1 mm long, lenticular, yellowish.

Disturbed tropical hammocks and pinelands. Rare; Hendry and Palm Beach Counties, southern peninsula. Escaped from cultivation. Florida and California; West Indies, Mexico, Central America, and South America; Africa, Asia, Australia, and Pacific Islands. Native to tropical America. Spring–fall.

MALVACEAE Juss., nom. cons. 1789. MALLOW FAMILY

Trees, shrubs, or herbs. Leaves alternate, the blade simple, pinnate- , palmate- , or pinnipalmate-veined, petiolate, stipulate or estipulate. Flowers axillary and/or terminal racemes, umbels, cymes, heads, paniculiform, or reduced to 1- to few-flowered fascicles, actinomorphic, bisexual or unisexual; sepals 2–5, free or connate; petals 4 or 5 or absent, free or adnate to the staminal tube; stamens 5-many, free or connate in bundles or into a staminal tube, sometimes on an androgynophore, the anthers medifixed, 2- or 3-thecate, dehiscent by longitudinal slits, introrse or extrorse; ovary superior, 1- to 40-carpellate. Fruit a capsule, schizocarp, berry, or follicle.

A family of about 240 genera and about 4,350 species; nearly cosmopolitan.

Byttneriaceae R. Br., nom. cons. (1814); *Sterculiaceae* Vent., nom. cons. (1807); *Tiliaceae* Juss., nom. cons. (1789).

Selected references: Brizicky (1965, 1966); Fryxell (1997).

1. Petals absent; carpels expanding into 5 radiating, stalked follicles, each opening before maturity to become leaflike and bearing 1–4 exposed marginal seeds ....**Firmiana**
1. Petals present; carpels other than as above.
    2. Calyx not subtended by involucral bracts.
        3. Stamens 5.
            4. Flowers axillary, solitary or in fascicles of 2–3; petals hooded, caducous; capsules with excrescences ....**Ayenia**
            4. Flowers in dense capitate or umbellate clusters; petals flat, persistent; capsules lacking excrescences.
                5. Ovary 5-carpellate; capsules 4- to 10-seeded; styles 5; corolla pink, purple, or white ....**Melochia**
                5. Ovary 1-carpellate; capsule 1-seeded; style 1; corolla orange-yellow ....**Waltheria**
        3. Stamens numerous (more than 10).
            6. Stamens free, the anthers 2-loculate.
                7. Peduncle adnate to half its length to a winglike, leafy bract ....**Tilia**
                7. Peduncle free from the subtending leafy bract.
                    8. Fruit with hooked spines...**Triumfetta**
                    8. Fruit without spines.
                        9. Tree or shrub; fruit a drupe...**Grewia**
                        9. Herb; fruit a capsule ....**Corchorus**
             6. Stamens monadelphous, the staminal column adnate to the corolla, the anthers 1-loculate.
                10. Carpels 2- to 5-seeded.
                    11. Fruit bladderlike; mericarps not beaked..**Herissantia**
                    11. Fruit not bladderlike; mericarps with a prominent beak.
                        12. Mericarps 3–5, each divided into 2 cells by a medial constriction, the lower cell indehiscent, 1-seeded, the upper cell dehiscent, 2-seeded......................**Wissadula**
                        12. Mericarps 5 or more, undivided, 3- to 5-seeded ....**Abutilon**
                10. Carpels 1-seeded.
                    13. Stigmas twice as many as the carpels; staminal column bearing anthers near the middle ....**Malachra**
                    13. Stigmas as many as the carpels; staminal column bearing anthers at the apex.
                        14. Corolla purple; lateral walls of the carpels fragile, soon breaking up; seed with an aril-like covering ....**Anoda**
                        14. Corolla yellow; lateral walls of the carpel firm, persistent; seed without an aril-like covering ....**Sida**
    2. Calyx subtended by involucral bracts.
        15. Fruit of several radially disposed, 1- to several-seeded, dry carpels, which split apart at maturity.
            16. Involucral bracts 5 or more.
                17. Leaves with 1–3 narrow, split glands on the medial veins of the lower surface; carpels with numerous spines..**Urena**
                17. Leaves not as above; carpels with 1–3 spines or lacking spines............................**Pavonia**
            16. Involucral bracts 2–3.
                18. Leaves ovate to elliptic-lanceolate...**Malvastrum**
                18. Leaves orbicular or suborbicular.
                    19. Leaves shallowly lobed or unlobed ....**Malva**
                  19. Leaves deeply palmately divided.

20. Petals erose-margined, purple, 2–4 cm long; fruit rounded dorsally ..... **Callirhoe**
20. Petals entire, orange or red, to 1 cm long; fruit with a dorsal bifurcate spine ....
.... **Modiola**
15. Fruit a loculicidal capsule or leathery and indehiscent or berrylike, but separating at maturity.
   21. Fruit leathery and indehiscent or berrylike, but separating at maturity.
      22. Corolla remaining closed at anthesis; staminal column longer than the petals; bracts
      7–12, persistent; fruit berrylike, but separating at maturity ....**Malvaviscus**
      22. Corolla open at anthesis; staminal column shorter than the petals; bracts 3–5, deciduous;
      fruit leathery, indehiscent .... **Thespesia**
   21. Fruit a loculicidal capsule.
      23. Calyx spathaceous, deciduous in fruit ....**Abelmoschus**
      23. Calyx of 3 equal lobes, persistent in fruit.
         24. Style branches clavate, erect.
            25. Involucrate bracts large, laciniate ....**Gossypium**
            25. Involucrate bracts small, entire ....**Cienfuegosia**
         24. Styles branches capitate, spreading.
            26. Fruit much depressed; carpels 1-seeded .... **Kosteletzkya**
            26. Fruit as long as broad or longer; carpels 2- to many-seeded.
               27. Stipular scar inconspicuous and not encircling the stem; leaf margin evidently
               serrate or crenate ....**Hibiscus**
               27. Stipular scar a conspicuous ring encircling the stem; leaf margin entire or ob-
               scurely denticulate or crenulate... **Talipariti**

Abelmoschus Medik. 1787. OKRA

Annual herbs. Leaves alternate, simple, the blade palmate-veined, petiolate, stipulate. Flowers axillary, solitary or in racemes, actinomorphic, bisexual; involucral bracts present; sepals 5, basally connate, spathaceous; petals 5, free; androgynophore anther-bearing near the base; ovary 5-carpellate, the style 5-branched, the stigmas peltate. Fruit a capsule, loculicidally dehiscent.

A genus of about 6 species; nearly cosmopolitan. [From the Arabic *abu-al-misk*, father of musk, in reference to the scented seeds.]

Selected reference: Bates (2015).

Abelmoschus esculentus (L.) Moench [Edible, in reference to the fruit.] OKRA.

Hibiscus esculentus Linnaeus, Sp. Pl. 696. 1753. *Abelmoschus esculentus* (Linnaeus) Moench, Methodus 617. 1794.

Herbs, to 2 m; stems erect, sparsely hirsute. Leaves with the blade broadly ovate to suborbicular, slightly 3- , 5- , or 7-lobed to palmate-divided, 10–25 cm long and wide, the apex acute to obtuse or rounded, the base cordate or hastate, the margin coarsely serrate-dentate, the upper and lower surfaces sparsely setose to hirsute, the petiole 10–30 cm long; stipules filiform to falcate. Flowers axillary, solitary or racemose; involucral bracts 8–12, linear, to 2.5 cm long, free or sometimes basally connate; calyx slitting vertically and spathaceous, then circumscissile dehiscent; petals broadly obcordate, to 8 cm long, white to yellow, purple or maroon at the base. Fruit cylindric, 8–30 cm long, slightly 5-angled, beaked, bristly and then glabrate; seeds numerous, obovoid-reniform, the surface strigose.

Disturbed sites. Rare; Miami-Dade County, Monroe County keys. Escaped from cultivation. Connecticut south to Florida, west to Michigan, Illinois, Mississippi, and Louisiana; West Indies, Mexico, Central America, and South America; Europe, Africa, Asia, Australia, and Pacific Islands. Native to Africa. Summer.

Our *Abelmoschus esculentus* is a cultigen apparently domesticated in Africa, where it has been grown for 4,000 years.

EXCLUDED TAXON

Abelmoschus manihot (Linnaeus) Medikus—Reported by Small (1903, as *Hibiscus manihot* Linnaeus; 1933) as naturalized in Florida. No Florida specimens known.

Abutilon Mill. 1754. INDIAN MALLOW

Herbs or shrubs. Leaves simple, palmate-veined, petiolate, stipulate. Flowers axillary or terminal, solitary or in racemes, panicles, or umbels, actinomorphic, bisexual; sepals 5, basally connate; petals 5, free; styles 5–25, the stigmas capitate. Fruit a schizocarp, the mericarps 5–25.

A genus of about 200 species; North America, Africa, Asia, and Australia. [From the Arabic *abu*, father of, and the Persian *tula* or *tulha*, mallow.]

Selected reference: Fryxell and Hill (2015a).

1. Carpels 10 or fewer.
    2. Stem with short, spreading stellate-pubescent trichomes .... **A. permolle**
    2. Stem with short, spreading stellate-pubescent and long, erect simple trichomes..............................
    .... **A. grandifolium**
1. Carpels 11 or more.
    3. Corolla solid orange-yellow..**A. theophrasti**
    3. Corolla salmon-pink to blackish maroon or orange-yellow with a purple base.
        4. Corolla salmon-pink to blackish maroon; seeds hirsute...**A. hulseanum**
        4. Corolla orange-yellow with a purple base; seeds minutely stellate-pubescent ....**A. hirtum**

Abutilon grandifolium (Willd.) Sweet [With large leaves.] HAIRY INDIAN MALLOW.

Sida grandifolia Willdenow, Enum. Pl. 724. 1809. *Abutilon grandifolium* (Willdenow) Sweet, Hort. Brit. 53. 1826.

Shrub, to 3 m; stem densely viscid, stellate-pubescent and hispid with simple trichomes. Leaves with the blade broadly ovate to lanceolate or suborbicular, 3–18 cm long, 2–13 cm wide, sometimes 3-angular, the apex acuminate to obtuse, the base deeply cordate, the margin coarsely serrate-crenate, the upper and lower surfaces densely stellate-pubescent, the petiole 4–10 cm long; stipules linear-lanceolate, 8–15 mm long, stellate-pubescent. Flowers axillary, solitary or 2–3 on a common peduncle equaling or longer than the petiole, articulate distally; sepals lanceolate to ovate, 1–1.5 cm long, connate ca. ½ their length, densely stellate-pubescent on both surfaces, also with simple trichomes toward the base within; petals broadly obovate, 1–1.5 cm long, yellow, the margin ciliate; staminal column stellate-pubescent; styles 10. Fruit ovoid to subglobose, 1–1.5 cm long, stellate-pubescent, the mericarps 10, 5–7 mm wide, short-beaked; seeds 2–3 per mericarp, ca. 2 mm long.

Disturbed sites. Hillsborough County. Escaped from cultivation. Florida; South America; Asia, Australia, and Pacific Islands. Native to South America.

Abutilon hirtum (Lam.) Sweet [Hairy, with long trichomes.] FLORIDA KEYS INDIAN MALLOW.

> *Sida hirta* Lamarck, Encycl. 1: 7. 1783. *Abutilon hirtum* (Lamarck) Sweet, Hort. Brit. 53. 1826. *Beloere cistiflora* Shuttleworth ex A. Gray, Smithsonian Contr. Knowl. 3(5): 21. 1852, nom. illegit. *Abutilon indicum* (Linnaeus) Sweet var. *hirtum* (Lamarck) Grisebach, Fl. Brit. W.I. 78. 1859. *Abutilon graveolens* (Roxburgh ex Hornemann) Wight & Arnott var. *hirtum* (Lamarck) Masters, in Hooker f., Fl. Brit. India 1: 327. 1874.

Herb, to 1 m; stem viscid-hirsute. Leaves with the blade ovate to suborbicular, 5–10 cm long and wide, the apex acuminate, the base cordate, the margin finely serrate, the upper and lower surfaces tomentose, the petiole equaling or exceeding the blade; stipules lanceolate, 7–9 mm long, recurved. Flowers solitary or in a terminal panicle; calyx 12–17 mm long, the lobes ovate; petals obovate, ca. 2 mm long, orange-yellow with a dark red center; styles 20–25. Fruit oblate, 12–14 mm long, ca. 20 mm wide, the mericarps 20–25, the apex obtuse to acute, stellate-hirsute; seeds 2–3 mm long, minutely scabridulous.

Disturbed sites. Rare; peninsula. Florida; West Indies, Mexico, Central America, and South America; Africa, Asia, and Australia. Native to Africa and Asia. All year.

Abutilon hulseanum (Torr. & A. Gray) Torr. & A. Gray [Commemorates Gilbert White Hulse (1808–1883), New York physician who served in the U.S. Army and collected in Florida during the Second Seminole War.] MAUVE.

> *Sida hulseana* Torrey & A. Gray, Fl. N. Amer. 1: 233. 1838. *Abutilon hulseanum* (Torrey & A. Gray) Torrey ex A. Gray, Mem. Amer. Acad. Arts, ser. 2. 4: 23. 1849. TYPE: FLORIDA: Hillsborough Co.: Tampa Bay, s.d., *Hulse s.n.* (holotype: NY).

Perennial herb or subshrub, to 2 m; stem stellate-tomentulose. Leaves with the blade ovate, 6–10 cm long and wide, the apex rounded-acute, the base cordate, the margin serrate-dentate, the upper and lower surfaces tomentulose, the petiole subequaling the blade; stipules filiform, ca. 8 mm long. Flowers axillary, solitary; calyx 12–15 mm long, the lobes basally overlapping; the petals yellowish fading pinkish, ca. 2 cm long; styles ca. 12. Fruit 12–15 mm long, 20–25 mm wide, the mericarps 12–15, the apex apiculate, the surface hirsute; seeds 4–6 per carpel, ca. 2 mm long, minutely pubescent.

Pinelands and disturbed sites. Occasional; Alachua County, central peninsula. Florida, Louisiana, and Texas; West Indies and Mexico. All year.

Abutilon permolle (Willd.) Sweet [Very soft, in reference to the short-pubescent parts.] COASTAL INDIAN MALLOW.

Sida permollis Willdenow, Enum. Pl. 723. 1809. *Abutilon permolle* (Willdenow) Sweet, Hort. Brit. 53. 1826.

Abutilon peraffine Shuttleworth ex A. Gray, Smithsonian Contr. Knowl. 3(5): 20. 1852. TYPE: FLORIDA: Monroe Co.: Key West, Feb 1846, *Rugel 95b* (holotype: GH?; isotype: BM, NY, P).

Shrub, to 2 m; stem stellate-pubescent. Leaves with the blade ovate, 7–12 cm long, 3.5–6 cm wide, the apex acuminate, the base cordate, the margin serrate-crenate, the upper and lower surfaces tomentose, the petiole 0.5–1 times the blade length; stipules lanceolate, 5–7 mm long. Flowers solitary or in a terminal panicle; sepals 1–1.5 cm long, the lobes lanceolate-ovate, basally overlapping; petals obovate, 10–18 mm long, yellow or yellow-orange; styles 10–12. Fruit 3–10 mm long and wide, hirsute, the mericarps 10–12, the apex acute; seeds ca. 1 mm long, papillate.

Disturbed sites. Rare; Manatee County, southern peninsula. Florida; West Indies, Mexico, and Central America. All year.

Abutilon theophrasti Medik. [Commemorates Theophrastus (370–285 BC), Greek philosopher and botanist.] VELVETLEAF; BUTTERPRINT.

Sida abutilon Linnaeus, Sp. Pl. 685. 1753. *Abutilon theophrasti* Medikus, Malvenfam. 28. 1787. *Abutilon avicennae* Gaertner, Fruct. Sem. Pl. 2: 251. 1791, nom. illegit. *Abutilon pubescens* Moench, Methodus 620. 1794, nom. illegit. *Abutilon abutilon* (Linnaeus) Huth, Helios 11: 132. 1893, nom. inadmiss. *Malva abutilon* (Linnaeus) E.H.L. Krause, in Sturm, Fl. Deutschl., ed. 2. 6: 237. 1902.

Annual herb, to 1 m; stem stellate-tomentose. Leaves with the blade broadly ovate to suborbicular, 8–15 cm wide and long, the apex acuminate, the base cordate, the margin crenulate, the upper and lower surfaces stellate-tomentose, the petiole subequaling the blade; stipules lanceolate. Flowers solitary or in a raceme; sepals ca. 1 cm long, the lobes ovate; petals 8–13 mm long, obovate, pale yellow; styles 13–15. Fruit ca. 1.5 cm long, ca. 2 cm wide, the mericarps 13–15, the apex spinose, the spines divergent, 3–6 mm long, the surface hirsute; seeds 3–4 mm long, minutely puberulent.

Disturbed sites. Occasional; central and southern peninsula, central and western panhandle. Nearly throughout North America; West Indies; Europe, Africa, Asia, and Australia. Native to Europe, Africa, and Asia. Spring–summer.

EXCLUDED TAXA

Abutilon abutiloides (Jacquin) Garcke ex Britton & P. Wilson—Reported for Florida by Chapman (1860, as *A. jacquinii* G. Don), the name misapplied to material of *A. permolle*.

Abutilon indicum (Linnaeus) Sweet—Reported as naturalized in Florida by Bailey et al. (1976). No Florida specimens known.

Abutilon pedunculata Kunth—Reported for Florida by Small (1903), who misapplied the name to material of *A. hulseanum*.

Anoda Cav. 1785.

Herbs. Leaves simple, petiolate, stipulate. Flowers solitary, actinomorphic, bisexual; sepals 5, basally connate; petals 5, free; ovary 10- to 19-carpellate, the styles 10–19, the stigmas capitate. Fruit a schizocarp, the mericarps 10–19.

A genus of 23 species; North America, Mexico, Central America, South America, Africa, Asia, and Australia. [Ceylonese vernacular name for a species of *Abutilon*.]

Selected references: Fryxell (1987); Fryxell and Hill (2015b).

Anoda cristata (L.) Schtldl. [Crested, in reference to the fruit.] CRESTED ANODA.

Sida cristata Linnaeus, Sp. Pl. 685. 1753. *Anoda lavateroides* Medikus, Malvenfam. 19. 1787, nom. illegit. *Anoda cristata* (Linnaeus) Schlechtendal, Linnaea 11: 210. 1837.

Anoda arizonica A. Gray var. *digitata* A. Gray, Proc. Amer. Acad. Arts. 22: 198. 1887. *Anoda triangularis* (Willdenow) de Candolle var. *digitata* (A. Gray) B. L. Robinson, in A. Gray & S. Watson, Syn. Fl. N. Amer. 1(1): 319. 1897. *Anoda cristata* (Linnaeus) Schlechtendal var. *digitata* (A. Gray) Hochreutiner, Annuaire Conserv. Jard. Bot. Genève 20: 47. 1916.

Decumbent to suberect herb; stem with spreading or retrorse trichomes. Leaves with the blade ovate to triangular, 3–9 cm long and wide, 3- to 5-lobed, the apex acute, the base cuneate to truncate, the margin crenate to subentire, the upper and lower surfaces sparsely pubescent, the trichomes mostly simple and appressed, the petiole 1.5–4 cm long, hispid. Flowers long-pedicellate; sepals 5–10 mm long, lobed ca. ½ to the base, the segments lanceolate, hispid; petals obovate, 8–26 mm long, purplish or lavender (rarely white); staminal column pubescent; styles with a dorsal spine 2–4 mm long; seed solitary in each mericarp.

Disturbed sites. Rare; Orange County. New York and Massachusetts south to Florida, west to California; West Indies, Mexico, Central America, and South America; Africa, Asia, and Australia. Native to southwestern United States and tropical America. Summer–fall.

Ayenia L. 1756.

Perennial herbs or subshrubs. Leaves alternate, simple, pinnipalmate-veined, petiolate, stipulate. Flowers axillary, solitary, actinomorphic, bisexual; sepals 5, basally connate; petals 5, free, with slender claws, the apex hooded, inflexed, and adnate to the staminal tube, with a subapical appendage, the entire corolla umbrella-like with a central protruding style; androgynophore present; stamens 5, alternating with the 5 antisepalous staminodia, connate into a staminal tube, the fertile anthers 3-thecate; ovary 5-carpellate and -loculate, the style simple, the stigma capitate. Fruit a schizocarpic capsule separating into 5 1-seeded, 2-valved mericarps covered with short excrescences.

A genus of about 80 species; North America, West Indies, Mexico, Central America, and South America. [Commemorates Louis de Noailles (1713–1793), duc d'Ayen, later duc de Noailles, French botanist.]

Ayenia is sometimes placed in the Sterculiaceae.

Selected references: Brizicky (1966); Cristóbal (1960); Dorr (2015b); Whetstone (1983).

Ayenia euphrasiifolia Griseb. [With leaves like *Euphrasia* (Orobanchaceae).] EYEBRIGHT AYENIA.

Ayenia euphrasiifolia Grisebach, Cat. Pl. Cub. 29. 1866. *Byttneria euphrasiifolia* (Grisebach) M. Gómez de la Maza y Jiménez, Anales Soc. Esp. Hist. Nat. 19: 217. 1890.

Decumbent or prostrate perennial herb or subshrub; stems to 4 dm long, wiry, reddish brown, appressed stellate-pubescent. Leaves with the blade orbicular to ovate, 4–14 mm long and wide, the apex cuspidate, the base truncate to rounded, the margin dentate, with bristle-tipped teeth, the upper and lower surfaces reddish brown mottled, stellate-pubescent, the petiole 2–3 mm long, pubescent; stipules subulate, 1 mm long. Flowers axillary, solitary, the pedicel 5–6 mm long; sepal-lobes ovate-lanceolate to elliptic, ca. 2 mm long, the outer surface sparsely pubescent; petals reddish, ca. 6 mm long, the claw 2–4 mm long, the blade rhombic, ca. 1 mm long, the apex emarginate to cleft, with an erect, abaxial appendage; androgynophore 2–3 mm long, glabrous; stamen filaments connate for nearly their length; style included but the stigma exserted. Fruit subglobose, 4–5 mm long, the excrescences muricate, stellate bristle-tipped; seeds slightly conic, 2 mm long, warty, brown with reddish glands.

Pinelands, pine rocklands, and scrub. Rare; central and southern peninsula. Florida; West Indies. All year.

EXCLUDED TAXON

Ayenia pusilla Linnaeus—Reported for Florida by Chapman (1860, 1883, 1897) and Small (1903, 1913a, 1913b, 1913d, 1933), but excluded from our flora by Cristóbal (1960), Brizicky (1966), and Whetstone (1983). The Chapman and Small reports were based on misidentifications of *A. euphrasiifolia*.

Callirhoe Nutt. 1822. POPPYMALLOW

Perennial herbs. Leaves alternate, simple, petiolate, stipulate. Flowers axillary, solitary or racemose, involucral bracts present or absent; sepals 5, basally connate; petals 5, free; ovary 10- to 28-carpellate, the style branches stigmatic along adaxial side. Fruit a schizocarp, the mericarps 10–28.

A genus of 9 species; North America and Mexico. [From the Greek, *kallos*, beautiful, and *rhoias*, flowing, named for an Athenian fountain and several mythological females.]

Selected references: Dorr (1990, 2015c).

1. Leaves triangular or hastate; peduncles several-flowered...**C. triangulata**
1. Leaves suborbicular, 3- to 7-divided or -cleft; peduncles 1-flowered.
    2. Stem mostly hirsute, at least some of the trichomes spreading ....**C. involucrata**
    2. Stem mostly appressed-pubescent or glabrate ....**C. papaver**

Callirhoe involucrata (Nutt. ex Torr.) A. Gray var. **lineariloba** (Torr. & A. Gray) A. Gray [With an involucre; with narrow lobes, in reference to the leaves.] PURPLE POPPYMALLOW.

Malva involucrata Torrey & A. Gray var. *lineariloba* Torrey & A. Gray, Fl. N. Amer. 1: 226. 1838. *Callirhoe involucrata* (Torrey & A. Gray) A. Gray var. *lineariloba* (Torrey & A. Gray) A. Gray, Proc.

Acad. Nat. Sci. Philadelphia 14: 161. 1862. *Malva lineariloba* (Torrey & A. Gray) M. Young, Fl. Texas 180. 1873. *Callirhoe lineariloba* (Torrey & A. Gray) A. Gray, Proc. Amer. Acad. Arts 19: 74. 1883.

Decumbent to weakly erect perennial herb, to 8 dm; stem pubescent with stellate and simple trichomes. Leaves with the blade suborbicular to ovate, 1–8 cm long, 1–9(12) cm wide, 3- to 5-cleft, the lobes oblong, obovate, or oblanceolate, the sinuses between the lobes on the cauline leaves extending to within 2–5 mm of the petiole, the margin entire, the upper and lower surfaces with sparse stellate and simple trichomes, the petiole 0.7–13(23) cm long, pubescent with stellate and simple trichomes; stipules ovate to ovate-lanceolate, 3–15(23) mm long, somewhat auriculate. Flowers in a raceme; involucral bracts 3, linear to ovate, 5–10(13) mm long; sepals ca. 2 cm long, free nearly to the base, the lobes divergent; petals obovate, 2.5–3 cm long, reddish purple with a white basal spot, entirely white, or mauve with a white margin. Fruit 7–11 mm in diameter, the mericarps 10–28, 2–3 mm long, indehiscent, the collars weakly developed, 2-lobed, pubescent with simple trichomes or glabrous.

Disturbed sites. Rare; Hillsborough County. Not recently collected. Florida, Kansas south to Texas, west to Colorado and New Mexico; Mexico. Native to the west of Florida. Spring–summer.

Callirhoe papaver (Cav.) A. Gray [To resemble *Papaver* (Papaveraceae).] WOODLAND POPPYMALLOW.

Malva papaver Cavanilles, Diss. 2: 64, t. 15(3). 1785. *Callirhoe papaver* (Cavanilles) A. Gray, Mem. Amer. Acad. Arts, ser. 2. 4: 17. 1849. *Sesquicella papaver* (Cavanilles) Alefeld, Oesterr. Bot. Z. 12: 256. 1862. *Malva nuttallioides* Croom, Amer. J. Sci. Arts 26: 313. 1834. TYPE: "Florida, and the southern parts of Georgia."

Weakly erect, ascending, or decumbent perennial herb, to 10 dm; stem pubescent with stellate or simple trichomes or glabrate. Leaves with the blade hastate, cordate, triangular, or ovate, 3–11 cm long, 3.5–13 cm wide, 3- to 5(7)-cleft, the lobes narrowly lanceolate, linear, or linear-falcate, the upper and lower surfaces with stellate and simple trichomes, the petiole 2–25 cm long; stipules oblong, ovate, or rhombic-ovate, 5–10(12) mm long. Flowers in a raceme; involucral bracts 3 or rarely absent, narrowly linear, 2–10 mm long; sepals ca. 1.5 cm long, free nearly to the base, the lobes lanceolate; petals narrowly obovate, 3–3.5 cm long, reddish purple. Fruit 8–11 mm in diameter, the mericarps 12–20, 3–4 mm long, indehiscent, the collar only slightly differentiated, sparsely pubescent.

Disturbed, dry hammocks. Rare; Alachua County, central panhandle. Georgia south to Florida, west to Texas. Spring.

Callirhoe papaver is listed as endangered in Florida (Florida Administrative Code, Chapter 5B-40).

Callirhoe triangulata (Leavenw.) A. Gray [Triangle-shaped, in reference to the leaves.] CLUSTERED POPPYMALLOW.

Malva triangulata Leavenworth, Amer. J. Sci. Arts 7: 62. 1824. *Nuttallia triangulata* (Leavenworth) Hooker, J. Bot. (Hooker) 1: 197. 1834. *Callirhoe triangulata* (Leavenworth) A. Gray, Mem. Amer.

Acad. Arts, ser. 2. 4: 16. 1849. *Sesquicella triangulata* (Leavenworth) Alefeld, Oesterr. Bot. Z. 12: 256. 1862.

Decumbent, ascending, or weakly erect perennial herb, to 5 dm; stem densely stellate-pubescent. Leaves with the blade triangular or ovate-lanceolate, unlobed or shallowly 3- to 5-cleft, (3)6–15(20) cm long, 3.3–10.5(14) cm wide, the upper and lower surfaces stellate-pubescent, the petiole 0.5–30 cm long; stipules ovate to lanceolate, 4–7 mm long. Flowers in a panicle; involucral bracts 3, spatulate or obovate, 3–8 mm long; sepals 2–3 mm long, free nearly to the base, the lobes lanceolate; petals obovate, 2–2.5 cm long, red with a white basal spot. Fruit 6–7 mm in diameter, the mericarps 10–13, 2–4 mm long, the collar absent, pubescent with 2-rayed trichomes.

Sandhills. Rare; locality unknown. Not recently collected. North Carolina south to Florida, west to Wisconsin, Iowa, Missouri, and Mississippi. Spring.

EXCLUDED TAXON

Callirhoe involucrata (Torrey & A. Gray) A. Gray—Because infraspecific categories were not recognized, the typical variety was reported for Florida by implication by Wunderlin (1982, 1998) and Wunderlin and Hansen (2003). Florida material is of var. *lineariloba*.

Cienfuegosia Cav. 1787. FLYMALLOW

Perennial herbs or subshrubs. Leaves alternate, simple, pinnate-veined, petiolate, stipulate. Flowers axillary, solitary, pedicellate; involucral bracts 6–9; sepals 5, basally connate; petals 5, free; stamens numerous, connate into a column; ovary 3-carpellate and -loculate, the style 1, the stigmas 3. Fruit a capsule; seeds several per carpel.

A genus of 26 species; North America, West Indies, Mexico, Central America, South America, and Asia. [Commemorates Bernard Cienfuegos (1580–1640), Spanish botanist.]

Selected references: Fryxell (1969); Fryxell and Hill (2015c).

Cienfuegosia yucatanensis Millsp. [Of Yucatan.] YUCATAN FLYMALLOW; YELLOW HIBISCUS.

Cienfuegosia yucatanensis Millspaugh, Publ. Field Columb. Mus., Bot. Ser. 2: 74. 1900.

Herb or subshrub, to 5 dm; stem glabrous. Leaves with the blade narrowly oblong-lanceolate, 2–4 cm long, 0.5–1 cm wide, the petiole 0.5–1.5 cm long; the apex acute, the base rounded, the margin entire, the upper and lower surfaces glabrous, minutely and obscurely punctate; stipules subulate, 1–2 mm long. Flowers solitary in the leaf axil, the pedicel 1–5 cm long, surmounted by 3 involucral nectaries or the nectaries absent; involucral bracts subulate, 1–2 mm long; sepals 8–12 mm long, the lobes lanceolate, punctate, glabrous; petals broadly obovate, 1–2 cm long, yellow; staminal column ½ the petal length. Fruit ovoid, 6–8 mm long, glabrous externally, prominently ciliate internally; seeds 2–3 mm long, densely pubescent with brownish trichomes 4–5 mm long.

Salt marshes. Rare; Monroe County keys. Florida; West Indies and Mexico. All year.

Cienfuegosia yucatanensis is listed as endangered in Florida (Florida Administrative Code, Chapter 5B-40).

EXCLUDED TAXON

Cienfuegosia heterophylla (Ventenat) Garcke—Reported for Florida by Chapman (1897, as *Fugosia heterophylla* Spach) and Small (1903, 1913a, 1913d, 1933), who misapplied the name to material of *C. yucatanensis*. *Cienfuegosia heterophylla* is excluded from Florida by Fryxell (1969).

Corchorus L. 1753.

Annual or perennial herbs, subshrubs, or shrubs. Leaves alternate, simple, pinnipalmate-veined, petiolate, stipulate. Flowers solitary or in umbellate axillary cymes, actinomorphic, bisexual, bracteate; sepals 5, free; petals 5, free; extrastaminal disk (gonophore) present or absent; stamens numerous, free or connate, the anthers introrse, dorsifixed; ovary 2- or 4-carpellate and -loculate. Fruit a loculicidal capsule.

A genus of about 100 species; North America, West Indies, Central America, South America, Africa, Asia, Australia, and Pacific Islands. [From the Greek *kore*, eye pupil, and *koreo*, to purge or clear, in reference to the medicinal use of the leaves.]

Corchorus is sometimes placed in the Tiliaceae.

Selected reference: Nesom (2015a).

1. Fruit angular, winged; leaves with 1 or 2 basal teeth with a setaceous appendage .... **C. aestuans**
1. Fruit subterete, not winged; leaves not appendaged.
    2. Plant densely stellate-pubescent to tomentose; fruit to 1.5 cm long.................................**C. hirsutus**
    2. Plant hirsute to glabrate, with simple trichomes only; fruit 2.5–6.5 cm long.
        3. Leaves and calyx hirsute; capsule apex attenuate-beaked; seeds separated by incomplete transverse partitions ....**C. hirtus**
        3. Leaves and calyx glabrous or glabrate; capsule apex truncate with 4 toothlike appendages; seeds not separated by transverse partitions ....**C. siliquosus**

Corchorus aestuans L. [Moving to and fro, apparently in reference to its wand-like growth habit.] JUTE.

Corchorus aestuans Linnaeus, Syst. Nat., ed. 10. 1079. 1759.
Corchorus acutangulus Lamarck, Encycl. 2: 104. 1786.

Erect or decumbent, annual herb, to 6 dm; stem puberulent to pilose-hirsute with simple trichomes. Leaves with the blade lanceolate-ovate to ovate-suborbicular, 1–7 cm long, 1–3.5 cm wide, the apex acute to acuminate, the base rounded to subcordate, the margin crenate-serrate, ciliate, the lowermost serration on 1 or both sides often prolonged into a hairlike bristle, the upper surface glabrate, the lower surface sparsely pilose hirsute, the petiole 0.5–2(2.5) cm long, pilose; stipules subulate, 3–10 cm long. Flowers 1–3 in the leaf axils, subsessile; sepals oblong, 3–4 mm long, the apex cucullate, glabrous; petals obovate, 3–4 mm long; gonophore evident, apically expanded into a collar that surrounds the base of the stamens; stamens free. Fruit narrowly oblong, 0.8–2 cm long, 3-locular, each valve with 2 longitudinally winged ridges that merge and project beyond the apex into a single bifid, recurved beak, glabrous; seeds 1–2 mm long.

Disturbed sites. Rare; peninsula, central panhandle. Georgia, Alabama, Florida, and Louisiana; West Indies, Mexico, Central America, and South America; Africa, Asia, and Australia. Native to tropical America. Spring–fall.

Corchorus hirsutus L. [With rough, course, ascending trichomes.] JACKSWITCH; WOOLLY CORCHORUS; CADILLO.

Corchorus hirsutus Linnaeus, Sp. Pl. 530. 1753.

Erect shrub, to 1(2) m; stem densely stellate-pubescent to -tomentose. Leaves with the blade ovate to oblong-ovate or oblong-lanceolate, 1.2–3.4 cm long, 0.9–1.9 cm wide, the apex obtuse to acute, the base rounded, the margin crenate to crenate serrate, the upper and lower surfaces densely stellate-pubescent to -tomentose, the petiole 2–7 mm long, densely stellate-pubescent; stipules linear-filiform to subulate, 2–3 mm long. Flowers 2–8 in an umbelliform cyme; sepals oblong to ovate, 5–7 mm long, the outer surface densely stellate-pubescent to -tomentose; petals obovate, 5–7 mm long, glabrous; gonophore short but evident; stamens free. Fruit oblong, to 1.4 cm long, 4-locular, densely stellate-pubescent; seeds 1–2 mm long.

Disturbed sites. Rare; Hillsborough and Miami-Dade Counties. Florida; West Indies, Mexico, Central America, and South America; Africa. Native to tropical America. Summer.

Corchorus hirtus L. [Hairy.] RED JUTE.

Corchorus hirtus Linnaeus, Sp. Pl., ed. 2. 747. 1762.
Corchorus pilolobus Link, Enum. Hort. Berol. Alt. 2: 72. 1822. *Corchorus hirtus* Linnaeus var. *pilolobus* (Link) K. Schumann, in Martius, Fl. Bras. 12(3): 128. 1886.
Corchorus orinocensis Kunth, in Humboldt et al., Nov. Gen. Sp. 5: 337. 1822. *Corchorus hirtus* Linnaeus var. *orinocensis* (Kunth) K. Schumann, in Martius, Fl. Bras. 12(3): 127. 1886.
Corchorus hirtus Linnaeus var. *glabellus* A. Gray, Syn. Fl. N. Amer. 1(2): 342. 1897.

Erect or spreading perennial herb or subshrub, to 6 dm; stem hirsute, usually with a more dense sublayer of hirsute-arachnoid trichomes in a longitudinal row. Leaves with the blade ovate or ovate-lanceolate to lanceolate, 1.5–7 cm long, 0.6–2.5 cm wide, the apex acute to acuminate, the base cuneate to rounded, the margin crenate-serrate to serrate, the upper and lower surfaces sparsely hirsute to glabrate, the petiole 3–15 mm long; stipules linear-lanceolate, 3–7 mm long, ciliate-hirsute. Flowers 1–2 in the axil of the leaves; sepals linear-lanceolate, 5–9 mm long, the outer surface sparsely hirsute; petals spatulate, 4–8 mm long, glabrous; gonophore absent; stamens free or connate to ca. ⅓ their length. Fruit linear, 3–5.5 cm long, 2- or 3-locular, hirsute to glabrate; seeds ca. 1 mm long.

Disturbed sites. Rare; Miami-Dade County. Florida, Alabama, Mississippi, Texas, and Arizona; West Indies, Mexico, Central America, and South America. Spring–fall.

Corchorus siliquosus L. [With a flattened fruit.] SLIPPERY BURR.

Corchorus siliquosus Linnaeus, Sp. Pl. 529. 1753.

Perennial herb or subshrub, to 1(2) m; stem glabrous except for 1–2 rows of simple trichomes. Leaves with the blade ovate, 0.7–5.2 cm long, 0.4–2.4 cm wide, the apex acute to short acuminate, the base rounded, the margin serrate, the upper and lower surfaces glabrous or nearly so, the petiole 0.5–2 cm long, the upper surface puberulent; stipules linear-lanceolate, 2–6 mm long, glabrous. Flowers 1–3 in the axil of the leaves; sepals oblong-lanceolate, 5–7 mm long, glabrous; petals obovate, 4–6 mm long, glabrous; gonophore absent; stamens free. Fruit linear, 3.5–6.5 cm long, 2-locular, the apex with 4 minute, toothlike projections, inconspicuously puberulous but with a slightly longer row of trichomes along the suture; seeds 1–2 mm long.

Disturbed sites. Occasional; southern peninsula. Florida; West Indies, Mexico, Central America, and South America. Native to North America, West Indies, Mexico, and Central America. Spring–fall.

Firmiana Marsili 1786. PARASOLTREE

Trees. Leaves simple, palmate-veined, petiolate, stipulate. Flowers in a terminal panicle, unisexual, bracteate; sepals 5, basally connate; petals absent; nectariferous disk present; androgynophore present. Staminate flowers with stamens 15, the anthers 2-celled, longitudinally dehiscent; pistillode present. Carpellate flowers with the ovary 5-carpellate and -loculate, the base ringed with indehiscent anthers, the styles 5, basally connate, the stigmas 5. Fruit a follicle, foliaceous, dehiscent before maturity and exposing the seeds adhering to the margins.

A genus of about 12 species; North America, Europe, Asia, Australia, and Pacific Islands. [Commemorates Karl Gotthard von Firmian (Carlo Giuseppe di Firmian) (1716–1782), Austrian statesman and governor-general of Lombardy.]

Firmiana is sometimes placed in the Sterculiaceae.

Selected references: Dorr (2015a); Kostermans (1957).

Firmiana simplex (L.) Wight [Simple, apparently in reference to the flowers.] CHINESE PARASOLTREE.

Hibiscus simplex Linnaeus, Sp. Pl., ed. 2. 977. 1763. *Firmiana simplex* (Linnaeus) W. Wight, Bull. Bur. Pl. Industr., U.S.D.A. 142: 67. 1909.

Sterculia platanifolia Linnaeus f., Suppl. Pl. 423. 1782 ("1781"). *Firmiana platanifolia* (Linnaeus f.) Marsili, Saggi Sci. Lett. Acad. Padova 1: 106. 1786. *Caucanthus platanifolius* (Linnaeus f.) Rafinesque, Sylva Tellur. 72. 1838.

Single or multistemmed tree, to 15(20) m; bark green with greenish white vertical stripes, becoming gray or chalky white. Leaves with the blade broadly ovate, (8)12–40 cm long, (12)20–50 cm wide, membranaceous, palmately 3- to 5-lobed (rarely unlobed), the lobes often constricted at the base, the apex acute or acuminate, the base cordate, the margin entire, the upper surface glabrous, the lower surface minutely stellate-pubescent, with domatia in the axil of the primary and secondary veins, the petiole 15–30(40) cm; stipules caducous. Flowers in a terminal panicle

20–50 cm long, the pedicel 2–3(5) mm long, articulated; sepals 7–9(10) mm long, divided nearly to the base, the tube cupuliform, the lobes lanceolate-oblong, reflexed, the outer surface puberulent, the inner surface villous in the throat. Staminate flowers with the androgynophore to 1 cm long; the anthers 15, irregularly fascicled; pistillode obscured by the anthers. Carpellate flowers with the androgynophore to 0.5 cm long; ovary densely stellate-pubescent, the base ringed with the 15 unopened anthers. Fruit ovate-lanceolate, 6–11 cm long, the margin (1)2–4-seeded, the outer surface stellate-pubescent or glabrous; seeds globose, 6–7 mm long, smooth, wrinkled when dry, glabrous.

Disturbed hammocks. Occasional; northern peninsula, central and western panhandle. Escaped from cultivation. Virginia south to Florida, west to Texas; Europe and Asia. Native to Asia. Spring.

Gossypium L. 1753. COTTON

Shrubs. Leaves alternate, simple, the blade palmate-veined, petiolate, stipulate. Flowers axillary, solitary or in a sympodial inflorescence; involucral bracts 3; sepals 5, basally connate; petals 5, free; stamens numerous, connate into a column; ovary 3- to 5-carpellate and -loculate, the styles 3–5. Fruit a 3- to 5-loculidal capsule, dehiscent; seeds comose.

A genus of about 50 species; nearly cosmopolitan. [From the Greek *gossypion*, cotton, or the Arabic *goz* or *gothn*, a silky or soft substance.]

Selected references: Fryxell (1992); Fryxell and Hill (2015d).

Gossypium hirsutum L. [With long erect trichomes.] UPLAND COTTON; WILD COTTON.

Gossypium hirsutum Linnaeus, Sp. Pl., ed. 2. 975. 1763. *Gossypium barbadense* Linnaeus var. *hirsutum* (Linnaeus) Triana & Planchon, Ann. Sci. Nat., Bot., ser. 4. 17: 171. 1862; non Hooker & Bentham, 1849. *Gossypium herbaceum* Linnaeus var. *hirsutum* (Linnaeus) Masters, in Hooker f., Fl. Brit. India 1: 347. 1874. *Hibiscus barbadensis* (Linnaeus) Kuntze var. *hirsutus* (Linnaeus) Kuntze, Revis. Gen. Pl. 1: 68. 1891.

Gossypium punctatum Schumacher & Thonning, in Schumacher, Beskr. Guin. Pl. 308. 1827. *Gossypium nigrum* Buchanan-Hamilton var. *punctatum* (Schumacher & Thonning) Hooker f. & Webb, in Hooker, Niger Fl. 107. 1849. *Gossypium punctatum* Schumacher & Thonning var. *nigeria* Watt, Wild Cult. Cotton 170. 1907, nom. inadmiss. *Gossypium purpurascens* Poiret var. *punctatum* (Schumacher & Thonning) Harland, Empire Cotton Growing Conf. 2: 24. 1934. *Gossypium hirsutum* Linnaeus forma *punctatum* (Schumacher & Thonning) Roberty, Candollea 7: 330. 1938. *Gossypium hirsutum* Linnaeus var. *punctatum* (Schumacher & Thonning) Roberty, Ann. Inst. Bot.-Geol. Colon. Marseille, ser. 6. 3: 42. 1945. *Gossypium hirsutum* Linnaeus subvar. *punctatum* (Schumacher & Thonning) Wouters, Publ. Inst. Natl. Etude Agron. Congo Belge, Ser. Sci. 34: 99. 1948. *Gossypium hirsutum* Linnaeus subsp. *punctatum* (Schumacher & Thonning) Mauer, Trudy Sredne-Aziatsk. Gosud. Univ. Lenina 18: 22. 1950.

Shrub, to 2 m; stem stellate-pubescent. Leaves with the blade ovate, 4–10 cm long and wide, shallowly 3- to 5-lobed, the lobes broadly ovate, the apex acute to acuminate, the base cordate, the margin entire, the petiole ½–1 times the blade length; stipules subulate to falcate, 0.5–1.5(2) cm long. Flowers usually in a sympodial inflorescence, the involucral bracts foliaceous, 2–4.5 cm long, the margin laciniate; sepals 5–6 mm long, truncate or 5-dentate; petals obovate, 2–5 cm long, cream-colored; staminal column ca. 1.5 cm long; style somewhat exceeding the androecium. Fruit broadly ovoid or subglobose, 2–4 cm long, glabrous, smooth; seeds ovate, 8–10 mm long, comose with white trichomes.

Coastal hammocks, beaches, and disturbed sites. Occasional; Gilchrist County, central and southern peninsula. Massachusetts south to Florida, west to Illinois, Missouri, Texas, and New Mexico, also California; West Indies, Mexico, Central America, and South America; Pacific Islands. Widely cultivated in the tropics and subtropics of the Old and New World and probably naturalized elsewhere. Native to North America, Mexico, and Central America. All year.

Gossypium hirsutum is listed as threatened in Florida (Florida Administrative Code, Chapter 5B-40).

EXCLUDED TAXA

Gossypium herbaceum Linnaeus—Reported for Florida as escaped from cultivation by Small (1903, 1913a, 1933), who apparently misapplied the name to material of *G. hirsutum*. No specimens known.

Gossypium barbadense Linnaeus—Reported for Florida by Small (1903, 1913a, 1913e, 1933), apparently based on a misapplication of the name to material of *G. hirsutum*.

Grewia L. 1753.

Shrubs or trees. Leaves alternate, simple, pinnipalmate-veined, petiolate, stipulate. Flowers in axillary cymes, actinomorphic, bisexual; sepals 5, free; petals 5, free; stamens numerous, free, the anthers dorsifixed, longitudinally dehiscent; ovary 2- to 4-carpellate and -loculate, the style 1, the stigma peltate, 4-lobed. Fruit a drupe.

A genus of about 90 species; North America, Africa, Asia, and Australia. [Commemorates Nehemiah Grew (1641–1712), English plant anatomist and physiologist.]

Grewia is sometimes placed in the Tiliaceae.

Grewia asiatica L. [Of Asia.] PHALSA.

Grewia asiatica Linnaeus, Mant. Pl. 122. 1767.

Large shrub or small tree, to 8 m; bark grayish white, the branchlets stellate-tomentose. Leaves with the blade elliptic-lanceolate or broadly ovate to suborbicular, 5–15 cm long, 5–15 cm wide, the apex acute to obtuse, the base broadly cuneate to rounded or obliquely shallow cordate, the margin serrate, the upper surface scabrous, the lower surface gray-tomentose, the petiole 1–1.5 cm long; stipules oblique-lanceolate or falcate, 1–1.3 cm long, stellate-pubescent. Flowers 2–6(10) in an axillary, dichasial cyme, the peduncle 2–3.5(5) cm long, the pedicel 3–4 mm long; sepals oblong, ca. 1.2 cm long, orange-yellow, the outer surface stellate-pubescent; petals narrowly obovate, 5–7 mm long, orange-yellow, with a basal whitish gland surrounded by a ring or trichomes, the apex irregularly lobed; stamens numerous, 4–6 mm long, orange-yellow. Fruit subglobose, 5–12 mm long, dark purple, pyrenes 4.

Disturbed pine rocklands. Rare; Miami-Dade County. Escaped from cultivation. Florida; Asia. Native to Asia. Summer.

Herissantia Medik. 1788.

Annual or perennial herbs. Leaves alternate, simple, pinnipalmate-veined, petiolate, stipulate. Flowers solitary in the leaf axils; sepals 5, basally connate; petals 5, free; stamens ca. 20, in a staminal column; ovary 10- to 12-carpellate and -loculate, the styles 10–12. Fruit a schizocarp.

A genus of 5 species; North America, West Indies, Mexico, Central America, South America, Africa, Asia, Australia, and Pacific Islands. [Commemorates Louis Antoine Prospère Hérissant (1745–1769), French physician, naturalist, and poet.]

Gayoides (Endl.) Small (1903).

Selected reference: La Duke (2015).

Herissantia crispa (L.) Brizicky [Irregularly waved and twisted, in reference to the capsule.] BLADDERMALLOW.

Sida crispa Linnaeus, Sp. Pl. 685. 1753. *Abutilon crispum* (Linnaeus) Medikus, Malvenfam. 29. 1787. *Napaea crispa* (Linnaeus) Moench, Suppl. Meth. 207. 1802. *Gayoides crispa* (Linneus) Small, Fl. S.E. U.S. 764, 1335. 1903. *Pseudobastardia crispa* (Linnaeus) Hassler, Bull. Soc. Bot. Genève, ser. 2. 1: 211. 1909. *Bogenhardia crispa* (Linnaeus) Kearney, Leafl. W. Bot. 7: 120. 1954. *Herissantia crispa* (Linnaeus) Brizicky, J. Arnold Arbor. 49: 279. 1968.

Sida imberbis de Candolle, Prodr. 1: 469. 1824. *Abutilon imberbe* (de Candolle) G. Don, Gen. Hist. 1: 502. 1831. *Gayoides imberbis* (de Candolle) Small, Fl. S.E. U.S. 764, 1335. 1903.

Diffuse or sprawling, annual or short-lived perennial, to 6 dm; stem soft tomentose with stellate and simple trichomes. Leaves with the blade ovate to elongate-triangular, 1–8 cm long, 1–4 cm wide, the apex acute to acuminate, the base cordate, the margin crenate to dentate, the upper and lower surfaces softly tomentose with stellate and simple trichomes, the petiole 0.5–1.5 cm long, the uppermost ones becoming subsessile. Flowers axillary, solitary, the peduncle 2–3 cm long, articulate distally; sepals 3–7 mm long, the lobes ovate to ovate-lanceolate, softly pubescent and often villous; petals obcordate, 6–11 mm long, pale yellow, whitish, or yellowish-orange. Fruit subglobose, 1–2 cm long, becoming papery and inflated in fruit, hirsute; seeds (2)3(6) per carpel, glabrous or minutely setulose.

Coastal hammocks and disturbed sites. Occasional; central and southern peninsula. Florida, Texas to California; West Indies, Mexico, Central America, and South America; Africa, Asia, Australia, and Pacific Islands. Native to North America and tropical America. All year.

Hibiscus L., nom. cons. 1753. ROSEMALLOW

Herbs, subshrubs, or shrubs. Leaves alternate, simple, palmate-veined, petiolate, stipulate. Flowers axillary, solitary, the pedicels jointed or not; involucrate bracts 4–12; sepals 5, basally connate; petals 5, basally connate to the staminal sheath; stamens numerous, forming a column, this basally adnate to the petals; ovary 5-carpellate and -loculate, the stigmas 5, capitate or discoid. Fruit a loculicidal 5-valved capsule.

A genus of about 350 species; nearly cosmopolitan. [From the Greek *hibiscos* or *ibiscos*, in reference to the cohabitation with the *ibis*, stork, in the marshes.]

Recent molecular data indicate that *Hibiscus*, as traditionally defined, is paraphyletic, which may result in either expanding the concept of the genus or breaking it up into numerous small genera. The resolution of this must await further study.

Trionum Schaeff. (1760).

Selected reference: Blanchard (2015a).

1. Lower leaf surface tomentose or scabrous.
    2. Involucrate bracts forked.
        3. Stem and leaves closely stellate-tomentose, velvety to the touch .... **H. furcellatus**
        3. Stem and leaves aculeate-scabrate, very rough to the touch .... **H. aculeatus**
    2. Involucrate bracts entire.
        4. Corolla 2–3 cm long; lower leaf surface sparsely stellate-pubescent .... **H. poeppigii**
        4. Corolla 6–12 cm long; lower leaf surface densely white-tomentose.
            5. Corolla pink; leaf base cordate .... **H. grandiflorus**
            5. Corolla white to pale yellow with a dark red eye; leaf base rounded to broadly cuneate
            .... **H. moscheutos**
1. Lower leaf surface glabrous or nearly so.
    6. Involucrate bracts bifurcate.
        7. Nectaries lacking on the leaves and the calyx lobes .... **H. radiatus**
        7. Nectaries present on the lower surface of the leaves near the base of the midrib and on the calyx lobes near the middle of the midrib.
            8. Corolla red with darker veins and a very dark center; leaves reddish; involucrate bracts with an appendage .... **H. acetosella**
            8. Corolla pale yellow, often with a dark purple center; leaves green or sometimes with red veins; involucrate bracts lacking an appendage .... **H. cannabinus**
    6. Involucrate bracts entire.
        9. Leaves (at least some) hastate-lobed .... **H. laevis**
        9. Leaves palmately divided or ovate to elliptic-lanceolate.
            10. Leaves ovate to elliptic-lanceolate .... **H. rosa-sinensis**
            10. Leaves palmately divided.
                11. Corolla 1.5–4 cm long, yellow or whitish with the base and one lateral edge dark purple; fruiting calyx inflated .... **H. trionum**
                11. Corolla 7–10 cm long, bright red; fruiting calyx not inflated .... **H. coccineus**

Hibiscus acetosella Welw. ex Hiern [Leaves with an acidity like that of sorrel, *Rumex acetosella* (Polygonaceae).] AFRICAN ROSEMALLOW.

Hibiscus acetosella Welwitsch ex Hiern, Cat. Afr. Pl. 1: 73. 1896.

Subshrub, to 4 m; stem glabrous or subglabrous, rarely sparsely pubescent, with a line of fine, curved trichomes. Leaves with the blade broadly ovate, 4–10 cm long, 3.5–10 cm wide, unlobed or 3- to 5-lobed or -fid, the lobes elliptic to obovate, usually dark red, the apex acute to acuminate, the base cuneate to truncate, the margin crenate or crenate-serrate, the upper and lower surfaces glabrous or subglabrous, the lower surface with a slit-like nectary on the midvein near the base, the petiole ½ to subequaling the blade length, with fine, curved trichomes adaxially; stipules linear-lanceolate, (8)10–15 mm long. Flowers solitary in the axil of the distal leaves, sometimes appearing racemose by the reduction of the subtending leaves, horizontal, the pedicel to 1.2 cm long, jointed near the middle; involucrate bracts 8–10, terete, 0.6–1.6 cm long, the apex 2-fid or appendaged, the margin setose, calyx funnelform-campanulate, 1.2–2 cm long, lobed nearly ⅔ its length, the lobes triangular, each with 1 median and 2 marginal ribs, the median one bearing a gland-like nectary, the veins setose with pustular-based simple trichomes; corolla funnelform-rotate, 3–5.5 cm long, the petals asymmetrically obovate, cream, yellow, or dull pink to dull red, maroon at the base, pubescent where exposed in bud; staminal column 1.5–2.5 cm long, dark-colored, bearing filaments nearly its length, the filaments 2–3 mm long; styles and stigmas dark-colored. Fruit ovoid, 1.6–2.5 cm long, reddish-brown, weakly antrorsely hispid with loose, scattered, simple trichomes; seeds angulately reniform-ovate, 3–4 mm long, olivaceous brown, papillose-pectinate scaly.

Disturbed sites. Occasional; central and southern peninsula. Escaped from cultivation. Florida; West Indies, Mexico, Central America, and South America; Africa and Asia. Native to Africa. Fall–winter.

Hibiscus aculeatus Walter [Spiny, in reference to the calyx.] COMFORTROOT.

Hibiscus aculeatus Walter, Fl. Carol. 177. 1788.
Hibiscus scaber Michaux, Fl. Bor.-Amer. 2: 45. 1803; non Lamarck, 1789. TYPE: "Carolina et Floridae."

Herb or subshrub, to 2 m; stem with scattered, stout, pustular-based trichomes and with a line of fine, curved trichomes. Leaves with the blade broadly ovate, 3.5–9.5 cm long, 4.5–13.5 cm wide, 3- to 5-fid or -lobed, the lobes obovate to oblanceolate, the apex acute to short-acuminate, the base cordate to cuneate, the margin coarsely and irregularly crenate-serrate, the upper and lower surfaces scabrous, the pubescence of scattered, stout, pustular-based trichomes, the lower surface with a slit-like nectary at or near the base of the midvein, the petiole of the lower leaves ⅔ to equaling the blade, much shorter in the inflorescence, with a fine pubescence of scattered, stout, pustular-based trichomes and adaxially with curved trichomes; stipules linear or linear-filiform, 2–6 mm long. Flowers solitary in the axil of the distal leaves, or appearing racemose by reduction of the subtending leaf, horizontal or declinate, the pedicel to 1.5 cm long, inconspicuously jointed at the base; involucrate bracts 8–11, linear-subulate, 1–1.6 cm long, the apex 2-fid, with bristly trichomes, calyx campanulate, 1.6–2.8 cm long, lobed ca. ½ its length, the lobes triangular, with 3 prominent, often reddish, 1 median, 2 marginal, the median with a prominent nectary, the veins and sometimes the space between them with conspicuous simple or stellate trichomes; corolla funnelform, 5–8 cm long, the lobes obovate, pale yellow to white, dark red at the base, pubescent where exposed in bud; staminal column 2–4.5 cm long, dark red, bearing filaments nearly throughout, the filaments 2–3 mm long; styles 9–22 mm long, dark red or white or intermediate, the stigmas dark red or white or intermediate. Fruit ovoid, 1.2–2 cm long, medium brown to stramineous, variously antrorsely hispid and minutely stellate-pubescent; seeds angulately reniform-ovate, 3–5 mm long, reddish brown to dark brown, sometimes with raised, pale concentric lines sparingly to moderately papillose-verrucose.

Seepage bogs, swamps, open, wet woods, and disturbed sites. Frequent; northern counties, Lake County. North Carolina south to Florida, west to Texas. Spring–summer.

Hibiscus cannabinus L. [Resembling *Cannabis* (Cannabaceae), in reference to the leaves.] BROWN INDIANHEMP; KENAF.

> *Hibiscus cannabinus* Linnaeus, Syst. Nat., ed. 10. 1149. 1759. *Ketmia glandulosa* Moench, Suppl. Meth. 202. 1802, nom. illegit. *Hibiscus cannabifolius* Stokes, Bot. Mat. Med. 3: 542. 1812, nom. illegit. *Hibiscus cannabinus* Linnaeus var. *genuinus* Hochreutiner, Annuaire Conserv. Jard. Bot. Genève 4: 115. 1900, nom. inadmiss. *Hibiscus sabdariffa* Linnaeus subsp. *cannabinus* (Linnaeus) Panagrahi & Murti, Fl. Bilaspur Distr. 1: 127. 1989.

Annual or perennial herb, to 3 m; stem sparsely spiny, glabrous. Leaves with the blade ovate, the proximal ones entire, the distal ones 3- to 7-lobed, the lobes lanceolate, 2–12 long, 0.6–2 cm wide, the apex acuminate, the base cordate or rounded, the margin serrate, the upper and lower surfaces glabrous, the lower surface glandular near the midrib base. Flowers solitary, in the axil of the distal leaves, horizontal, nearly sessile, the pedicel 1–2 cm long; involucral bracts 7–10, filiform, 6–8 mm long, sparsely spiny; calyx subcampanulate, 2–4 cm long, lobed ca. ½ its length, the lobes lanceolate, spiny, and white-tomentose; corolla broadly funnelform, ca. 6 cm long, the lobes oblong-ovate, yellow with a red center; staminal column 1.5–2 cm long; styles glabrous. Fruit globose, ca. 1.5 cm long, densely spiny; seeds reniform, subglabrous.

Disturbed sites. Rare; Miami-Dade County. Escaped from cultivation. Florida; West Indies; Asia and Africa. Native to Africa and Asia. Spring–summer.

Hibiscus cannabinus is widely cultivated in the Old World as a fiber crop.

Hibiscus coccineus Walter [Deep red, in reference to the petals.] SCARLET ROSEMALLOW.

> *Hibiscus coccineus* Walter, Fl. Carol. 177. 1778. TYPE: FLORIDA: St. Johns Co.: Fl 16, 0.4 mi. E of St. Johns River, between Orangedale and Green Cove Springs, 14 Jul 1968, *Blanchard 173* (neotype: GH; isoneotype: CAS). Neotypified by Ward (2008: 478).
>
> *Hibiscus coccineus* Walter var. *integrifolius* Chapman, Bot. Gaz. 3: 3. 1878. *Hibiscus semilobatus* Chapman, Fl. South. U.S., ed. 3. 52. 1897. *Hibiscus integrifolius* (Chapman) Small, Bull. Torrey Bot. Club 25: 135. 1898, nom. illegit. TYPE: FLORIDA: Duval Co.: marshes near Jacksonville, 1871, *Chapman s.n.* (holotype: NY).

Perennial herb, to 3 m, stem glabrous, glaucous. Leaves with the blade, suborbicular to transversely elliptic, 10–19 cm long, 13–25 cm wide, 3- to 5-parted or -divided, the segments linear-lanceolate, the apex acuminate, the base cordate, the margin irregularly serrate, the upper and lower surfaces glabrous, the petiole ⅓ to equaling the blade; stipules linear-subulate, 1–3 mm long. Flowers solitary in the axil of distal leaves, horizontal or ascending, the pedicel 3–14 cm long, the joint distal; involucral bracts 9–15, linear-subulate, 2.5–4 cm long; calyx rotate, 3.5–6 cm long, lobed for ½ its length, the lobes narrowly triangular; corolla rotate, (6)7.5–10 cm long, the lobes spatulate-obovate, deep red (rarely white), minutely pubescent where exposed in bud; staminal column 6.5–7 cm long, bearing filaments in the distal ⅓, the filaments 4–8 mm long; styles and stigmas red. Fruit ovoid to globose, 2.8–3.5 cm long, brown, glabrous; seeds reniform-globose, 3–4 mm long, brown, brownish to reddish pubescent.

Swamps. Occasional; nearly throughout. Virginia south to Florida, west to Texas. Summer.

Hibiscus furcellatus Desr. [Minutely or slightly forked, in reference to the calyx lobes.] LINDENLEAF ROSEMALLOW.

Hibiscus furcellatus Desrousseaux, in Lamarck, Encycl. 3: 358. 1789. *Hibiscus furcellatus* Desrousseaux var. *genuinus* Gürke, in Martius, Fl. Bras. 12(3): 562. 1892, nom. inadmiss.

Herb or subshrub, to 4 m; stem densely stellate-tomentulose, sometimes also with longer stiff, simple trichomes. Leaves with the blade broadly ovate, 6.5–11 cm long, 6–12 cm wide, unlobed or 3- to 7-lobed, the lobe apex acute to short-acuminate, the base cordate, the margin irregularly serrate or crenate-serrate, the upper and lower surfaces densely stellate-tomentulose, the lower surface with a slit-like nectary at or near the base of the midvein, the petiole ⅔ to equaling the blade, shorter in the inflorescence; stipules linear, 3–10 mm long. Flowers solitary in the axil of the distal leaves, horizontal or declinate, sometimes appearing racemose by reduction of the subtending leaves, the pedicel to 2.5 cm long, jointed at the base; the involucrate bracts 9–11, 0.8–1.6 cm long, 2-fid or appendaged at the apex; calyx campanulate, 1.5–2.4 cm long, lobed ½–⅔ its length, the lobes triangular, with 3 prominent ribs, 1 median and 2 marginal, the median one with a nectary, variously with small stellate trichomes and larger simple or stellate, pustular-based trichomes; corolla narrow funnelform, 5.5–9.5 cm long, the lobes obliquely obovate, pink with maroon at the base, pubescent where exposed in the bud; staminal column 3–3.5 cm long, maroon, bearing filaments throughout, the filaments ca. 1 mm long, those on the distal half of the column in subverticillate clusters; styles 1–3 mm long, dark maroon, the stigmas dark maroon. Fruit ovoid, 2–2.5 cm long, with dense, simple, antrorsely appressed, pale trichomes; seeds angulately reniform-ovate, 3–4 mm long, olivaceous-brown to reddish or purplish brown, smooth, glabrous.

Dry, open areas. Occasional; central peninsula, Broward County. Florida; West Indies, Mexico, Central America, and South America; Pacific islands. Native to North America and tropical America. All year.

Hibiscus grandiflorus Michx. [Large-flowered.] SWAMP ROSEMALLOW.

Hibiscus grandiflorus Michaux, Fl. Bor.-Amer. 2: 46. 1803. TYPE: "Georgiae et Floridae et in regione Natchez ad Mississipi," *Michaux s.n.* (lectotype: P; isolectotypes: P). Lectotypified by Blanchard (2008: 5).

Perennial herb, to 3 m; stem glabrous or stellate pubescent on the younger parts. Leaves with the blade broadly ovate to transversely broadly elliptic, 7–30 cm long, 8–29 cm wide, 3-lobed, the lateral lobes acuminate to broadly rounded, the apex acute to acuminate, the base cordate to truncate, the margin crenate to crenate-dentate, the upper and lower surfaces velvety-pubescent, sometimes the upper scabridulous, the petiole ½–¾ the blade length; stipules linear-subulate, 2–4 mm long. Flowers solitary in the axil of the distal leaves, horizontal or ascending, the pedicel 2–10 cm long, jointed near the middle or distally; involucral bracts 9–13, linear-subulate, 1.3–2.7 cm long, velvety-pubescent; calyx broadly campanulate, 2.9–6 cm long, lobed to about the middle, the lobes triangular, velvety-pubescent; corolla funnelform, 8.5–14 cm long, the lobes narrowly obovate, pink to white, red at the base, minutely pubescent where exposed in bud; staminal column 6.2–9.5 cm long, bearing filaments throughout its length, the filaments 3–9 mm long; the style white, the stigmas yellow. Fruit ovoid to subglobose, 2.2–3.5 cm long, hispid with yellowish brown to reddish brown simple trichomes; seeds reniform-globose, 2–3 mm long, brown to reddish brown, verrucose-papillose.

Swamps and marshes. Frequent; nearly throughout. South Carolina south to Florida, west to Texas; West Indies. Spring–summer.

Hibiscus laevis All. [Smooth.] HALBERDLEAF ROSEMALLOW.

Hibiscus laevis Allioni, Auct. Syn. 31. 1773.
Hibiscus militaris Cavanilles, Diss. 6: 352, t. 198(2). 1788.

Perennial herb, to 2.5 m; stem glabrous or nearly so, often glaucous. Leaves with the blade ovate to triangular-ovate, 6–18 cm long, 3–16 cm wide, hastately 3- to 5-lobed or unlobed, the apex acuminate, the base cordate to truncate, the margin crenate-serrate to serrate or serrate-dentate, the petiole ½ to somewhat exceeding the blade; stipules linear-subulate, 2–7 mm long. Flowers solitary in the axil of distal leaves, horizontal, the pedicel 1–10 cm long, jointed distally; involucral bracts (8)9–15(16), linear-subulate, 1–3 cm long; calyx broadly cylindric-campanulate, 2.5–3 cm long, somewhat inflated, lobed ⅓–½ its length, the lobes broadly triangular; corolla broadly funnelform, 5–8 cm long, the lobes obovate, pink to white, red at the base, minutely pubescent where exposed in bud; staminal column 2.5–4 cm long, bearing filaments nearly throughout, the filaments 2–4 mm long; styles and stigmas pink. Fruit ovoid, 1.8–3 cm long, brown, glabrous; seeds reniform-globose, 3–5 mm long, reddish pubescent.

Swamps and floodplain forests. Occasional; central panhandle, Escambia County. New York south to Florida, west to Ontario, Minnesota, Nebraska, Kansas, Oklahoma, and Texas. Summer.

Hibiscus moscheutos L. [*Moschatus*, musk-scented.] CRIMSONEYED ROSEMALLOW.

Hibiscus moscheutos Linnaeus, Sp. Pl. 693. 1753. *Hibiscus petioliflorus* Stokes, Bot. Mat. Med. 3: 543. 1812, nom. illegit.
Hibiscus lasiocarpos Cavanilles, Diss. 3: 159, t. 70(1). 1787. *Hibiscus moscheutos* Linnaeus subsp. *lasiocarpos* (Cavanilles) O. J. Blanchard, Novon 18: 4. 2008. *Hibiscus moscheutos* Linnaeus var. *lasiocarpos* (Cavanilles) B. L. Turner, Phytologia 90: 379. 2008.
Hibiscus incanus J. C. Wendland, Bot. Beob. 54. 1798. *Hibiscus moscheutos* Linnaeus subsp. *incanus* (J. C. Wendland) Ahles, J. Elisha Mitchell Sci. Soc. 80: 173. 1964.

Perennial herb, to 2.5 m; stem glabrous or variously pubescent. Leaves with the blade lanceolate to suborbicular, 8–20 cm long, 3–13 cm wide, sometimes 3-lobed or -angulate, the apex acute to acuminate, the base cordate or cuneate, the margin crenate to dentate or serrate, the upper surface pubescent or glabrous, the lower surface pubescent, the petiole ¼–¾ the blade length; stipules subulate, 1–4 mm long. Flowers solitary in the axil of the distal leaves, horizontal, the pedicel 2–15 cm long, jointed distally; involucral bracts 10–14, linear-lanceolate, pubescent; calyx broadly campanulate, 1.5–4 cm long, lobed for ½–⅔ its length, the lobes triangular or tri-angular-ovate, pubescent; corolla funnelform, 4–12 cm long, the lobes obliquely obovate, pink or white, sometimes with a basal red spot, minutely pubescent where exposed in bud; staminal column 1.2–5 cm long, bearing filaments nearly throughout, the filaments 2–8 mm long; styles and stigmas creamy white to yellow. Fruit ovoid to subglobose, 1.4–3.5 cm long, dark brown, glabrous or pubescent; seeds reniform-globose, 2–3 mm long, brown, verrucose-papillose.

Marshes and swamps. Occasional; northern counties south to Lake County. New York and Massachusetts south to Florida, west to Ontario, Wisconsin, and New Mexico, also California; Mexico; Europe and Asia. Native to North America and Mexico. Summer–fall.

Hibiscus poeppigii (Spreng.) Garcke [Commemorates Eduard Friedrich Poeppig (1798–1868), German botanist, zoologist, and explorer.] POEPPIG'S ROSEMALLOW.

> *Achania poeppigii* Sprengel, Syst. Veg. 3: 100. 1826. *Malvaviscus poeppigii* (Sprengel) G. Don, Gen. Hist. 1: 475. 1831. *Pavonia poeppigii* (Sprengel) Steudel, Nomencl. Bot., ed. 2, 2: 279. 1841. *Hibiscus poeppigii* (Sprengel) Garcke, Jahresber. Naturwiss. Vereins Halle 2: 133. 1850. *Malache poeppigii* (Sprengel) Kuntze, Revis. Gen. Pl. 1: 71. 1891.
>
> *Malvaviscus floridanus* Nuttall, J. Acad. Nat. Sci. Philadelphia 7: 89. 1834. *Achania floridana* (Nuttall) Rafinesque, New Fl. 1: 49. 1836. *Hibiscus floridanus* (Nuttall) Shuttleworth ex A. Gray, Smithsonian Contr. Knowl. 3(5): 22. 1852. TYPE: FLORIDA: Monroe Co.: Key West, s.d., *Peale s.n.* (holotype: PH).

Subshrub, to 1.8 m; stem appressed stellate-pubescent and with a fine line of curved simple trichomes internodally and decurrent from the leaf base, older branches brown or greenish brown to gray. Leaves with the blade broadly ovate to transversely ovate, 1.2–4.5 cm long, 1.2–4.3 cm wide, often 3-lobed, the apex obtuse to acute, the base cordate to truncate or broadly cuneate, the margin irregularly crenate to serrate, the upper and lower surfaces with appressed stellate trichomes, the lower surface with a nectary at the base of the midvein, the petiole usu-ally ¼–¾ the blade length, with a line of fine curved trichomes adaxially; stipules narrowly triangular, 2–4 mm long. Flowers solitary in the axil of the distal leaves, nodding or pendu-lous, the pedicel 2–5 cm long, jointed beyond the middle; involucrate bracts 9–11, linear to narrowly oblanceolate, 0.5–0.9 cm long; calyx narrowly campanulate, 0.7–1.2 cm long, lobed to ½ its length, the lobes narrowly triangular, without nectaries; corolla narrowly funnelform, 1.6–2.8 cm long, the lobes broadly oblanceolate, red, pubescent where exposed in bud; stami-nal column exceeding the corolla, bearing filaments on the distal ½, the filaments 2–3 mm long, red; styles 1–3 mm long, red, the stigmas red. Fruit subglobose, ca. 1 cm long, apiculate, dull brown, coarsely stellate-pubescent; seeds angulately reniform-ovate, 2–3 mm long, pale silky-pubescent.

Rockland hammocks. Rare; Miami-Dade and Monroe Counties. Florida; West Indies, Mex-ico, and Central America. All year.

Hibiscus poeppigii is listed as endangered in Florida (Florida Administrative Code, Chapter 5B-40).

Hibiscus radiatus Cav. [Spreading out from a common point, in reference to the palmate-lobed leaves.] MONARCH ROSEMALLOW.

Hibiscus radiatus Cavanilles, Diss. 3: 150, t. 54(2). 1787.

Herb or subshrub, to 2 m; stem glabrous or with a few prickles and/or simple trichomes, always with a line of fine curved trichomes. Leaves with the blade broadly triangular-ovate, 4.5–15 cm long, 6–17 cm wide, 3- to 5-fid, the segments lanceolate, the apex acuminate, the base truncate or cordate, the margin serrate, the upper surface glabrous, the lower surface with retrorse prickles on the veins, a nectary sometimes present on the midvein near the base, the petiole ⅔ to equaling the blade, with fine, curved trichomes adaxially; stipules linear-lanceolate, 10–16 mm long, ciliate. Flowers solitary in the axil of the distal leaves, horizontal or ascending, the petiole to 1.5 cm long, jointed near the middle, prickly distal to the joint; involucrate bracts 9–10, flattened or canaliculate, 1–1.8 cm long, simple or apically 2-fid or appendaged, setose; calyx cylindric-campanulate, 2–2.5 cm long, divided ⅔ its length, the lobes narrowly triangular, each with 3 prominent ribs, 1 median and 2 marginal, setose; corolla rotate, 3.5–7 cm long, the lobes asymmetrically obovate, rose-purple or yellow with a dark purple base, pubescent where exposed in bud; staminal column 2.4–3.5 cm long, dark purple, bearing filaments nearly throughout, the filaments 2–3 mm long; styles and stigmas dark purple. Fruit ovoid, 1.8–2.5 cm long, antrorsely hispid; seeds angulately reniform-ovate, dark olivaceous, with fine, raised concentric lines, ca. 4 mm long, moderately verrucose-lepidote, the scales striate-fimbriate.

Disturbed sites. Occasional; peninsula. Escaped from cultivation. Florida and Texas; West Indies, Mexico, Central America, and South America; Africa, Asia, and Australia. Native to Asia. All year.

Hibiscus rosa-sinensis L. [Vernacular name, meaning "China rose."]

Shrub or tree, to 5 m; stem sparingly with stellate or simple trichomes. Leaves with the blade ovate or ovate-lanceolate, 3.5–12 cm long, 1.5–8.5 cm wide, the apex acute to short acuminate, the base rounded to cuneate, the margin coarsely serrate in the distal ⅔–¾, the upper and lower surfaces glabrate, the lower surface with a nectary on the midvein near the base, the petiole to ⅓ the blade length, with minute sinuous trichomes in a groove, adaxially; the stipules linear to lanceolate, 8–16 mm long or triangular 1–3 mm long. Flowers solitary in the axil of the distal leaves, horizontal or declinate, the pedicel 4–9.5 cm long, jointed close to the flower; involucral bracts 6–8, narrowly triangular, 1.2–2.2 cm long or less than ¹⁄₁₆ the calyx length; calyx narrowly campanulate to funnelform, (1)2–3 cm long, lobed ⅛–¾ its length, the lobes triangular, often minutely and sparingly pubescent; corolla funnelform to rotate or slightly reflexed, 6–10.5 cm long, the lobes obovate, red, pink, or white (other colors in horticultural forms), usually darker at the base, pubescent where exposed in bud; staminal column 6.5–11.5 cm long, usually colored as in the petals, bearing filaments in the distal ½, the filaments 3–9 mm long; the styles and stigmas colored as the petals. Fruit ovate, ca. 2.5 cm long, glabrous; seeds reniform, 2–3 mm long.

1. Petals entire; involucrate bracts ⅓ to nearly equaling the calyx length .... var. **rosa-sinensis**
1. Petals laciniate; involucrate bracts less than ¹⁄₁₆ the calyx length .... var. **schizopetalus**

Hibiscus rosa-sinensis var. **rosa-sinensis** GARDEN ROSEMALLOW; SHOE-BLACK-PLANT.

> *Hibiscus rosa-sinensis* Linnaeus, Sp. Pl. 694. 1753. *Hibiscus festalis* Salisbury, Prodr. Stirp. Chap. Allerton 383. 1796, nom. illegit. *Hibiscus rosiflorus* Stokes, Bot. Mat. Med. 3: 543. 1812, nom. illegit. *Hibiscus rosiflorus* Stokes var. *simplex* Stokes, Bot. Mat. Med. 3: 543. 1812, nom. inadmiss. *Hibiscus rosa-sinensis* Linnaeus var. *genuinus* Hochreutiner, Annuaire Conserv. Jard. Bot. Genève 4: 134. 1900, nom. inadmiss.

Leaves with the blade ovate, 5–12 cm long, 3–8.5 cm wide; stipules linear to lanceolate, 8–16 mm long. Flowers with the involucrate bracts ⅓ to nearly equaling the calyx length; petals entire.

Disturbed sites. Rare; Brevard, Indian River, St. Lucie, and Miami-Dade Counties. Escaped from cultivation. Florida and California; West Indies, Mexico, Central America, and South America; Africa, Asia, Australia, and Pacific Islands. Native to Asia. Summer–fall.

Hibiscus rosa-sinensis var. **schizopetalus** Dyer [From the Greek *schizo*, split or cut, in reference to the deeply cut petals.] FRINGED ROSEMALLOW.

> *Hibiscus rosa-sinensis* Linnaeus var. *schizopetalus* Dyer, Gard. Chron., ser. 2. 11: 568. 1879. *Hibiscus schizopetalus* (Dyer) Hooker f., Bot. Mag. 106: t. 6524. 1880.

Leaves with the blade ovate-lanceolate, 3.5–10.5 cm long, 1.5–4 cm wide; stipules triangular 1–3 mm long. Flowers with the involucrate bracts less than ¹⁄₁₆ the calyx length; petals laciniate.

Disturbed sites. Rare; Lee and Palm Beach Counties, southern peninsula. Escaped from cultivation. Florida; West Indies, Mexico, and Central America; Africa, Asia, Australia, and Pacific Islands. Native to Africa.

Hibiscus trionum L. [Pre-Linnaean name for the plant, in reference to the usually 3-lobed or -divided leaves.] FLOWER-OF-AN-HOUR.

> *Hibiscus trionum* Linnaeus, Sp. Pl. 697. 1753. *Ketmia trionum* (Linnaeus) Scopoli, Fl. Carniol., ed. 2. 2: 44. 1772. *Trionum annuum* Medikus, Malvenfam. 47. 1787. *Trionum diffusum* Moench, Methodus 618. 1794, nom. illegit. *Hibiscus trionicus* Saint-Lager, Ann. Soc. Bot. Lyon 7: 127. 1880, nom. illegit. *Trionum trionum* (Linnaeus) Wooten & Standley, Contr. U.S. Natl. Herb. 19: 417. 1915, nom. inadmiss.

Annual herb, to 6 dm; stem with coarse, simple or few-armed stellate trichomes and fine, many-armed stellate trichomes, additionally with a line of fine, curved trichomes extending from node to node. Leaves with the blade broadly ovate, 2.5–6.5 cm long, 3–7 cm wide, 3- to 5-lobed or divided, the apex rounded to truncate, the base cuneate, the margin pinnately lobed, the upper and lower surfaces with coarse, simple or few-armed stellate trichomes and fine, many-armed stellate trichomes, the petiole ½ to subequaling the blade, with fine, curved trichomes adaxially; stipules narrowly triangular, 2–5(8) mm long. Flowers solitary in the axil of the distal leaves, ascending or erect, the pedicel 1–4 cm long, jointed distally; involucral bracts 10–15, 1–2.5 cm long, ciliate; calyx campanulate, 0.8–1.8 cm long, lobed to about the middle; corolla funnelform, 1.5–3(4) cm long, spreading, the lobes obovate, yellow with a purple-brown basal spot, purplish and pubescent where exposed in bud; staminal column 4–7(11) mm long, purplish, bearing filaments in the distal ½, the filaments purplish; styles yellowish, the stigmas purplish. Fruit ellipsoid to ovoid, 1.2–1.5 cm long, dark brown-black, enclosed by the inflated calyx, the surface somewhat conforming to the enclosed seed mass, pubescent; seeds reniform-ovate, somewhat laterally compressed, 2–3 mm long, dark gray-brown, sparingly and minutely papillose.

Dry, disturbed sites. Rare; Franklin County. Nearly throughout North America; West Indies; Europe, Africa, Asia, Australia, and Pacific Islands. Native to Europe, Africa, Asia, Australia, and Pacific Islands. Summer–fall.

EXCLUDED TAXA

Hibiscus bifurcatus Cavanilles—Reported as occurring in the Florida Keys by Menzel et al. (1983). No Florida specimens known.

Hibiscus mutabilis Linnaeus—Reported as escaped in Florida by Small (1933). No Florida specimens known.

Hibiscus pilosus (Swartz) Fawcett & Rendle—Reported for Florida by Small (1933) and Long and Lakela (1971), this name has been shown by Fryxell (1988) to apply to *Malvaviscus arboreus* var. *mexicanus* Schlechtendal. Florida material with this name is *H. poeppigii*.

Hibiscus sabdariffa Linnaeus—Reported for Florida by Small (1913a, 1913b, 1933), but no specimens are known to document this cultivated plant as a legitimate escape from cultivation.

Hibiscus spiralis Cavanilles—Reported for Florida by Small (1913a, 1913d, 1913e), who misapplied the name of this Mexican endemic to material of *H. poeppigii*.

Hibiscus syriacus Linnaeus—Reported for Florida by Small (1903, 1913a, 1913c, 1913d, 1933), but no specimens have been seen to document this species as an escape from cultivation in Florida.

Kosteletzkya C. Presl, nom. cons. 1835.

Perennial herbs or subshrubs. Leaves alternate, simple, palmate-veined, petiolate, stipulate. Flowers solitary in the axils of the upper leaves or forming open panicles by upper leaf reduction, the pedicel jointed; involucre 6- to 10-segmented, the segments free or nearly so; sepals 5, basally connate; petals 5, basally adnate to the stamen column; stamens numerous, the filaments connate into a column nearly the length of the androgynophore; ovary 5-carpellate and -loculate, the style 1, apically 5-fid, the stigmas capitate. Fruit a capsule.

A genus of 17 species; North America, West Indies, Mexico, Central America, South America, Europe, Africa, and Asia. [Commemorates Vincenz Franz Kosteletzky (1801–1887), Czech physician and botanist.]

Selected reference: Blanchard (2015b).

1. Petals pink, 2–4 cm long; capsule with slender trichomes fairly well distributed over the surface, the sides of capsule also often with some short trichomes; seeds glabrous and with evident striations .... **K. pentacarpos**
1. Petals white or yellow (fading pink), 0.8–1 cm long; capsule with stout, often uncinate, trichomes confined to the suture margins, the sides of capsule finely stellate-pubescent; seeds sparsely stellate, lacking evident striations .... **K. depressa**

Kosteletzkya depressa (L.) O. J. Blanch. et al. [Flattened, in reference to the growth habit.] WHITE FENROSE; WHITE FEN.

Melochia depressa Linnaeus, Sp. Pl. 674. 1753. *Riedlea depressa* (Linnaeus) de Candolle, Prodr. 1: 491. 1824. *Melochia diffusa* Bertero ex Colla, Hortus Ripul. 88. 1824, nom. illegit. *Visenia depressa* (Linnaeus) Sprengel, Syst. Veg. 3: 30. 1826. *Kosteletzkya depressa* (Linnaeus) O. J. Blanchard et al., Gentes Herb. 11: 357. 1978.

Hibiscus pentaspermus Bertero ex de Candolle, Prodr. 1: 447. 1824; non Nuttall, 1822. *Pavonia berteroi* Sprengel, Syst. Veg. 3: 99. 1826. *Kosteletzkya pentasperma* Grisebach, Fl. Brit. W.I. 83. 1859, nom. illegit.

Herbs or subshrubs, to 2 m; stem hispid. Leaves with the blade ovate, sometimes palmately or sagittately 3- to 5-lobed, 3–8 cm long, 1.5–6.5 cm wide, the apex acute to acuminate, the base rounded to truncate, the margin irregularly serrate to crenate, the upper and lower surfaces hispid, the petiole of the lower leaves ¼–⅔ times the blade length; stipules linear subulate, 3–6 mm long. Flowers axillary or in a leafy, open panicle, the pedicel exceeding the petiole and sometimes the blade; involucre bracts 6–7, 2–3 mm long; sepals 3–6 mm long, divided ca. ½ their length, the segments triangular-ovate; petals asymmetrically ovate to suborbicular, 5–14 mm long, white, sometimes with a pinkish tinge, sometimes drying yellow; staminal column 4–10 mm long, yellow, bearing the filaments in the distal ½–⅔, the anthers yellow; styles white or pink, the stigmas pink. Fruit depressed globose, 8–11 mm wide, olivaceous to brown, the valve margin 5-angulate, with short trichomes and larger curved or hooked simple trichomes on the sutures; seeds 2–3 mm long, reniform-ovate, brown, puberulent.

Hammocks. Rare; Miami-Dade and Monroe Counties. Florida and Texas; West Indies, Mexico, Central America, and South America. All year.

Kosteletzkya depressa is listed as endangered in Florida (Florida Administrative Code, Chapter 5B-40).

Kosteletzkya pentacarpos (L.) Ledeb. [From the Greek, *pentos*, five, and *carpos*, fruit, in reference to the plant; a pre-Linnaean name.] VIRGINIA SALTMARSH MALLOW.

Hibiscus pentacarpos Linnaeus, Sp. Pl. 697. 1753. *Pavonia pentacarpos* (Linnaeus) Poiret, in Lamarck, Encycl., Suppl. 4: 335. 1816. *Pavonia veneta* Sprengel, Syst. Veg. 3: 98. 1826, nom. illegit. *Kosteletzkya pentacarpos* (Linnaeus) Ledebour, Fl. Ross. 1: 437. 1842. *Pentagonocarpus zannichellii* Parlatore, Fl. Ital. 5: 106. 1873, nom. illegit.

Hibiscus virginicus Linnaeus, Sp. Pl. 697. 1753. *Pavonia virginica* (Linnaeus) Sprengel, Syst. Veg. 3: 98. 1826. *Kosteletzkya virginica* (Linnaeus) C. Presl ex A. Gray, Gen. Amer. Bor. 80. 1848.

Pentagonocarpus virginicus (Linnaeus) Parlatore, Fl. Ital. 3: 106. 1873. *Kosteletzkya virginica* (Linnaeus) C. Presl ex A. Gray var. *typica* Fernald, Rhodora 43: 608. 1941, nom. inadmiss.

Hibiscus pentaspermus Nuttall, Amer. J. Sci. Arts 5: 298. 1822. TYPE: FLORIDA: "E. Flor.," s.d., *Ware s.n.* (lectotype: PH). Lectotypified by Alexander et al. (2012: 115).

Kosteletzkya virginica (Linnaeus) C. Presl ex A. Gray var. *althaeifolia* Chapman, Fl. South. U.S. 57. 1860. *Kosteletzkya althaeifolia* (Chapman) A. Gray ex S. Watson, Bibliogr. Index N. Amer. Bot. 1: 136. 1878. TYPE: FLORIDA: Manatee Co.: along the Manatee River, s.d., *Rugel 102* (holotype: ?).

Kosteletzkya virginica (Linnaeus) C. Presl ex A. Gray var. *smilacifolia* Chapman, Fl. South. U.S. 57. 1860. *Kosteletzkya smilacifolia* (Chapman) Chapman, Fl. South. U.S., ed. 2. 610. 1883. *Kosteletzkya pentacarpos* (Linnaeus) Ledebour var. *smilacifolia* (Chapman) S. N. Alexander, in S. N. Alexander et al., Castanea 77: 115. 2012. TYPE: FLORIDA: Manatee Co.: between the Manatee River and Sarasota Bay, Aug. 1845, *Rugel 103* (lectotype: US; isolectotype: NY). Lectotypified by Alexander et al. (2012: 115).

Herb or subshrub, to 3 m; stem stellate-pubescent to subglabrous. Leaves with the blade ovate, often palmately or hastately 3- to 5-lobed, 5.5–17 cm long, 3.5–16 cm wide, the apex acute to acuminate, the base cordate to rounded or truncate, the margin coarsely to finely serrate or doubly serrate to subentire, the upper and lower surfaces stellate-pubescent, the petiole of the lower leaves ca. ½ times the blade length; stipules linear-subulate, 3–10 mm long. Flowers in the axil of the upper leaves or axillary on the lateral branches, the pedicel exceeding the petiole, the involucral bracts 8–10, 4–10 mm long; sepals 7–11 mm long, divided ca. ¾ their length, the segments triangular-ovate; petals asymmetrically obovate, 1.5–4.5 cm long, pink or rarely white, yellow at the base; staminal column 1.5–3.5 cm long, bearing filaments nearly to the base, the filaments 1–3 mm long, the anthers yellow; styles 3–9 mm long, deep pink, the stigmas pink. Fruit depressed globose, 10–14 mm wide, 5-angled, dark brown, the valves hispid, the margin rounded; seeds reniform-ovate, 3–4 mm long, brown with pale concentric lines, glabrous.

Brackish and freshwater marshes. Common; nearly throughout. New York south to Florida, west to Texas; West Indies; Europe and Asia. Spring–fall.

EXCLUDED TAXON

Kosteletzkya tubiflora (de Candolle) O. J. Blanchard & McVaugh—Reported for Florida by Chapman (1897, as *Hibiscus tubiflorus* de Candolle) and Small (1903, as *Hibiscus tubiflorus* de Candolle), the name misapplied to *Hibiscus poeppigii*.

Malachra L. 1767. LEAFBRACT

Herbs or subshrubs. Leaves alternate, simple, palmate-veined, petiolate, stipulate. Flowers subtended by sessile, cordate bracts in head-like, axillary or terminal, racemes; involucral bracts absent; sepals 5, basally connate; petals 5, free; ovary 5-carpellate and -loculate, the styles 10, the stigmas capitate. Fruit a schizocarp.

A genus of about 8 species; North America, West Indies, Mexico, Central America, Africa, Asia, and Australia. [From the Greek *malache*, mallow.]

Selected reference: Fryxell and Hill (2015e).

1. Flowering heads on peduncles to 8 cm long; large trichomes on the stem and leaves usually stellate, small stellate trichomes numerous and evenly distributed on the stem and leaves; calyx 6–8 mm long ....**M. capitata**
1. Flowering heads sessile or on short peduncles to 2 cm long; large trichomes on the stem and leaves mostly simple and spreading, small stellate trichomes lacking except in bands on the stem and petioles; calyx 4–5 mm long ....**M. fasciata**

Malachra capitata (L.) L. [Head-like, in reference to the inflorescence.] MALVA DE CABALLO.

Sida capitata Linnaeus, Sp. Pl. 685. 1753. *Malachra capitata* (Linnaeus) Linnaeus, Syst. Nat., ed. 12. 458. 1767. *Urena capitata* (Linnaeus) M. Gómez de la Maza y Jiménez, Anales Soc. Esp. Hist. Nat. 19: 219. 1890.

Perennial herb or subshrub, to 1.5 m; stem densely stellate-pubescent, sometimes with longer trichomes. Leaves with the blade ovate, lyrate, or 3- to 5-lobed, the apex acute or obtuse, the base cordate, the margin crenate, the upper and lower surfaces stellate-pubescent, the petiole ½ or less the blade length; stipules filiform, 9–15 mm long. Flowers in a head-like, axillary or terminal inflorescence, these subtended by cordate foliaceous bracts with prominent white markings at the base, the apex acute, stellate-pubescent, the peduncle exceeding the petiole; sepals 6–8 mm long, deeply divided, hispid; petals 7–10 mm long, yellow. Fruit depressed globose, the mericarps 5, obovoid, brownish, 3 mm long, glabrous; seed solitary in each mericarp, glabrous.

Disturbed sites. Rare; Broward and Miami-Dade Counties. Florida, Louisiana, and Texas; West Indies, Mexico, Central America, and South America; Africa and Asia. Native to North America and tropical America. Spring–fall.

Malachra fasciata Jacq. [Grown together, in reference to the flowers appearing to be fasciate.] ROADSIDE LEAFBRACT.

Malachra fasciata Jacquin, Collectanae 2: 352. 1789. *Malachra alceifolia* Jacquin var. *fasciata* (Jacquin) A. Robyns, Ann. Missouri Bot. Gard. 52: 527. 1966.
Malachra urens Poiteau, in Ledebour & Alderstam, Diss. Bot. Pl. Doming. 22. 1805. *Urena urens* (Poiteau) M. Gómez de la Maza y Jiménez, Anales Soc. Esp. Hist. Nat. 19: 220. 1892.

Annual herb or subshrub, to 5 dm; stem sparingly hispid with simple trichomes, sometimes with stellate ones. Leaves with the blade ovate to ovate-lanceolate, 4–6 cm long, 1–4 cm wide, unlobed or 3(5)-lobed, the apex obtuse to acute, the base rounded to truncate, the margin sharply dentate, the petiole ca. ½ the blade length; stipules 5–15 mm long. Flowers in a subsessile head in the leaf axil, subtended by triangular-lanceolate bracts 1–1.5 cm long, these strongly veiny, the base cordate; sepals 4–5 mm long, the lobes lanceolate, aristate; petals obovate, ca. 1.5 cm long, red to orange or yellow. Fruit depressed globose, the mericarps 5, obovoid, brownish, 3 mm long, retorsely hispidulous; seed solitary in each mericarp, glabrous.

Disturbed sites. Rare; Lee County, southern peninsula. Florida; West Indies, Mexico, Central America, and South America; Africa, Asia, and Australia. Native to North America and tropical America. Spring–fall.

EXCLUDED TAXON

Malachra alceifolia Jacquin—Reported for Florida by Small (1933), Long and Lakela (1971), Wunderlin (1998), and Wunderlin and Hansen (2003), all of whom misapplied the name to material of *M. fasciata.*

Malva L. 1753. MALLOW

Annual herbs. Leaves alternate, simple, palmate-veined, petiolate, stipulate. Flowers solitary or in fascicles in the leaf axils; involucral bracts 3, free; sepals 5, basally connate; petals 5, free; stamens numerous, connate into a tube; ovary 10- to 11-carpellate and -loculate, the style branches 10–11. Fruit a schizocarp.

A genus of about 25 species; nearly cosmopolitan. [From the Greek *malos*, tender, in reference to the vegetative qualities.]

Selected reference: Hill (2015a).

Malva parviflora L. [Small-flowered.] CHEESEWEED MALLOW.

Malva parviflora Linnaeus, Demonstr. Pl. 18. 1753.

Erect or ascending annual herb, to 8 dm; stem glabrous or sparsely stellate-pubescent distally. Leaves with the blade suborbiculate-cordate or reniform, 2–8 cm long and wide, shallowly 5- to 7-lobed, the lobes deltate or rounded, the apex rounded, the margin evenly crenulate, the upper and lower surfaces glabrous or pubescent with stellate and simple trichomes, the petiole 2–3(4) times the blade length; stipules broadly lanceolate, 4–5 mm long. Flowers axillary, solitary or 2–4 in a fascicle, the pedicel to 10 mm long; involucrate bracts linear to filiform, (1)2–3 mm long, glabrous or sparsely ciliate; sepals 3–5 mm long, the lobes orbiculate-deltate, glabrous or stellate-pubescent; petals obovate, 3–5 mm long, white to pale lilac, shallowly 2-notched or -lobed; staminal tube 1–2 mm long. Fruit 6–7 mm in diameter, the mericarps 2–3 mm long, the dorsal face strongly reticulate-wrinkled, the side strongly, radially ribbed, the margin sharp-edged, toothed, and narrowly winged, the surface glabrous or pubescent; seeds reniform-suborbicular, ca. 2 mm long, black or dark brown, notched, finely roughened or smooth, glabrous.

Disturbed sites. Occasional; northern and central peninsula west to central panhandle. Nearly cosmopolitan. Native to Europe, Africa, and Asia. Spring–summer.

EXCLUDED TAXA

Malva rotundifolia Linnaeus—Cited by Small (1903, 1913a, 1933) as "throughout North America," which would presumably include Florida. No Florida specimens known.

Malva sylvestris Linnaeus—Cited by Small (1903) as "In waste places, in British America, the United States and Mexico," which would presumably include Florida, and as occurring in Florida by Radford et al. (1968). No Florida specimens known.

Malvastrum A. Gray, nom. cons. 1849. FALSE MALLOW

Herbs and subshrubs. Leaves alternate, simple, pinnipalmate-veined, petiolate, stipulate. Flowers solitary, axillary or in terminal spikes or heads, pedicellate; involucrate bracts 3, adnate basally to the calyx; sepals 5, basally connate; petals 5, free; stamens 16–50, the filaments connate into a tube; ovary 5- to 18-carpellate and -loculate, the style branches 9–18. Fruit a schizocarp; seed 1 per mericarp.

A genus of about 15 species; North America, West Indies, Mexico, Central America, South America, Africa, Asia, Australia, and Pacific Islands. [Genus *Malva* and *astrum*, an incomplete resemblance to *Malva*.]

Selected references: Hill (1982, 2015b).

1. Stem with erect, short-armed, stellate trichomes ....**M. americanum**
1. Stem with appressed, long-armed, stellate trichomes.
  2. Leaves with simple and stellate trichomes on the lower surface; mature carpels with dorsal spines .... **M. coromandelianum**
  2. Leaves with only stellate trichomes on the lower surface; mature carpels without spines.................. .... **M. corchorifolium**

Malvastrum americanum (L.) Torr. [Of America.] INDIAN VALLEY FALSE MALLOW.

Malva americana Linnaeus, Sp. Pl. 687. 1953. *Malva hispida* Moench, Methodus 609. 1794, nom. illegit. *Malvastrum americanum* (Linnaeus) Torrey, in Emory, Rep. U.S. Mex. Bound. 2(1): 38. 1858 ("1859"). *Malveopsis americana* (Linnaeus) Kuntze, Revis. Gen. Pl. 1: 72. 1891. *Sphaeralcea americana* (Linnaeus) Metz, Catholica Univ. Amer., Biol. Ser. 16: 142. 1934.

Erect perennial herb or subshrub, to 2 m; stem with erect, short-armed, stellate trichomes. Leaves with the blade ovate to ovate-lanceolate, 5–12 cm long, 4–10 cm wide on the lower leaves, 2–4 cm long, 1.5–3 cm wide on the upper leaves, often shallowly 3-lobed in the upper ½ or unlobed, the margin dentate to denticulate, the upper and lower surfaces stellate-pubescent, the petiole 3.5–8 cm long on the lower leaves, 1–1.5 cm long on the upper leaves; stipules lanceolate, subfalcate, 3–5 mm long. Flowers solitary in the axil of the upper 1–2 leaves, the remainder in a dense terminal spike 3–10 cm long, these terminating each branch, the flowers subtended by a bifid bract 2–5 mm long, the pedicel to 3 mm long; involucral bracts lanceolate, subfalcate, 5–7 mm long, the base adnate to the calyx for ca. 2 mm; sepals 5–6 mm long, broadly campanulate, united ¼–⅓ their length, the outside surface densely appressed-hirsute and with appressed stellate trichomes; corolla wide spreading, the petals obovate, 6–10 mm long, asymmetrically lobed, orange-yellow; staminal tube 2–3 mm long, stellate-puberulent; style branches (9)10–15(18). Fruit 4–6 mm in diameter, the mericarps 2–3 mm long, indehiscent to weakly dehiscent, but not separating, narrowly notched, with a proximal-apical mucro, minutely spreading hirsute on the top, the sides radially ribbed; seeds reniform, 1–2 mm long, glabrous.

Disturbed sites. Rare; Franklin and Lee Counties. Florida, Alabama, and Texas; West Indies,

Mexico, Central America, and South America; Africa, Asia, Australia, and Pacific Islands. All year.

Malvastrum corchorifolium (Desr.) Britton ex Small [With leaves like *Corchorus.*] FALSE MALLOW.

Malva scoparia Jacquin, Collectanea 1: 59. 1787; non L'Héritier de Brutelle, 1786. *Malva corchorifolia* Desrousseaux, in Lamarck, Encycl. 3: 755. 1792. *Malvastrum corchorifolium* (Desrousseaux) Britton ex Small, Fl. Miami 119, 200. 1913.

Malvastrum rugelii S. Watson, Proc. Amer. Acad. Arts 17: 367. 1882. TYPE: FLORIDA: Manatee Co.: mouth of the Manatee River, 1845, *Rugel 90* (holotype: GH; isotypes: BM, CU, G, UC, US).

Erect annual or perennial herb, to 1.5 m; stem with swollen-based, scattered, appressed stellate trichomes. Leaves with the blade ovate, (1.6)3–5(8) cm long, (0.8)2.5–4(7.5) cm wide, the apex acute to rounded, the base broadly cuneate to rounded, the margin dentate, the upper and lower surfaces sparsely stellate pubescent, with a few minute, marginal simple trichomes, the petiole 1.5–3.5(7) cm long; stipules lanceolate, subfalcate, 3–5 mm long. Flowers at first solitary, axillary, later in a congested or loose terminal spike 1–2 cm long, these in the axil of the upper leaves or terminating each branch, flowers subtended by a bifid bract or the leaf and stipules, the pedicel 1–2 mm long; involucral bracts lanceolate, subfalcate, 4–6 mm long, the base adnate to the calyx for about 1 mm; calyx broadly campanulate, the sepals 5–6 mm long, united ¼–⅓ their length, the outside surface moderately pubescent with rigid, simple and stellate trichomes; corolla campanulate to wide spreading, 12–17 mm in diameter, the petals obovate, 6–7 mm long, asymmetrically lobed, yellow to pale yellow-orange; staminal tube 2–3 mm long sparsely stellate-puberulent; style branches (9)11–13(16). Fruit 4–7 mm in diameter, the mericarps indehiscent to weakly dehiscent but separating, narrowly notched, with 3 minute, apical cusps, 1 at the proximal-apical surface, 2 at the distal-apical surface, moderately pubescent on the upper ⅓ with minute simple or stellate, erect trichomes, the sides ribbed; seeds reniform, ca. 2 mm long, glabrous.

Disturbed coastal hammocks. Occasional; central and southern peninsula, Escambia County. Alabama and Florida; West Indies, Mexico, and Central America; Africa. Native to North America, West Indies, Mexico, and Central America. All year.

Malvastrum coromandelianum (L.) Garcke [Of the Coromandel coast, India.] THREELOBE FALSE MALLOW.

Malva coromandeliana Linnaeus, Sp. Pl. 687. 1753. *Malva tricuspidata* R. Brown, in W. T. Aiton, Hortus Kew. 4: 210. 1812, nom. illegit. *Malvastrum tricuspidatum* A. Gray, Smithsonian Contr. Knowl. 3(5): 16. 1852, nom. illegit. *Malvastrum coromandelianum* (Linnaeus) Garcke, Bonplandia 5: 295. 1857. *Malveopsis coromandeliana* (Linnaeus) Morong, Ann. New York Acad. Sci. 7: 55. 1892. *Malveopsis coromandeliana* (Linnaeus) Morong var. *normalis* Kuntze, Revis. Gen. Pl. 3(2): 20. 1898, nom. inadmiss. *Malvastrum tricuspidatum* A. Gray var. *normale* Stuckert, Anales Soc. Ci. Argent. 114: 28. 1932, nom. inadmiss.

Erect or decumbent annual or perennial herb, to 6(10) dm; stem with scattered, appressed stellate trichomes. Leaves with the blade ovate to sublanceolate, (1.7)3–4(6.5) cm long,

(0.6)1.5–3(5.5) cm wide, the apex acute, the base truncate to rounded or cuneate, the margin dentate to serrate, the upper and lower surfaces sparsely pubescent with stellate or simple trichomes, the petiole 1–2(4) cm long; stipules lanceolate, falcate, 3–6 mm long. the margin ciliate. Flowers solitary, axillary, or sometimes congested toward the branch tips, the pedicel 1–2 mm long; involucrate bracts lanceolate, subfalcate, 4–6 mm long, the base adnate to the calyx for ca. 1 mm; calyx broadly campanulate, the sepals 5–7 mm long, united ½ their length, the outer surface sparsely appressed stellate-pubescent, the lobe margin ciliate; corolla wide-spreading, 9–21 mm in diameter, the petals obovate, 8–10 mm long, yellow to orange-yellow, asymmetrically 2-lobed; staminal tube 2–3 mm long, glabrous; style branches (9)10–14(15). Fruit 6–7 mm in diameter, the mericarps ca. 4 mm long, indehiscent, narrowly notched, with 3 conspicuous, apical cusps ca. 1 mm long, 1 at the medial-apical surface, 2 at the distal-apical surface, these moderately pubescent with stellate or simple trichomes on top, the sides glabrous, radially ribbed; seed reniform, ca. 2 mm long, glabrous.

Disturbed coastal hammocks. Occasional; Volusia, Brevard, and Miami-Dade Counties, central panhandle, Escambia County. Florida, Louisiana, and Texas; West Indies, Mexico, Central America, and South America; Africa, Asia, Australia, and Pacific Islands. Native to tropical America. All year.

Malvaviscus Fabr. 1759. WAXMALLOW

Shrubs. Leaves alternate, simple, pinnipalmate-veined, petiolate, stipulate. Flowers solitary in the leaf axils or several in apical cymes; involucral bracts 8–9; sepals 5, basally connate; petals 5, free; stamens numerous, the filaments connate into a column; ovary 5-carpellate and -loculate, the style branches 10. Fruit a schizocarp, the mericarps 5; seeds 1 per mericarp.

A genus of 5 species; North America, West Indies, Mexico, Central America, and South America. [*Malva*, mallow, and *viscum*, sticky, in reference to the sticky sap].

Selected references: Mendenhall and Fryxell (2015); Turner and Mendenhall (1993).

1. Leaves broadly ovate to cordate, 3- to 5-lobed, conspicuously stellate-pubescent on the lower surface. ....**M. arboreus**
1. Leaves ovate-lanceolate to lanceolate, unlobed or shallowly lobed, glabrate on the lower surface ....**M. penduliflorus**

Malvaviscus arboreus Cav. var. **drummondii** (Torr. & A. Gray) Schery [Treelike; commemorates Thomas Drummond (1793–1835), Scottish-born botanist who collected in the southwestern United States] TEXAS WAXMALLOW

> *Malvaviscus drummondii* Torrey & A. Gray, Fl. N. Amer. 1: 230. 1838. *Pavonia drummondii* (Torrey & A. Gray) Torrey & A. Gray, Fl. N. Amer. 1: 682. 1840. *Hibiscus drummondii* (Torrey & A. Gray) M. J. Young, Familiar Lessons Bot. 186. 1873. *Malvaviscus arboreus* Cavanilles var. *drummondii* (Torrey & A. Gray) Schery, Ann. Missouri Bot. Gard. 29: 215. 1942.

Shrub, to 3 m; stem stellate-pubescent to glabrate. Leaves with the blade broadly ovate to cordate, 4–9 cm long, 4–12 cm wide, 3- to 5-lobed, the apex obtuse to acute, the base cordate, the margin crenate-dentate, the upper and lower surfaces stellate-pubescent, the petiole 1–6(10) cm long; stipules subulate. Flowers solitary or several in the axil of the leaves or sometimes in an apical cyme, ascending; involucrate bracts 5–8(11), linear to linear-spatulate; sepals 8–15 mm long, connate for ½–⅔ their length, pubescent with simple or stellate trichomes; petals 2–3.5 cm long, red, rarely white; staminal column exserted. Fruit obloid, 8–14 mm in diameter, red, smooth, glabrous, the mericarps each a broadly ellipsoidal wedge; seeds reniform, glabrous.

Disturbed sites. Occasional; peninsula west to central panhandle. Escaped from cultivation. South Carolina south to Florida, west to Texas; Mexico. Native to Texas and Mexico. All year.

Malvaviscus penduliflorus DC. [With pendulent flowers.] MAZAPAN; TURKSCAP MALLOW.

> *Malvaviscus penduliflorus* de Candolle, Prodr. 1: 445. 1824. *Malvaviscus arboreus* Cavanilles var. *penduliflorus* (de Candolle) Schery, Ann. Missouri Bot. Gard. 29: 233. 1942. *Malvaviscus arboreus* subsp. *penduliflorus* (de Candolle) Hadac, Folia Geobot. Phytotax. 5: 432. 1970.

Shrub, to 3 m; stem sparsely stellate-pubescent to glabrate. Leaves with the blade narrowly ovate to lanceolate, 4–9(20) cm long, 4–8(10) cm wide, unlobed or rarely slightly 3-lobed, the apex acute, the base rounded to cuneate, the margin crenate-dentate, the upper and lower surfaces sparsely stellate-pubescent to glabrate, the petiole 1–6 cm long; stipules subulate. Flowers solitary or several in the axil of the leaves or sometimes in an apical cyme, ascending; involucrate bracts 5–8(11), linear to linear-spatulate; sepals 1.5–2 cm long, connate for ⅔ their length, sparsely pubescent with simple or stellate trichomes; petals 4–7 cm long, red; staminal column usually included. Fruit obloid, 8–14 mm in diameter, red, smooth, glabrous, the mericarps each a broadly ellipsoidal wedge; seeds reniform, glabrous.

Disturbed sites. Frequent; central and southern peninsula, Franklin County. Escaped from cultivation. Florida and Texas; West Indies, Mexico, Central America, and South America. Native to tropical America. Spring–fall.

EXCLUDED TAXON

> *Malvaviscus arboreus* Cavanilles—Reported for Florida by Correll and Correll (1982), the name misapplied to material of *M. penduliflorus*. The specimen cited by Turner and Mendenhall (1993) is of cultivated material.

Melochia L., nom. cons. 1753.

Herbs or shrubs. Leaves alternate, simple, pinnipalmate-veined, petiolate, stipulate. Flowers in axillary or terminal umbellate, spiciform, or cymose glomerules; bracteolate; sepals 5, basally connate; petals 5, free; stamens 5, basally connate; ovary 5-carpellate and -loculate, the styles 5. Fruit a capsule, dehiscence loculicidal and septicidal; seeds 1–2 per locule.

A genus of about 60 species; North America, West Indies, Central America, South America, Africa, Asia, Australia, and Pacific Islands. [*Melóchich*, an Arabic name for *Corchorus olitorius*, widely cultivated as a potherb in the eastern Mediterranean since ancient times.]

Melochia is sometimes placed in the Sterculiaceae.

Moluchia Medik. (1789); *Riedlea* Vent. (1807).

Selected references: Goldberg (1967, 2015).

1. Fruit pyramidal, 5-winged; cymes loose, not subtended by an involucel of bracts.
   2. Inflorescences mostly leaf-opposed; capsule with sharply acute angles, sparingly stellate-pubescent; stem and leaves sparsely pubescent with simple, forked, and stellate trichomes ......**M. pyramidata**
   2. Inflorescences mostly axillary; capsule with rounded angles or lobes, densely stellate-pubescent; stem and leaves densely stellate-pubescent ....**M. tomentosa**
1. Fruit subglobular or ovoid, not winged; cymes compact, subtended by an involucel of bracts.
   3. Cymes terminal on the primary and lateral branches; petiole of the mature leaves more than 1.5 cm long .... **M. corchorifolia**
   3. Cymes in the upper leaf axils; petiole of the mature leaves less than 1 cm long.
      4. Bracts and calyx lobes linear-lanceolate, longer than the mature fruit, glandular-stipitate
      .... **M. spicata**
      4. Bracts and calyx lobes lanceolate to deltate, shorter than the mature fruit, eglandular ....
      ....**M. nodiflora**

Melochia corchorifolia L. [With leaves like *Corchorus.*] CHOCOLATEWEED.

Melochia corchorifolia Linnaeus, Sp. Pl. 675. 1753. *Riedlea corchorifolia* (Linnaeus) de Candolle, Prodr. 1: 491. 1824.

Annual or perennial herbs or shrubs, to 2 m; stem pubescent to subglabrous. Leaves with the blade ovate to lanceolate, 1.3–9 cm long, 1–7 cm wide, the apex acute, the base rounded, the margin serrate, the upper and lower surfaces subglabrous, with a few simple trichomes on the veins; stipules subulate. Flowers 10–40 in a terminal, cymose glomerule, sessile or subsessile, sometimes with a few at the apex of a short axillary branchlet, each flower with 3–4 subtending bracteoles; sepals 2–3 mm long, the lobes distant and the sinuses between them rounded; petals 3–6 mm long, purple, pink, or white with a yellow base; stamens 2–3 mm long; ovary 2–3 mm long. Fruit subglobose, 4–5 mm long, dehiscence loculicidal and then septicidal with the fruit fragmenting; seed 1 per locule, obovoid, with 2 flat sides and 1 rounded side.

Disturbed sites. Frequent; peninsula, central and western panhandle. North Carolina south to Florida, west to Arkansas and Texas; West Indies, Central America and South America; Africa, Asia, Australia, and Pacific Islands. Native to Africa, Asia, Australia, and Pacific Islands. Spring–fall.

Melochia nodiflora Sw. [With flowers at the axillary nodes.] BRETONICA PRIETA.

Melochia nodiflora Swartz, Prodr. 97. 1788. *Mougeotia nodiflora* (Swartz) Kunth, in Humboldt et al., Nov. Gen. Sp. 5: 330. 1823. *Riedlea nodiflora* (Swartz) de Candolle, Prodr. 1: 491. 1824. *Visenia nodiflora* (Swartz) Sprengel, Syst. Veg. 3: 30. 1826.

Perennial herb, subshrub, or shrub, to 3 m; branchlets with simple and stellate trichomes, soon glabrate. Leaves with the blade ovate to lanceolate-ovate, (2.5)3–6.7(8.6) cm long, 1.3–4.2(5.5) cm wide, the apex acuminate, the base rounded to truncate or subcordate, the margin serrate, the upper and lower surfaces with sparse, adpressed, simple trichomes, the petiole 0.4–2.7 cm long, pilose. Flowers in an axillary, sessile or subsessile, dense bracteate head, the peduncle to

3 mm long; sepals 3–5 mm long, the lobes lanceolate to deltate, the outer surface stellate-puberulent and pilose; petals oblong-obovate, 4–5 mm long, pink, occasionally white with pink venation, rarely yellow, glabrous; staminal tube 2–3 mm long, the anthers sessile; styles free. Fruit obloid to subglobose, septicidal, schizocarpic capsule, 5-lobed, 2–4 mm long, stellate-puberulent and pilose; seeds ca. 2 mm long.

Dry, disturbed sites. Rare; Miami-Dade County. Florida; West Indies, Mexico, Central America, and South America. Native to tropical America. All year.

Melochia pyramidata L. [Pyramid-shaped, in reference to the inflorescence.] PYRAMIDFLOWER.

Melochia pyramidata Linnaeus, Sp. Pl. 674. 1753. *Moluchia herbacea* Medikus, Malvenfam. 10. 1787, nom. illegit. *Melochia pyramidata* Linnaeus var. *normalis* Kuntze, Revis. Gen. Pl. 3(2): 25. 1898, nom. inadmiss. *Moluchia pyramidata* (Linnaeus) Britton, Brooklyn Bot. Gard. Mem. 1: 69. 1918.

Annual or perennial herb or shrub, to 2 m; stem glabrous or subglabrous, the trichomes mostly simple. Leaves with the blade lanceolate to ovate, 1–6 cm long, 0.5–3 cm wide, the apex acute, the base rounded, the margin serrate, the upper and lower surfaces glabrous or subglabrous, the trichomes mostly simple. Flowers in an axillary umbellate cyme, the peduncle 1–12 mm long, the bracteoles at the base of the pedicel, the pedicel 1–5 mm long; sepals 3–5 mm long, the lobes linear-lanceolate; petals obovate, 6–8 mm long, purple, pink, or violet, sometimes yellow basally; stamens 4–5 mm long; pistil 3–6 mm long. Fruit subglobose, 4–5 mm long, dehiscence loculicidal and then septicidal, the fruit fragmenting; seeds 1 per locule, obovoid, with 2 flat sides and 1 rounded side.

Disturbed sites. Occasional; southern peninsula. Florida, Louisiana, and Texas; West Indies, Mexico, Central America, and South America. Spring–fall.

Melochia spicata (L.) Fryxell [Flowers in spikes.] BRETONICA PELUDA.

Malva spicata Linnaeus, Syst. Nat., ed. 10. 1146. 1759. *Malvastrum spicatum* (Linnaeus) A. Gray, Mem. Amer. Acad. Arts, ser. 2. 4: 22. 1849. *Malveopsis spicata* (Linnaeus) Kuntze, Revis. Gen. Pl. 1: 72. 1891. *Malveopsis spicata* (Linnaeus) Kuntze var. *normalis* Kuntze, Revis. Gen. Pl. 1: 72. 1891, nom. inadmiss. *Melochia spicata* (Linnaeus) Fryxell, Monogr. Syst. Bot. 25: 457. 1988.

Sida villosa Miller, Gard. Dict., ed. 8. 1768. *Melochia villosa* (Miller) Fawcett & Rendle, Fl. Jamaica 5: 165. 1926.

Melochia hirsuta Cavanilles. Diss. 6: 323, pl. 175(1). 1788. *Mougeotia hirsuta* (Cavanilles) Kunth, in Humboldt et al., Nov. Gen. Sp. 5: 331. 1823. *Riedlea hirsuta* (Cavanilles) de Candolle, Prodr. 1: 492. 1824. *Visenia hirsuta* (Cavanilles) Sprengel, Syst. Veg. 3: 30. 1826.

Riedlea serrata Ventenat, Choix Pl. pl. 37. 1807. *Visenia serrata* (Ventenat) Sprengel, Syst. Veg. 3: 30. 1826. *Melochia serrata* (Ventenat) A. Saint-Hilaire & Naudin, Ann. Sci. Nat., Bot., ser. 2. 18: 36. 1842.

Riedlea serrata Ventenat var. *glabrescens* C. Presl, Reliq. Haenk. 2: 147. 1845. *Melochia hirsuta* Cavanilles var. *glabrescens* (C. Presl) A. Gray, Syn. Fl. N. Amer. 1(1): 340. 1897. *Riedlea glabrescens* (C. Presl) Small, Fl. S.E. U.S. 780, 1335. 1903.

Perennial herb or shrub, to 2 m; stem sericeous to subglabrate. Leaves lanceolate to ovate, 1–5 cm long, 0.5–2 cm wide, the apex acute, the base rounded to truncate, the margin serrate, the upper and lower surfaces sericeous to subglabrate with simple trichomes, the petiole 0.3–5.5 cm long; stipules linear-lanceolate. Flowers in a terminal, interrupted spiciform, 1–3(12)-flowered cyme, sometimes also an axillary cluster, sessile or subsessile, the bracteoles 2–3 subtending each flower; sepals 4–5 mm long, the lobes linear-lanceolate; petals obovate, 8–11 mm long, pink, purple, or violet; stamens 4–8 mm long; pistil 4–8 mm long. Fruit subglobose, 2–4 mm long, obtusely 5-angled, dehiscence loculicidal, tardily septicidal, the fruit fragmenting; seeds 1(2) per locule, obovoid, with 2 flat sides and 1 rounded side.

Sandhills and disturbed sites. Occasional; peninsula. Georgia, Florida, and Louisiana; West Indies, Mexico, Central America, and South America. All year.

Melochia tomentosa L. [Thickly and evenly covered with matted trichomes.] WOOLLY PYRAMIDFLOWER; TEABUSH; BROOMWOOD.

Melochia tomentosa Linnaeus, Syst. Nat., ed. 10. 1140. 1759. *Melochia tomentosa* Linnaeus var. *typica* K. Schumann, in Martius, Fl. Bras. 12(3): 1886, nom. inadmiss. *Moluchia tomentosa* (Linnaeus) Britton, Brooklyn Bot. Gard. Mem. 1: 69. 1918.

Perennial herb, subshrub, or shrub, to 2.5 m; stems stellate-tomentose. Leaves with the blade ovate to lanceolate, 1.5–6 cm long, 0.8–4 cm wide, the apex acute, the base cordate, the margin serrate, the upper and lower surfaces stellate-tomentose, the petiole 0.3–2.5 cm long; stipules lanceolate. Flowers 3–5 in an axillary umbellate cyme, the peduncle 6–8 mm long, the bracteoles at the base of the pedicel, the pedicel 1–5 mm long; sepals 5–7 mm long, the lobes lanceolate; petals (6)8–11 mm long, purple, pink, or violet; stamens 4–10 mm long; pistil 6–10 mm long. Fruit pyramidal, 6–9 mm long, 5-winged, dehiscence loculicidal; seeds 1–2 per locule, obovoid, with 2 flat sides and 1 rounded side.

Pine rocklands and disturbed sites. Rare; Hillsborough, St. Lucie, and Miami-Dade Counties. Florida and Texas; West Indies, Mexico, Central America, and South America. All year.

Modiola Moench 1794. BRISTLEMALLOW

Perennial herbs. Leaves alternate, simple, palmate-nerved, petiolate, stipulate. Flowers solitary, axillary; involucrate bracts 3, free; sepals 5, basally connate; petals 5, free; stamens numerous, the filaments connate into a column; ovary 16- to 22-carpellate, the number of styles and stigmas equaling the carpel number. Fruit a schizocarp; seeds 1 per carpel.

A monotypic genus; nearly cosmopolitan. [*Modiolus*, hub of a wheel, in reference to the fruit shape.]

Selected reference: Hill (2015c).

Modiola caroliniana (L.) G. Don [Of Carolina.] CAROLINA BRISTLEMALLOW.

Malva caroliniana Linnaeus, Sp. Pl. 688. 1753. *Modiola multifida* Moench, Methodus 620. 1794, nom. illegit. *Modiola caroliniana* (Linnaeus) G. Don, Gen. Hist. 1: 466. 1831. *Modanthos caroliniana* (Linnaeus) Alefeld, Oesterr. Bot. Z. 13: 12. 1863. *Abutilodes caroliniana* (Linnaeus) Kuntze, Revis. Gen. Pl. 1: 65. 1891.

Perennial herb; stem to 5 dm long, trailing, rooting at the nodes, hirsute. Leaves with the blade suborbicular, 1.5–4 cm long and wide, usually palmately 5- to 7-lobed, the apex acute, the base cordate, the margin irregularly serrate, the upper and lower surfaces sparsely hirsute, the petiole 1–2 times the blade length; stipules ovate, 3–4 mm long. Flowers solitary, axillary, the pedicel usually shorter than the subtending petioles, hirsute; involucral bracts lanceolate, 4–5 mm long; sepals 5–7 mm long, the lobes ovate-lanceolate, hirsute; petals 6–8 mm long, dark orange or salmon-orange, often with a darker center, erect; staminal column yellowish, the anthers crowded at the apex; styles and stigmas 16–22. Fruit with 16–22 mericarps, these 5–6 mm long, with 2 apical spines 2–3 mm long, each mericarp divided into 2 cells by a partial septum, each cell 1-seeded, the lower cell strongly rugose, the upper cell nearly smooth, hirsute; seeds ca. 2 mm long, sparsely puberulent.

Disturbed sites. Occasional; northern counties, central peninsula. Massachusetts south to Florida, west to Oregon and California; West Indies, Mexico, Central America, and South America; Europe, Africa, Asia, Australia, and Pacific Islands. Native to South America. Spring–fall.

Pavonia Cav., nom. cons. 1787. SWAMPMALLOW

Shrubs or subshrubs. Leaves alternate, simple, palmate-veined, petiolate, stipulate. Flowers solitary and axillary or in terminal racemes; involucral bracts 5–7, free or basally connate; sepals 5, basally connate; petals 5; stamens connate into a column surmounted by 5 apical teeth; ovary 5-carpellate and -loculate, the styles 10-branched, the stigmas capitate. Fruit a schizocarp, the mericarps 5; seed 1 per mericarp.

A genus of about 250 species; North America, West Indies, Mexico, Central America, South America, Africa, Asia, Australia, and Pacific Islands. [Commemorates José Antonio Pavón (1754–1844), Spanish physician and botanist].

Malache Vogel (1772).

Selected references: Fryxell (1999); Fryxell and Hill (2015f).

1. Carpels with 3 apical, retrorsely barbed awns to 1 cm long...**P. spinifex**
1. Carpels lacking awns.
    2. Petals deep rose; involucrate bracts 5, 4–6 mm long..**P. hastata**
    2. Petals greenish white or greenish yellow; involucrate bracts 6–10, ca. 10 mm long......**P. paludicola**

Pavonia hastata Cav. [With equal more or less triangular basal lobes directed outward, in reference to the leaves.] SPEARLEAF SWAMPMALLOW.

Pavonia hastata Cavanilles, Diss. 3: 138, t. 47(2). 1787. *Malache hastata* (Cavanilles) Kuntze, Revis. Gen. Pl. 1: 70. 1891. *Pavonia hastata* Cavanilles forma *longifolia* Gürke, in Martius, Fl. Bras. 12(3): 500. 1892, nom. inadmiss. *Pavonia hastata* Cavanilles var. *pubescens* Gürke, in Martius, Fl. Bras. 12(3): 500. 1892, nom. inadmiss. *Lass hastata* (Cavanilles) Kuntze, Revis. Gen. Pl. 3(2): 19. 1898. *Pavonia hastata* Cavanilles var. *genuina* Hassler, Bull. Herb. Boissier, ser. 2. 7: 737. 1907, nom. inadmiss.

Erect subshrub, to 5 dm; stem stellate-pubescent. Leaves with the blade ovate-triangular to hastate-oblong, to 2 cm long, the apex acute or subobtuse, the base cordate, the margin coarsely crenate, the upper and lower surfaces minutely stellate-pubescent, the lower more densely so, the petiole ⅓ or less the length of the blade; stipules subulate, ca. 2 mm long. Flowers solitary, axillary; involucral bracts 5, alternate with the sepal lobes, ovate, 4–6 mm long; sepals 6–8 mm long; petals 15–22 mm long, lavender. Fruit ca. 6 mm in diameter, the mericarps dorsally keeled, ca. 4 mm long, reticulate-costate, minutely pubescent.

Disturbed sites. Rare; Levy and Citrus Counties. Escaped from cultivation. Georgia, Florida, and Texas; Mexico and South America; Australia. Native to South America. Spring–fall.

Pavonia paludicola Nicolson ex Fryxell [*Palud*, marshes, and *cola*, dweller.] SWAMPBUSH.

Malache scabra Vogel, in Trew, Pl. Select. 9: 50, t. 90. 1772. *Pavonia spicata* Cavanilles, Diss. 3: 136. 1787, nom. illegit. *Althaea racemosa* Swartz, Prodr. 102. 1788, nom. illegit. *Pavonia racemosa* Swartz, Fl. Ind. Occid. 2: 1215. 1800, nom. illegit. *Malache spicata* Kuntze, Revis. Gen. Pl. 1: 70. 1891, nom. illegit. *Pavonia scabra* (Vogel) Ciferri, Atti Ist. Bot. "Giovanni Briosi" 8: 321. 1936; non C. Presl, 1835; nec Bentham ex Garcke, 1881. *Pavonia paludicola* Nicolson ex Fryxell, in R. A. Howard, Fl. Less. Antill., Dicot. 2: 241. 1989.

Semiscandent shrub, to 4 m; stem minutely stellate-pubescent. Leaves with the blade broadly ovate, the apex acuminate, the base subcordate, the margin obscurely dentate to subentire, the upper and lower surfaces sparsely stellate-pubescent, the petiole 4–5 cm long. Flowers in a terminal, leafless raceme, the pedicel 1–3 cm long; involucral bracts lanceolate, 9–10 mm long; sepal lobes ovate, 8–11 mm long, minutely stellate-pubescent; petals 12–18 mm long, yellow or yellow-green. Fruit 10–13 mm in diameter, woody, glabrous, usually with a crown of 5 points, the mericarps narrowed basally, usually 3-pointed apically.

Hammocks. Rare; southern peninsula. Florida; West Indies, Central America, and South America. All year.

Pavonia paludicola is listed as endangered in Florida (Florida Administrative Code, Chapter 5B-40).

Pavonia spinifex (L.) Cav. [Thorn-bearing.] GINGERBUSH.

Hibiscus spinifex Linnaeus, Syst. Nat., ed. 10. 1149. 1759. *Pavonia spinifex* (Linnaeus) Cavanilles, Diss. 3: 133. 1787. *Pavonia spinifex* (Linnaeus) Cavanilles var. *ovalifolia* de Candolle, Prodr. 1: 443. 1824, nom. inadmiss. *Typhalea spinifex* (Linnaeus) C. Presl, Bot. Bemerk. 19. 1844. *Malache spinifex* (Linneus) Kuntze, Revis. Gen. Pl. 1: 70. 1891. *Pavonia spinifex* (Linnaeus) Cavanilles subsp. *genuina* Gürke, in Martius, Fl. Bras. 12(3): 481. 1892, nom. inadmiss. *Lass spinifex* (Linnaeus) Kuntze, Revis. Gen. Pl. 3(2): 20. 1898.

Erect subshrub, to 2 m; stem minutely pubescent to glabrate, the trichomes often in well-defined rows. Leaves with the blade ovate, the apex acute, the base truncate to subcordate, the margin irregularly dentate, the upper and lower surfaces glabrate, the petiole 2–3 cm long. Flowers solitary, axillary, the pedicel 1–3 cm long; involucral bracts 5–7, lanceolate to spatulate; sepals 8–11 mm long, the lobes ciliate; petals 2–2.5 cm long, yellow. Fruit 8–10 mm in diameter, the mericarps 3-spined, the spines 6–7 mm long, retrorsely barbed, the central spine erect, the lateral spines divergent.

Hammocks. Occasional; northern and central peninsula. Florida; West Indies and South America. Spring–fall.

Sida L. 1753. FANPETALS

Perennial herbs or subshrubs. Leaves alternate, simple, pinnipalmate-veined, petiolate, stipulate. Flowers axillary, solitary, in glomerules, or open terminal inflorescences, actinomorphic, bisexual; sepals 5, basally connate; petals 5, basally adnate to the staminal tube; ovary 5- to 14-carpellate and -loculate, the styles 5–14, the stigmas capitate. Fruit a schizocarp, the mericarps 5–14; seed 1 per carpel.

A genus of about 150 species; nearly cosmopolitan. [*Side*, name used by Theophastos (ca. 373–287 BC) for *Nymphaea alba* (Nymphaeaceae).]

There is some disagreement on the interpretation of several *Sida* species, especially the weedy ones, and our treatment is considered tentative.

Selected references: Fryxell (1985); Fryxell and Hill (2015g); Krapovickas (2003).

1. Flowers congested into a head at the end of the branches; peduncle adnate to the petiole of a leaflike bract .... **S. ciliaris**
1. Flowers solitary or variously clustered; peduncle not adnate to the petiole of a leaflike bract.
  2. Leaf bases (at least the larger leaves) cordate or subcordate.
    3. Stem with a small spinelike tubercle just below the petiole .... **S. spinosa**
    3. Stem lacking a small spinelike tubercle.
      4. Flowers and fruits crowded into a terminal paniculate inflorescence.
        5. Leaves thinly stellate-pubescent; petals red-blotched at the base; carpels 5 .... **S. urens**
        5. Leaves densely stellate-pubescent, velvety; petals uniformly yellow-orange; carpels 8–14 .... **S. cordifolia**
      4. Flowers and fruits axillary.
        6. Leaf blades 2–7 cm long, the apex acuminate; stem viscid-pilose, usually erect .... **S. glabra**
        6. Leaf blades to 1.5(2) cm long, the apex rounded to obtuse; stem not viscid-pilose, usually procumbent or spreading .... **S. abutifolia**
  2. Leaf bases cuneate or truncate.
    7. Carpels uniformly 5; stem usually with a small, spinelike tubercle just below the petiole .... **S. spinosa**
    7. Carpels 7–10; stem lacking a small, spinelike tubercle just below the petiole.
      8. Leaves linear to oblong .... **S. elliottii**
      8. Leaves ovate to lanceolate.
        9. Flowers in axillary, subterminal, or terminal clusters at the end of short axillary branches .... **S. planicaulis**
        9. Flowers solitary or few in the leaf axils.
          10. Petals yellow with a red spot at their base .... **S. santaremensis**
          10. Petals solid yellow-orange.
            11. Leaves densely tomentose on the lower surface; flowers on slender peduncles to 4 cm long .... **S. rhombifolia**
            11. Leaves sparsely pubescent on the lower surface; flowers subsessile or with the pedicels less than 1 cm long .... **S. ulmifolia**

Sida abutifolia Mill. [With leaves like *Abutilon*.] SPREADING FANPETALS.

Sida abutifolia Miller, Gard. Dict., ed. 8. 1768.
Sida procumbens Swartz, Prodr. 101. 1788.
Sida supina L'Héritier de Brutelle, Stirp. Nov. 109bis, t. 52bis. 1789.
Sida diffusa Kunth, in Humboldt et al., Nov. Gen. Sp. 5: 257. 1822.
Sida filicaulis Torrey & A. Gray, Fl. N. Amer. 1: 232. 1838. *Sphaeroma filicaule* (Torrey & A. Gray) Kuntze, Revis. Gen. Pl. 1: 74. 1891.

Procumbent perennial herb; stem stellate-pubescent and usually also with simple trichomes. Leaves with the blade ovate to oblong, 0.5–1.5 cm long, ⅕–3 times as long as wide, the apex obtuse to acute, the base cordate, the margin crenate, the upper and lower surfaces pubescent, the petiole ½ to equaling or exceeding the blade length; stipules inconspicuous. Flowers solitary, axillary, the pedicel 1–2.5 cm long; calyx angulate, 4–5(7) mm long, hirsute, the lobes ovate; petals 5–6(10) mm long, white; staminal column minutely pubescent; styles 5. Fruit conic, pubescent, the mericarps 5, apically 2-spined, these antrorsely pubescent.

Dry, open, often disturbed sites. Rare; Miami-Dade County, Monroe County keys. Florida, Oklahoma and Texas west to Arizona; West Indies, Mexico, Central America, and South America. All year.

Sida ciliaris L. [With cilia, in reference to the stipules.] BRACTED FANPETALS; FRINGED FANPETALS.

Sida ciliaris Linnaeus, Syst. Nat., ed. 10. 1145. 1759. *Sida ciliaris* Linnaeus var. *typica* K. Schumann, in Martius, Fl. Bras. 12(3): 284. 1891, nom. inadmiss. *Pseudomalachra ciliaris* (Linnaeus) Monteiro, Portugaliae Acta Biol., Sér. B, Sist. 12: 133. 1973.

Procumbent perennial herb; stem appressed stellate-pubescent. Leaves with the blade elliptic, 1–2 cm long, usually 2–3 times as long as wide, the apex acute to obtuse, the base cuneate, the margin few-toothed apically, the upper surface glabrous, the lower surface stellate-pubescent, the petiole ¼–½ the blade length; stipules linear, 4–12 mm long, partly adnate to the petiole. Flowers crowded at the branch apex, subsessile; calyx 4–6 mm long, hirsute; petals 5–11 mm long, rose or yellowish; staminal column pubescent; styles 5–8. Fruit conic, the mericarps 5–8.

Dry, open sites. Rare; Pinellas and Hillsborough Counties, southern peninsula. Florida and Texas; West Indies, Mexico, Central America, and South America; Pacific Islands. Native to North America and tropical America. All year.

Sida cordifolia L. [With heart-shaped leaves.] LLIMA.

Sida cordifolia Linnaeus, Sp. Pl. 684. 1753.

Perennial subshrub or shrub, to 1.5 m; stem stellate-tomentose. Leaves with the blade broadly cordate to lance-ovate, to 6 cm long, 1–2 times as long as wide, the apex acute, the base cordate, the margin dentate, the upper and lower surfaces stellate-tomentose, the petiole ½ or less the blade length; stipules linear, 5–8 mm long. Flowers axillary, but usually terminally aggregated into a paniculate or corymbiform inflorescence; calyx 5–7 mm long, stellate-tomentose; petals 8–11 mm long, yellow-orange; staminal column pubescent; styles 8–14. Fruit 6–7 mm in diameter, apically stellate-tomentose, the mericarps 8–14, apically 2-spined, these retrorsely barbed.

Disturbed sandhills and hammocks. Frequent; peninsula, west to eastern panhandle. Florida, Alabama, and Texas; West Indies, Mexico, Central America, and South America; Africa, Asia, Australia, and Pacific Islands. Native to South America, Africa, and Asia. Spring–fall.

Sida elliottii Torr. & A. Gray [Commemorates Stephen Elliott (1771–1830), South Carolina legislator, banker, and botanist.] ELLIOTT'S FANPETALS.

Sida gracilis Elliott, Sketch Bot. S. Carolina 2: 159. 1822; non Richard, 1792; nec Salisbury, 1796; nec R. Brown, 1814. *Sida elliottii* Torrey & A. Gray, Fl. N. Amer. 1: 231. 1838.

Sida rubomarginata Nash, Bull. Torrey Bot. Club 23: 102. 1896. TYPE: FLORIDA: Hillsborough Co.: Ballast Point, Tampa, 24 Aug 1895, *Nash 2472* (holotye: NY; isotypes: F, GH, K, MASS, MO, OS, US).

Sida carpinifolia Linnaeus f. var. *parviflora* Chapman, Fl. South. U.S., ed. 3. 47. 1897. TYPE: FLORIDA: Collier Co.: Roberts Key in Caximbas [Caxambas] Bay, s.d. *Chapman s.n.* (holotype: ?; isotypes: NY, US).

Sida elliottii Torrey & A. Gray var. *parviflora* Chapman, Fl. South. U.S., ed. 3. 48. 1897. TYPE: FLORIDA: Monroe Co.: Key West, s.d., *without collector s.n.* (neotype: NY). Neotypified by Siedo (1999: 107).

Sida flabellata Gandoger, Bull. Soc. Bot. France 71: 631. 1924. TYPE: FLORIDA: Collier Co.: Marco [Island], Jul–Aug 1900, *Hitchcock 23* (holotype: LY?; isotype: US).

Sida floridana Siedo, Lundellia 4: 69, f. 1–3. 2001; non Gandoger, 1924). *Sida littoralis* Siedo, Phytoneuron 2014–75: 1. 2014. TYPE: FLORIDA: Lee Co.: Middle Captiva Island, 29 Apr 1967, *Brumbach 5819* (holotype: GH; isotype: US).

Perennial herb or subshrub, to 1 m; stem minutely puberulent to glabrate. Leaves with the blade narrowly elliptic to subrhombic, 2–8 cm long, 2.5–5 times as long as wide, the apex acute, the base truncate, the margin crenate or dentate, the upper surface glabrous, the lower surface minutely pubescent, the petiole 2–9 mm long; stipules subulate, 5–11 mm long. Flowers solitary or paired, axillary, sometimes somewhat congested distally, the pedicel 0.5–2.5(10) cm long; calyx 5–10 mm long, minutely pubescent and hirsute basally; petals 12–15 mm long, yellow-orange; staminal column pubescent; styles 8–12. Fruit 6–8 mm in diameter, the mericarps 8–12, apically 2-spined, laterally reticulate.

Sandhills. Occasional; peninsula, central panhandle. Virginia south to Florida, west to Texas; Mexico and Central America. Summer.

Sida glabra Mill. [Smooth.] SMOOTH FANPETALS.

Sida glabra Miller, Gard. Dict., ed. 8. 1768.
Sida glutinosa Cavanilles, Diss. 1: 16, t. 2(8). 1785.

Erect or reclining perennial subshrub, to 1 m; stem glandular-puberulent and with simple trichomes. Leaves with the blade cordate-ovate, 3–6 cm long, 1.5–3.5 cm wide, the apex acute, the base cordate, the margin dentate, the upper and lower surfaces stellate-pubescent, the petiole ½–3.4 times the blade length; stipules subulate, 1–2 mm long. Flowers solitary, axillary, the pedicel 1–1.5 cm long; calyx 4–5 mm long, with stellate and glandular trichomes; petals 5–6 m long, yellow; staminal column pubescent; styles 5. Fruit 4–5 mm in diameter, the mericarps 5, apically 2-spined, these antrorsely pubescent.

Disturbed sites. Rare; Miami-Dade County. Florida; West Indies, Mexico, Central America, and South America. Native to tropical America. Fall–spring.

Sida planicaulis Cav. [*Planities*, flat surface, and *caulis*, stem, in reference to the distichous leaves.]

Sida planicaulis Cavanilles, Diss. 1: 24, t. 3(11). 1785.
Sida carpinifolia Linnaeus f., Suppl. Pl. 307. 1782; non Miller, 1768. *Malvinda carpinifolia* Medikus, Malvenfam. 24. 1787. *Malvastrum carpinifolium* (Medikus) A. Gray, Mem. Amer. Acad. Arts, ser. 2. 4(1): 22. 1849. *Sida acuta* Burman f. var. *carpinifolia* (Medikus) K. Schumann, in Martius, Fl. Bras. 12(3): 326. 1891. *Sida carpinifolia* Linnaeus f. var. *genuina* Stehle, Boissiera 7: 347. 1943, nom. inadmiss. *Sida acuta* Burman f. subsp. *carpinifolia* (Medikus) Borssum Waalkes, Blumea 14: 188. 1966.

Ascending or erect perennial herb or subshrub, to 4 dm; stem hirsute and sparsely stellate-puberulent. Leaves with the blade broadly lanceolate-elliptic to elliptic, (2.5)6–9 cm long, (1)3–4 cm wide, the apex acute, the base cuneate to subrotund, the margin serrate, the upper surface hirsute and minutely stellate-puberulent, the lower surface minutely stellate-puberulent, frequently ciliate along the margin, the petiole 3–7 mm long, glabrate; stipules filiform, 4–8 mm long, straight or subfalcate. Flowers solitary, axillary, sometimes congested on short axillary branches, the pedicel 1–3 mm long; calyx 5–6 mm long; petals 8–10 mm long; staminal column minutely glandular-puberulent; styles ca. 8. Fruit 3–5 mm wide, the mericarps ca. 8, with 2 apical spines.

Disturbed oak hammocks. Occasional; central peninsula. Florida; South America; Africa and Pacific Islands. Native to South America. All year.

Sida rhombifolia L. [With diamond-shaped leaves.] CUBAN JUTE; INDIAN HEMP.

Sida rhombifolia Linnaeus, Sp. Pl. 684. 1753. *Napaea rhombifolia* (Linnaeus) Moench, Methodus 621. 1794. *Sida rhombifolia* Linnaeus var. *typica* K. Schumann, in Martius, Fl. Bras. 12(3): 339. 1891, nom. inadmiss. *Sida rhombifolia* Linnaeus var. *normalis* Kuntze, Revis. Gen. Pl. 3(2): 22. 1898, nom. inadmiss. *Malva rhombifolia* (Linnaeus) E.H.L. Krause, in Sturm, Deutschl. Fl., ed. 2. 6: 238. 1901.
Sida bilobulata Gandoger, Bull. Soc. Bot. France 71: 632. 1924. TYPE: FLORIDA: Lake Co.: Eustis, s.d., *Hitchcock s.n.* (holotype: LY?).
Sida floridana Gandoger, Bull. Soc. Bot. France 71: 633. 1924. TYPE: FLORIDA: Manatee Co.: Perico Island, s.d., *Tracy 6639* (holotype: LY?).
Sida laevefacta Gandoger, Bull. Soc. Bot. France 71: 631. 1924. TYPE: FLORIDA: Walton Co.: De Funiak [Springs], s.d., *Rolfs 175* (holotype: LY?).
Sida parietariifolia Gandoger, Bull. Soc. Bot. France 71: 631. 1924. TYPE: FLORIDA: Escambia Co.: Pensacola, 29 Oct 1903, *Tracy 8688* (holotype: LY?; isotype: US).
Sida recticaulis Gandoger, Bull. Soc. Bot. France 71: 631. 1924. TYPE: FLORIDA: Lee Co.: s.d., *Willis s.n.* (holotype: LY?).

Erect perennial subshrub, to 1 m; stem stellate-puberulent. Leaves with the blade elliptic to rhombic, 2.5–9 cm long, 2–4 times as long as wide, the apex acute to subobtuse, the base cuneate, the margin serrate distally, the upper surface stellate-puberulent to glabrate, the lower surface stellate-puberulent, the petiole to 7 mm long; stipules subulate, 5–6 mm long. Flowers solitary, axillary, the pedicel 1–3 cm long; calyx 5–6 mm long, puberulent; petals 7–9 mm long, yellow; staminal column pubescent; styles 10–14. Fruit 4–5 mm in diameter, the mericarps 10–14, glabrous, laterally reticulate, apically 2-spined to muricate.

Disturbed sites. Frequent; nearly throughout. Nearly cosmopolitan. Native to North America, tropical America, Africa, and Asia. Summer–fall.

Sida santaremensis Monteiro [Of Santarem, Brazil.] MOTH FANPETALS.

Sida santaremensis Monteiro, Monogr. Malv. Bras. 44, t. 8. 1936.

Erect perennial subshrub, to 1 m; stem sparsely stellate-pubescent. Leaves with the blade broadly elliptic to subrhombic, to 5.5 cm wide, to 3.5 cm wide, the apex acute, the base cuneate, the margin dentate, the upper and lower surfaces stellate-pubescent, the petiole 3–10 mm long; stipules linear, ca. 7 mm long. Flowers usually solitary, the pedicel to 2 cm long; calyx 6–7 mm long, stellate-pubescent; petals ca. 1 cm long, cream-colored with a dark spot at the base; staminal column cream-colored; styles ca. 11. Fruit 5–6 mm in diameter, the mericarps ca. 11, the apex short apiculate.

Disturbed sites. Occasional; Pasco, Hillsborough, Pinellas, and Collier Counties. Florida; South America. Native to South America. Spring–fall.

Sida spinosa L. [With spines, in reference to the nodal spinelike spur.] PRICKLY FANPETALS.

Sida spinosa Linnaeus, Sp. Pl. 683. 1753. *Malvinda spinosa* (Linnaeus) Moench, Methodus 619. 1794. *Malva spinosa* (Linnaeus) E.H.L. Krause, in Sturm, Deutschl. Fl., ed. 2. 6: 237. 1902. *Sida spinosa* Linnaeus var. *typica* Rodrigo, Revista Mus. La Plata, Secc. Bot. 6: 122. 1944, nom. inadmiss.

Erect perennial subshrub or shrub, to 1 m; stem stellate-pubescent, usually with a nodal spinelike spur. Leaves with the blade ovate, lanceolate, or narrowly oblong, 2–6 cm long, 2–5 times as long as wide, the apex acute, the base cuneate to subcordate, the margin serrate, the upper surface glabrate, the lower surface stellate-tomentulose, the petiole ¼–½ the blade length; stipules subulate, 3–6 mm long. Flowers solitary or few, axillary, the pedicel to 1 cm long; calyx angulate, 5–7 mm long, minutely tomentose; petals ca. 5 mm long, yellow or white; staminal column glabrous; styles 5. Fruit 4–5 mm in diameter, the mericarps 5, apically 2-spined, these antrorsely pubescent.

Disturbed sites. Occasional; peninsula, central panhandle. Maine south to Florida, west to Ontario and California; West Indies, Mexico, Central America, and South America; Europe, Africa, Asia, Australia, and Pacific Islands. Native to North America, tropical America, Africa, Asia, and Australia. Spring–fall.

Sida ulmifolia Mill. [With leaves like *Ulmus* (Ulmaceae).] COMMON WIREWEED; COMMON FANPETALS.

Sida ulmifolia Miller, Gard. Dict., ed. 8. 1768.

Sida stipulata Cavanilles, Diss. 1: 22, t. 3(10). 1785. *Sida acuta* Burman f. var. *stipulata* (Cavanilles) K. Schumann, in Martius, Fl. Bras. 12(3): 327. 1891.

Sida glabra Nuttall, J. Acad. Nat. Sci. Philadelphia 7: 90. 1834; non Miller, 1768. TYPE: FLORIDA: "East Florida," s.d., *Peale s.n.* (holotype: PH).

Sida carpinifolia Linnaeus f. var. *brevicuspidata* Grisebach, Fl. Brit. W.I. 73. 1859.

Sida antillensis Urban, Symb. Antill. 5: 418. 1908.

Sida hitchcockii Gandoger, Bull. Soc. Bot. France 71: 628. 1924. TYPE: FLORIDA: Lee Co.: Fort Myers, Jul–Aug 1900, *Hitchcock 24* (holotype: LY?; isotype: US).

Sida incerta Gandoger, Bull. Soc. Bot. France 71: 633. 1924. TYPE: FLORIDA: Duval Co.: Jacksonville, s.d., *Williamson s.n.* (holotype: LY?).

Sida laxifolia Gandoger, Bull. Soc. Bot. France 71: 629. 1924. TYPE: FLORIDA: Manatee Co.: Sneed's Rod [Sneed Island Road], Palmetto, 5 Aug 1900, *Tracy 6637* (holotype: LY?; isotype: US).

Sida persicarioides Gandoger, Bull. Soc. Bot. France 71: 631. 1924. TYPE: FLORIDA: Lake Co.: near Lake Dorr, Tavares, 29 Jun 1893, *Rolfs 68* (holotype: LY?; isotype: FLAS).

Sida vicina Gandoger, Bull. Soc. Bot. France 71: 629. 1924. TYPE: FLORIDA: Duval Co.: Mayport, s.d., *Keeler s.n.* (holotype: LY?).

Erect perennial subshrub or shrub, to 1 m; stem minutely stellate-pubescent. Leaves with the blade lanceolate to ovate, 3–9 cm long, 2–4 times as long as wide, the apex acute to obtuse, the base cuneate to truncate, the margin serrate or dentate at least distally, the upper and lower surfaces hirsute to glabrate, the petiole 1–5(8) mm long; stipules linear-lanceolate, falcate, often exceeding the petiole. Flowers solitary or paired, axillary, the pedicel subequaling the petiole; calyx 6–8 mm long, often ciliate; petals 7–10 mm long, yellow; staminal column glabrous or pubescent; styles 8–10. Fruit glabrous, the mericarps 8–10, apically 2-spined, laterally reticulate.

Disturbed sites. Common; nearly throughout. New Jersey and Pennsylvania south to Florida, west to Texas; West Indies, Mexico, Central America, and South America; Africa, Asia, Australia, and Pacific Islands. All year.

Sida urens L. [Stinging.] TROPICAL FANPETALS.

Sida urens Linnaeus, Syst. Nat., ed. 10. 1145. 1759.

Erect or reclining subshrub, to 1.5 m; stem hispid with simple trichomes and short, stellate trichomes. Leaves with the blade cordate, the apex acuminate, the base cordate, the margin crenate-serrate, the upper surface glabrous, the lower surface stellate-pubescent and with simple trichomes, the petiole 1–4 cm long; stipules linear. Flowers in an axillary glomerule or a pedunculate head, the pedicel shorter than the calyx; calyx 5–8 mm long, the lobes with a dark green margin, setose; petals orange, often with a dark red center; staminal column glabrous; styles 5. Fruit 3–4 mm in diameter, the mericarps 5, apically 2-mucronate.

Disturbed sites. Rare; Broward County. Florida and Alabama; West Indies, Mexico, Central America, and South America; Africa. Native to tropical America and Africa. All year.

EXCLUDED TAXA

Sida acuta Burman f.—Reported for Florida by most authors, including Chapman (1897), Small (1903), Long and Lakela (1971), Correll and Correll (1982), Wunderlin (1982, 1998), Wunderlin and

Hansen (2003), Wilhelm (1984), and Clewell (1985). According to Krapovickas (2003), *S. acuta* is misapplied to our plants and all Florida material so named belongs in *S. ulmifolia*.

Sida lindheimeri Engelmann & A. Gray—Reported for Florida by Chapman (1860), probably based on a misidentification of material of *S. elliottii*.

Talipariti Fryxell 2001.

Shrubs or trees. Leaves alternate, simple, palmate-veined, petiolate, stipulate. Flowers solitary, axillary, actinomorphic, bisexual; involucrate bracts 8–12; sepals 5, basally connate; petals 5, basally adnate to the staminal column; stamens numerous, the filaments connate into a column; ovary 5-carpellate and -loculate, the styles 5, the stigmas capitate. Fruit a capsule.

A genus of 22 species; North America, West Indies, Central America, South America, Africa, Asia, Australia, and Pacific Islands. [*Tali*, mucilaginous, and *pariti Hibiscus*.]

Selected reference: Fryxell (2001); Fryxell and Hill (2015h).

Talipariti tiliaceum (L.) Fryxell [Resembling *Tilia*, in reference to the leaves.] MAHOE.

Tree or shrub, to 8 m; branchlets minutely stellate-pubescent to glabrescent. Leaves with the blade broadly ovate, 6–13 cm long, slightly wider than long, the apex acuminate, the base deeply cordate, the margin entire or obscurely denticulate or crenulate, the upper surface glabrate, the lower surface densely stellate-pubescent to subglabrous, palmately 7- to 9-nerved, the main veins each with a basal nectary 2–7 mm long, the petiole 4–12 cm long, with pubescence like the stem but often more dense; stipules oblong-lanceolate, 1.5–4 cm long, sessile, enclosing the apical bud, stellate-puberulent externally, deciduous. Flowers solitary, axillary, or sometimes several aggregated distally with the leaves reduced, the pedicel 0.5–3 cm long; involucel cupuliform, to ½ the calyx length, 8- to 12-dentate, the teeth triangular to lanceolate, 1–6(20) mm long, the sinuses broadly rounded; calyx 1.5–2 cm long, densely stellate-pubescent, the lobes lanceolate, 12–15 mm long, each with a medial nectary; petals 4–6(8) cm long, yellow, with or without a red center, the outer surface pubescent; staminal column 2.5–3 cm long, pallid, the filaments 1–3 mm long; stigmas purplish. Fruit subglobose to subobovoid, 1.5–2 cm long, densely yellowish or brownish, antrorsely pubescent; seeds numerous, ca. 4 mm long, minutely papillate.

1. Corolla yellow with a red center; branchlets and petioles glabrate (rarely with a few stellate trichomes); involucrate bracts 8–10 mm long, ca. ⅓ the length of the calyx, the lobes 3–5 mm long, deltoid ....var. **tiliaceum**
1. Corolla yellow throughout; branchlets and petioles stellate-pubescent; involucrate bracts 12–17 mm long, ca. ½ the length of the calyx, the lobes 6–10 mm long, lanceolate...............var. **pernambucense**

Talipariti tiliaceum var. **tiliaceum** SEA HIBISCUS; MAHOE.

Hibiscus tiliaceus Linnaeus, Sp. Pl. 694. 1753. *Hibiscus tiliifolius* Salisbury, Prodr. Stirp. Chap. Allerton 383. 1796, nom. illegit. *Pariti tiliaceum* (Linnaeus) A. Jussieu, in A. Saint-Hilaire, Fl. Bras. Merid. 1: 256. 1827. *Hibiscus tiliaceus* Linnaeus var. *genuinus* Hochreutiner, Annuaire Conserv. Jard. Bot.

Genève 4: 63, nom. inadmiss. *Talipariti tiliaceum* (Linnaeus) Fryxell, Contr. Univ. Michigan Herb. 23: 258. 2001.

Hibiscus abutiloides Willdenow, Enum. Pl. 736. 1809. *Pariti abutiloides* (Willdenow) G. Don, Gen. Hist. 1: 485. 1831. *Pariti elatum* (Swartz) G. Don var. *abutiloides* (Willdenow) Grisebach, Fl. Brit. W.I. 87. 1859. *Hibiscus tiliaceus* Linnaeus var. *abutiloides* (Willdenow) Hochreutiner, Nova Guinea 14: 163. 1914.

Pariti grande Britton ex Small, Man. S.E. Fl. 859. 1933. TYPE: FLORIDA.

Branchlets and petioles glabrate (rarely with a few stellate trichomes). Involucrate bracts 8–10 mm long, ca. ⅓ the length of the calyx, the lobes 3–5 mm long, deltoid; corolla yellow with a red center.

Disturbed coastal sites. Occasional; St. Lucie and Manatee Counties southward. Escaped from cultivation. Florida; West Indies, Mexico, Central America, and South America; Africa, Asia, Australia, and Pacific Islands. Native to Africa, Asia, Australia, and Pacific Islands. All year.

Talipariti tiliaceum is listed as a Category II invasive species in Florida by the Florida Exotic Pest Plant Council (FLEPPC, 2015).

Talipariti tiliaceum var. **pernambucense** (Arruda) Fryxell [Of Pernambuco, Brazil.] YELLOW MAHOE.

Hibiscus pernambucensis Arruda da Cámara, Diss. Pl. Brazil 44. 1810. *Pariti pernambucense* (Arruda da Cámara) G. Don, Gen. Hist. 1: 485. 1831. *Hibiscus tiliaceus* Linnaeus var. *pernambucensis* (Arruda da Cámara) I. M. Johnston, Sargentia 8: 196, 1949. *Hibiscus tiliaceus* Linnaeus subsp. *pernambucensis* (Arruda da Cámara) A. Castellanos, Sellowia 19: 50. 1967. *Talipariti tiliaceum* (Linnaeus) Fryxell var. *pernambucense* (Arruda da Cámara) Fryxell, Contr. Univ. Michigan Herb. 23: 262. 2001.

Branchlets and petioles stellate-pubescent. Involucrate bracts 12–17 mm long, ca. ½ the length of the calyx, the lobes 6–10 mm long, lanceolate; corolla yellow throughout.

Disturbed sites. Rare; Lee and Miami-Dade Counties, Monroe County keys. Escaped from cultivation. Florida; West Indies, Mexico, Central America, and South America. Native to South America. All year.

Thespesia Sol. ex Corrêa, nom. cons. 1807.

Trees. Leaves alternate, simple, palmate-veined, petiolate, stipulate. Flowers solitary, axillary, actinomorphic, bisexual; involucral bracts 3; calyx 5, basally connate; petals 5, basally adnate to the staminal column; stamens numerous, the filaments connate into a column; ovary 3- to 5-carpellate and -loculate, the style 1, the stigmas 3–5. Fruit a capsule.

A genus of 17 species; North America, West Indies, Mexico, Central America, South America, Africa, Asia, Australia, and Pacific Islands. [From the Greek *Thespesios*, divine, wondrous, or excellent.]

Selected reference: Fryxell and Hill (2015i).

Thespesia populnea (L.) Sol. ex Corrêa [Resembling *Populus* (Salicaceae), in reference to the leaves.] PORTA TREE.

Hibiscus populneus Linnaeus, Sp. Pl. 694. 1753. *Malvaviscus populneus* (Linnaeus) Gaertner, Fruct. Sem. Pl. 2: 253. 1791. *Hibiscus populifolius* Salisbury, Prodr. Stirp. Chapm. Allerton 383. 1796, nom. illegit. *Thespesia populnea* (Linnaeus) Solander ex Corrêa da Serra, Ann. Mus. Natl. Hist. Nat. 9: 290. 1807. *Bupariti populnea* (Linnaeus) Rothmaler, Feddes Repert. Spec. Nov. Regni Veg. 53: 6. 1944.

Tree, to 12 m; branchlets lepidote to glabrate when young. Leaves with the blade ovate, 6–13 cm long, 4–10 cm wide, the apex acute to acuminate, the base deeply cordate, the margin entire, the upper and lower surfaces glabrate, the lower surface with a nectariferous zone near the base of the midrib, the petiole ⅔ to subequaling the blade length; stipules lanceolate or falcate, 3–7 mm long. Flower solitary, axillary, the pedicel 2–4 cm long, shorter than the subtending petiole; involucral bracts ligulate, 1–2 cm long, irregularly inserted; calyx cupulate, 8–10 mm long, the apex truncate, minutely lepidote; petals 4–6 cm long, yellow with a maroon basal spot, punctate; staminal column ca. ½ the length of the petals, glabrous; style exceeding the androecium. Fruit oblate, 3–3.5 cm long and wide, leathery, indehiscent; seeds 8–9 mm long, short pubescent.

Coastal hammocks and beaches. Occasional; central and southern peninsula. Escaped from cultivation. Florida; West Indies, Mexico, Central America, and South America; Africa, Asia, Australia, and Pacific Islands. Native to Africa, Asia, Australia, and Pacific Islands. All year.

Thespesia populnea is listed as a Category II invasive species in Florida by the Florida Exotic Pest Plant Council (FLEPPC, 2015).

Tilia L. 1753. BASSWOOD

Trees. Leaves alternate, pinnipalmate-veined, petiolate, stipulate. Flowers in axillary, corymbiform cymose panicles, actinomorphic, bisexual, bracteate, the peduncle adnate to half its length to a membranaceous winglike bract, bisexual; sepals 5, free, each with a basal nectary on the inner surface; petals 5, free; stamens numerous, in 5 fascicles, the innermost stamen of each group modified into a spatulate, petaloid staminode, the filaments sometimes forked distally and each branch bearing a half anther, the anthers dorsifixed, extrorse, 1-loculate; ovary 5-carpellate and -loculate. Fruit dry, nutlike, indehiscent.

A genus of about 40 species; North America, Mexico, Europe, Asia. [From the Greek *Ptilon*, wing, in reference to the winglike bract of the inflorescence.]

Tilia is sometimes placed in the Tiliaceae.

Selected references: Brizicky (1965); Jones (1968); Strother (2015b).

Tilia americana L. BASSWOOD.

Tree, to 20 m; bark gray or reddish brown, longitudinally furrowed, the branchlets silver-gray, yellow-brown, to reddish, glabrous or pubescent, the axillary buds ovoid, reddish, the outer scales ciliate. Leaves deciduous, the blade ovate to ovate-oblong or suborbicular, 5–20 cm long, 5–12 cm wide, the apex acute to abruptly acuminate, the base asymmetrically truncate to subcordate or cordate, the margin serrate to serrate-dentate, the upper surface dark green, glabrous, the lower surface densely stellate-pubescent or glabrate, the petiole to 4 cm long, glabrous; stipules 1–2 mm long. Flowers 10–20(30) in a pendulous cyme, the peduncle stellate-pubescent, the basal bract with the apex obtuse, the base cuneate, asymmetrical, 8–15 cm long, 2–3 cm wide, adnate ca. ½ its length to the peduncle, glabrous or stellate-pubescent; sepals 5, lanceolate-ovate, 4–6 mm long, glabrous or stellate-pubescent; petals 5, lanceolate, 6–8 mm long, white to pale yellow; stamens numerous, arranged in 5 clusters with the filaments connate at the base, with 1 petaloid staminodium opposite each petal and ca. ⅔ as long; ovary villous. Fruit ellipsoid, rusty-brown tomentose; seed 1(2), subglobose, ca. 4 mm long, glabrous.

1. Lower leaf surface glabrate to tawny or rusty stellate-puberulent ....var. **caroliniana**
1. Lower leaf surface white stellate-tomentose ....var. **heterophylla**

Tilia americana var. **caroliniana** (Mill.) Castigl. [Of Carolina.] CAROLINA BASSWOOD.

> *Tilia caroliniana* Miller, Gard. Dict., ed. 8. 1768. *Tilia pubescens* Aiton, Hort. Kew. 2: 229. 1789, nom. illegit. *Tilia americana* Linnaeus var. *caroliniana* (Miller) Castiglioni, Viagg. Stati Uniti 2: 389. 1790. *Tilia grata* Salisbury, Prodr. Stirp. Chap. Allerton 367. 1796, nom. illegit. *Tilia americana* Linnaeus var. *pubescens* Loudon, Arbor. Frutic. Brit. 1: 374. 1838, nom. illegit. *Tilia americana* Linnaeus subsp. *caroliniana* (Miller) E. Murray, Kalmia 13: 31. 1983.
>
> *Tilia americana* Linnaeus var. *walteri* A. W. Wood, Fl. Atl. 64. 1879. TYPE: "Va. to Fla."
>
> *Tilia floridana* Small, Fl. S.E. U.S. 761, 1335. 1903. *Tilia caroliniana* Miller var. *floridana* (Small) V. Engler, Monogr. Gatt. Tilia 132. 1909. *Tilia caroliniana* Miller subsp. *floridana* (Small) E. Murray, Kalmia 15: 11. 1984 ("1985"). *Tilia americana* Linnaeus subsp. *floridana* (Small) E. Murray, Kalmia 13: 31. 1983. *Tilia americana* Linnaeus var. *floridana* (Small) E. Murray, Kalmia 13: 31. 1983. TYPE: FLORIDA: Jackson Co.: Jun, *Curtis 401** (holotype: NY; isotypes: A, F, KANU, NEB, NY, US).
>
> *Tilia caroliniana* Miller var. *vagans* V. Engler, Monogr. Gatt. Tilia 132. 1909. TYPE: FLORIDA: Columbia Co.: s.d., *Hitchcock s.n.*(?) (holotype: WRSL?).
>
> *Tilia pubescens* Aiton var. *aitonii* V. Engler, Monogr. Gatt. Tilia 129. 1909. SYNTYPE: FLORIDA.
>
> *Tilia pubescens* Aiton forma *heteromorpha* V. Engler, Monogr. Gatt. Tilia 129. 1909. TYPE: FLORIDA: Columbia Co.: Lake City, s.d., *Nash 2188* (holotype: WRSL?; isotype: NY).
>
> *Tilia crenoserrata* Sargent, Bot. Gaz. 66: 430. 1918. TYPE: FLORIDA: Seminole Co.: Oviedo.
>
> *Tilia floridana* Small var. *oblongifolia* Sargent, Bot. Gaz. 66: 435. 1918. *Tilia leucocarpa* Ashe var. *oblongifolia* (Sargent) Bush, Bull. Torrey Bot. Club 54: 248. 1927.
>
> *Tilia georgiana* Sargent, Bot. Gaz. 66: 510. 1918.
>
> *Tilia porracea* Ashe, Quart. Charleston Mus. 1: 31. 1925. *Tilia floridana* Small var. *porracea* (Ashe) Coker & Totten, Trees S.E. States 331. 1937. TYPE: FLORIDA: Okaloosa Co.: Hickory Head, s.d., *Ashe s.n.* (holotype: NCU).
>
> *Tilia ashei* Bush, Bull. Torrey Bot. Club 54: 241. 1927. TYPE: FLORIDA: Okaloosa Co.: Hickory Head, 23 Apr 1923, *Ashe 3* (holotype: NCU; isotype: NCU).
>
> *Tilia crenoserrata* Sargent var. *acuminata* Ashe, Bull. Torrey Bot. Club 55: 465. 1928. TYPE: FLORIDA: Alachua Co.: near Gainesville, 9 Jun 1925, *Cody & Ashe s.n.* (holotype: NCU).

Lower leaf surface glabrate to tawny or rusty stellate-puberulent.

Moist hammocks. Frequent; northern counties, central peninsula. North Carolina south to Florida, west to Oklahoma and Texas. Spring.

Tilia americana var. **heterophylla** (Vent.) Loudon [From the Greek *heteros* and *phyllon*, the leaves of more than one form.] WHITE BASSWOOD.

Tilia heterophylla Ventenat, Anales Hist. Nat. 2: 68. 1800. *Tilia americana* Linnaeus var. *heterophylla* (Ventenat) Loudon, Arbor. Frutic. Brit. 1: 375. 1838. *Tilia americana* Linnaeus subsp. *heterophylla* (Ventenat) E. Murray, Kalmia 13: 32. 1983.

Tilia heterophylla Ventenat var. *amphibola* Sargent, Bot. Gaz. 66: 507. 1918. TYPE: FLORIDA: Gadsden County: River Junction, s.d., *Harbison 1484* (holotype: A).

Tilia heterophylla Ventenat var. *nivea* Sargent, Bot. Gaz. 66: 507. 1918. TYPE: FLORIDA: Gadsden Co.: River Junction, 19 Apr & 20 Jun 1917, *Harbison 29* (holotype: A).

Tilia lasioclada Sargent, Bot. Gaz. 66: 502. 1918. *Tilia eburnea* Ashe var. *lasioclada* (Sargent) Ashe, Bull. Torrey Bot. Club 53: 31. 1926.

Lower leaf surface white stellate-tomentose.

Moist hammocks. Occasional; Jefferson County, central panhandle. New Brunswick south to Florida, west to Ontario, Iowa, Missouri, Arkansas, and Louisiana. Spring.

EXCLUDED TAXON

Tilia americana Linnaeus—Because infraspecific categories were not recognized, the typical variety was reported for Florida by implication by Kurz and Godfrey (1962) and Godfrey (1988), who lumped all variation into a single species. All Florida material is either var. *caroliniana* or var. *heterophylla*.

Triumfetta L. 1753. BURRBARK

Annual or perennial herbs or subshrubs. Leaves alternate, simple, palmate-veined, petiolate, stipulate. Flowers in axillary and terminal, umbellate cymes, actinomorphic, bisexual, pedicellate, bracteate; sepals 5, free, each with a subapical appendage; petals 5, free; androgynophore usually present, with 5 antipetalous glands and sometimes an apical extrastaminal disk; stamens 10–15+, inserted at the apex of the androgynophore, free, the anthers dorsifixed, introrse; ovary 2- to 5-carpellate, 2- or 3-loculate, the stigma 2- or 3-lobed. Fruit nutlike, indehiscent.

A genus of about 160 species; North America, Mexico, Central America, South America, Africa, Asia, Australia, and Pacific Islands. [Commemorates Giovanni Battista Triumfetti (1658–1708), Italian physician and botanist.]

Triumfetta is sometimes placed in the Tiliaceae.

Selected reference: Brizicky (1965); Nesom 2015b).

1. Capsule glabrous, the prickles retrorsely hispidulous; lower leaf surface velutinous .... **T. semitriloba**
1. Capsule hispid, the prickles pubescent throughout or pubescent only at the base; lower leaf surface not velutinous.
    2. Leaves 3-nerved from the base; branchlets with both simple (strigose) and stellate trichomes; capsule prickles plumose-pubescent throughout ....**T. pentandra**
    2. Leaves 5- to 7-nerved from the base; branchlets with stellate trichomes only; capsule prickles pubescent only at the base.. **T. rhomboidea**

Triumfetta pentandra A. Rich. [From the Greek *pentos* and *andros*, with five stamens.] FIVESTAMEN BURRBARK.

Triumfetta pentandra A. Richard, in Guillemin et al., Fl. Seneg. Tent. 93, t. 19. 1831.

Erect annual herb, to 6 dm; stem subscabrous, stellate-pubescent. Leaves with the blade rhomboid-ovate, 4.5–9 cm long, 3.5–6 cm wide, palmately 3-lobed or unlobed, the apex acute to acuminate, the base cuneate, the margin unequally coarsely serrate, the upper surface sparsely pubescent with simple trichomes to glabrate, the lower surface stellate-pubescent, the petiole 1–5 cm long, pubescent; stipules subulate, 5–6 mm long. Flowers in a short pedicellate and pedunculate cyme, sometimes forming a terminal, racemose inflorescence due to the reduction of the upper leaves; bracts linear, 2–3 mm long, stellate-pubescent; sepals lanceolate, 2–3 mm long, yellow, cucullate and awned at the apex, sparsely stellate-puberulent on the outer surface; petals spatulate, 2–3 mm long, yellow, glabrous; stamens 5(10), the filaments pubescent at the base; ovary 2-carpellate and -loculate. Fruit ovoid, 6–7 mm long, the spines ca. 2 mm long, with a line of spreading trichomes on the upper surface.

Disturbed sites. Occasional; northern and central peninsula. Florida; South America, Africa, Asia, and Australia. Native to Africa. All year.

Triumfetta rhomboidea Jacq. [Diamond-like, in reference to the leaf shape.] DIAMOND BURRBARK.

Triumfetta rhomboidea Jacquin, Enum. Syst. Pl. 22. 1760. *Triumfetta rhombeifolia* Swartz, Prodr. 76. 1788, nom. illegit.

Bartramia indica Linnaeus, Sp. Pl. 389. 1753. *Triumfetta bartramia* Linnaeus, Syst, Nat., ed 10. 1044. 1759, nom. illegit.

Perennial herb or subshrub; branchlets gray-brown-tomentose. Leaves with the blade broadly ovate-orbicular, rhomboid, or broadly ovate, 3–9.5 cm long, 2–8.5 cm wide, 3-lobed, the apex acute, the base broadly cuneate to rounded, the margin irregularly bluntly serrate, the upper and lower surfaces stellate-pilose, the petiole 1–5 cm long, the uppermost blades oblong-lanceolate, unlobed; stipules linear-lanceolate to lanceolate. Flowers 3–5 in an axillary cyme, the peduncle to 2 mm long, the pedicel less than 1 mm long; calyx cylindrical in bud, the sepals narrowly oblong, 5–6 mm long, villous, each with a subapical appendage; petals 6–7 mm long, yellow, pubescent along the margin; stamens 10. Fruit globose, ca. 3 mm long, indehiscent, spiny, the spines ca. 2 mm long, gray-yellow puberulent, the apex hooked; seeds 2–6.

Disturbed sites. Rare; DeSoto and Broward Counties. Alabama and Florida; West Indies, Mexico, Central America, and South America; Africa, Asia, Australia, and Pacific Islands. Native to Africa, Asia, and Australia. Summer–fall.

Triumfetta semitriloba Jacq. [Somewhat three-lobed.] SACRAMENTO BURRBARK.

Triumfetta semitriloba Jacquin, Enum. Syst. Pl. 22. 1760. *Triumfetta semitriloba* Jacquin var. *typica* K. Schumann, in Martius, Fl. Bras. 12(3): 135. 1886, nom. inadmiss.

Perennial herb or subshrub, to 1 m; branchlets pubescent with simple and stellate trichomes. Leaves with the blade broadly ovate, 4–8 cm long, 3–6 cm wide, unlobed or 3-lobed, the apex acuminate, the base rounded to shallowly cordate, the margin irregularly serrate-dentate, the upper surface pubescent with simple trichomes, the lower surface stellate-pubescent, the petiole 0.5–6 cm long, the uppermost blades usually oblong; stipules linear-lanceolate to lanceolate. Flowers 2–3 in an axillary cyme, the peduncle 1–2(3) mm long, the pedicel 2–3 mm long; calyx cylindrical in bud, the sepals linear, 5–6 mm long, stellate-pubescent, the subapical appendage ca. 1 mm long, pubescent; petals narrowly elliptic-obovate, subequaling the sepals; stamens 15+, the filaments free, glabrous; ovary 3-loculate, the style 3-lobed. Fruit globose, 3–5 mm long, indehiscent, spiny, the spines ca. 2 mm long, retrorsely pilosulose proximally, the apex hooked; seeds 1 or 2 per locule.

Disturbed sites. Occasional; Manatee and Okeechobee Counties southward. Georgia and Florida; West Indies, Mexico, Central America, and South America; Asia and Pacific Islands. Native to tropical America. All year.

Urena L. 1753.

Perennial herbs, subshrubs, or shrubs. Leaves alternate, simple, palmate-veined, petiolate, stipulate. Flowers solitary or few in axillary fascicles, actinomorphic, bisexual; involucrate bracts 5, basally connate; sepals 5, basally connate; petals 5, basally adnate to the staminal column; stamens numerous, connate into a column; ovary 5-carpellate and -loculate, the styles 10. Fruit a schizocarp.

A genus of 6 species; North America, West Indies, Mexico, Central America, South America, Africa, Asia, Australia, and Pacific Islands. [Malayalam *ooren*, to loosen or soak, in reference to the retting process to extract fiber from the stems of *U. lobata*.]

Selected reference: Hill (2015d).

Urena lobata L. [Lobed, in reference to the leaves.] CAESARWEED.

Urena lobata Linnaeus, Sp. Pl. 682. 1753.

Perennial herb, subshrub, or shrub, to 1 m; stem minutely stellate-puberulent. Leaves with the blade ovate, 3–9 cm long and wide, the apex obtuse, the base cordate, the margin shallowly 3- to 5-lobed or -angled, crenate, the upper and lower surfaces stellate-puberulent, the petiole of the lower leaves subequaling the blade length, progressively reduced upward to ¼ the blade length; stipules subulate, 2–3 mm long. Flowers axillary, solitary or few in a fascicle, subsessile or the pedicel to 7 mm long; involucral bracts linear, 5–6 mm long; calyx 5–9 mm long, basally connate ½ its length, stellate-puberulent, the lobes ciliate; petals 15–20 mm long, pink, the outer surface stellate-puberulent; staminal column glabrous, the anthers subsessile. Fruit oblate, ca. 8 mm long, stellate-puberulent and with numerous glochidiate spines; seeds 1 per carpel, obovoid to reniform, 3–5 mm long, glabrous.

Disturbed sites. Common; peninsula, Leon, Gadsden, and Escambia Counties. South

Carolina, Florida, and Louisiana; West Indies, Mexico, Central America, and South America; Africa, Asia, Australia, and Pacific Islands. Native to Asia and Australia. All year.

Urena lobata is listed as a Category I invasive species in Florida by the Florida Exotic Pest Plant Council (FLEPPC, 2015).

Waltheria L. 1753.

Perennial herbs. Leaves alternate, simple, pinnate-veined, petiolate, stipulate. Flowers in axillary compound cymes, actinomorphic, bisexual, the bracts 4, subtending a floral pair; sepals 5, basally connate; petals 5, free; stamens 5, the filaments connate into a column; ovary 1-loculate, the style 1. Fruit a capsule.

A genus of about 50 species; North America, West Indies, Mexico, Central America, South America, Africa, Asia, Australia, and Pacific Islands. [Commemorates Friedrich Walther (1688–1746), German physician and botanist.]

Waltheria is sometimes placed in the Sterculiaceae.

Selected references: Brizicky (1966); Saunders (2015).

Waltheria indica L. [Of India.] SLEEPY MORNING.

Waltheria indica Linnaeus, Sp. Pl. 673. 1753. *Waltheria arborescens* Cavanilles, Diss. 6: 316. 1788, nom. illegit. *Waltheria americana* Linnaeus var. *indica* (Linnaeus) K. Schumann, in Engler. Monogr. Afrik. Pflanzen-Fam. 5: 47. 1900.
Waltheria americana Linnaeus, Sp. Pl. 673. 1753. *Waltheria indica* Linnaeus var. *americana* (Linnaeus) R. Brown ex Hosaka, Occas. Pap. Bernice Pauahi Bishop Mus. 13: 224. 1937.
Waltheria bahamensis Britton, Torreya 3: 105–106. 1903.

Suffrutescent perennial herb, to 12 dm; stem with stellate and simple trichomes. Leaves with the blade elliptic-ovate, 3–7(10) cm long, 1–4 cm wide, the apex acute to obtuse, the base rounded to cordate, the margin serrate, the upper and lower surfaces pubescent, the petiole to 5 mm long. Flowers in an axillary glomerulate, sessile or pedunculate, compound cyme; bract subulate, 1 cm long; sepals ca. 1 cm long, the lobes subulate, ca. 5 mm long, the outer surface villous-hirsute; petals ca. 6 mm long, yellow. Fruit 2–3 mm long, chartaceous, apically pubescent; seed dark brown.

Sandhills and hammocks. Frequent; central and southern peninsula. Florida, Alabama, Texas, and Arizona; West Indies, Mexico, Central America, and South America; Africa, Asia, Australia, and Pacific Islands. Native to North America and tropical America. All year.

Waltheria indica s.l. is a pantropical complex and is in need of revision.

Wissadula Medik. VELVETLEAF

Shrubs. Leaves alternate, simple, pinnipalmate-veined, petiolate, stipulate. Flowers in open terminal panicles; sepals 5, basally connate, petals 5, basally adnate to the staminal tube; stamens numerous, the filaments connate into a tube; styles 3–5, the stigmas capitate. Fruit a septicidally dehiscent schizocarp.

A genus of about 25 species; North America, West Indies, Mexico, Central America, South America, Africa, and Asia. [Vernacular name from the Sinhala language.]

Wissadula hernandioides (L'Hér.) Garke [Commemorates Hernán Cortés (1485–1547), Spanish conquistador.] BIG YELLOW VELVETLEAF.

> *Sida hernandioides* L'Héritier de Brutelle, Stirp. Nov. 121, t. 58. 1789. *Abutilon hernandioides* (L'Héritier de Brutelle) Sweet, Hort. Brit. 53. 1836. *Wissadula periplocifolia* (Linnaeus) C. Presl ex Thwaites var. *hernandioides* (L'Héritier de Brutelle) Grisebach, Cat. Pl. Cub. 25. 1866. *Wissadula rostrata* (Schumacher) Hooker f. var. *hernandioides* (L'Héritier de Brutelle) M. Gómez de la Maza y Jiménez, Anales Soc. Esp. Hist. Nat. 19: 219. 1890. *Wissadula hernandioides* (L'Héritier de Brutelle) Garck, Naturwiss. 63: 124. 1890.

Shrub, to 2 m; stem with stipitate-stellate trichomes and minutely puberulent with simple trichomes. Leaves with the blade ovate, 4–12 cm long, 5.5–8 cm wide, the apex acuminate, the base deeply cordate, the margin entire, the petiole 4–6 cm long; stipules subulate, ca. 3 mm long. Flowers in an open, terminal, usually leafless panicle, long-pedunculate; calyx ca. 3 mm long, lobed to ca. ½; petals ca. 5 mm long, yellow; stamens yellowish, the free filaments longer than the basal column. Fruit 6–7 mm long, with 3–5 mericarps, these constricted below, bulbous-apiculate distally, minutely puberulent; seeds 2–3 mm long, patchy pubescent.

Disturbed sites. Rare; Collier County. Florida, Louisiana, and Texas; West Indies, Mexico, Central America, and South America; Africa. Native to tropical America. All year.

EXCLUDED TAXON

> *Wissadula amplissima* (Linnaeus) R. E. Fries—Reported for Florida by Wunderlin and Hansen (2011), the name misapplied to *W. hernandioides* (Krapovickas, 1996).

EXCLUDED GENUS

> *Alcea rosea* Linnaeus—Reported for Florida by Wunderlin (1998, as *Althaea rosea* (Linnaeus) Cavanilles) and Wunderlin and Hansen (2003), but the voucher specimen is from a cultivated plant.

THYMELAEACEAE Juss., nom. cons. 1789. MEZEREON FAMILY

Shrubs. Leaves alternate, simple, pinnate-veined, petiolate, estipulate. Flowers in axillary racemes, actinomorphic, bisexual; sepals 4, connate; petals 8, free; stamens 8, free, the anthers basifixed, introrse, 2-thecate, longitudinally dehiscent; nectariferous disk present; ovary superior, 1-carpellate. Fruit a berry.

A family of about 50 genera and about 890 species; nearly cosmopolitan.

Daphnaceae Vent. (1799).

Selected reference: Nevling (1962).

Dirca L. 1753. LEATHERWOOD

Shrubs. Leaves alternate, simple, pinnate-veined, petiolate, estipulate. Flowers in axillary racemes; sepals 4, connate into a tube; petal 8, inserted on the calyx tube between the stamens; stamens 8 in 2 whorls, inserted on the calyx tube; nectariferous disk annular, surrounding the ovary base; ovary 1, 1-carpellate and -loculate, the style 1, eccentric, the stigma punctate. Fruit a 1-seeded berry.

A genus of 4 species; North America and Mexico. [From Dirce of Greek mythology, the second wife of Lycus who was transformed by Dionysus into a fountain.]

Selected reference: Floden and Nevling (2015).

Dirca palustris L. 1753. [Of swamps.] EASTERN LEATHERWOOD.

Dirca palustris Linnaeus, Sp. Pl. 358. 1753.

Shrub, to 2 m; stem smooth, glabrous. Leaves with the blade ovate, oblanceolate, elliptic, or obovate, the apex obtuse to rounded, the base cuneate to rounded, margin entire, the upper surface glabrous, the lower surface pale green, glabrous or sparsely short-pubescent, the petiole ca. 2 mm long, calyptrate over the axillary bud. Flowers in a short axillary raceme; calyx funnelform, 5–6 mm long, yellowish green, shallowly notched at the apex, the petals minute and inconspicuous, inserted at the base of the calyx tube between the filaments; stamens in 2 whorls of 4 each of unequal length, these slightly exserted from the calyx tube; ovary with a minute nectar ring surrounding the base, the style eccentric on the ovary, exserted from the tube, minutely capitate. Fruit subelliptic, 5–7 mm long, slightly asymmetric, yellow or greenish yellow; seed with a fleshy outer and a hard inner coat.

Bluff forests and riverbanks. Rare; Gadsden, Liberty, and Jackson Counties. Quebec south to Florida, west to Ontario, North Dakota, Iowa, Kansas, Oklahoma, and Louisiana. Spring.

Dirca palustris is listed as endangered in Florida (Florida Administrative Code, Chapter 5B-40).

CISTACEAE Juss., nom. cons. 1789. ROCKROSE FAMILY

Herbs or subshrubs. Leaves alternate, simple, pinnate-veined, petiolate or epetiolate, estipulate. Flowers in terminal or axillary cymes, actinomorphic, bisexual; sepals 3 or 5, basally connate; petals 3 or 5, free, or absent; stamens 3–10 or numerous, free, the anthers basifixed, 2-locular, longitudinally dehiscent; ovary 3-carpellate, 1- to 3-loculate, the style simple, the stigma 1. Fruit a loculicidal capsule.

A family of 8 genera and about 175 species; North America, West Indies, Mexico, Central America, South America, Europe, Africa, and Asia.

Selected reference: Brizicky (1964).

1. Flowers 10–20 mm wide; petals 5, yellow, or absent..**Crocanthemum**
1. Flowers 1–2 mm wide; petals 3, greenish or purplish ....**Lechea**

Crocanthemum Spach 1836. FROSTWEED

Herbs or subshrubs. Leaves alternate, simple, pinnate-veined, petiolate or epetiolate, estipulate. Flowers in paniculate, spiciform, umbellate, thyrsiform, or scorpioid cymes; bracteate, chasmogamous and cleistogamous or chasmogamous only. Chasmogamous: sepals 5, basally connate, the calyx bilaterally symmetric; petals 5, free, radially symmetric; stamens 10–50, adnate to the petals at the corolla throat; staminodes absent; ovary 1-locular. Cleistogamous: sepals 5, basally connate, the calyx bilaterally symmetric; petals absent; stamens 4–6; staminodes absent; ovary 1-locular. Fruit a capsule.

A genus of 21 species; North America, West Indies, Mexico, Central America, and South America. [From the Greek *krokos*, saffron, and *anthemon*, flower, in reference to the flower color.]

Recent molecular data (Arrington, 2004, Guzmán and Vargas, 2009) show that the New World taxa of *Helianthemum* are best placed in *Crocanthemum*.

Selected references: Daoud and Wilbur (1965); Sorrie (2015).

1. Lower leaf surface easily visible; lower stem short pilose; basal rosette of leaves usually present ....
.... C. carolinianum
1. Lower leaf surface densely stellate-tomentose and concealing the surface; lower stem pubescent; basal rosette of leaves absent.
   2. Ovary and/or fruit densely stellate-pubescent, at least in the upper ½.
      3. Flowers in terminal umbellate clusters; fruit 2-valved .... C. arenicola
      3. Flowers in leafy thyrses; fruit 3-valved ....C. nashii
   2. Ovary and/or fruit glabrous.
      4. Flowers in short glomerulate spikes from the leaf axils; leaves without evident secondary veins
....C. rosmarinifolium
      4. Flowers in terminal or subterminal umbellate cymes; leaves with evident secondary veins.
         5. Flowers many in dense cymes...C. corymbosum
         5. Flowers few in scattered clusters or solitary..C. georgianum

Crocanthemum arenicola (Chapm.) Barnhart [*Arena*, sandy place, and *-cola*, dweller, growing in sandy areas.] COASTALSAND FROSTWEED.

Helianthemum arenicola Chapman, Fl. South. U.S. 35. 1860. *Halimium arenicola* (Chapman) Grosser
   in Engler, Pflanzenr. 4(Heft 14): 49. 1903. *Crocanthemum arenicola* (Chapman) Barnhart, in Small,
   Man. S.E. Fl. 879. 1933. TYPE: FLORIDA: Franklin Co.: Apalachicola, s.d., *Chapman s.n.* (lectotype: US; isolectotypes, GH, MO, NY). Lectotypified by Daoud and Wilbur (1965: 264).
Helianthemum canadense (Linnaeus) Michaux var. *obtusum* A. W. Wood, Class-Book Bot., ed. 1861.
   246. 1861. TYPE: FLORIDA.

Spreading or erect perennial herb, to 1.5(2) dm; stem stellate-pubescent. Leaves all cauline, the blade oblanceolate to lanceolate, 0.8–2(2.8) cm long, 2–4(10) mm wide, the apex acute, the base cuneate, the margin entire, the upper and lower surfaces stellate-pubescent, the lateral veins obscure on the lower surface, the petiole 0.5–2 mm long. Flowers in a terminal or occasionally lateral in an umbellate cyme; bracts 1–4 mm long. Chasmogamous flowers 2–10, the pedicel (2)5–10(15) mm long; inner sepals lanceolate, 4–8 mm long, the outer sepals linear, 2–4 mm long; petals obovate, 8–9 mm long, yellow; stamens 17–29; ovary stellate-pubescent distally. Fruit subglobose, 3–5 mm long, stellate-pubescent distally; seeds 14–17, dark brown, ovoid, short-papillate. Cleistogamous flowers with the pedicel 1–2 mm long; inner sepals ovate, 3–5 mm long, the outer sepals linear, 1–2 mm long; stamens 4–8; ovary stellate-pubescent in the upper ½. Fruit subglobose, 3–5 mm long, stellate-pubescent distally; seeds (6)9–14, similar to those of the chasmogamous flowers.

Sandhills, scrub, and dunes. Occasional; central and western panhandle. Florida, Alabama, and Mississippi. Spring.

Crocanthemum carolinianum (Walter) Spach [Of Carolina.] CAROLINA FROSTWEED.

Cistus carolinianus Walter, Fl. Carol. 152. 1788. *Helianthemum carolinianum* (Walter) Michaux, Fl. Bor.-Amer. 1: 307. 1803. *Crocanthemum carolinianum* (Walter) Spach, Ann. Sci. Nat., Bot., ser. 2. 6: 370. 1836. *Halimium carolinianum* (Walter) Grosser, in Engler, Pflanzenr. 4(Heft 14): 44. 1903.

Erect perennial herb, to 3(3.8) dm; stem stellate-pubescent. Leaves basal and cauline, the basal ones rosette-forming, the blade spatulate to obovate or elliptic, 1–3.5(6) cm long, 0.5–1.8(2.8) cm wide, the apex obtuse, the base cuneate, the margin entire, revolute, the upper and lower surfaces stellate-pubescent, the lateral veins prominent on the lower surface, the petiole 1–4 mm long, the cauline ones similar to the basal, but the margin not revolute. Flowers in a terminal, scorpioid cyme; bracts 3–6 mm long. Chasmogamous flowers: 1–3, the pedicel 4–15(20) mm long; inner sepals 7–12 mm long, the outer ones lanceolate, 3–6(8) mm long; petals broadly spatulate, 8–18 mm long, yellow; stamens 20–35(50); ovary glabrous. Fruit subglobose, 6–9 mm long, glabrous; seeds 80–92(135), ovoid, red-brown to dark brown, short-papillate. Cleistogamous flowers rarely produced, the pedicel 0.5–4 mm long; inner sepals 3–5 mm long, the outer sepals 2–3 mm long; stamens 4–6; ovary glabrous. Fruit as in the chasmogamous flowers.

Flatwoods and dry hammocks. Occasional; northern counties, central peninsula. North Carolina south to Florida, west to Texas. Summer.

Crocanthemum corymbosum (Michx.) Britton [Cluster, in reference to the flowers and/or fruits.] PINEBARREN FROSTWEED.

Helianthemum corymbosum Michaux, Fl. Bor.-Amer. 1: 307. 1803. *Cistus corymbosus* (Michaux) Poiret, in Lamarck, Encycl., Suppl. 2: 272. 1811. *Heteromeris cymosa* Spach, Ann. Sci. Nat., Bot., ser. 2. 6: 370. 1836, nom. illegit. *Halimium corymbosum* (Michaux) Grosser, in Engler, Pflanzenr. 4(Heft 14): 50. 1903. *Crocanthemum Corymbosum* (Michaux) Britton, in Britton & A. Brown, Ill. Fl. N. U.S., ed. 2. 2: 541. 1913.

Erect perennial herb or subshrub, to 3(5) dm; stem stellate-pubescent, becoming glabrate. Leaves all cauline, the blade obovate-elliptic to elliptic-lanceolate, 1.2–3.5(4.7) cm long, 3–10(13) mm wide, the apex obtuse, the base cuneate, the margin entire, the upper and lower surfaces stellate-pubescent, the lateral veins prominent on the lower surface, the petiole 1–5 mm long. Flowers in a terminal compound cyme; bracts 2–7 mm long, the pedicel with simple and stellate trichomes. Chasmogamous flowers 1–6, the pedicel 6–15 mm long; inner sepals 3–7 mm long, the apex acute, the outer ones spatulate-linear, 2.5–4.5 mm long, the apex obtuse; petals obovate, 6–11 mm long, yellow; stamens 20–30; ovary glabrous. Fruit subglobose, 3.6–5.4 mm long, glabrous; seeds 15–30, ovoid, light to dark brown, smooth. Cleistogamous flowers 10–45, the pedicel 1–3 mm long; inner sepals ovate, 2–5 mm long, the apex acute, the outer ones linear, 2–3 mm long; stamens 4–6; ovary glabrous. Fruit subglobose, 2–4 mm long, glabrous; seeds 4–10, similar to those of the chasmogamous flowers.

Sandhills, dunes, and dry, open hammocks. Common, nearly throughout. North Carolina south to Florida, west to Mississippi. Spring–summer.

Crocanthemum georgianum (Chapm.) Barnhart [Of Georgia.] GEORGIA FROSTWEED.

Helianthemum georgianum Chapman, Fl. South. U.S., ed. 3. 36. 1897. *Halimium georgianum* (Chapman) Grosser, in Engler, Pflanzenr. 4(Heft 14): 51. 1903. *Crocanthemum georgianum* (Chapman) Barnhart, in Small, Man. S.E. Fl. 879. 1933.

Erect perennial herb or subshrub, to 3(4) dm; stem stellate-tomentose. Leaves basal and cauline, the basal ones rosette-forming or absent, the blade spatulate to oblanceolate, 1–2.8 cm long, 5–11 mm wide, the upper and lower surfaces stellate-tomentose, the petiole 1–2 mm long, the cauline with the blade oblanceolate proximally, narrowly elliptic distally, 2–4 cm long, (2)5–6(12) mm wide, the upper surface stellate-pubescent, the lower surface stellate-tomentose, the lateral veins prominent. Flowers in a terminal racemiform cyme, the bracts 1–5 mm long. Chasmogamous flowers 2–7, the pedicel 5–12 mm long; inner sepals 4–7 mm long, the apex acute, the outer ones linear, 2–4 mm long; petals obovate, 6–10(12) mm long, yellow; stamens 15–25; ovary glabrous. Fruit subglobose, 4–6 mm long, glabrous; seeds 20–35, ovoid, dark brown, smooth. Cleistogamous flowers 2–7, the pedicel 0.5–3 mm long; inner sepals ovate, 3–4 mm long, the apex acute, the outer ones linear, ca. 2 mm long; stamens (3)5(8); ovary glabrous. Fruit subglobose, 3–4 mm long, glabrous; seeds 12–20, similar to those of the chasmogamous flowers.

Sandhills, dunes, and dry hammocks. Occasional; northern counties south to Hernando County. North Carolina south to Florida, west to Oklahoma and Texas. Spring.

Crocanthemum nashii (Britton) Barnhart [Commemorates George Valentine Nash (1864–1921), Agrostologist and head gardener at the New York Botanical Garden.] FLORIDA SCRUB FROSTWEED.

Helianthemum nashii Britton, in Nash, Bull. Torrey Bot. Club 22: 147. 1895. *Halimium nashii* (Britton) Grosser, in Engler, Pflanzenr. 4(Heft 14): 49. 1903. *Crocanthemum nashii* (Britton) Barnhart, in Small, Man. S.E. Fl. 879. 1933. TYPE: FLORIDA: Lake Co.: vicinity of Eustis, 16–31 Mar 1894, *Nash 815* (lectotype: NY: isolectotypes: CU, F, GH, MIN, MO, MSC, UC, US). Lectotypified by Daoud and Wilbur (1965: 266).

Helianthemum thyrsoideum Barnhart, in Small, Fl. S.E. U.S. 797, 1335. 1903. *Crocanthemum thyrsoideum* (Barnhart) Janchen, in Engler & Prantl, Nat. Pflanzenfam., ed. 2. 21: 307. 1925. TYPE: FLORIDA: Pinellas Co.: Sutherland, 1 May 1899, *Barnhart 2740* (holotype: NY; isotypes: NY).

Erect to ascending perennial herb, to 3.5(4) dm; stem stellate-pubescent. Leaves all cauline, the blade oblanceolate-elliptic to lanceolate-elliptic, 1.5–3(3.8) cm long, 3–6(8) mm wide, the apex acute, the base cuneate, the margin entire, the upper and lower surfaces stellate-tomentose, the lateral veins slightly to moderately elevated on the lower surface, the petiole 1–2(3) mm long. Flowers in a terminal, elongate, leafy thyrsiform cyme; bracts linear-lanceolate, 1–3 mm long. Chasmogamous flowers 1–8, the pedicel 3–8(10) mm long; inner sepals ovate-elliptic, 3–5 mm long, the apex acute, stellate-pubescent or with simple trichomes, the other ones linear, 1–3 mm long; petals broadly cuneate, 6–10 mm long, yellow; stamens 12–18; ovary stellate-pubescent. Fruit subglobose, 3–5 mm long, 2-valved, stellate-pubescent distally; seeds 6–10, dark brown, short papillate. Cleistogamous flowers 8–10, the pedicel 1–3 mm long; inner sepals ovate-lanceolate, 1–3 mm long, the apex acute, the outer ones linear, 1–2 mm long; stamens 5(7); ovary stellate-pubescent. Fruit 3–4 mm long, 2-valved, stellate-pubescent distally; seeds 5–8, similar to those of the chasmogamous fruit.

Scrub. Frequent; peninsula. North Carolina and Florida. Spring–summer.

Crocanthemum rosmarinifolium (Pursh) Janch. [With leaves resembling those of rosemary, *Rosmarinus officinalis* (Lamiaceae).] ROSEMARY FROSTWEED.

Helianthemum rosmarinifolium Pursh, Fl. Amer. Sept. 364. 1814. *Halimium rosmarinifolium* (Pursh) Grosser, in Engler, Pflanzenr. 4(Heft 14): 49. 1903. *Crocanthemum rosmarinifolium* (Pursh) Janchen, in Engler & Prantl, Nat. Pflanzenfam., ed. 2. 21: 307. 1925.

Erect perennial herb, to 4(5) dm; stem stellate-tomentose. Leaves cauline and sometimes also basal, the basal ones with the blade obovate, 1–2.2 cm long, 3–5 mm wide, the cauline oblanceolate to narrowly lanceolate, 1–4(5) cm long, 2–6 mm wide, the apex attenuate, the margin entire, the upper surface stellate-pubescent, the lower surface stellate-tomentose, the lateral veins obscure, the petiole 1–4 mm long. Flowers in a terminal umbelliform cyme; bracts 3–7 mm long. Chasmogamous flowers solitary, terminal, the pedicel 10–22 mm long; inner sepals 2–4 mm long, the apex acute, the outer ones linear, 1–3 mm long; petals obovate, 4–6 mm long, yellow; stamens 15–24; ovary glabrous. Fruit subglobose, 2–3 mm long, glabrous; seeds 1–3(6), ovoid, reddish brown, smooth. Cleistogamous flowers 2–7, in axillary and terminal glomerules subtending the chasmogamous flowers, the pedicel 1–3 mm long; inner sepals ovate, 1–2 mm long, the apex acute, the outer ones linear, ca. 1 mm long; stamens 3–5; ovary glabrous. Fruit subglobose, 1–2 mm long, glabrous; seeds 1–2, similar to those of the chasmogamous fruit.

Sandhills. Rare; Putnam County, panhandle. North Carolina south to Florida, west to Oklahoma and Texas; West Indies. Summer.

EXCLUDED TAXA

Crocanthemum canadense (Linnaeus) Britton—Reported for Florida by Chapman (1860, as *Helianthemum canadense* (Linnaeus) Michaux), the name misapplied to material of *C. rosmarinifolium*.

Lechea L. 1753. PINWEED

Herbs or subshrubs. Leaves alternate, opposite, or whorled, simple, pinnate-veined, sessile or short-petiolate. Flowers in leafy panicles or racemiform inflorescences; sepals 5, basally connate, dimorphic, the outer 2 differing from the inner 3; petals 3, free; stamens 3–25; ovary

3-carpellate and -loculate, the style minute or absent, the stigmas 3, fimbriate-plumose. Fruit a 3-locular capsule or indehiscent.

A genus of 18 species; North America, West Indies, Mexico, and Central America. [Commemorates Johan Leche (1704–1764), Swedish botanist.]

Selected references: Hodgdon (1938); Lemke (2015); Wilbur and Daoud (1961); Spaulding (2013).

1. Fertile stems spreading pubescent.
    2. Leaves frequently more than 1.5 cm long; fruit thin-walled, readily splitting into 3 valves at maturity, subglobose, about equal to the calyx in length ....**L. mucronata**
    2. Leaves all less than 1 cm long; fruit thick-walled, indehiscent, ellipsoid, exserted from the calyx ⅓–½ its length.. **L. divaricata**
1. Fertile stems appressed-pubescent or glabrous.
    3. Outer narrow sepals equal in length or longer than the inner wider sepals.
        4. Outer sepals longer than the inner sepals, equaling or exceeding the fruit .... **L. minor**
        4. Outer sepals about equal to the inner sepals, shorter than the fruit .... **L. sessiliflora**
    3. Outer narrow sepals shorter than the inner wider sepals.
        5. Leaves pubescent on both surfaces; flowers and/or fruits clustered in 2s or 3s .... **L. cernua**
        5. Leaves glabrous on the upper surface; flowers and/or fruits borne separately.
            6. Fruit exserted ⅓–½ its length; plants suffruticose .... **L. deckertii**
            6. Fruit not exserted more than ⅕ its length; plant herbaceous.
                7. Leaves abruptly tapering at the apex into a hardened, shiny, conical callosity; inner sepals 3-nerved; fruit slightly exceeding the sepals ....**L. pulchella**
                7. Leaves merely acute, not differentiated into a callosity; inner sepals 1-nerved; fruit subequaling the sepals.
                    8. Calyx, stems, and the leaf undersurface (at least the midvein) appressed pilose; fruit equaling or slightly shorter than the inner sepals...**L. torreyi**
                    8. Calyx, stems, and leaves glabrous; fruit slightly exceeding the inner sepals...**L. lakelae**

Lechea cernua Small [Slightly drooping, in reference to the fruits.] NODDING PINWEED; SCRUB PINWEED.

Lechea cernua Small, Bull. Torrey Bot. Club 51: 384. 1924. TYPE: FLORIDA: Indian River Co.: near Sebastian, 6 Sep 1922, *Small et al. s.n.* (holotype: NY).

Perennial herb or subshrub, to 6 dm; stem finely appressed-pubescent. Leaves of the flowering stems alternate, the blade elliptic, 4–6 mm long, 2–4 mm wide, the apex acute, the base cuneate, the margin entire, the upper and lower surfaces pale-gray, short, shaggy, somewhat viscid pubescent, those of the basal shoots suborbicular, oval, or elliptic, 4–12 cm long, the upper and lower surfaces densely gray-pubescent, sessile or short-petiolate. Flowers with the bracts similar to the branch leaves, the pedicel 2–3 mm long, appressed-pubescent; outer sepals minute, scalelike, the inner ones suborbicular, ca. 2 mm long, appressed-pubescent. Fruit obovoid, ca. 2 mm long, slightly exserted, glabrous, smooth, thick-walled, 3-valvate; seeds 1–2.

Scrub. Frequent; central and southern peninsula. Endemic. Summer–fall.

Lechea cernua is listed as threatened in Florida (Florida Administrative Code, Chapter 5B-40).

Lechea deckertii Small [Commemorates Richard F. Deckert (fl. 1925), Miami naturalist.] DECKERT'S PINWEED.

> *Lechea deckertii* Small, Torreya 27: 103. 1927. TYPE: FLORIDA: Miami-Dade Co.: Lemon City, NW 64th St., Miami, Jul 1925, *Deckert s.n.* (holotype: NY; isotype: NY).
> *Lechea myriophylla* Small, Man. S.E. Fl. 884, 1505. 1933. TYPE: FLORIDA: Highlands Co.: E of Sebring, 13 Dec 1920, *Small & DeWinkeler 9772* (holotype: NY).

Erect or ascending perennial, suffruticose herb, to 1.5 cm; stem appressed-pubescent. Leaves of the flowering stems alternate, the blade linear to linear-subulate, 1–3 mm long, the apex acute, the base narrowly cuneate, the margin entire, the upper and lower surfaces glabrous, sessile or short-petiolate. Flowers few, the pedicel ca. 1 mm long, glabrous; outer sepals subulate, the apex acute, ca. 0.5 mm long, the inner ones oval, ca. 1 mm long, the apex obtuse; petals oval or suborbicular, ca. 1 mm long. Fruit subglobose, 1–2 mm long, exserted ⅓–½ its length, glabrous, exceeding the sepals, thin-walled, 3-valvate; seeds 1(2).

Sandhills and scrub. Frequent; peninsula, central panhandle. Georgia and Florida. Summer–fall.

Lechea divaricata Shuttlew. ex Britton [Spreading at an angle, in reference to the branches.] DRYSAND PINWEED; SPREADING PINWEED.

> *Lechea divaricata* Shuttleworth ex Britton, Bull. Torrey Bot. Club 21: 249. 1894. *Lechea major* Linnaeus var. *divaricata* (Shuttleworth ex Britton) A. Gray, Syn. Fl. N. Amer. 1(1): 192. 1895. TYPE: FLORIDA: Manatee Co.: s.d., *Rugel s.n.* (lectotype: GH). Lectotypified by Hodgdon (1938: 97).

Erect perennial herb, to 5 dm; stem densely villous-pubescent. Leaves of the flowering stems alternate, the blade ovate, oblong, or linear-oblong, 4–8 mm long, the apex acute, the base cuneate, the upper and lower surfaces villous-pubescent, short-petiolate. Flowers with the pedicel ca. 2 mm long; outer sepals shorter than the inner. Fruit globose, ca. 2 mm long, exserted from calyx, thick-walled, indehiscent.

Flatwoods. Occasional; central and southern peninsula. Endemic. Summer–fall.

Lechea divaricata is listed as endangered in Florida (Florida Administrative Code, Chapter 5B-40).

Lechea lakelae Wilbur [Commemorates Olga Korhonen Lakela (1890–1980), Finnish-born Minnesota and Florida botanist, curator of the herbarium at the University of South Florida (1960–1973).] LAKELA'S PINWEED.

> *Lechea lakelae* Wilbur, Rhodora 76: 481. 1974. TYPE: FLORIDA: Collier Co.: Marco Island, 7 Aug 1967, *Lakela 30953* (holotype: DUKE; isotype: USF).

Erect perennial herb, to 3(4) dm; stem glabrous. Leaves of the flowering stems alternate, the blade linear-elliptic, the apex acute, the base cuneate, the upper and lower surfaces glabrous, sessile or subsessile. Flowers on a pedicel ca. 1 mm long; inner sepals ca. 2 mm long, the outer ones linear to narrowly lanceolate, ca. 1 mm long. Fruit broadly cylindric to ovoid, 1–2 mm long, slightly exceeding the sepals, thin-walled, 3-valvate; seeds 2–3.

Scrub. Rare; Collier County. Endemic. Summer–fall.

Lechea lakelae is listed as endangered in Florida (Florida Administrative Code, Chapter 5B-40).

Lechea minor L. [Small.] THYMELEAF PINWEED.

Lechea minor Linnaeus, Sp. Pl. 90. 1753. *Lechea revoluta* Rafinesque, New Fl. 1: 94. 1836, nom. illegit. *Lechea thymifolia* Michaux, Fl. Bor.-Amer. 1: 77. 1803. *Lechea cinerea* Rafinesque, New Fl. 1: 91. 1836, nom. illegit.

Perennial or biennial herb, to 5 dm; stem sparsely appressed-pubescent. Basal leaves in whorls of 3 or opposite, the blade ovate to elliptic-ovate, 4–7 mm long, 2–4 mm wide, the apex mucronate, the base cuneate, the margin entire, the upper surface sparsely pilose, the lower surface spreading pilose to subvillous, subsessile; leaves of the flowering stem in whorls of 3–4 or opposite, the blade elliptic to lanceolate, 0.5–1.5 cm long, 2–7 mm wide, the apex acute, the base cuneate, the margin entire, the upper surface sparsely pilose to glabrous, the lower surface pilose on the midvein and near the margin, the petiole not well differentiated from the blade. Flowers with the bracts similar to the leaves, the pedicel 1–2 mm long; outer sepals lanceolate to linear, ca. 2 mm long, subequaling or longer than the elliptic ovate inner sepals, subappressed-pilose. Fruit ellipsoid to ovoid, ca. 2 mm long, thin-walled, 3-valvate; seeds 2–3.

Sandhills, flatwoods, and hammocks. Frequent; northern counties, central peninsula. Vermont and New Hampshire south to Florida, west to Ontario, Wisconsin, Illinois, Tennessee, and Texas. Summer.

Lechea mucronata Raf. [Mucronate, in reference to the leaf apex.] HAIRY PINWEED.

Lechea mucronata Rafinesque, Précis Découv. Somiol. 37. 1814. *Lechea mucronata* Rafinesque var. *simplex* Rafinesque, New Fl. 1: 93: 1836, nom. inadmiss.
Lechea villosa Elliott, Sketch Bot. S. Carolina 1: 184. 1816. *Lechea major* Linnaeus var. *villosa* (Elliott) Grosser, Pflanzenr. 4(Heft 14): 136. 1903. *Lechea villosa* Elliott var. *typica* Hodgdon, Rhodora 40: 53. 1938, nom. inadmiss. *Lechea minor* Linnaeus var. *villosa* (Elliott) B. Boivin, Naturaliste Canad. 93: 433. 1966.
Lechea drummondii Spach, in Hooker, Companion Bot. Mag. 2: 284. 1837.

Perennial or biennial herb, to 1 dm; stem densely villous. Basal leaves whorled, the blade elliptic-ovate, 0.8–1.5 cm long, 0.5–1.2 cm wide, the apex abruptly tapering to a hardened, conical callosity, the base cuneate, the margin entire, the upper surface glabrous, the lower surface villous on the midrib and margin; leaves of the flower stem opposite or whorled, the blade elliptic to ovate, 1–3 cm long, 3–4 mm wide, the apex acute to obtuse, the base cuneate, the margin entire, the upper surface glabrous, the lower surface villous. Flowers with the pedicel ca. 1 mm long; outer sepals narrowly lanceolate to linear, 1–1.5 mm long, shorter than the inner ones, the inner lanceolate-elliptic, 1.5–2 mm long. Fruit depressed globose to obovoid, ca. 2 mm long, subequaling the calyx, thin-walled, 3-valvate; seeds (2)3(5).

Sandhills and dry hammocks. Frequent; northern counties, central peninsula. Vermont and

New Hampshire south to Florida, west to Ontario, Wisconsin, Nebraska, Kansas, and New Mexico. Summer.

Lechea pulchella Raf. var. **ramosissima** (Hodgdon) Sorrie & Weakley [Beautiful and little; much branching.] LEGGETT'S PINWEED.

> *Lechea leggettii* Britton & Hollick var. *ramosissima* Hodgdon, Rhodora 40: 119, pl. 491(3). 1938. *Lechea pulchella* Rafinesque var. *ramosissima* (Hodgdon) Sorrie & Weakley, J. Bot. Res. Inst. Texas 1: 370. 2007.

Perennial or biennial herb, to 8 dm; stem appressed-pubescent to glabrate. Basal leaves opposite or whorled, the blade lanceolate to oblanceolate or elliptic, 0.5–1 cm long, 1–3 cm wide, the apex acute to obtuse, the base cuneate, the margin entire, the upper surface glabrous, the lower surface pilose on the midrib and margin; leaves of the flowering stem opposite or whorled, the blade, elliptic to ovate, 1–2 cm long, 2–4 mm wide, the apex acute to obtuse, mucronate, the base cuneate, the margin entire, the upper surface glabrous, the lower surface short-pilose. Flowers with the pedicel 1–2 mm long; outer sepals ca. 1.5 m long, shorter than the inner ones, the inner ones ca. 2 mm long. Fruit globose to obovoid, ca. 1.5 mm long, thin-walled, 3-valvate; seeds 2(3).

Sandhills and flatwoods. Occasional; northern counties south to Hernando County. Virginia south to Florida, west to Texas. Summer.

Lechea sessiliflora Raf. [With sessile flowers.] PINELAND PINWEED.

> *Lechea sessiliflora* Rafinesque, New Fl. 1: 97. 1836. TYPE: "Florida and Alabama."
> *Lechea floridana* Rafinesque, New Fl. 1: 95. 1836. TYPE: FLORIDA.
> *Lechea patula* Leggett, Bull. Torrey Bot. Club 6: 251. 1878. TYPE: FLORIDA: Duval Co.: Aug-Sep, *Curtiss 232*** (lectotype: NY). Lectotypified by Hodgdon (1938: 61).
> *Lechea exserta* Small, Man. S.E. Fl. 883, 1505. 1922. TYPE: FLORIDA: Broward Co.: W of Hallandale, 16 Dec 1923, *Small et al. 11089* (holotype: NY).
> *Lechea prismatica* Small, Man. S.E. Fl. 882, 1505. 1933. TYPE: FLORIDA: Highlands Co.: E of Sebring, 13 Dec 1920, *Small & DeWinkeler 9787* (holotype: NY).

Perennial herb, to 7 dm; stem appressed-pubescent. Basal leaves alternate, narrowly elliptic to oblong, 3–5 mm long, ca. 1 mm wide, the apex acute, the base cuneate, the margin entire, the upper surface glabrous, the lower surface finely pilose; leaves of the flowering stem alternate, the blade linear to narrowly elliptic, 0.5–1.5 cm long, 0.5–1.5 mm wide, the apex acute, the base cuneate, the margin entire, the upper surface glabrous, the lower surface pilose on the midvein and the margin, the petiole not well differentiated from the blade. Flowers with the pedicel 1–2 mm long; outer sepals subequaling the inner ones, the inner ones ca. 1 mm long. Fruit ellipsoid to obovoid, 1–2 mm long, thin-walled, 3-valvate; seeds 1(2).

Sandhills and flatwoods. Common; nearly throughout. Virginia south to Florida, west to Mississippi. Summer–fall.

Lechea torreyi (Chapm.) Legg. ex Britton [Commemorates John Torrey (1796–1873), of Columbia University, distinguished American botanist.] PIEDMONT PINWEED.

Lechea cinerea Rafinesque var. *torreyi* Chapman, Fl. South. U.S., ed. 2, suppl. 2. 678. 1892. *Lechea torreyi* (Chapman) Leggett ex Britton, Bull. Torrey Bot. Club 21: 251. 1894. TYPE: FLORIDA: St. Johns Co.: St. Augustine vicinity, Mar 1872, *Torrey s.n.* (lectotype: NY). Lectotypified by Hodgdon (1938: 104).

Lechea torreyi (Chapman) Leggett ex Britton var. *congesta* Hodgdon ex D. E. Lemke, Phytoneuron 2014–33: 1. 2014. TYPE: FLORIDA: Duval Co.: without locality, Aug, *Curtiss 232** (holotype: NY; isotype: NY).

Perennial herb, to 5 dm; stem appressed-pilose. Basal leaves alternate, linear, 5–8 mm long, ca. 1 mm wide, the apex acute, the base narrowly cuneate, the margin entire, the upper surface glabrous, the lower surface sparsely villous, the leaves of the flowering stem alternate, the blade linear to elliptic, 1–2 cm long, 1–2 mm wide, the apex acute, the base narrowly cuneate, the margin entire, the upper surface glabrous, the lower surface pilose, the petiole not well differentiated from the blade. Flowers with the pedicel ca. 1 mm long; outer sepals ca. 2 mm long, shorter than the inner ones. Fruit ovoid, ca. 1.5 mm long, subequaling or shorter than the calyx, thin-walled, 3-valvate; seeds 3(6).

Flatwoods. Common; nearly throughout. North Carolina south to Florida, west to Mississippi; Central America. Summer–fall.

EXCLUDED TAXA

Lechea pulchella Rafinesque—Because infraspecific categories were not recognized, the typical variety was reported for Florida by implication by Small (1933, as *L. leggettii* Britton & Hollick), Radford Ahles & Bell (1964, as *L. leggettii* Britton & Hollick), Wilhelm (1984), Clewell (1985), Wunderlin (1998), and Wunderlin and Hansen (2003). All Florida material is of var. *ramosissima*.

Lechea major Linnaeus—Reported for Florida by Chapman (1860), who misapplied this name, which is a synonym of *Crocanthemum canadense* (Linnaeus) Britton, to material of *L. mucronata*.

Lechea racemulosa Michaux—Reported for Florida by Small (1903, 1913, 1933). Excluded from the state by Hodgdon (1938) and by Wilbur and Daoud (1961).

Lechea tenuifolia Michaux—Reported for Florida by Small (1903, 1913, 1933) and Radford et al. (1968). Excluded from the state by Hodgdon (1938) and Wilbur and Daoud (1961).

MORINGACEAE Martinov, nom. cons. 1820.
HORSERADISHTREE FAMILY

Shrubs or trees. Leaves alternate, odd-bipinnately compound, petiolate, estipulate. Flowers in axillary panicles, bracteate, zygomorphic, bisexual; sepals 5, free; petals 5, free; nectariferous disk present; stamens 10, 5 fertile, 5 staminodial, the fertile anthers dorsifixed, 1-locular, longitudinally dehiscent, the staminodes 5; ovary weakly perigynous, 3-carpellate, 1-loculate, stipitate, the style 1. Fruit a capsule.

A monogeneric family with about 11 species; North America, West Indies, Mexico, Central America, Africa, and Asia.

Selected reference: Olson (2010).

Moringa Adans. 1763. HORSERADISHTREE

Shrubs or trees. Leaves alternate, odd-bipinnately compound, petiolate, estipulate. Flowers in axillary panicles, bracteate, zygomorphic, bisexual; sepals 5, free, petaloid; petals 5, free; stamens 10, 5 fertile, 5 staminodial, the fertile anthers dorsifixed, 1-locular, longitudinally dehiscent; ovary 3-carpellate, 1-loculate, on a short gynophore. Fruit 3-valved capsules.

A genus with about 11 species; North America, West Indies, Mexico, Central America, Africa, and Asia. [Tamil language vernacular name *murungai*, twisted pod, in reference to the young fruit.]

Moringa oleifera Lam. [*Oleum*, oil, and *fer*, bearing.] HORSERADISHTREE.

Guilandina moringa Linnaeus, Sp. Pl. 381. 1753. *Moringa oleifera* Lamarck, Encycl. 1: 398. 1785. *Anoma moringa* (Linnaeus) Loureiro, Fl. Cochinch. 279. 1790. *Hyperanthera moringa* (Linnaeus) Vahl, Symb. Bot. 1: 30. 1790. *Moringa pterygosperma* Gaertner, Fruct. Sem. Pl. 2: 314. 1791, nom. illegit. *Moringa erecta* Salisbury, Prodr. Stirp. Chap. Allerton 326. 1796, nom. illegit. *Moringa zeylanica* Persoon, Syn. Pl. 1: 461. 1805, nom. illegit. *Moringa polygona* de Candolle, Prodr. 2: 478. 1825, nom. illegit. *Moringa moringa* (Linnaeus) Millspaugh, Publ. Field Columb. Mus., Bot. Ser. 1: 490. 1902, nom. inadmiss.

Large shrub or small tree, to 9 m; bark rough corky, pale gray or tan, the branches puberulent or glabrate. Leaves with the blade odd-bipinnately compound, 3–6 dm long, the pinnae 4–8, opposite, the leaflets 75–150, opposite, the blade oblong to obovate, 1–2 cm long, 0.5–1.5 cm wide, membranaceous, the apex rounded to emarginate, the base rounded to cuneate, asymmetrical, the margin entire, short-petiolulate, the petiole to ca. 1.5 dm long. Flowers in axillary panicles 10–25 cm long, the peduncle 5–10 mm long, the pedicel 5–10 mm long, each flower subtended by a glandular bract; sepals linear to linear-oblong, 9–13 mm long, white, unequal, reflexed; petals linear to linear-oblong, 10–15 mm long, white with a yellow base, unequal; stamens and staminodes ca. 6 mm long, white, puberulent at the base; ovary puberulent, the style ca. 4 mm long, the gynophore 2–3 mm long. Fruit linear, 2–4.5 dm long, 1–2 cm wide, obtusely trigonous, beaked, tan, the valves strongly ribbed; seeds numerous, globular, ca. 10 mm long, the wings 2.5–3 cm long, membranous.

Disturbed sites. Rare; Manatee County, southern peninsula. Escaped from cultivation. Florida; West Indies, Mexico, Central America; Asia and Africa. Native to Asia. Spring.

CARICACEAE Dumort., nom. cons. 1829. PAPAYA FAMILY

Trees. Leaves alternate, simple, palmate-lobed, petiolate, estipulate. Flowers axillary, paniculate, actinomorphic, bisexual or unisexual (plants dioecious or polygamous); sepals 5, basally connate; petals 5, basally connate; stamens 10 or reduced to staminodia in the carpellate flowers, the anthers 2-locular; ovary superior, 5-carpellate, 1-loculate, the stigmas 5. Fruit a berry.

A family of 6 genera and about 30 species; North America, West Indies, Mexico, Central America, South America, Africa, Asia, and Pacific Islands.

Selected reference: Holmes (2010).

Carica L. 1753. PAPAYA

Trees. Leaves alternate, simple, palmate-lobed, petiolate, estipulate. Flowers axillary, paniculate, actinomorphic, bisexual or unisexual (plants dioecious or polygamous). Staminate flowers: sepals 5, basally connate; petals 5, funnelform; stamens 10 in 2 series, the anthers 2-locular; ovary vestigial or absent. Carpellate flowers: sepals 5, basally connate; petals 5, basally connate; staminodia absent; ovary 5-carpellate, 1-loculate, the style 1, the stigmas 5. Bisexual flowers: petals 5, basally connate or funnelform; stamens 10; ovary 5-carpellate, 1-loculate. Fruit a berry.

A genus of about 20 species; North America, West Indies, Mexico, Central America, South America, Africa, Asia, and Pacific Islands. [Resembling the fig, *Ficus carica* (Moraceae).]

Carica papaya L. [Pre-Linnaean name.] PAPAYA.

Carica papaya Linnaeus, Sp. Pl. 1036. 1753. *Papaya carica* Gaertner, Fruct. Sem. Pl. 2: 191. 1791. *Papaya vulgaris* Poiret, in Lamarck, Encycl. 5: 2. 1804, nom. illegit. *Papaya sativa* Tussac, Fl. Antill. 3: 45. 1824–1826, nom. illegit. *Papaya papaya* (Linnaeus) H. Karsten, Deut. Fl. 894. 1882, nom. inadmiss.

Trees, to 6 m; stems glabrous, succulent, hollow, with conspicuous leaf scars, with a milky latex. Leaves the blade suborbicular, 2–5 dm long and wide, the lobes 7(11), pinnatifid or the basal ones entire, the lobe apex acute to acuminate, the base deeply cordate, the margin entire or irregularly lobed, the upper and lower surfaces glabrous. Staminate flowers many in a panicle to 50 cm long, the peduncle to 21 cm long, the axis glabrous or puberulent; calyx 1–1.5 mm long, the lobes ovate to triangular, ca. 1 mm long, glabrate; corolla salverform, 2.2–4 cm long, cream-white or yellow, the tube 1–2.4 cm long, the lobes oblong, 1.1–1.8 cm long, convolute; 5 longer stamens with the filaments 3–4 mm long, densely pilose, the anther ca. 2 mm long, the connective not expanded, the 5 shorter stamens with the filament to 1 mm long, pilose, the anther 2–3 mm long, the connective prolonged beyond the theca; ovary 3–7 mm long. Carpellate flowers few, the peduncle less than 2 cm long, the rachis glabrous or puberulent; calyx as in the staminate, but often slightly larger; petals nearly free, lanceolate, 2–5 mm long, white to yellow; staminodia absent; ovary ovoid to elliptic, 1–3 cm long, the style less than 1 mm long, the stigmas to 1 cm long, usually irregularly branched. Fruit ovoid to ellipsoid or globose, 2–10 cm long, 1.5–6 cm wide, yellow to orange, the pulp yellow to orange; seeds numerous, ellipsoid, 4–7 mm long, black, surrounded by a membranous aril.

Coastal hammocks, shell middens, and disturbed sites. Occasional; peninsula. Florida; West Indies, Mexico, Central America, and South America; Africa and Pacific Islands. Native to North America and tropical America. All year.

In plants recently cultivated, the flowers may be bisexual, strongly sympetalous, with 10 stamens, a slightly 5-lobed ovary, and fruits 50–60 cm long, 20–30 cm wide.

Carica papaya is believed to have originated in Mexico and Central America where its closest relatives occur. Widespread by pre-Columbian aboriginals, it apparently arrived in Florida no later than 300 A.D. and thus is considered native (Ward, 2011).

BATACEAE Mart. ex Perleb, nom. cons. 1838. SALTWORT FAMILY

Subshrubs. Leaves simple, opposite, epetiolate, stipulate. Flowers in axillary catkins, actinomorphic, unisexual (plants dioecious). Staminate flowers: tepals (staminodes?) 4(5), free; stamens 4(5), free, the anthers dorsifixed, versatile, dehiscing by longitudinal slits; ovary absent. Carpellate flowers: bracteate; tepals absent; staminodes absent; ovary 2 carpellate, falsely 4-loculate. Fruit a syncarp.

A unigeneric family with 2 species; North America, West Indies, Mexico, Central America, South America, Australia, and Pacific Islands.

Selected references: Rogers (1982); Thorne (2010).

Batis P. Browne 1756. SALTWORT

Subshrubs. Leaves simple, opposite, stipulate. Flowers unisexual (plants dioecious). Staminate flowers in axillary catkins, bracteate, tepals (staminodes?) 4(5); stamens 4(5), the anthers dorsifixed, dehiscing by longitudinal slits. Carpellate flowers in axillary spikes; perianth and staminodes absent; gynoecium 2 carpellate, 4-loculate, stigmas 2, sessile. Fruit a drupaceous syncarp.

A genus of 2 species; North America, West Indies, Mexico, Central America, South America, Australia, and Pacific Islands. [From the Greek *batos*, bramble, apparently in reference to the habit and fruit.]

Batis maritima L. [Growing by the sea]. SALTWORT; TURTLEWEED.

Batis maritima Linnaeus, Syst. Nat., ed. 10. 1289. 1759.

Creeping or prostrate succulent subshrub, to 1.5 m; stem initially quadrangular in cross section, becoming terete, glabrous, rooting at the nodes, the bark light gray, flaking. Leaves with the blade narrowly elliptic to narrowly oblanceolate, somewhat falcate, somewhat triangular to terete in cross section, the lower surface flattened, the apex acuminate to mucronate, with a small deltoid appendage below the insertion, the base narrowly cuneate, the margin entire, the upper and lower surfaces glabrous, epetiolate; stipules paired, minute, succulent, inserted on the stem, glandular. Staminate flowers in axillary ellipsoid, turbinate, or subglobose catkins, 4-ranked, 5–10 mm long, these sessile or short-pedunculate, sometimes clustered on short branches, each flower subtended by an ovate, succulent bract, the apex of this obtuse, usually mucronate, the margin membranaceous, erose, at first enclosed in a sheath (spathella) that dehisces to form a 2-lipped cup; tepals white, without venation or with an unbranched midvein, ungulate; stamens longer than the tepals and alternate with them, exserted beyond the subtending bract, glabrous. Carpellate flowers 2–14(24), decussately arranged in a fleshy spike 5–10 mm long, often clustered on short branchlets, sessile or short-pedunculate, the bracts ovate to orbicular, peltate, sometimes 3-lobed, glabrous, the margin membranaceous, erose, the apex rounded to acute or apiculate, the base rounded to cordate; the stigmas fimbriate.

Fruit a subcylindric-obconic syncarp 1–2 cm long, this with rounded protuberances representing each flower, the apical flowers not developing; seeds surrounded by a hard endocarp (pyrenes), narrowly deltoid, laterally compressed, white.

Coastal strands, salt flats, marshes, and mangroves. Frequent; peninsula, west to central panhandle. North Carolina south to Florida, west to Texas, also California; West Indies, Mexico, Central America, and South America; Pacific Islands. Spring–summer.

BRASSICACEAE Burnett, nom. cons. 1835. MUSTARD FAMILY

Herbs, shrubs, or trees. Leaves alternate or whorled, simple or palmately compound, pinnate-veined, the margin sometimes lobed and the leaves appearing pinnately compound, petiolate or epetiolate, stipulate or estipulate. Flowers in terminal and axillary racemes, corymbs, or solitary, pedunculate, pedicellate, bracteate or ebracteate, ebracteolate, actinomorphic or zygomorphic, bisexual (sometimes appearing unisexual due to incomplete development); sepals 4, free or basally connate; petals 4, adnate to the receptacle, free, equal or unequal, sometimes absent; intrastaminal nectariferous disk or glands present or absent; stamens 2–120, the filaments free or basally adnate to the gynophore or androgynophore, the anthers 2-thecal, dehiscing by longitudinal slits; ovary 1, 1- or 2-carpellate, 2-loculate, the style superior, the stigma 1, capitate, lobed or unlobed, the gynophore present or absent. Fruit a capsule dehiscent by 2 valves, a dehiscent silique, a schizocarp, berry, or nutlike; seeds 1-many, arillate or not.

A family of about 360 genera and about 4,500 species; nearly cosmopolitan.

Recent molecular and morphological data reveal that the Capparaceae as traditionally circumscribed is paraphyletic with the Brassicaceae and embedded within it (see Tucker, 2010a). This has resulted in some workers recognizing three monophyletic lineages as families: Capparaceae, Cleomeaceae, and Brassicaceae s.s. We follow the broad circumscription of the Brassicaceae in combining all three families into one.

Capparaceae Juss., nom. cons. (1789); *Cleomaceae* Bercht. & J. Presl (1825); *Cruciferae* Juss, nom. cons. et nom. alt. (1789).

Selected references: Al-Shehbaz (2010a); Tucker (2010a).

1. Trees or shrubs.
    2. Upper stem and the lower leaf surface glabrous ....**Cynophalla**
    2. Upper stem and the lower leaf surface lepidote..**Quadrella**
1. Herbs.
    3. Leaves palmately 3- to 7-foliolate (upper ones often simple).
        4. Gynophore long, filiform ....**Cleome**
        4. Gynophore short, stout.
            5. Flowers pale to bright yellow or orange-yellow ....**Cleome**
            5. Flowers white to pale pink..**Polanisia**
    3. Leaves simple, the margin sometimes lobed or dissected and appearing pinnately compound or rarely palmately divided.
        6. Fruit borne on a conspicuous gynophore more than 1 cm long ....**Warea**
        6. Fruit sessile or borne on a gynophore less than 3 mm long.

7. Fruit flattened contrary to the septum, the septum therefore narrower than the width of the fruit.

    8. Fruit obcordate or obdeltoid...**Capsella**

    8. Fruit oval, rotund, or elliptical.

        9. Seeds several in each locule..**Thlaspi**

        9. Seeds 1 in each locule..**Lepidium**

7. Fruit terete, prismatic, or flattened parallel to the septum, the septum therefore as wide as the fruit.

    10. Fruit flattened or compressed.

        11. Fruit 1-seeded ....**Lobularia**

        11. Fruit several-seeded.

            12. Leaves palmately divided into 4–7 linear segments ....**Cardamine**

            12. Leaves not as above.

                13. Leaves entire or with merely dentate margins.

                    14. Fruit 7–10 cm long ....**Boechera**

                    14. Seeds to 3 cm long.

                        15. Seeds in 2 rows in each locule..**Draba**

                        15. Seeds in 1 row in each locule.. **Cardamine**

                13. Leaves lyrate pinnatifid or bipinnatifid.

                    16. Flowers white to pink.

                        17. Stem glabrous; fruit replum flat; seeds winged...................... **Cardamine**

                        17. Stem pubescent near the base; fruit replum rounded; seeds not winged.

                            .... **Planodes**

                  16. Flowers yellow.

                      18. Plant with simple trichomes.. **Erucastrum**

                      18. Plant with stellate trichomes ....**Descurainia**

    10. Fruit terete or prismatic.

        19. Fruit prominently beaked.

            20. Beak conic, terete.

                21. Seeds in 1 row in each locule.. **Brassica**

                21. Seeds in 2 rows in each locule...**Diplotaxis**

            20. Beak flat, 2-edged.

                22. Fruit dehiscing into 2 valves.. **Sinapis**

                22. Fruit ilndehiscent or transeversely dividing into 1-seeded segments ....

                    .... **Raphanus**

        19. Fruit merely tipped with the short or obsolete persistent style.

            23. Fruit transversely 2-jointed, dividing into two segments at maturity ....**Cakile**

            23. Fruit not transversely 2-jointed and dividing into two segments at maturity.

                24. Fruit 4-sided .... **Erysimum**

                24. Fruit terete.

                    25. Seeds in 2 rows in each locule.

                      26. Petals yellow or absent ....**Rorippa**

                      26. Petals white ....**Nasturtium**

                    25. Seeds in 1 row in each locule.

                      27. Leaves entire, cordate-clasping.. **Conringia**

                      27. Leaves coarsely toothed or pinnatifid, not cordate-clasping.

28. Petals pale yellow ....**Sisymbrium**
28. Petals white .... **Cardamine**

Boechera Á. Löve & D. Löve 1976. ROCKCRESS

Herbs. Leaves basal and cauline, alternate, simple, pinnate-veined, epetiolate, estipulate. Flowers in terminal racemes, ebracteate, actinomorphic, bisexual; sepals 4, free, equal; petals 4, equal; stamens 6, tetradynamous, the filaments free; intrastaminal nectar glands confluent, subtending the stamen bases. Fruit a silique, dehiscent, the replum rounded, the septum complete; seeds uniseriate, winged.

A genus of about 110 species; North America, Mexico, and Asia. [Commemorates Tyge Wittrock Böcher (1909–1983), Danish cytogeneticist who worked on subarctic plants.]

Selected reference: Al-Shehbaz and Windham (2010).

Boechera canadensis (L.) Al-Shehbaz [Of Canada]. SICKLEPOD.

Arabis canadensis Linnaeus, Sp. Pl. 665. 1753. *Erysimum canadense* (Linnaeus) Kuntze, Revis. Gen. Pl. 1: 28. 1891. *Turritis canadensis* (Linnaeus) Nieuwland, Amer. Midl. Naturalist 4: 39. 1915. *Boechera canadensis* (Linnaeus) Al-Shehbaz, Novon 13: 384. 2003. *Borodina canadensis* (Linnaeus) P. J. Alexander & Windham, in P. J. Alexander et al., Syst. Bot. 38: 203. 2013.

Biennial herb, to 10(12.5) dm; stem sparsely to densely pubescent with simple trichomes. Basal leaves with the blade obovate to oblanceolate, 6–10 cm long, 1–3 cm wide, the apex acute to obtuse, the base cuneate, the margin dentate, ciliate proximally, the upper and lower surfaces simple and stellate, the petiole 3–4 cm long, the cauline leaves with the blade 3–6 cm long, 0.5–1 cm wide, the proximalmost ones with basal auricles 1–2(5) mm long, the upper and lower surfaces glabrous or sparsely pubescent. Flowers 15–65(80) in a simple or branched raceme, the pedicel 6–12(15) mm long, reflexed or horizontal, curved, pubescent or glabrous; sepals 3–5 mm long, white, glabrous. Fruit linear, 4–10 cm long, 3–4 mm wide, curved, usually secund, glabrous; seeds 40–60, 2–3 mm long, the wing to 1 mm wide.

Wet hammocks and riverbanks. Rare; Liberty and Jackson Counties. Quebec south to Florida, west to Ontario, North Dakota, South Dakota, Nebraska, Kansas, Oklahoma, and Texas. Spring.

Boechera canadensis is listed as endangered in Florida (Florida Administrative Code, Chapter 5B-40).

Brassica L. 1753. MUSTARD

Annual or biennial herbs. Leaves basal and cauline, alternate, simple, pinnate-veined, petiolate or epetiolate, estipulate. Flowers in terminal corymbose racemes, pedicellate, ebracteate, actinomorphic, bisexual; sepals 4 free, equal; petals 4, free, equal; stamens 6, tetradynamous; intrastaminal nectar glands present. Fruit a silique, dehiscent, 2-segmented, the terminal segment beak-like, seedless or few-seeded, the lower segment valvular, several-seeded, the replum rounded, the septum complete; seeds uniseriate.

A genus of 35 species; nearly cosmopolitan. [Latin name for cabbage].
Selected reference: Warwick (2010a).

1. Upper leaves clasping the stem.
   2. Sepals to 6.5 mm long; petals to 10 mm long; beak of the fruit more than 9 mm long, often more than 12 mm long...**B. rapa**
   2. Sepals more than 6.5 mm long; petals more than 10 mm long, beak of the fruit rarely more than 9 mm long, never more than 12 mm long...**B. napus**
1. Upper leaves not clasping the stem.
   3. Plant glabrous or glabrate; pedicels elongate; silique terete, loosely ascending, 3–6 cm long............
   ....**B. juncea**
   3. Plant hirsute; pedicels short and thick; silique somewhat 4-angled, closely appressed, 1–2 cm long
   ....**B. nigra**

Brassica juncea (L.) Czern. [*Junceus*, rush-like.] INDIA MUSTARD; LEAF MUSTARD.

Sinapis juncea Linnaeus, Sp. Pl. 668. 1753. *Raphanus junceus* (Linnaeus) Crantz, Cl. Crucif. Emend. 110. 1769. *Brassica juncea* (Linnaeus) Czernajew, Consp. Pl. Charc. 8. 1859. *Brassica arvensis* (Linnaeus) Rabenhorst var. *juncea* (Linnaeus) Kuntz, Revis. Gen. Pl. 1: 19. 1891. *Caulis junceus* (Linnaeus) E.H.L. Krause, Bot. Centralbl. 81: 207. 1900. *Crucifera juncea* (Linnaeus) E.H.L. Krause, in Sturm, Fl. Deutschl., ed. 2. 6: 133. 1902. *Brassica juncea* (Linnaeus) Czernajew subsp. *eujuncea* Thellung, Verh. Bot. Vereins Prov. Brandenburg 50: 152. 1908, nom. inadmiss.

Sinapis japonica Thunberg, Fl. Jap. 262. 1784. *Brassica japonica* (Thunberg) Siebold ex Miquel, Ann. Mus. Bot. Lugduno-Batavi 2: 74. 1866. *Brassica nigra* (Linnaeus) W.D.J. Koch var. *japonica* (Thunberg) O. E. Schulz, in Engler, Pflanzenr. 4(Heft 70): 79. 1919. *Brassica juncea* (Linnaeus) Czernajew var. *japonica* (Thunberg) L. H. Bailey, Gentes Herb. 1: 93. 1922.

Annual herb, to 1 m; stem glabrate, somewhat glaucous. Basal leaves with the blade oblanceolate, (4)6–30(80) cm long, 1.5–15(28) cm wide, the margin pinnatifid to pinnately lobed, the lobes 1–3 on each side, the upper and lower surfaces glabrate, somewhat glaucous, the petiole (1)2–8(15) cm long, the cauline leaves with the blade oblong or lanceolate, the apex acute to obtuse, the base cuneate, the margin dentate to lobed, the upper and lower surfaces glabrate, somewhat glaucous, short-petiolate or sessile. Flowers in a terminal, corymbose raceme, the pedicel (0.5)1–1.5(2) cm long, spreading to divaricately ascending; sepals oblong, 4–6(7) mm long; petals ovate to obovate, (7)9–13 mm long, pale yellow, the claw 3–6 mm long, the apex rounded or emarginate; stamens 4–7 mm long. Fruit subcylindric, (2)3–5(6) cm long, 2–5 mm wide, torulose, the terminal segment conic, (4)5–10(15) mm long, seedless, the valvular segment (1.5)2–4.5 cm long, with 6–15(20) seeds per locule; seeds globose, 1–2 mm long, brown or yellow, the surface finely reticulate-alveolate.

Disturbed sites. Occasional; nearly throughout. Escaped from cultivation. Nearly cosmopolitan. Native to Europe, Africa, and Asia. Spring–summer.

Brassica napus L. [Pre-Linnaean vernacular name.] RAPE.

Brassica napus Linnaeus, Sp. Pl. 666. 1753. *Rapa napus* (Linnaeus) Miller, Gard. Dict., ed. 8. 1768. *Raphanus napus* (Linnaeus) Crantz, Cl. Crucif. Emend. 113. 1769. *Brassica asperifolia* Lamarck,

Encycl. 1: 746. 1784, nom. illegit. *Brassica asperifolia* Lamarck var. *sylvestris* Lamarck, Encycl. 1: 746. 1984, nom. inadmiss. *Brassica oleifera* Moench, Methodus 253. 1794, nom. illegit. *Brassica napus* Linnaeus var. *sylvestris*, Schkuhr, Bot. Handb. 2: 258. 1796, nom. inadmiss. *Sinapis napus* (Linnaeus) Brotero, Fl. Lusit. 1: 586. 1804. *Sinapis rapa* (Linnaeus) Brotero var. *sylvestris* Stokes, Bot. Mat. Med. 3: 479. 1812. *Brassica napus* Linnaeus var. *oleifera* de Candolle, Syst. Nat. 2: 592. 1821, nom. inadmiss. *Brassica oleracea* Linnaeus var. *napus* (Linnaeus) Sprengel, Syst. Veg. 2: 909. 1825. *Napus oleifera* Schimper & Spenner, in Spenner, Fl. Friburg 3: 940. 1829. *Brassica napus* Linnaeus subsp. *oleifera* Metzger, Syst. Beschr. Kohlart 40. 1833, nom. inadmiss. *Brassica napus* Linnaeus var. *biennis* Metzger, Syst. Beschr. Kohlart 40. 1833, nom. inadmiss. *Brassica napus* Linnaeus var. *leptorhiza* Spach, Hist. Nat. Veg. 6: 367. 1838, nom. inadmiss. *Brassica campestris* Linnaeus subsp. *napus* (Linnaeus) Hooker f. & T. Anderson, in Hooker f., Fl. Brit. India 1: 156. 1872. *Brassica campestris* Linnaeus var. *napus* (Linnaeus) Babington, Man. Brit. Bot., ed. 8. 121. 1881. *Brassica sativa* Clavaud subsp. *asperifolia* Clavaud, Acta Soc. Linn. Bordeaux 35: 297. 1881. *Brassica napus* Linnaeus var. *typica* Pospichal, Fl. Oesterr. Küstenl. 1: 497. 1897, nom. inadmiss.

Annual herb, 1.3 m; stem glabrous, glabrescent, or pubescent. Basal leaves with the blade oblanceolate, 0.5–2.5(4) dm long, 2–7(10) dm wide, the margin lyrate-pinnatifid or pinnately lobed, the lobes 0–6 on each side, the upper and lower surfaces glaucous, glabrous or pubescent, the petiole to 15 cm long, often winged, the cauline leaves with the blade oblong or lanceolate, the apex obtuse, the base auriculate or amplexicaul, the margin entire, the upper and lower surfaces glabrous, glabrate, or pubescent, glaucous, epetiolate. Flowers in a corymbose raceme, the pedicel 1–3 cm long, spreading to ascending; sepals oblong, (5)6–10 mm long; petals broadly obovate, 10–16 mm long, yellow, the claw 5–9 mm long, the apex rounded; stamens (5)7–10 mm long. Fruit subcylindric, (3.5)5–10(11) cm long, 3–5 mm wide, terete, sometimes slightly torulose, the terminal segment attenuate-conic, (5)9–16 mm long, usually seedless or rarely 1- or 2-seeded, the valvular segment (3)4–8.5(9.5) cm long, with 12–20(30) seeds per locule; seeds globose, 2–3 mm long, brown to black or reddish, the surface finely reticulate-alveolate.

Disturbed sites. Rare; Leon, Lee, and Miami-Dade Counties. Escaped from cultivation. Nearly throughout North America; Mexico, Central America, and South America; Europe, Africa, Asia, Australia, and Pacific Islands. Native to Europe. Spring–summer.

Brassica nigra (L.) W.D.J. Koch [Black, in reference to the seed color.] ABYSSINIAN MUSTARD; BLACK MUSTARD.

Sinapis nigra Linnaeus, Sp. Pl. 668. 1753. *Raphanus sinapis-officinalis* Crantz, Cl. Crucif. Emend. 109. 1769, nom. illegit. *Melanosinapis communis* Spenner, Fl. Friburg. 945. 1829, nom. illegit. *Brassica sinapioides* Roth, Man. Bot. 2: 947. 1830, nom. illegit. *Brassica nigra* (Linnaeus) W.D.J. Koch, in Röhling, Deutchl. Fl., ed. 3. 4: 713. 1833. *Brassica sinapis* Noulet, Fl. Bass. Sous-Pyren. 32. 1837, nom. illegit. *Brassica melanosinapis* Endlicher, Cat. Horti Vindob. 2: 259. 1842, nom. illegit. *Brassica nigra* (Linnaeus) W.D.J. Koch var. *vulgaris* Alefeld, Landw. Fl. 233. 1866, nom. inadmiss. *Sisymbrium nigrum* (Linnaeus) Prantl, Exkurs.-Fl. Bayern 222. 1884. *Crucifera sinapis* E.H.L. Krause, in Sturm, Fl. Deutschl. ed. 2. 6: 131, pl. 38. 1902, nom. illegit. *Melanosinapis nigra* (Linnaeus) Calestani, Nuovo Giorn. Bot. Ital., ser. 2. 15: 384. 1908.

Annual herb, to 20 dm; stem sparsely to densely hirsute-hispid, (at least basally). Basal leaves with the blade oblanceolate, 6–30 cm long, 1–10 cm wide, the margin lyrate-pinnatifid to sinuate-lobed, the lobes 1–3 on each side, the terminal lobe ovate, the apex obtuse, the upper and lower surfaces sparsely to densely hirsute-hispid, the petiole to 10 cm long, the cauline leaves with the blade ovate-elliptic to lanceolate, similar to the basal, reduced distally and less divided, the margin entire to sinuate-serrate, sessile or subsessile. Flowers in a corymbose raceme, the pedicel (2)3–5(6) mm long, erect, straight; sepals oblong, 4–6(7) mm long; petal obovate, 7–11(13) mm long, the claw 3–6 mm long, the apex rounded; filaments 3–5 mm long. Fruit subcylindric, 1–2.5 cm long, 2–3 mm wide, erect-ascending, somewhat appressed to the rachis, somewhat 4-angled, smooth, the terminal segment linear, (1)2–5(6) mm long, seedless, the valvular segment (0.4)0.8–2(2.5) cm long, with 2–5(8) seeds per locule; seeds globose, 1–2 mm long, brown to black, the surface coarsely reticulate, minutely alveolate.

Disturbed sites. Rare; Leon County. Escaped from cultivation. Nearly throughout North America; Mexico, Central America, and South America; Europe, Africa, Asia, Australia, and Pacific Islands. Native to Europe. Spring.

Brassica rapa L. [Pre-Linnaean generic name.] RAPE; TURNIP.

Brassica rapa Linnaeus, Sp. Pl. 666. 1753. *Raphanus rapa* (Linnaeus) Crantz, Cl. Crucif. Emend. 113. 1769. *Brassica tuberosa* Salisbury, Prodr. Stirp. Chap. Allerton 272. 1796, nom. illegit. *Sinapis tuberosa* Poiret, in Lamarck, Encycl. 4: 346. 1797, nom. illegit. *Sinapis rapa* (Linnaeus) Brotero, Fl. Lusit. 1: 586. 1804. *Sinapis rapa* (Linnaeus) Brotero var. *rotunda* Stokes, Bot. Mat. Med. 3: 480. 1812, nom. inadmiss. *Brassica sphaerorrhiza* S. Gray, Nat. Arr. Brit. Pl. 2: 683. 1821, nom. illegit. *Napus rapa* (Linnaeus) Schimper & Spenner, in Spenner, Fl. Friburg. 941. 1829. *Brassica rapa* Linnaeus subsp. *rapifera* Metzger, Syst. Beschr. Kohlart. 52. 1833, nom. inadmiss. *Brassica rapa* Linnaeus var. *communis* Schübler & G. Martens, Fl. Würtemberg 438. 1834, nom. inadmiss. *Brassica rapa* Linnaeus var. *sarcorhiza* Spach, Hist. Nat. Vég. 6: 371. 1838, nom. inadmiss. *Brassica campestris* Linnaeus subsp. *rapa* (Linnaeus) Babington, Man. Brit. Bot., ed. 3. 24. 1843. *Brassica campestris* Linnaeus var. *rapa* (Linnaeus) Hartman, Handb. Skand. Fl., ed. 6. 110. 1854. *Brassica rapa* Linnaeus var. *rapifera* Alefeld, Landw. Fl. 24. 1866, nom. inadmiss. *Brassica sativa* Clavaud subsp. *rapa* (Linnaeus) Clavaud, Acta Soc. Linn. Bordeaux 35: 297. 1881. *Brassica rapa* Linnaeus var. *typica* Pospichal, Fl. Oesterr. Küstenl. 1: 496. 1897. *Crucifera rapa* (Linnaeus) E.H.L. Krause, in Sturm, Fl. Deutschl., ed. 2. 6: 137. 1902. *Brassica rapifera* Dalla Torre & Sarntheim, Fl. Tirol. 6(2): 341. 1909, nom. illegit.

Brassica campestris Linnaeus, Sp. Pl. 666. 1753. *Raphanus campestris* (Linnaeus) Crantz, Cl. Crucif. Emend. 113. 1769. *Erysimum campestre* (Linnaeus) Scopoli, Fl. Carniol., ed. 2. 2: 27. 1772. *Gorinkia campestris* (Linnaeus) J. Presl & C. Presl, Fl. Cech. 141. 1819. *Brassica campestris* Linnaeus var. *oleifera* de Candolle, Syst. Nat. 2: 589. 1821, nom. inadmiss. *Napus campestris* (Linnaeus) Schimper & Spenner, in Spenner, Fl. Friburg 3: 941. 1829. *Brassica rapa* Linnaeus subsp. *oleifera* Metzger, Syst. Beschr. Kohlart 49. 1833. *Brassica rapa* Linnaeus var. *annua* Metzger, Syst. Beschr. Kohlart 51. 1833. *Brassica rapa* Linnaeus var. *leptorhiza* Spach, Hist. Nat. Vég. 6: 371. 1838, nom. illegit. *Brassica rapa* Linnaeus var. *campestris* (Linnaeus) W.D.J. Koch, Syn. Fl. Germ. Helv., ed. 2. 59. 1843, nom. illegit. *Brassica asperifolia* Lamarck var. *oleifera* (Metzger) Godron, in Grenier & Godron, Fl. France 1: 77. 1848, nom. illegit. *Brassica rapa* Linnaeus var. *oleifera* (Metzger) Alefled, Landw. Fl. 246. 1866, nom. illegit. *Brassica campestris* Linnaeus var. *genuina* Celakovsky, Prodr. Fl. Böhmen 3: 469. 1875, nom. inadmiss. *Brassica sativa* Clavaud, Acta Soc. Linn. Bordeaux 35: 291. 1881, nom. illegit. *Brassica sativa* Clavaud var. *campestris* (Linnaeus) Clavaud, Acta Soc. Linn. Bordeaux 35: 292. 1881, nom. inadmiss. *Brassica rapa* Linnaeus forma *campestris* (Linnaeus) Bogenhard ex Thellung, in Hegi, Ill. Fl. Mitt.-Eur. 4(1): 259. 1918. *Brassica rapa* Linnaeus subsp. *campestris* (Linnaeus) Clapham, in Clapham et al., Fl. Brit. Isles 153. 1952, nom. illegit.

Annual or biennial herb, to 1 m; stem glabrous or sparsely pubescent, glaucous. Basal leaves with the blade ovate-oblanceolate, (5)10–40(60) cm long, 3–10(20) cm wide, the margin lyrate-pinnatifid to pinnate or pinnatisect, sinuate-dentate, sometimes ciliate, the lobes 2–4(6) on each side, the terminal lobe oblong-obovate, the apex obtuse, the upper and lower surfaces usually setose, the petiole (1)2–10(17) cm long, winged, the cauline leaves with the blade base auriculate to amplexicaul, the margin subentire, epetiolate. Flowers in a corymbose raceme, the pedicel (0.5)1–2.5(3) cm long, ascending to spreading; sepals oblong, (3)4–6(8) mm long; petals obovate, 6–11(13) mm long, yellow, the apex rounded; stamens 4–6(7) mm long. Fruit terete, (2)3–8(11) cm long, 2–4(5) mm wide, torulose, the terminal segment 8–22 mm long, seedless, the valvular segment 2–5(7) cm long, with 8–15 seeds per locule; seeds globose, 1–2 mm long, black, brown, or reddish, the surface finely reticulate-alveolate.

Disturbed sites. Occasional; peninsula, central and western panhandle. Escaped from cultivation. Nearly cosmopolitan. Native to Europe and Asia. Spring–summer.

Cakile Mill. 1754. SEAROCKET

Annual or perennial herbs. Leaves cauline, alternate, simple, pinnate-veined, petiolate or epetiolate, estipulate. Flowers in terminal racemes, pedicellate, ebracteate, actinomorphic, bisexual; sepals 4, free; petals 4, free, equal; stamens 6, tetradynamous; intrastaminal nectary glands 4, free, median glands present. Fruit a silique, indehiscent, the segments 2, each segment falsely 1-loculate, the replum pressed to 1 side, not distinguishable, the terminal segment deciduous by a transverse articulation, the proximal segment remaining attached to the petiole; seeds uniseriate.

A genus of 7 species; nearly cosmopolitan. [From the Arabic *qaqulleh*.]

Selected references: Rodman (1974, 2010).

1. Beak of the fruit somewhat flattened, the apex retuse or truncate..**C. edentula**
1. Beak of the fruit conical, the apex acute ....**C. lanceolata**

Cakile edentula (Bigelow) Hook subsp. **harperi** (Small) Rodman [*E*- without, and *dentulata*, very small teeth or prongs, in reference to the fruit apex; commemorates Roland McMillan Harper (1878–1966), southeastern U.S. botanist.] AMERICAN SEAROCKET.

> *Cakile harperi* Small, Fl. S.E. U.S. 478, 1331. 1903. *Cakile edentula* (Bigelow) Hooker subsp. *harperi* (Small) Rodman, Contr. Gray Herb. 205: 125. 1974.

Erect or prostrate, succulent annual or perennial herb, to 8 dm; stem glabrous. Leaves with the blade ovate or spatulate, the apex acute, the base cuneate, the margin sinuate, the upper and lower surfaces glabrous. Flowers in a congested raceme 10–20 cm long, the pedicel 1–8 mm long; sepals ovate or oblong, 4–5 mm long; petals 5–9 mm long, white to pale lavender, the claw indistinct. Fruit cylindric, 20–29 mm long, 5–9 mm wide, 8-ribbed, the terminal segment somewhat cylindric, the beak somewhat flattened, blunt, the apex retuse or truncate, the proximal segment terete; seeds flattened, brown, smooth.

Coastal strands. Occasional; eastern peninsula from Duval County south to St. Lucie County. North Carolina south to Florida. Spring–summer.

Cakile lanceolata (Willd.) O. E. Schulz [Lance-shaped, in reference to the leaves.] COASTAL SEAROCKET.

Raphanus lanceolatus Willdenow, Sp. Pl. 3: 562. 1800. *Sinapis lanceolata* (Willdenow) de Candolle, Syst. Nat. 2: 611. 1821. *Brassica willdenowii* Boissier, Ann. Sci. Nat., Bot., ser. 2. 17: 88. 1842, nom. illegit. *Brassica lanceolata* (Willdenow) C. Wright, in Sauvalle, Anales Acad. Ci. Med. Habana 5: 199. 1868; non Lange, 1856. *Cakile lanceolata* (Willdenow) O. E. Schulz, in Urban, Symb. Antill. 3: 504. 1903.

Cakile aequalis L'Héritier de Brutelle ex de Candolle, Syst. Nat. 2: 430. 1821. *Cakile maritima* Scopoli var. *aequalis* (L'Héritier de Brutelle ex de Candolle) Chapman, Fl. South. U.S. 31. 1860.

Cakile americana Nuttall var. *cubensis* de Candolle, Syst. Nat. 2: 429. 1821. *Cakile cubensis* (de Candolle) Kunth, in Humboldt et al., Nov. Gen. Sp. 5: 75. 1821. *Cakile maritima* Scopoli var. *cubensis* (de Candolle) Chapman, Fl. South. U.S., ed. 2. 606. 1883.

Cakile fusiformis Greene, Pittonia 3: 346. 1898. *Cakile lanceolata* (Willdenow) O. E. Schulz subsp. *fusiformis* (Greene) Rodman, Contr. Gray Herb. 205: 114. 1974. TYPE: FLORIDA: Monroe Co.: Mangrove Key (Salt Pond Key), 10 Mar 1898, *Pollard et al. 19* (holotype: NDG; isotypes: F, NY, US).

Cakile alacranensis Millspaugh, Publ. Field Columb. Mus., Bot. Ser. 2: 130. 1900. *Cakile lanceolata* (Willdenow) O. E. Schulz var. *alacranensis* (Millspaugh) O. E. Schulz, in Urban, Symb. Antill. 3: 506. 1903. *Cakile edentula* (Bigelow) Hooker var. *alacranensis* (Millspaugh) O. E. Schulz, in Engler, Pflanzenr. 4(Heft 84): 27. 1923. *Cakile lanceolata* (Willdenow) O. E. Schulz subsp. *alacranensis* (Millspaugh) Rodman, Contr. Gray Herb. 205: 116. 1974.

Cakile chapmanii Millspaugh, Publ. Field Columb. Mus., Bot. Ser. 2: 130. 1900. TYPE: FLORIDA: Miami-Dade Co.: Miami, Jul 1877, *Garber 6472* (Lectotype: US; isolectotypes: F, PH). Lectotypified by Rodman (1974: 114).

Cakile contricta Rodman, Contr. Gray Herb. 205: 131. 1974. TYPE: FLORIDA: Bay Co.: St. Andrews, 2 May 1901, *Tracy 7303* (holotype: US; isotypes: BM, F, MO, NY).

Cakile lanceolata (Willdenow) O. E. Schulz subsp. *pseudoconstricta* Rodman, Contr. Gray Herb. 105: 116. 1974. TYPE: FLORIDA: Hillsborough Co.: 6 Mar 1904, *Fredholm 6324* (holotype: GH; isotypes: MO, US).

Erect or prostrate, succulent annual herb, to 5 dm; stem glabrous. Leaves with the blade ovate to ovate-lanceolate or oblanceolate, the apex acute, the base cuneate, the margin serrate, dentate, or pinnatisect, the upper and lower surfaces glabrous, short petiolate or sessile. Flowers in a raceme to 30 cm long, the pedicel 2–4 mm long; sepals 4–5 mm long, the lateral pair saccate; petals 5–9 mm long, white or lavender, distinctly clawed. Fruit fusiform or lanceolate, 1.5–3 cm long, 3–4 mm wide, 4-angled to terete, striate or sulcate, the terminal segment slenderly conical, 9–18 mm long, the apex acute, the proximal segment terete, 5–10 mm long; seeds flattened, brown, smooth.

Coastal strands. Common; peninsula, central and western panhandle. Florida and Texas; West Indies and Mexico. Spring–summer.

EXCLUDED TAXA

Cakile edentula (Bigelow) Hooker—The typical species was reported for Florida by Small (1903), Radford et al. (1968), and Long and Lakela (1971), misapplied to *C. edentula* var. *harperi*. Also reported for Florida by Schulz (1923, as *C. edentula* var. *americana* (Nuttall) Torrey & A. Gray), who apparently misapplied the name to material of *C. lanceolata*.

Cakile geniculata (B. L. Robinson) Millspaugh—Reported for Florida by Correll and Johnston (1970). According to Rodman (1974), material from that area is all *C. constricta* (=*C. lanceolata*).

Cakile maritima Scopoli—Reported for Florida by Small (1933, as *C. cakile* (Linnaeus) H. Karsten). Not found in Florida according to Rodman (1974).

Capsella Medik., nom. cons. 1792.

Annual or perennial herbs. Leaves basal and cauline, alternate, simple, pinnate-veined, petiolate or epetiolate, estipulate. Flowers in corymbose racemes, pedicellate, actinomorphic, bisexual; sepals 4, free; petals 4, free; stamens 6, tetradynamous; intrastaminal nectar glands 4. Fruit a silique, dehiscent, 2-valved, the replum rounded, the septum complete; seeds uniseriate.

A genus of 4 species; nearly cosmopolitan. Native to Europe, Africa, and Asia. [*Capsa*, box or case, and -*ellus*, diminutive, in reference to the fruit resembling a small medieval purse.]

Bursa Boehmer (1760).

Selected reference: Al-Shehbaz (2010b).

Capsella bursa-pastoris (L.) Medik. [*Bursa*, bag, and *pascuum*, of the pasture, in reference to the shape of the fruit resembling a shepherd's purse.] SHEPHERD'S PURSE.

Thlaspi bursa-pastoris Linnaeus, Sp. Pl. 647. 1753. *Iberis bursa-pastoris* (Linnaeus) Crantz, Stirp. Austr. Fasc. 1: 21. 1762. *Bursa pastoris* Weber, in F. H. Wiggers, Prim. Fl. Holsat. 47. 1780, nom. illegit. *Thlaspi bursa* Thunberg, Fl. Jap. 259. 1784, nom. illegit. *Lepidium bursa-pastoris* (Linnaeus) Willdenow, Fl. Berol. Prodr. 211. 1787. *Nasturtium bursa-pastoris* (Linnaeus) Roth, Tent. Fl. Germ. 1: 281. 1788. *Capsella bursa-pastoris* (Linnaeus) Medikus, Pfl.-Gatt. 85. 1792. *Thlaspi infestum* Salisbury, Prodr. Stirp. Chap. Allerton 267. 1796, nom. illegit. *Rodshieldia bursa-pastoris* (Linnaeus) P. Gaertner et al., Oekon. Fl. Wetterau 2: 435. 1800. *Capsella polymorpha* Cavanilles, Descr. Pl. 411, 614. 1802, nom. illegit. *Thlaspi cuneatum* Stokes, Bot. Mat. Med. 3: 434. 1812, nom. illegit. *Capsella bursa* Rafinesque, New Fl. 2: 28. 1837 ("1836"), nom. illegit. *Capsella pastoralis* Dulac, Fl. Hautes-Pyrénées 189. 1867, nom. illegit. *Capsella pastoris* Ruprecht, Fl. Caucasi 128. 1869, nom. illegit. *Capsella triangularis* Saint-Lager, Ann. Soc. Bot. Lyon 7: 69. 1880, nom. illegit. *Bursa bursa-pastoris* (Linnaeus) Britton, Mem. Torrey Bot. Club 5: 172. 1894. *Crucifera capsella* E.H.L. Krause, in Sturm, Fl. Deutschl., ed. 2. 6: 144, pl. 23, nom. illegit.

Capsella rubella Reuter, Compt.-Rend. Trav. Soc. Haller. 2: 18. 1853–1854. *Crucifera rubella* (Reuter) E.H.L. Krause, Beih. Bot. Centralbl. 32(2): 330. 1914.

Annual or biennial herb, to 5(7) dm; stem sparsely to densely pubescent with sessile, 3- to 5-rayed, stellate and simple, longer trichomes. Basal leaves with the blade oblong or oblanceolate, (0.5)1.5–10(15) cm long, 0.2–2.5(5) cm wide, the apex acute or acuminate, the base cuneate or attenuate, the margin pinnately lobed, rarely entire or dentate, the upper and lower surfaces sparsely to densely pubescent with sessile, 3- to 5-rayed, stellate and simple longer trichomes, the petiole 0.5–4(6) cm long, the cauline leaves with the blade narrowly oblong, lanceolate, or linear, 1–5.5(8) cm long, 1–15(20) mm wide, the base sagittate, amplexicaul, or auriculate. Flowers in a corymbose raceme, the pedicel to 1.5(2) cm long in fruit, divaricate, glabrous; sepals oblong, ca. 2 mm long, green or reddish; petals obovate, 2–4(5) mm long, white or pinkish, clawed; stamens 1–2 mm long; nectar glands 4, lateral, 1 on each side of the lateral stamen.

Fruit obdeltoid to obdeltoid-obcordiform, (3)4–9(10) mm long, (2)3–7(9) mm wide, strongly flattened, the apex emarginate or truncate, the base cuneate, the valves papery, with prominent, lateral, subparallel veins, glabrous; seeds oblong, ca. 1 mm long, brown, the surface reticulate.

Disturbed sites. Occasional; nearly throughout. Nearly cosmopolitan. Native to Europe, Africa, and Asia. Spring.

Cardamine L. 1753. BITTERCRESS

Annual, biennial, or perennial herbs. Leaves cauline, rhizomal, or basal, simple or palmately or pinnately compound, alternate, opposite, or whorled, pinnate-veined, petiolate or epetiolate, estipulate. Flowers in corymbose or paniculate racemes, ebracteate, pedicellate, actinomorphic, bisexual; sepals 4, free; petals 4, free, or absent; stamens 6(5–4), equal, or if unequal, 2 shorter than the other 4; intrastaminal nectary glands present. Fruit a silique, dehiscent, 2-valved, the replum flattened, the septum complete; seeds uniseriate.

A genus of about 200 species; nearly cosmopolitan. [From the Greek *kardamon*, vernacular name for a cress.]

Dentaria L. (1753).

Selected reference: Al-Shehbaz, Marhold, and Lihová (2010).

1. Leaves palmately compound or divided .... **C. concatenata**
1. Leaves simple or pinnately compound or divided.
    2. Leaves simple, the margin entire or sinuate-crenate; roots tuberous................................**C. bulbosa**
    2. Leaves (at least the basal ones) pinnately compound or deeply pinnately lobed; roots fibrous.
        3. Petiole of the lower leaves sparsely ciliate ....**C. hirsuta**
        3. Petiole of the lower leaves eciliate.
            4. Middle and upper cauline leaves with the leaf segments uniformly linear, the terminal segment 1–2 mm wide ....**C. parviflora**
            4. Middle and upper cauline leaves with the leaf segments obovate to orbicular, the terminal segment 6–25 mm wide.. **C. pensylvanica**

Cardamine bulbosa (Schreb. ex Muhl.) Britton et al. [Bulb-like, in reference to the tuberous rhizomes.] BULBOUS BITTERCRESS; SPRING CRESS.

Arabis bulbosa Schreber ex Muhlenberg, Trans. Amer. Philos. Soc. 3: 174. 1793. *Cardamine bulbosa* (Schreber ex Muhlenberg) Britton et al., Prelim. Cat. 4. 1888. *Dracamine bulbosa* (Schreber ex Muhlenberg) Nieuwland, Amer. Midl. Naturalist 4: 40. 1915.
Arabis rhomboidea Persoon, Syn. Pl. 2: 204. 1806. *Cardamine rhomboidea* (Persoon) de Candolle, Syst, Nat. 2: 246. 1821. *Dentaria rhomboidea* (Persoon) Greene, Pittonia 3: 124. 1896.

Erect perennial herb, to 6 dm; stem glabrous or sparsely pubescent distally, with fleshy, subglobose, tuberous rhizomes at the stem base. Rhizomal leaves simple, the blade reniform to cordate, ovate, or oblong, (1)2–4(6) cm long and wide, the apex rounded to obtuse, the base obtuse to cordate, the margin repand, entire, or shallow dentate, the petiole (1.5)2.5–10(13) cm long, the cauline leaves simple, the blade ovate to oblong or oblong-linear to lanceolate, 3–6(9) cm long, 1–3(4.5) cm wide, the apex rounded or obtuse, the base cuneate or rounded, short-petiolate or sessile, the margin entire or sinuate-crenate, minutely pubescent. Flowers in a raceme, the pedicel ascending to divaricate, (1)1.5–2(3) cm long in fruit; sepals oblong, 3–5 mm long, glabrous; Petals obovate, (6)7–12(16) mm long, white or pale pink, the apex rounded, the base short-clawed; stamens 6, the medial pairs 4–7 mm long, the lateral pair 2–4 mm long. Fruit linear, 2–3.5(4) cm long, ca. 2 mm wide, the style 2–4(5) mm long, glabrous; seeds 14–24, oblong or globose, ca. 2 mm long, dark orange or greenish yellow.

Floodplain forests and the margin of streams and spring runs. Occasional; northern and central peninsula, central panhandle. Quebec south to Florida, west to Manitoba, North Dakota, South Dakota, Nebraska, Kansas, Oklahoma, and Texas. Spring.

Cardamine concatenata (Michx.) O. Schwarz [Linked or connected, in reference to the palmately divided leaves.] CUTLEAF TOOTHCUP.

> *Dentaria concatenata* Michaux, Fl. Bor.-Amer. 2: 30. 1803. *Cardamine concatenata* (Michaux) O. Schwarz, Repert. Spec. Nov. Regni Veg. 46: 188. 1939.
>
> *Dentaria laciniata* Muhlenberg ex Willdenow, Sp. Pl. 3: 479. 1800. *Cardamine laciniata* (Muhlenberg ex Willdenow) A. W. Wood, Amer. Bot. Fl. 38. 1870; non (Hooker) Steudel, 1840; nec F. Mueller, 1855.

Erect perennial herb, to 4(5.5) dm; stem glabrous or pubescent distally, with fleshy, moniliform, tuberous rhizomes at the stem base. Rhizomal leaves 3-foliolate, (7)10–20(30) cm long, the leaflets sessile, the terminal leaflet blade oblong, lanceolate, oblanceolate, or linear, the apex acute, the base cuneate, the margin coarsely dentate to incised, or 3-lobed, rarely entire, the upper and lower surfaces puberulent or glabrous, the lateral leaflets similar to the terminal, the petiole (4)7–18(25) cm long, the cauline leaves 2 or 3, 3-foliolate, whorled or opposite, rarely alternate, similar to the rhizomal leaves, petiolate, the leaflets petiolulate or subsessile, the petiole (1)1.5–6(7.5) cm long, the terminal leaflet with the blade lanceolate, linear, or oblanceolate, (3)4–10(12) cm long, (3)5–20(25) mm wide, the margin coarsely dentate to incised, rarely subentire, minutely puberulent, subsessile or with the petiole to 3 cm long, the lateral leaflets sessile, the blade similar to the terminal one, sometimes smaller. Flowers in a raceme, the pedicel ascending to divaricate, (6)10–27(33) mm long in fruit; sepals oblong, (4)5–8 mm long, the lateral pair slightly saccate basally; petals oblanceolate, (8)10–20 mm long, white or pale pink, the apex rounded, the base short-clawed; stamens 6, the median pairs 8–12 mm long, the lateral pair 6–8 mm long. Fruit linear-lanceolate, (2)2.5–3.8(4.8) cm long, 1.5–3 mm wide, the style (2)5–9(12) mm long, glabrous or sparsely pubescent; seeds 10–14, oblong, brown.

Moist hammocks. Rare; Gadsden, Liberty, and Jackson Counties. Quebec south to Florida, west to Ontario, North Dakota, South Dakota, Nebraska, Kansas, Oklahoma, and Texas. Spring.

Cardamine hirsuta L. [With coarse trichomes.] HAIRY BITTERCRESS.

> *Cardamine hirsuta* Linnaeus, Sp. Pl. 655. 1753. *Ghinia hirsuta* (Linnaeus) Bubani, Fl. Pyren. 3: 162. 1901. *Crucifera cardamine* E.H.L. Krause, in Sturm, Fl. Deutschl., ed. 2. 6: 112. 1902, nom. illegit.

Erect, ascending, or decumbent annual herb, to 3.5(3.5) dm; stem sparsely hirsute basally, often glabrous distally. Basal leaves pinnately (5)8- to 15(22)-foliolate, (2)3.5–15(17) cm long, the leaflets petiolulate, the terminal leaflet with the blade reniform or suborbicular, 0.4–2 cm long, 0.6–3 cm wide, the margin dentate or 3- or 5-lobed, ciliate, the lateral leaflets with the blade oblong, ovate, obovate, or orbicular, ciliate, smaller than the terminal one, the margin entire, crenate, or 3-lobed, the petiole 0.5–5 cm long, the cauline leaves similar to the basal ones, but smaller. Flowers in a raceme, the pedicel erect to ascending, (2)3–10(14) mm long in fruit; sepals oblong, 2–3 mm long; petals spatulate (sometimes absent), 3–5 mm long, white; stamens usually 4, (lateral pair often absent), rarely 5 or 6, 2–3 mm long. Fruit linear, (0.9)1.5–2.5(2.8) cm long, 1–2 mm wide, the style to 1 mm long, torulose, glabrous; seeds 14–40, oblong or subquadrate, ca. 1 mm long, light brown.

Wet, disturbed sites. Occasional; northern counties, central peninsula. Quebec south to Florida, British Columbia south to California; Central America and South America; Europe, Africa, Asia, and Australia. Native to Europe and Asia. Spring.

Cardamine parviflora L. [With small flowers.] SAND BITTERCRESS.

Cardamine parviflora Linnaeus, Syst. Nat., ed. 10. 1131. 1759.
Cardamine arenicola Britton, Bull. Torrey Bot. Club 19: 220. 1892. Cardamine parviflora Linnaeus var. arenicola (Britton) O. E. Schulz, Bot. Jahrb. Syst. 32: 485. 1903.

Erect annual herb, to 3(4) dm; stem glabrous or pubescent. Basal leaves pinnately (5)7- to 13(17-foliolate, (2)4–10 cm long, the leaflets sessile or petiolulate, the terminal leaflet with the blade linear to oblong, oblanceolate to obovate or suborbicular, (1)3–10 mm long, 1–7 mm wide, the apex obtuse, the base cuneate, the margin entire to 3(5)-toothed or -lobed, the upper and lower surfaces glabrous or pubescent, the lateral leaflets similar to the terminal one, sometimes smaller, the petiole 0.5–2.5(4.5) cm long, the cauline leaves (5)9- to 15(17)-foliolate, the leaflets sessile, sometimes smaller, the terminal leaflet with the blade filiform, linear, or narrowly oblong, 3–1(1.5) cm long, 1–3 mm wide, the margin entire or 1- to 3-lobed, the upper and lower surfaces glabrous or pubescent, glabrous, the lateral leaflets similar to the terminal one, the petiole 3–10 mm long. Flowers in a raceme, the pedicel divaricate or ascending, 4–10 mm long in fruit; sepals oblong, 1–2 mm long; petals oblanceolate, 2–3 mm long, white; stamens 6, 2–3 mm long. Fruit linear, (0.5)1–2(2.5) cm long, to 1 mm wide, the style to 1 mm long, torulose; seeds 20–50, oblong-ovoid, ca. 1 mm long, pale brown.

Wet hammocks. Rare; Alachua County, central and western panhandle. Nearly throughout North America; Europe and Asia. Spring.

Cardamine pensylvanica Muhl. ex Willd. [Of Pennsylvania.] PENNSYLVANIA BITTERCRESS.

Cardamine pensylvanica Muhlenberg ex Willdenow, Sp. Pl. 3: 486. 1800. Cardamine hirsuta Linnaeus forma pensylvanica (Muhlenberg ex Willdenow) E. L. Rand & Redfield, Fl. Mt. Desert. Isl. 78. 1894. Cardamine flexuosa Withering subsp. pensylvanica (Muhlenberg ex Willdenow) O. E. Schulz, in Urban, Symb. Antill. 3: 521. 1903. Dracamine pensylvanica (Muhlenberg ex Willdenow) Nieuwland, Amer. Midl. Naturalist 4: 40. 1915. Cardamine hirsuta Linnaeus var. pensylvanica (Muhlenberg ex Willdenow) P. W. Graff, Castanea 10: 95. 1945.
Cardamine pensylvanica Muhlenberg ex Willdenow var. brittoniana Farwell, Asa Gray Bull. 2: 46. 1894.

Erect annual or biennial herb, to 5(7) dm; stem sparsely hirsute basally. Basal leaves similar to the proximalmost cauline ones, 4–15 cm long, the cauline leaves pinnately 7- to 13(19)-foliolate, 2–11 cm long, the terminal leaflet with the blade suborbicular, obovate to oblanceolate or elliptic, 1.3–3(4) cm long, 0.6–2.5 cm wide, the apex rounded, the base cuneate, the margin entire or obscurely 3- or 5-lobed, short-petiolulate to 1 cm long or subsessile, the lateral ones similar to the terminal, but narrower and smaller than the terminal one, the margin entire, crenate, or few-lobed, short petiolulate or sessile, the petiole (0.4)1–3.5(4.5) cm long. Flowers in a raceme, the pedicel divaricate-ascending, (3)4–10(13) mm long in fruit; sepals oblong, 1–2 mm long; petals narrowly spatulate to oblanceolate, 2–4 mm long, white; filaments 6, the median pairs 2–3 mm long, the lateral pair 1–2 mm long. Fruit linear, 1.5–3 cm long, ca. 1 mm wide, the style ca. 1 mm long, torulose; seeds 40–80, oblong to ovoid, ca. 1 mm long, brown.

Wet hammocks along streams and spring runs. Common; nearly throughout. Nearly throughout North America. Spring.

EXCLUDED TAXA

Cardamine flexuosa With.—This European species was reported for Florida by Small (1933, as *C. debilis* D. Don), but no Florida specimen is known. The name is probably misapplied to material of *C. hirsuta*.

Cleome L., nom. cons. 1753. SPIDERFLOWER

Annual or perennial herbs. Leaves alternate, palmately compound, the leaflets pinnate-veined, petiolate, stipulate or estipulate. Flowers in terminal or axillary racemes, bracteate or ebracteate, zygomorphic, bisexual; sepals 4, free or basally connate; petals 4, free, unequal; intrastaminal nectaries present or absent; stamens 6 or 14–25, the filaments free or adnate and forming an androgynophore. Fruit a capsule, dehiscent or only partly so, the replum round; seeds biseriate, arillate or not.

A genus of about 150 species; nearly cosmopolitan. [From the Greek *kleos*, glory, or after Kleo, Greek muse of history.]

Cleome is distributed as several lineages among other recognized genera, but a modern comprehensive revision is lacking.

Cleome is sometimes placed in the Capparaceae (e.g., Wunderlin, 1998) or Cleomaceae by (e.g., Tucker and Vanderpool, 2010).

Arivela Raf. (1838); *Cleoserrata* Iltis (2007); *Corynandra* Schrad. ex Spreng. (1827); *Gynandropsis* DC., nom. cons. (1824); *Neocleome* Small (1933); *Tarenaya* Raf. (1838).

Selected references: Cochrane and Iltis (2014); Iltis and Cochrane (2007); Tucker (2010c, 2010d, 2010e, 2010f); Tucker and Iltis (2010).

1. Plant armed with prickles.
   2. Petals colored; fruit glabrous... **C. hassleriana**
   2. Petals white; fruit pubescent... **C. spinosa**

1. Plant unarmed.

    3. Stamens 14–25; fruit sessile on the pedicel (without a distinct gynophore) .... **C. viscosa**

    3. Stamens 6; fruit borne on a gynophore distinct from the pedicel.

        4. Bracts simple; petals 1.5–4 cm long..**C. speciosa**

        4. Bracts 3-foliolate; petals 7–14 mm long.

            5. Androgynophore much shorter than the petals; fruit glabrous.................... **C. rutidosperma**

            5. Androgynophore subequaling or longer than the petals; fruit glandular-pubescent ....

            .... **C. gynandra**

Cleome gynandra L. [From the Greek *gyne*, female, and *andro*, male, in reference to the female and male flowers on an androgynophore.] SPIDERWISP.

Cleome gynandra Linnaeus, Sp. Pl. 671. 1753. *Cleome pentaphylla* Linnaeus, Sp. Pl., ed. 2. 938. 1763, nom. illegit. *Pedicellaria pentaphylla* Schrank, Bot. Mag. (Römer & Usteri) 3(8): 11. 1790, nom. illegit. *Gynandropsis pentaphylla* de Candolle, Prodr. 1: 238. 1824, nom. illegit. *Podogyne pentaphylla* Hoffmannsegg, Verz. Pfl.-Kult. 185. 1824, nom. illegit. *Gynandropsis gynandra* (Linnaeus) Briquet, Annuaire Conserv. Jard. Bot. Genève 17: 382. 1914.

Erect annual herb, to 1.5 m; stem glandular-pubescent. Leaves with 3 or 5 leaflets, the leaflet blade oblanceolate to rhombic, 2.5–4.5 cm long, 1.2–2.5 cm wide, the apex acute, the base cuneate, the margin serrulate-denticulate, the upper and lower surfaces glabrate to glandular-pubescent, the petiole 3.5–4.5(8) cm long, glandular pubescent; estipulate. Flowers in a raceme 5–20 cm long, the bracts 3-foliolate, 10–25 mm long, the pedicel 8–15 mm long, glabrous, purple; sepals free, lanceolate, 4–5 mm long, green, glandular-pubescent; petals oblong to ovate, 7–14 mm long, purple or white; stamens 6, 8–30 mm long, purple, the filaments basally adnate to the gynophore, glabrous; gynophore 10–14 mm long, purple, the ovary 6–10 mm long, the style ca. 1 mm long. Fruit linear-oblong, 4.5–9.5 cm long, 3–4 mm wide, glandular-pubescent, dehiscent; seeds 10–20, subglobose, 1–2 mm long, reddish brown to black, the surface rugose to tuberculate, not arillate.

Disturbed sites. Occasional; peninsula west to central panhandle. Nearly throughout. Native to Africa, Asia, Australia, and Pacific Islands. Summer–fall.

Cleome hassleriana Chodat [Commemorates Emil Hassler (1864–1937), Swiss physician, naturalist, and botanist known for his work on the flora of Paraguay.] PINKQUEEN SPIDERFLOWER.

Cleome hassleriana Chodat, Bull. Herb. Boissier 6(App. 1): 12. 1898. *Tarenaya hassleriana* (Chodat) Iltis, in Iltis & Cochrane, Novon 17: 450. 2007.

Erect annual herb, to 2 m; stem glandular-pubescent. Leaves 5 or 7 leaflets, the leaflet blade elliptic to oblanceolate, 2–6(12) cm long, 1–3 cm wide, the apex acute, the base cuneate, the margin serrulate-denticulate, the upper surface glandular, the lower surface glandular-pubescent, the petiole 2.5–7.5 cm long, glandular-pubescent, with scattered spines 1–3 mm long; stipules spinescent, 1–3 mm long. Flowers in a raceme 5–30 cm long, the bracts 1-foliolate, ovate, 10–25 mm long, the pedicel 20–45 mm long, glandular-pubescent; sepals free, linear-lanceolate, 5–7 mm long, green, glabrous; petals pink or purple, oblong-ovate, 2–3(4.5) cm long; stamens 6,

3–5 cm long, purple, the filaments inserted on a discoid or conic androgynophore, glabrous; gynophore 4.5–8 cm long, the ovary 6–10 mm long, the style less than 1 mm long. Fruit oblong, (2.5)4–8 cm long, 3–4 mm wide, glabrous, dehiscent; seeds 10–20, triangular to subglobose, ca. 2 mm long, tuberculate, not arillate.

Disturbed sites. Occasional; northern counties, central peninsula. Escaped from cultivation. Nearly cosmopolitan. Native to South America. Summer–fall.

Cleome rutidosperma DC. [*Rutilus*, reddish-orange, and *sperma*, seed.] FRINGED SPIDERFLOWER.

Cleome rutidosperma de Candolle, Prodr. 1: 241. 1824.

Erect to decumbent annual or perennial herb, to 1 m; stem glabrous or glabrescent to glandular-pubescent. Leaves with 3 leaflets, the leaflet blade oblanceolate to rhombic-elliptic, 1–3.5 cm long, 0.5–1.7 cm wide, the apex acute to acuminate or obtuse, the base cuneate, the margin entire or serrulate-ciliate, the upper surface glabrous, the lower surface with curved trichomes, the petiole 0.5–3.5 cm long, winged proximally; stipules absent or scalelike and to 0.5 mm long. Flowers in a raceme 2–4 cm long, the bracts 3-foliolate, 1–3.5 cm long, the pedicel 1–2 cm long; sepals basally connate ½ their length, lanceolate, 3–4 m long, yellow, the margin denticulate, ciliate, glabrous; petals oblong to narrowly ovate, 7–10 mm long, white or purple-spotted, the 2 central ones with a yellow transverse band; stamens 6, 4–7 mm long, the filaments inserted on a discoid or conic androgynophore, yellow; gynophore 4–12 mm long, the ovary 2–3 mm long, glabrous, the style ca. 1 mm long. Fruit oblong, 4–7 cm long, glabrous, dehiscent; seeds 4–25, reniform, reddish brown to black, arillate.

Disturbed sites. Occasional; central and southern peninsula, Bay County. South Carolina and Florida; West Indies, Mexico, Central America, and South America; Africa, Asia, and Australia. Native to Africa. Summer–fall.

Cleome speciosa Raf. [Showy.] GARDEN SPIDERFLOWER.

Cleome speciosa Rafinesque, Fl. Ludov. 86. 1817. *Cleoserrata speciosa* (Rafinesque) Iltis, in Iltis & Cochrane, Novon 17: 448. 2007.
Cleome speciosa Kunth, in Humboldt et al., Nov. Gen. Sp. 5: 84, t. 436. 1821; non Rafinesque, 1817.
Gynandropsis speciosa de Candolle, Prodr. 1: 238. 1824.

Erect annual herb, to 1.5 m; stem glabrous sparsely glandular-pubescent. Leaves with 5–9 leaflets, the leaflet blade narrowly lanceolate-elliptic, the apex subobtuse, the base cuneate, 6–15 cm long, 1–5 cm wide, the margin entire or serrulate, the upper and lower surfaces glandular-pubescent, the petiole 2–12 cm long, glandular-pubescent; estipulate. Flowers in a raceme 15–50 cm long, the rachis glandular-pubescent, the bracts 1-foliolate, ovate-cordate, 3–18 mm long, the pedicel 1–5 cm long, glabrous; sepals free, lanceolate, 4–7 mm long, green, glandular-pubescent; petals ovate, 1.5–4 cm long, pink to purple, fading to pink or white, rarely initially white, clawed; stamens 6, 4–8.5 cm long, the filaments inserted on a discoid or conic androgynophore, green; gynophore 3–8.5 cm long, the ovary 6–10 mm long, the style ca. 1 mm long.

Fruit oblong, 6–15 cm long, 3–5 mm wide, glabrous, dehiscent; seeds 10–30, subglobose, 2–3 mm long, pale green to brown, tuberculate.

Disturbed sites. Rare; Duval County. Not recently collected. Escaped from cultivation. Florida; West Indies, Mexico, Central America, and South America. Native to tropical America. Summer–fall.

Cleome spinosa Jacq. [Spiny.] SPINY SPIDERFLOWER.

Cleome spinosa Jacquin, Enum. Syst. Pl. 26. 1760. *Tarenaya spinosa* (Jacquin) Rafinesque, Sylva Tellur. 111. 1838. *Neocleome spinosa* (Jacquin) Small, Man. S.E. Fl. 577, 1504. 1933.

Cleome pungens Willdenow, Enum. Pl. 689. 1809. *Cleome spinosa* Jacquin forma *pungens* (Willdenow) Eichler, in Martius, Fl. Bras. 13(1): 253. 1865.

Erect annual herb, to 2 m; stem glandular-pubescent. Leaves with (3)5 or 7 leaflets, the leaflet blade lanceolate to elliptic-oblanceolate, 4–10(12) cm long, 1–3 cm wide, the apex acute to acuminate, the base cuneate, the margin serrulate-denticulate, the upper surface glandular, the lower surface glandular-pubescent, the petiole 2.5–10 cm long, with scattered antrorse spines 1–3 mm long; stipules spinescent, 2–5 mm long. Flowers in a raceme 1–20 cm long, the bracts 1-foliolate, ovate to broadly elliptic, 8–25 mm long, the pedicel 1.5–3 cm long, glandular-pubescent; sepals free, lanceolate, 6–8 mm long, green, glandular-pubescent; petals oblong to ovate, 1–2(3) cm long, white or greenish white, somewhat reddish on the outer surface, somewhat glandular pubescent externally, especially apically; stamens 6, 2–5 cm long, the filaments inserted on a discoid or conic androgynophore, white; gynophore 4.5–7 cm long, the ovary 6–10 mm long, slightly glandular-pubescent, the style less than 1 mm long. Fruit oblong, 4–11(15) cm long, 3–5 mm wide, slightly glandular-pubescent, dehiscent; seeds 20–30, triangular to subglobose, ca. 2 mm long, smooth, not arillate.

Disturbed sites. Rare; locality unknown. Not recently collected. Pennsylvania, West Virginia, Alabama, and Florida; West Indies, Mexico, Central America, and South America; Pacific Islands. Native to South America. Summer–fall.

Cleome viscosa L. [Sticky, glutinous]. ASIAN SPIDERFLOWER.

Cleome viscosa Linnaeus, Sp. Pl. 672. 1753. *Sinapistrum viscosum* (Linnaeus) Moench, Suppl. Meth. 83: 1802. *Polanisia viscosa* (Linnaeus) de Candolle, Prodr. 1: 242. 1824. *Arivela viscosa* (Linnaeus) Rafinesque, Sylva Tellur. 110. 1838. *Corynandra viscosa* (Linnaeus) Cochrane & Iltis, Novon 23: 24. 2014.

Erect annual herb, to 1(1.5) m; stem glandular-pubescent. Leaves with 3 or 5 leaflets, the leaflet blade ovate to oblanceolate-elliptic, (0.6)2–6 cm long, 0.5–3.5 cm wide, the apex acute to obtuse, the base cuneate, the margin entire, glandular-ciliate, the upper and lower surfaces glandular-hirsute, the petiole 1.5–4.5(8) cm long, glandular-hirsute; estipulate. Flowers in a raceme 5–10 cm long, the rachis glandular-hirsute, the bracts 3-foliolate, 1–2.5 cm long, glandular-hirsute, the pedicel 0.6–3 cm long, glandular-hirsute; sepals free, lanceolate, 5–10 mm long, green, the outer surface, glandular-hirsute; petals oblong to ovate, 7–14 mm long, yellow, sometimes purplish basally; stamens 14–25, 5–9 mm long, the filaments inserted on a discoid or conic androgynophore, dimorphic, 4–10 much shorter and with a swelling near the anther, green, glandular-hirsute; the gynophore obsolete, the ovary 6–10 mm long, densely glandular, the style ca. 1 mm long. Fruit oblong, 3–10 cm long, glandular-hirsute, dehiscing on partway to the base; seeds 25–40, subglobose, 1–2 mm long, light brown, finely transversely ridged, not arillate.

Disturbed sites. Rare; Hillsborough and Pinellas Counties, panhandle. Pennsylvania and New Jersey south to Florida, west to Louisiana; West Indies, Mexico, Central America, and South America; Africa, Asia, Australia, and Pacific Islands. Native to Africa, Asia, and Australia. All year.

EXCLUDED TAXON

Cleome houtteana Schlechtendal—Reported for Florida by Radford et al. (1968), apparently based on a misidentification of material of *C. hassleriana*.

Conringia Heist. ex Fabr. 1759. HARE'S-EAR MUSTARD

Annual herbs. Leaves basal and cauline, simple, pinnate-veined, subsessile or sessile, estipulate. Flowers in corymbose racemes, pedicellate, ebracteate, actinomorphic, bisexual; sepals 4, free; petals 4, free, equal; stamens 6, tetradynamous; intrastaminal nectar glands present. Fruit a silique, dehiscent, the replum rounded, the septum complete; seeds uniseriate.

A genus of 6 species; North America, West Indies, Mexico, Europe, Africa, Asia, and Australia. [Commemorates Hermann Conring (1606–1681), German physician and professor of medicine.]

Selected reference: Warwick (2010e).

Conringia orientalis (L.) Dumort. [Of the Orient.] HARE'S-EAR MUSTARD.

Brassica orientalis Linnaeus, Sp. Pl. 666. 1753. *Erysimum orientale* (Linnaeus) Miller, Gard. Dict., ed. 8. 1768. *Erysimum glaucum* Moench, Methodus 255. 1794, nom. illegit. *Gorinkia orientalis* (Linnaeus) J. Presl & C. Presl, Fl. Cech. 141. 1819. *Conringia orientalis* (Linnaeus) Dumortier, Fl. Belg. 123. 1827. *Erysimum orientale* (Linnaeus) Miller var. *typicum* Lindeman, Bull. Soc. Imp. Naturalistes Moscou 49(3): 69. 1875, nom. inadmiss. *Sisymbrium tetragonum* Trautvetter, Trudy Imp. S.-Peterburgsk. Bot. Sada 4: 350. 1876. *Arabis orientalis* (Linnaeus) Prantl, Excurs.-Fl. Bayern 227. 1884. *Crucifera conringia* E.H.L. Krause, in Sturm, Deutschl. Fl., ed. 2. 6: 85, t. 3. 1902, nom. illegit.

Erect annual herb, to 7 dm; stem glabrous, usually glaucous. Basal leaves with the blade oblanceolate to obovate, 5–9 cm long, the apex obtuse, the base cordate to amplexicaul, the margin entire or nearly so, the upper and lower surfaces glabrous, the cauline leaves with the blade oblong to elliptic, or lanceolate, (1)3–10(15) cm long, (0.5)2–2.5(5) cm wide, the apex rounded, the base deeply cordate-amplexicaul, the margin entire, the upper and lower surfaces glabrous. Flowers in a corymbose raceme, the pedicel ascending, 1–1.5(2) cm in fruit; sepals oblong, 6–8 mm long, the median pair narrower than the lateral; petals narrowly obovate, 7–12 mm long, the claw usually as long as the sepals; stamens 5–7 mm long. Fruit 4-angled to subcylindric, (5)8–14 cm long, 2–3 mm wide, the style cylindric, 1–4 mm long, somewhat torulose; seeds oblong, 2–3 mm long, brown, the surface papillose.

Disturbed sites. Rare; Volusia County. Not recently collected. Nearly throughout North America; West Indies and Mexico; Europe, Africa, Asia, and Australia. Native to Europe, Africa, and Asia. Spring.

Cynophalla (DC.) J. Presl 1825. CAPERTREE

Shrubs or trees. Leaves alternate, simple, pinnate-veined, petiolate, stipules glandular or absent. Flowers in terminal or axillary racemes or corymbs, pedunculate, pedicellate, actinomorphic, bisexual (sometimes appearing unisexual); sepals 4, free, in 2 unequal pairs; petals 4, free, equal; stamens 18–150, the filaments free, inserted on an androgynophore; ovary on a gynophore, 1-carpellate, 2-loculate, the stigma subsessile. Fruit capsule, dehiscent or indehiscent; seeds arillate or not.

A genus of about 20 species; North America, West Indies, Mexico, Central America, and South America. [From the Greek *kynos*, dog, and *phallas*, penis, in reference to the brilliant red color inside the ruptured fruit resembling a dog's penis.]

Cynophalla is sometimes placed in the Capparaceae by various authors.

Selected reference: Tucker (2010a).

Cynophalla flexuosa (L.) J. Presl [Bent alternately in opposite directions, zigzag, in reference to the branches.] BAYLEAF CAPERTREE.

Morisonia flexuosa Linnaeus, Pl. Jamaic. Pug. 14. 1759. *Capparis flexuosa* (Linnaeus) Linnaeus, Sp. Pl., ed. 2. 722. 1762. *Cynophalla flexuosa* (Linnaeus) J. Presl, in Berchtold & J. Presl, Prir. Rostlin 2: 755. 1825.

Shrub, to 4 m; branches glabrous or glabrescent, rarely puberulent. Leaves with the blade oblong to obovate, 3–7(9) cm long, 2–4(5) cm wide, the apex emarginate, rounded, or acute, the base cuneate to rounded, the margin entire, the upper surface glabrous, the lower surface glabrescent, the petiole 4–10 mm long; stipules glandular. Flowers in a terminal or axillary raceme (1)2–6 cm long, the pedicel 7–12 mm long, puberulent; sepals suborbicular, of 2 unequal pairs, the proximal pair 5–7 mm long, the distal pair 4–5 mm long, glabrous; petals oblong-obovate, (1)1.5–2 cm long, white, becoming yellowish or reddish white; stamens 120–150, 4–6(12) cm long, white, inserted on discoid receptacle (androgynophore); gynophore 4–6(8) cm long. Fruit linear-oblong, 0.5–1.5(2.5) cm long, 9–13(17) mm wide, irregularly constricted between the seeds, reddish brown to yellowish, glabrous; seeds reniform to subglobose, 7–14 mm long, green, arillate, embedded in a red pulp.

Coastal hammocks and shell middens. Occasional; central and southern peninsula. Florida; West Indies, Mexico, Central America, and South America. Spring–summer.

EXCLUDED TAXON

Capparis cynophallophora Linnaeus—This name was misapplied to *Cynophalla flexuosa* by Chapman (1860, 1883, 1897) and small (1903, 1913a, 1913b, 1913d, 1913e).

Descurainia Webb & Berthel., nom. cons. 1836. TANSEYMUSTARD

Annual herbs. Leaves basal and cauline, alternate, pinnate-lobed, petiolate or epetiolate, estipulate. Flowers in racemes, pedicellate, ebracteate, actinomorphic, bisexual; sepals 4, free; petals 4, free, equal; stamens 6, tetradynamous; intrastaminal nectar glands present, confluent. Fruit a silique, the replum rounded, the septum complete.

A genus of about 45 species; North America, Mexico, South America, Europe, Africa, and Asia. [Commemorates François Descurain (1658–1740), French botanist and apothecary.]

Sophia Adans., nom. rej. (1763).

Selected reference: Goodson and Al-Shehbaz (2010).

Descurainia pinnata (Walter) Britton [Resembling a feather by having the leaflets arranged on each side of a common axis.] WESTERN TANSEYMUSTARD.

Erysimum pinnatum Walter, Fl. Carol. 174. 1788. *Sisymbium pinnatum* (Walter) Greene, Bull. Calif. Acad. Sci. 2: 390. 1887; non Barneoud ex C. Gay, 1846. *Hesperis pinnata* (Walter) Kuntze, Revis. Gen. Pl. 2: 936. 1891. *Descurainia pinnata* (Walter) Britton, Mem. Torrey Bot. Club 5: 173. 1894. *Sophia pinnata* (Walter) Howell, Fl. N.W. Amer. 1: 56. 1897. *Descurainia pinnata* (Walter) Britton subsp. *typica* Detling, Amer. Midl. Naturalist 22: 503. 1939, nom. inadmiss.

Cardamine multifida Pursh, Fl. Amer. Sept. 440. 1814. *Nasturtium multifidum* (Pursh) Sprengel, Syst. Veg. 2: 883. 1825. *Sisymbrium multifidum* (Pursh) MacMillan, Metasp. Minnesota Valley 258. 1892; non Willdenow ex Sprengel, 1825. *Descurainia multifida* (Pursh) O. E. Schulz, in Engler, Pflanzenr. 4(Heft 86): 328. 1924. TYPE: FLORIDA: s.d., *Bartram s.n.* (holotype: BM?).

Sisymbrium canescens Nuttall, Gen. N. Amer. Pl. 2: 68. 1818. *Descurainia canescens* (Nuttall) Prantl, in Engler & Prantl, Pflanzenfam. 3(2): 192. 1892. *Sisymbrium multifidum* (Pursh) MacMillan subsp. *canescens* (Nuttall) Thellung, in Hegi, Ill. Fl. Mitt.-Eur. 4: 153. 1916.

Erect annual herb, to 6(9) dm; stem sparsely to densely pubescent, glandular or eglandular. Basal leaves with the blade deeply pinnately lobed and appearing 1- or 2-pinnately compound, ovate to oblanceolate, 1–15 cm long, the lateral lobes 4–9 pairs, linear or oblanceolate to ovate, the margin entire or dentate, the petiole 0.5–3.6 cm long, the cauline leaves smaller distally, the distal lobes narrower, the surfaces densely pubescent, sessile or short petiolate. Flowers in a raceme, the pedicel divaricate to horizontal or descending, 0.5–1.5 cm long in fruit, the rachis sparsely to densely pubescent, glandular or eglandular; sepals oblong, 1–2 mm long, yellow or rose, pubescent; petals narrowly oblanceolate, 1–2 mm long, whitish or yellow; median stamens 1–3 mm long. Fruit clavate, rarely linear, 4–13 mm long, 1–2 mm wide, glabrous; seeds 16–40, oblong, 0.5–1 mm long, reddish brown, the surface minutely reticulate.

Disturbed sites. Common; northern counties, central peninsula. North Carolina south to Florida, west to Texas. Spring.

EXCLUDED TAXON

Descurainia sophia (Linnaeus) Webb ex Prantl—Cited by Correll and Johnston (1970) as "throughout the U.S.," which would presumably include Florida. No Florida specimens known.

Diplotaxis DC. 1821. WALLROCKET

Annual or perennial herbs. Leaves basal and cauline, alternate, simple, pinnate-veined, petiolate or epetiolate, estipulate. Flowers in racemes, pedicellate, bracteate basally or ebracteate, actinomorphic, bisexual; sepals 4, free; petals 4, free, equal; stamens 6, tetradynamous; intrastaminal nectary glands 4, the lateral cushion-like, the median cylindric. Fruit a silique, dehiscent, the segments 2, the replum rounded, the septum complete; seeds biserrate.

A genus of about 25 species; North America, West Indies, Mexico, South America, Europe, Africa, Asia, Australia, and Pacific Islands. [*Diplo*, double, and *taxis*, arrangement, in reference to the number of seed rows in each fruit locule.]

Selected reference: Martínez-Laborde (2010).

1. Perennials with adventitious buds on the roots; gynophore 0.5–3 mm long.......................**D. tenuifolia**
1. Annuals or short-lived perennials lacking adventitious buds on the roots; gynophore absent or to 0.5 mm long ....**D. muralis**

Diplotaxis muralis (L.) DC. [Growing on walls.] ANNUAL WALLROCKET.

> *Sisymbrium murale* Linnaeus, Sp. Pl. 658. 1753. *Brassica muralis* (Linnaeus) Hudson, Fl. Angl., ed. 2. 290. 1778. *Sinapis muralis* (Linnaeus) W. T. Aiton, Hortus Kew. 4: 128. 1812. *Diplotaxis muralis* (Linnaeus) de Candolle, Syst. Nat. 2: 634. 1821. *Crucifera diplotaxis* E.H.L. Krause, in Sturm, Fl. Deutschl., ed 2. 6: 127, pl. 33. 1902, nom. illegit.

Ascending to suberect annual or short-lived perennial herb, to 5(6) dm; roots lacking adventitious buds, the stem pubescent. Basal leaves with the blade elliptic to obovate, 2–9 cm long, 1–3.5 cm wide, sinuate to pinnatifid, lyrate, the apex acute to obtuse, the base cuneate, the margin and veins glabrescent to sparsely pubescent, petiolate or sessile, the cauline leaves with the margin entire or dentate, petiolate or sessile. Flowers in a raceme, the pedicel ascending or divaricate, (3)8–20(37) mm long in fruit; sepals oblong, 3–6 mm long, pubescent or glabrous; petals obovate, 5–8(10) mm long, yellow, the apex rounded or truncate; stamens 4–8 mm long; gynophore absent or to 0.5 mm long. Fruit linear-oblong, (1.5)2–4 cm long, 2–3 mm wide, the terminal segment beak-like (1)2–3 m long; seeds ovoid or ellipsoid, ca. 1 mm long, smooth or minutely reticulate.

Disturbed sites. Rare; Monroe County keys. Nearly throughout North America; West Indies, Mexico, and South America; Europe, Africa, Asia, Australia, and Pacific Islands. Native to Europe, Africa, and Asia. Spring.

Diplotaxis tenuifolia (L.) DC. [With slender leaves.] PERENNIAL WALLROCKET.

> *Sisymbrium tenuifolium* Linnaeus, Cent. Pl. 1: 18. 1755. *Sinapis tenuifolia* (Linnaeus) W. T. Aiton, Hortus Kew. 4: 128. 1812. *Diplotaxis tenuifolia* (Linnaeus) de Candolle, Syst. Nat. 2: 632. 1821. *Brassica tenuifolia* (Linnaeus) Baillon, Hist. Pl. 227. 1871.

Erect perennial herb, to 7(10) dm; roots with adventitious buds, the stem glabrescent or sparsely pubescent basally. Basal leaves with the blade elliptic to obovate, 2–15 cm long, 1–6(8) cm wide, sinuate to deeply pinnatifid with 2–5 lobes on each side, the apex acute to obtuse, the base cuneate, petiolate or sessile, the cauline leaves with the blade similar to the basal but with narrow segments, the surfaces usually glabrescent, petiolate. Flowers in a raceme, the pedicel erect, 0.8–3.5 cm long in fruit; sepals oblong, 4–6 mm long, glabrous or pubescent; petals oblong, 7–11(13) mm long, the apex rounded; stamens 4–8 mm long; gynophore 0.5–3 mm long. Fruit linear-oblong, 2–5 cm long, 2–3 mm wide, the terminal segment beak-like 2–3 m long; seeds ovoid or ellipsoid, ca. 1 mm long, smooth or minutely reticulate.

Disturbed sites. Rare; Escambia County, not recently collected. Widely distributed in eastern, southern, and western North America; South America; Europe, Africa, Asia, and Australia. Native to Europe, Africa, and Asia. Spring.

Draba L., nom. cons. 1753. WHITLOWGRASS

Annuals or perennials. Leaves basal and cauline, alternate, simple, pinnate-veined, petiolate or epetiolate, estipulate. Flowers in racemes, pedicellate, ebracteate, actinomorphic, bisexual; sepals 4, free; petals 4, free, equal; stamens 6, tetradynamous; intrastminal nectary glands present. Fruit a silique, dehiscent; seeds biseriate.

A genus of about 380 species; North America, Mexico, South America, Europe, Africa, and Asia. [From the Greek *drabe*, acrid, having the taste like the mustard plant.]

Selected reference: Al-Shehbaz, Windham, and Elven (2010).

1. Plant with numerous cauline leaves in flower or fruit; stem and leaves with 4-rayed trichomes; fruit glabrous ....**D. brachycarpa**
1. Plant scapose or subscapose in flower or fruit; stem with at least some simple trichomes; fruit with simple or 2-rayed trichomes ....**D. cuneifolia**

Draba brachycarpa Nutt. ex Torr. & A. Gray [From the Greek *brachys*, short, and *carpos*, fruit.] SHORTPOD WHITLOWGRASS.

Draba brachycarpa Nuttall ex Torrey & A. Gray, Fl. N. Amer. 1: 108. 1838. *Abdra brachycarpa* (Nuttall ex Torrey & A. Gray) Greene, Pittonia 4: 207. 1900.

Annual herb, to 2 dm; stem usually branched, pubescent, the trichomes cruciform. Basal leaves with the blade oblanceolate to obovate, 0.5–2 cm long, 2–8 mm wide, the apex acute, the base cuneate, the margin entire, the upper and lower surfaces pubescent with cruciform trichomes, the petiole to 5 mm long, the cauline leaves numerous to the inflorescence, the blade lanceolate to oblong or linear, the apex acute, the base cuneate, the margin entire, the upper and lower surfaces pubescent with cruciform trichomes, sessile. Flowers in a raceme, the axis pubescent with cruciform trichomes, the fruiting pedicel horizontal to divaricate-ascending, straight, (1)2–4(5) mm long; sepals oblong, ca. 1 mm long, green or pink, pubescent; petals spatulate, 2–3 mm long, white. Fruit elliptic to linear-elliptic, rarely ovate-elliptic, flattened, 2–6 mm long, 1–2 mm wide, glabrous; seeds ovoid, winged, to 1 mm long.

Dry hammocks and disturbed sites. Occasional; Suwannee County, central and western panhandle. Virginia south to Florida, west to Nebraska, Kansas, Oklahoma, and Texas, also Montana, Oregon, and Arizona. Spring.

Draba cuneifolia Nutt. ex Torr. & A. Gray [With cuneate leaves.] WEDGELEAF WHITLOWGRASS.

Draba cuneifolia Nuttall ex Torrey & A. Gray, Fl. N. Amer. 1: 108. 1838.

Annual herb, to 3 dm; stem scapose or subscapose, hirsute with 3- to 4-rayed trichomes and at least some simple ones. Basal leaves with the blade oblanceolate to spatulate or broadly obovate, 1–3.5 cm long, 0.5–2 cm wide, the apex obtuse, the base broadly cuneate, the margin dentate, the surfaces pubescent with simple or 2- to 4-rayed trichomes, subsessile, the cauline leaves absent or sometimes a few on the basal ⅓ of the stem, similar to the basal leaves. Flowers in a raceme, the axis densely pubescent with simple and 2- to 4-rayed trichomes, the fruiting pedicel horizontal to divaricate-ascending, straight, 2–7(10) mm long, pubescent as the rachis; sepals oblong, 2–3 mm long, green or pink, glabrous or pubescent with simple trichomes; petals spatulate, (2)3–5 mm long, white. Fruit oblong to linear or lanceolate, (5)7–12(16) mm long, flattened, with simple or rarely 2-rayed trichomes; seeds broadly ovoid, to 1 mm long.

Dry hammocks and disturbed sites. Rare; St. Johns and Jackson Counties. Pennsylvania south to Florida, west to California; Mexico. Spring.

Erucastrum (DC.) C. Presl, nom. cons. 1826. DOGMUSTARD

Annual or biennial herbs. Leaves basal and cauline, alternate, simple, pinnate-veined, petiolate or epetiolate, estipulate. Flowers in racemes, pedicellate, bracteate, actinomorphic, bisexual; sepals 4, free; petals 4, free, equal; stamens 6, tetradynamous; intrastaminal nectar glands present. Fruit a silique, the segments 2, dehiscent; seeds 1-seriate.

A genus of 19 species; North America, Europe, Africa, and Asia. [The genus *Eruca*, and -*astrum*, to resemble.]

Selected reference: Warwick (2010b).

Erucastrum gallicum (Willd.) O. E. Schulz [Of Gallia, France]. COMMON DOGMUSTARD.

Sisymbrium gallicum Willdenow, Enum. Pl. 678. 1809. *Erucastrum inodorum* Reichenbach, Fl. Germ. Excurs. 693. 1832, nom. illegit. *Sisymbrium irio* Linnaeus var. *xerophilum* E. Fournier, Rech. Anat. Crucif. 74. 1865. *Erucastrum gallicum* (Willdenow) O. E. Schulz, Bot. Jahrb. Syst. 54(Beibl. 119): 56. 1916. *Hirschfeldia gallica* (Willdenow) Fritsch, Excursionsfl. Oesterreich, ed. 3. 157. 1922.

Erect or ascending annual or perennial herb, to 6.5(8) dm; stem pubescent. Basal leaves with the blade oblanceolate, 3–28 cm long, 0.8–11 cm wide, the margin dentate to deeply lobed or pinnatifid, the lobes 3–10 on each side, smaller than the terminal one, the lobe margin crenate or dentate, the upper and lower surfaces sparsely pubescent, the cauline leaves similar to the basal, the distal ones short-petiolate or sessile, the blade smaller, the distalmost ones 1–2 cm long, passing into leaflike, linear bracts, the margin entire. Flowers in a raceme, the pedicel ascending to divaricate, (3)5–10(20) mm long in fruit; sepals oblong, 3–5 mm long, sparsely hispid apically; petals obovate to oblanceolate, 4–8 mm long, white to pale yellow, the apex rounded, the base short-clawed. Fruit linear, 1–4.5 cm long, 1–3 mm wide, subterete or

4-angled, slightly torulose, the terminal segment style-like, 2–4 mm long, seedless, the valves each with a prominent midvein, glabrous; seeds elliptic, ca. 1 mm long, reddish brown.

Disturbed sites. Rare; Palm Beach County, southern peninsula. Nearly throughout North America; Europe. Native to Europe. Spring–summer.

Erysimum L. 1753. WALLFLOWER

Annual herbs. Leaves basal and cauline, alternate, simple, pinnate-veined, petiolate or epetiolate, estipulate. Flowers in racemes, pedicellate, ebracteate, actinomorphic, bisexual; sepals 4, free; petals 4, free, equal; stamens 6, tetradynamous; intrastaminal nectar glands present. Fruit a silique, dehiscent, the segments 2, the replum rounded, the septum complete; seeds uniseriate.

A genus of about 150 species; North America, Mexico, Central America, Europe, Africa, Asia, and Australia. [From the Greek, *eryso*, to ward off or to cure, in reference to the supposed medicinal properties of some species.]

Selected reference: Al-Shehbaz (2010f).

Erysimum cheiranthoides L. [Resembling *Cheiranthus* (Brassicaceae).] WORMSEED WALLFLOWER.

Erysimum cheiranthoides Linnaeus, Sp. Pl. 661. 1753. *Cheiranthus turritoides* Lamarck, Encycl. 2: 716. 1788, nom. illegit. *Cheirinia cheiranthoides* (Linnaeus) Link, Enum. Hort. Berol. Alt. 2: 170. 1822. *Cuspidaria cheiranthoides* (Linnaeus) Link, Handbuch 2: 315. 1829. *Sisymbrium cheiranthoides* (Linnaeus) Eaton & J. Wright, Man. Bot., ed. 8. 429. 1840. *Cheiranthus cheiranthoides* (Linnaeus) A. Heller, Cat. N. Amer. Pl. 4. 1898. *Crucifera erysimum* E.H.L. Krause, in Sturm, Fl. Deutsch., ed. 2. 2: 75, pl. 1. 1902, nom. illegit.

Erect annual herb, to 1(1.5) m; stem sparsely stellate pubescent. Basal leaves similar to cauline, the cauline with the blade lanceolate, linear, or elliptic-oblong, (1)2–7(11) cm long, 0.5–1.2(2) cm wide, the apex acute or obtuse, the base cuneate, the margin subentire or denticulate or sinuate-dentate, the upper and lower surfaces sparsely stellate-pubescent, short-petiolate or sessile. Flowers in a raceme, the pedicel divaricate or ascending, 5–13(16) mm long in fruit; sepals oblong, 2–3 mm long; petals narrowly spatulate, 3–5 mm long, yellow, the apex rounded, the claw 2–3 mm long; stamens 2–3 mm long. Fruit linear, (1)1.5–2.5(4) cm long, straight, 4-angled, somewhat torulose, the valves with a prominent midvein, stellate-pubescent, the inside surface densely pubescent, the style cylindric; seeds oblong, ca. 1 mm long, brown, the surface minutely reticulate.

Disturbed sites. Rare; Jackson and Miami-Dade Counties. Nearly throughout North America; South America; Europe, Africa, Asia, and Pacific Islands. Native to Europe and Asia. Spring.

Lepidium L. 1753. PEPPERWEED

Annuals or biennials. Leaves basal and cauline, alternate, simple, pinnate-veined, petiolate or epetiolate, estipulate. Flowers in racemes, pedicellate, ebracteate, actinomorphic, bisexual;

sepals 4, free; petals absent or 4, free, equal, with or without a claw; stamens 2 and equal in length or 6 and tetradynamous; intrastaminal nectary glands 4 or 6, distinct. Fruit a dehiscent silique or a schizocarp, the replum rounded, the septum complete or perforate; seeds 2.

A genus of about 220 species; nearly cosmopolitan. [From the Greek *lepidion* or *leidos*, scale, in reference to the appearance of the fruit.]

Carara Medik. (1792); *Coronopus* Zinn, nom. cons. (1757).

Selected reference: Al-Shehbaz and Gaskin (2010).

1. Middle and upper leaves clasping the stem; fruit 5–6 mm long..**L. campestre**
1. Middle and upper leaves narrowly cuneate; fruit 1.5–3.5 mm long.
   2. Fruit a schizocarp, separating from the septum as 2 nutlets, wrinkled ....**L. didymum**
   2. Fruit a dehiscent silique, smooth.
      3. Fruit obovate to obovate-suborbicular, widest beyond the middle; petals shorter than the sepals or lacking..**L. densiflorum**
      3. Fruit orbicular or suborbicular, widest at the middle; petals as long as or slightly longer than the sepals..**L. virginicum**

Lepidium campestre (L.) W. T. Aiton [Of plains or flat areas.] FIELD PEPPERWEED.

Thlaspi campestre Linnaeus, Sp. Pl. 641. 1753. *Lepidium campestre* (Linnaeus) W. T. Aiton, Hortus Kew. 4: 88. 1812. *Lepia campestris* (Linnaeus) Desvaux, J. Bot. Agric. 3: 165. 1815. *Iberis campestris* (Linnaeus) Wallroth, Sched. Crit. 341. 1822. *Crucifera lepidium* E.H.L. Krause, in Sturm, Fl. Deutschl., ed. 2. 6: 152. 1902, nom. illegit. *Neolepia campestris* (Linnaeus) W. A. Weber, Phytologia 67: 427. 1989.

Erect annual herb, to 5(6) dm; stem densely hirsute. Basal leaves with the blade oblanceolate or oblong, (1)2–6(8) cm long, 0.5–1.5 cm wide, the margin entire, lyrate, or pinnatifid, the apex rounded, the base cuneate, the upper and lower surfaces densely hirsute, the petiole (0.5)1.5–6 cm long, the cauline leaves with the blade oblong, lanceolate or narrowly deltate-lanceolate, (0.7)1–4(6.5) cm long, (2)5–10(15) mm wide, the apex acute, the base sagittate or auriculate, the margin dentate or subentire, the upper and lower surfaces densely hirsute, sessile. The flowers in a raceme, the rachis hirsute, the pedicel horizontal or slightly recurved, (3)4–8(10) mm long in fruit, hirsute; sepals oblong, 1–2 mm long; petals spatulate, 2–3 mm long, white, the claw ca. 1 mm long; stamens 6, tetradynamous, ca. 2 mm long. Fruit a dehiscent silique, broadly oblong to ovate, (4)5–6 mm long, (3)4–5 mm wide, the apex broadly winged and notched, smooth; seeds ovoid, ca. 2 mm long, brown.

Disturbed sites. Rare; Washington County. Nearly throughout North America; South America; Europe, Africa, Asia, Australia, and Pacific Islands. Native to Europe and Asia. Spring.

Lepidium densiflorum Schrad. [Densely flowered.] PRAIRIE PEPPERWEED.

Lepidium densiflorum Schrader, Index Sem. Hort. Gott. 4. 1832. *Lepidium densiflorum* Schrader var. *typicum* Thellung, Bull. Herb. Boissier, ser. 2. 4: 706. 1904, nom. inadmiss.

Erect annual or biennial herb, to 5(6.5) dm; stem puberulent or glabrous. Basal leaves with the blade oblanceolate, spatulate, or oblong, (1.5)2.5–8(11) cm long, 0.5–1(2) cm wide, the apex obtuse, the base cuneate, the margin coarsely serrate or pinnatifid, the upper and lower surfaces puberulent or glabrous, the petiole 0.5–1.5(2) cm long, the cauline leaves with the blade narrowly oblanceolate or linear, (0.7)1.3–6.2(8) cm long, 1–10(18) mm wide, the apex obtuse, the base attenuate to cuneate, the margin entire or irregularly serrate to dentate, rarely pinnatifid, the upper and lower surfaces puberulent or glabrous, short-petiolate. The flowers in a raceme, the rachis puberulent, the pedicel divaricate-ascending to horizontal, straight or slightly recurved, 2–4 mm long in fruit, puberulent; sepals oblong, 0.5–1 mm long; petals absent or rudimentary, filiform, to 1 mm long; stamens 2, 1–2 mm long. Fruit a dehiscent silique, obovate to obovate-suborbicular, widest beyond the middle, 2–3 mm long, ca. 2 mm wide, the apex winged and notched, smooth, sparsely puberulent at least on the margin; seeds ovate, ca. 1 mm long.

Disturbed sites. Rare; Okaloosa and Escambia Counties. Nearly throughout North America; Mexico; Europe, Asia, and Pacific Islands. Native to North America to the north and west of our area and Mexico. Spring.

Lepidium didymum L. [In pairs, in reference to the flowers with 2 stamens and 2-seeded fruits.] LESSER SWINECRESS.

Lepidium didymum Linnaeus, Mant. Pl. 92. 1767. *Nasturtiolum castratum* Medikus, Pfl.-Gatt. 82. 1792, nom. illegit. *Senebiera pinnatifida* de Candolle, Mém. Soc. Hist. Nat. Paris 1: 144, t. 9. 1799. *Coronopus didymus* (Linnaeus) Smith, Fl. Brit. 691. 1800. *Nasturtiolum pinnatum* Moench, Suppl. Meth. 71. 1802, nom. illegit. *Cochlearia humifusa* Michaux, Fl. Bor.-Amer. 2: 27. 1803, nom. illegit. *Senebiera didyma* (Linnaeus) Persoon, Syn. Pl. 2: 185. 1806. *Coronopus pinnatus* Hornemann, Hort. Bot. Hafn. 599. 1815, nom. illegit. *Coronopus pinnatifidus* (de Candolle) Dulac, Fl. Hautes-Pyrénées 186. 1867. *Carara didyma* (Linnaeus) Britton, in Britton & A. Brown, Ill. Fl. N. U.S., ed. 2. 2: 167. 1913.

Erect to ascending or decumbent annual herb, to 5(7) dm; stem glabrous or pilose. Basal leaves with the blade 1- to 2-pinnatisect, 1–6(8) cm long, the apex obtuse, the base cuneate, the margin of the lobes entire or dentate to lobed, the upper and lower surfaces glabrous or pilose, the petiole 0.5–4(6) cm long, the cauline leaves with the blade similar to the basal ones, but smaller and less divided distally, 1.5–3.5(4.5) cm long, 5–12 cm wide, the lobes lanceolate to oblong or elliptic, the margin of the lobes entire, serrate, or incised, short-petiolate to subsessile. Flowers in a raceme, the rachis glabrous or pubescent, the pedicel divaricate or horizontal, straight or slightly recurved, 2–4 mm long in fruit, glabrous or sparsely pubescent; sepals ovate, 0.5–1 mm long; petals elliptic to linear, ca. 0.5 mm long, white; stamens 2, ca. 0.5 mm long. Fruit a schizocarp, separating from the septum as 2 nutlets, ca. 1.5 mm long, ca. 2 mm wide, the apex with a deep notch, rugose; seeds ovate, ca. 1 mm long.

Disturbed sites. Occasional; nearly throughout. Nearly cosmopolitan. Native to Europe. Spring.

Lepidium virginicum L. [Of Virginia.] VIRGINIA PEPPERWEED.

Lepidium virginicum Linnaeus, Sp. Pl. 645. 1753. *Thlaspi virginicum* (Linnaeus) Cavanilles, Descr. Pl. 413, 624. 1802. *Thlaspi virginianum* Poiret, in Lamarck, Encycl. 7: 544. 1806, nom. illegit. *Cyanocardamum virginicum* (Linnaeus) Webb & Berthelot, Nat. Hist. Iles Canaries 3(2(1)): 97. 1836.

Iberis virginica (Linnaeus) Fischer & C. A. Meyer, Index Sem. Hort. Petrop. 2: 13. 1836. *Nasturtium virginicum* (Linnaeus) Kuntze, Revis. Gen. Pl. 1: 35. 1891. *Crucifera virginica* (Linnaeus) E.H.L. Krause, in Sturm, Fl. Deutsch., ed 2. 6: 158. 1902. *Lepidium virginicum* Linnaeus subsp. *euvirginicum* Thellung, Neue Denkschr. Schweiz. Naturf. Ges. 41(1): 225. 1906, nom. inadmiss. *Lepidium virginicum* Linnaeus var. *typicum* Thellung, Neue Denkschr. Schweiz. Naturf. Ges. 41(1): 224. 1906, nom. inadmiss.

Erect annual herb, to 5.5(7) dm; stem puberulent. Basal leaves with the blade obovate, spatulate, or oblanceolate, (1)2.5–10(15) cm long, 0.5–3(5) cm wide, the apex obtuse, the base cuneate, the margin pinnatifid to lyrate or dentate, the upper and lower surfaces puberulent, the petiole 0.5–3.5 cm long, the cauline leaves with the blade oblanceolate or linear, 1–6 cm long, (1)3–10 mm wide, the apex acute, the base attenuate to subcuneate, the margin serrate or entire, the upper and lower surfaces puberulent, short-petiolate. Flowers in a raceme, the rachis puberulent or glabrous, the pedicel divaricate-ascending to nearly horizontal, straight or slightly recurved, 3–4(6) mm long in fruit, puberulent or glabrous; sepals oblong to ovate, ca. 1 mm long; petals spatulate to oblanceolate, 1–3 mm long, white; stamens 2, ca. 1 mm long. Fruit a dehiscent silique, orbicular or suborbicular, 3–4 mm long and wide, widest at the middle, the apex winged and notched, smooth, glabrous; seeds ovate, 1–2 mm long.

Disturbed sites. Common; nearly throughout. Nearly throughout North America; West Indies, Mexico, Central America, and South America; Europe, Africa, Asia, Australia, and Pacific Islands. Native to North America, Mexico, and Central America. Spring.

EXCLUDED TAXA

Lepidium coronopus (Linnaeus) Al-Shehbaz—Reported for Florida by Small (1903, 1913a, both as *Coronopus coronopus* (Linnaeus) H. Karsten; 1933, as *Carara coronopus* (Linnaeus) Medikus). No Florida specimens known.

Lepidium ruderale Linnaeus—Reported for Florida by Small (1933), Al-Shehbaz (1986), and Al-Shehbaz and Gaskin, 2010). No Florida specimens known.

Lobularia Desv., nom. cons. 1815.

Perennial herbs or subshrubs. Leaves alternate, simple, pinnate-veined, epetiolate, estipulate. Flowers in racemes, pedicellate, bracteate or ebracteate, actinomorphic, bisexual; sepals 4, free; petals 4, free, equal; stamens tetradynamous; intrastaminal nectary glands present. Fruit a silique, dehiscent, the replum rounded, the septum complete; seeds uniseriate.

A genus of 4 species; nearly cosmopolitan. [*Lobulus*, small lobe, in reference to the small siliques.]

Koniga R. Br. (1826).

Selected reference: Borgen (2010).

Lobularia maritima (L.) Desv. [Growing by the sea.] SEASIDE LOBULARIA; SWEET ALYSSUM.

Clypeola maritima Linnaeus, Sp. Pl. 652. 1753. *Alyssum maritimum* (Linnaeus) Lamarck, Encycl. 1: 98. 1783. *Lobularia maritima* (Linnaeus) Desvaux, J. Bot. Agric. 3: 169. 1815. *Adyseton maritimum* (Linnaeus) Link, Enum. Hort. Berol. Alt. 2: 157. 1822. *Glyce maritima* (Linnaeus) Lindley, Syn. Brit. Fl. 26. 1829. *Octadenia maritima* (Linnaeus) Fischer & C. A. Meyer, Index Sem. Hort. Petrop. 3: 41. 1837.

Erect, procumbent, or decumbent perennial herb or subshrub, to 2.5(4) dm tall; stem pubescent. Leaves with the blade linear or linear-lanceolate, (1)1.6–2.5(4.2) cm long, (1)2–3(7) mm wide, the apex acute, the base attenuate, the margin entire, the upper and lower surfaces pubescent. Flowers in a dense raceme, (1)4–8(16) cm long in fruit, the lower ones sometimes bracteate, the rachis pubescent, the pedicel ascending, pubescent; sepals oblong, 1–2 mm long, white tinged purplish; petals broadly obovate, 2–3 mm long, abruptly contracted into a claw; stamens ca. 2 mm long; intrastaminal nectary glands 8, the 4 lateral rudimentary, the 4 median cylindric. Fruit elliptic-suborbicular, 2–3 mm long and wide, flattened, smooth, sparsely pubescent; seeds ovate, ca. 1 mm long, light brown to reddish brown.

Disturbed sites. Rare; Levy, Volusia, and Broward Counties. Escaped from cultivation. Nearly cosmopolitan. Native to Europe. Spring–fall.

Nasturtium W. T. Aiton, nom. cons. 1812. WATERCRESS

Perennial herbs. Leaves alternate, simple, pinnate-veined, petiolate or epetiolate, estipulate. Flowers in racemes, pedicellate, ebracteate, actinomorphic, bisexual; sepals 4, free; petals 4, free, equal; stamens tetradynamous; intrastaminal nectary glands 2, lateral. Fruit a silique, dehiscent, the replum rounded, the septum complete; seeds uniseriate or biseriate.

A genus of 5 species; nearly cosmopolitan. [*Nasus*, nose, and *tortus*, distortion, alluding to the pungency of the plants.]

Selected reference: Al-Shehbaz (2010c).

1. Fruit 1–1.5 mm wide; seeds uniseriate, each seed face very finely alveolate-reticulate (400–500 alveoli on each side), dull yellow-brown ....**N. floridanum**
1. Fruit 2–3 mm wide; seeds biseriate, each seed face relatively coarsely alveolate-reticulate (25–50 alveoli on each side), shiny reddish brown ....**N. officinale**

Nasturtium floridanum (Al-Shehbaz & Rollins) Al-Shehbaz & R. A. Price [Of Florida]. FLORIDA WATERCRESS.

Cardamine curvisiliqua Shuttleworth ex Chapman, Fl. South. U.S., ed. 2. 605. 1883. *Nasturtium stylosum* Shuttleworth ex O. E. Schulz, in Engler & Prantl, Nat. Pflanzenfam., ed. 2. 17B: 553. 1936; non (de Candolle) O. E. Schulz ex Cheeseman, 1911. *Rorippa floridana* Al-Shehbaz & Rollins, J. Arnold Arbor. 69: 68. 1988. *Nasturtium floridanum* (Al-Shehbaz & Rollins) Al-Shehbaz & R. A. Price, Novon 8: 125. 1998. TYPE: FLORIDA: Wakulla Co.: "in uliginosis subsalsis ad fluv. St. Marks, prope St. Marks," Apr–May 1843, *Rugel s.n.* (lectotype: GH; isolectotype: GH). Lectotypified by Al-Shehbaz & Rollins (1988: 68–69).

Perennial herb, to 9 dm; stem glabrous or sparsely pubescent. Leaves with the blade 3- to 5(7)-pinnatifid, (1.5)3–7(10) cm long, the lateral segments often much smaller than that of the terminal, the terminal one subreniform, orbicular, ovate, or oblong, 0.5–3.5 cm long and wide, the apex obtuse or rounded, the base obtuse, subcordate, rounded, or cuneate, the margin repand, entire, or obtusely dentate, the upper and lower surfaces glabrous or rarely sparsely pubescent, petiolate. Flowers in a raceme, the pedicel divaricate, straight or slightly recurved, 5–15 mm long in fruit; sepals oblong, 2–3 mm long; petals spatulate, 4–5 mm long, white, the apex rounded; stamens 2–3 mm long. Fruit linear, 1.5–3 cm long, ca. 1 mm wide, terete, glabrous; seeds uniseriate, ovoid, ca. 1 mm long, yellow-brown, minutely alveolate-reticulate (400–500 alveoli on each side).

Spring runs and blackwater swamps. Frequent; peninsula west to central panhandle. Endemic. Spring.

Nasturtium officinale W. T. Aiton [Used in medicine.] EUROPEAN WATERCRESS.

Sisymbrium nasturtium-aquaticum Linnaeus, Sp. Pl. 657. 1753. *Cardamine fontana* Lamarck, Fl. Franc. 2: 499. 1779 ("1778"), nom. illegit. *Sisymbium nasturtium* Thunberg, Fl. Jap. (Thunberg) 260. 1784, nom. illegit. *Cardaminum nasturtium* Moench, Methodus 262. 1794, nom. illegit. *Sisymbrium amarum* Salisbury, Prodr. Stirp. Chapm. Allerton 270. 1796, nom. illegit. *Beaumerta nasturtium* P. Gaertner et al., Oekon Fl. Wetterau 2: 467. 1800, nom. illegit. *Nasturtium officinale* W. T. Aiton, Hortus Kew. 4: 110. 1812. *Nasturtium aquaticum* Wahlenberg, in Palmstruch, Svensk Bot. 9: t. 624. 1824, nom. illegit. *Nasturtium fontanum* Ascherson, Fl. Brandenburg 1: 32. 1860, nom. illegit. *Nasturtium officinale* W. T. Aiton var. *vulgare* Alefeld, Landw. Fl. 231. 1866, nom. inadmiss. *Nasturtium nasturtium-aquaticum* (Linnaeus) H. Karsten, Deut. Fl. 657. 1882, nom. inadmiss. *Cardamine nasturtium* Kuntze, Revis. Gen. Pl. 1: 22. 1891, nom. illegit. *Nasturtium nasturtium* Cockerell, Bull. Torrey Bot. Club 19: 95. 1892, nom. inadmiss. *Rorippa nasturtium* Beck, Fl. Nieder-Oesterreich 463. 1892, nom. illegit. *Rorippa nasturtium* Beck var. *typica* Beck, Fl. Nieder-Oesterreich 463. 1892, nom. inadmiss. *Cardamine nasturtium-aquaticum* (Linnaeus) Borbás, Balaton Fl. 390. 1900. *Crucifera fontana* E.H.L. Krause, in Sturm, Fl. Deutschl., ed. 2. 6: 109, pl. 9. 1902, nom. illegit. *Rorippa nasturtium-aquaticum* (Linnaeus) Hayek, Sched. Fl. Stiriac. 22. 1904. *Radicula officinalis* (W. T. Aiton) H. Groves, in Babington, Man. Brit. Bot., ed. 9. 26. 1904, nom. illegit. *Radicula nasturtium-aquaticum* (Linnaeus) Rendle & Britten, List Brit. Seed Pl. 3. 1907. *Cardamine aquatica* Nieuwland, Amer. Midl. Naturalist 4: 34. 1915, nom. illegit. *Rorippa officinalis* (W. T. Aiton) P. Royen, Alp. Fl. New Guinea 3: 2029, nom. illegit.

Perennial herb, to 11(20) dm; stem glabrous or sparsely pubescent. Leaves with the blade 3- to 9(13)-pinnatifid, (1)2–5(22) cm long, the lateral segments smaller than the terminal, the terminal one suborbicular to ovate, or oblong to lanceolate, (0.5)1–4(5) cm long, (3)0.7–2.5(4) cm wide, the apex obtuse, the base obtuse, cuneate, or subcordate, the margin entire or repand, the upper and lower surfaces glabrous or sparsely pubescent, petiolate. Flowers in a raceme, pedicel divaricate or ascending, straight or recurved, 5–17(24) mm long in fruit; sepals oblong, 2–4 mm long; petals spatulate or obovate, 3–5 mm long, white or pink, the apex rounded; stamens 2–4 mm long. Fruit linear, (0.6)1–1.8(2.5) cm long, 2–3 mm wide, terete, glabrous; seeds biseriate, ovoid, ca. 1 mm long, reddish brown, coarsely alveolate-reticulate (25–50 alveoli on each side).

Springs and fast-flowing shallow streams. Occasional; northern counties, central peninsula. Escaped from cultivation. Nearly cosmopolitan. Native to Europe, Africa, and Asia. Spring–summer.

EXCLUDED TAXA

Nasturtium microphyllum Boenninghausen ex Reichenbach—Reported for Florida by Godfrey and Wooten (1981), Wunderlin (1982), and Clewell (1985), based on misidentification of material of *N. floridanum.*

Nasturtium ×*sterile* (Airy-Shaw) Oefelein—This hybrid of *N. microphyllum* Boenninghausen ex Reichenbach and *N. officinale* was reported for Florida by Rollins (1978), based on a misidentification of Florida material of *N. floridanum.*

Planodes Greene 1912. WINGED ROCKCRESS

Annual herbs. Leaves basal and cauline, alternate, simple, pinnate-veined, petiolate, estipulate. Flowers in racemes, pedicellate, ebracteate, actinomorphic, bisexual; sepals 4, free; petals 4, free, equal; stamens 6, tetradynamous; intrastaminal nectar glands present. Fruit a silique, the replum rounded, the septum complete; seeds uniseriate.

A monotypic genus; North America and Mexico. [From the Greek *planis*, wanderer, and *odes*, resemblance, in reference to its various generic placements.]

Selected reference: Al-Shehbaz (2010d).

Planodes virginica (L.) Greene [Of Virginia.] VIRGINIA WINGED ROCKCRESS.

Cardamine virginica Linnaeus, Sp. Pl. 656. 1753. *Arabis virginica* (Linnaeus) Poiret, in Lamarck, Encycl., Suppl. 1: 413. 1810. *Cardamine hirsuta* Linnaeus var. *virginica* (Linnaeus) Torrey & A. Gray, Fl. N. Amer. 1: 85. 1838. *Cardamine parviflora* Linnaeus subsp. *virginica* (Linnaeus) O. E. Schulz, Bot. Jahrb. Syst. 32: 484. 1903. *Planodes virginica* (Linnaeus) Greene, Leafl. Bot. Observ. Crit. 2: 221. 1912. *Sibara virginica* (Linnaeus) Rollins, Rhodora 3: 481. 1941.

Cardamine ludoviciana Hooker, J. Bot. (Hooker) 1: 191. 1834. *Arabis ludoviciana* (Hooker) C. A. Meyer, in Fischer & C. A. Meyer, Index Sem. Hort. Petrop. 9: 60. 1843.

Erect, ascending, or decumbent annual herb, to 3.5(4.5) dm; stem hirsute to puberulent proximally, usually glabrous distally. Basal leaves with the blade oblong to oblanceolate, (1)1.5–7(10) cm long, 0.4–2(3) cm wide, the margin pinnatifid to pinnatisect, the lobes (4)6–12(15) per side, ovate or oblong to linear, the margin entire or dentate, often 1-toothed, the upper and lower surfaces pubescent to glabrate, the terminal lobe equal to or larger than the lateral, the petiole (0.3)0.8–1.5(2) cm long, the cauline leaves with the blade similar to the basal but smaller, short-petiolate. Flowers in a terminal and axillary raceme, the pedicel divaricate-ascending, straight, 2–6(8) mm long in fruit; sepals oblong, 1–2 mm long, glabrous or with a few trichomes apically; petals oblanceolate, 2–3 mm long, white; stamens ca. 2 mm long; nectar glands 1 on each side of the lateral stamen, minute. Fruit linear, (1)1.5–2.5(3) cm long, 1–2 mm wide, slightly torulose, glabrous; seeds orbicular or suborbicular, ca. 1 mm long, flattened, narrowly winged.

Disturbed sites. Rare; Leon County. Maryland south to Florida, west to Nebraska, Kansas, Oklahoma, and Texas, also California; Mexico. Native to eastern North America. Spring.

Polanisia Raf. 1819. CLAMMYWEED

Herbs. Leaves alternate, 3-foliolate, petiolate, estipulate. Flowers in terminal or axillary racemes, pedicellate, bracteate, zygomorphic, bisexual; sepals 4, free; petals 4, free, unequal (1 pair longer than the other pair); stamens 8–13, the filaments inserted on an androgynophore. Fruit a capsule, dehiscent in the distal ½; seeds uniseriate.

A genus of 5 species; North America and Mexico. [From the Greek *Polys*, many, and *anisos*, unequal, in reference to the stamens.]

Aldenella Greene (1900).

Selected reference: Tucker (2010b).

Polanisia tenuifolia Torr. & A. Gray [With slender leaves.] SLENDERLEAF CLAMMYWEED.

Polanisia tenuifolia Torrey & A. Gray, Fl. N. Amer. 1: 123. 1838. *Jacksonia tenuifolia* (Torrey & A. Gray) Greene, Pittonia 2: 175. 1891. *Aldenella tenuifolia* (Torrey & A. Gray) Greene, Pittonia 4: 212. 1900. *Cleome aldenella* W. R. Ernst, J. Arnold Arbor. 44: 89. 1963.

Erect annual herb, to 4.5(9) dm; stem with stalked glandular trichomes. Leaves 3-foliolate, the leaflet blade linear, 1–5 cm long, 1–2 mm wide, the apex obtuse, mucronulate, the base narrowly cuneate, the margin entire, the upper surface glabrous or sparsely glandular, the lower surface sparsely glandular, the petiole 1–3 cm long. Flowers in a raceme 1–3 cm long, the bracts 3-foliolate, 0.5–1.5 cm long, the pedicel 0.5–1.5 cm long; sepals lanceolate to deltoid, 1–3 mm long, pale yellow, clawed, glandular, reflexed; petals oblong-ovate, white, the upper pair 2–3 mm long, 6- to 7-lobed, the apex emarginate to lacerate, the lower pair 3–5 mm long, subsessile, 4- to 5-lobed, the apex emarginate to lacerate; nectary glands yellow; stamens 3–6 mm long, slightly exserted, the anthers yellow with a maroon apex; gynophore 1–2 mm long. Fruit linear, 4–6 cm long, 2–4 mm wide, dehiscent in the distal ½, glandular or glabrous, reticulate; seeds 18–36, spheroidal, ca. 1 mm long, reddish brown, the surface pebbled.

Sandhills and scrub. Frequent; peninsula, central panhandle. Georgia, Florida, Alabama, and Mississippi. Spring–fall.

Quadrella (Jacq.) J. Presl 1825. CAPERTREE

Shrubs or trees. Leaves alternate, simple, pinnate-veined, petiolate, stipules glandular or absent. Flowers in terminal or axillary racemes or corymbs, pedunculate, pedicellate, actinomorphic, bisexual (sometimes appearing unisexual); sepals 4, free, in 2 unequal pairs; petals 4, free, equal; stamens 18–150, the filaments free, inserted on an androgynophore; ovary on a gynophore, 1-carpellate, 2-loculate, the stigma subsessile. Fruit a capsule, dehiscent or indehiscent; seeds arillate or not.

A genus of about 25 species; North America, West Indies, Mexico, Central America, and South America. [*Quadra*, four, and *-ella*, diminutive, in reference to the perianth.]

Selected reference: Tucker (2010a).

Quadrella jamaicensis (Jacq.) J. Presl [Of Jamaica.] JAMAICAN CAPERTREE.

Capparis jamaicensis Jacquin, Enum. Syst. Pl. 23: 1760. *Quadrella jamaicensis* (Jacquin) J. Presl, in
  Berchtold & J. Presl, Prir. Rostlin 2: 621. 1825.

Shrub or tree, to 6 m; branches with peltate, lepidote scales. Leaves with the blade ovate to
ovate-elliptic, 5–15(21) cm long, 2.5–5(8) cm wide, the apex usually emarginate, sometimes
acute to acuminate, the base cuneate to rounded, the margin entire, the upper surface glabrous,
the lower surface lepidote, the petiole 6–21(25) mm long, lepidote; estipulate. Flowers in a terminal or axillary raceme or corymb (1)2–6 cm long, pedicel 7–18(35) mm long, lepidote; sepals
obovate, of equal length, (7)12–18 mm long, the outer surface reddish lepidote; petals elliptic-
oblanceolate, 12–18 mm long, creamy white to purple; stamens 18–30, 4–5 cm long, purplish,
woolly basally, inserted on a discoid androgynophore; gynophore (1)6–7(9) cm long. Fruit
linear-cylindric, (7)20–38 cm long, 0.8–1.2 cm wide, irregularly constricted between the seeds,
brown-lepidote; seeds reniform, (3)6–16(38) mm long, not arillate, embedded in a red pulp.

Coastal hammocks and shell middens. Frequent; central and southern peninsula. Florida;
West Indies, Mexico, and Central America. Spring–summer.

EXCLUDED TAXON

Capparis cynophallophora Linnaeus—This name was misapplied by Small (1933), Wunderlin (1982,
  1998), Correll and Correll (1982), and Wunderlin and Hansen (2003), to Florida material of *Quadrella jamaicensis* (fide Iltis and Cornejo, 2010).

Raphanus L. 1753. RADISH

Annuals or biennials. Leaves basal and cauline, alternate, simple, pinnate-veined, petiolate
or subsessile, estipulate. Flowers in racemes, pedicellate, ebracteate, actinomorphic, bisexual;
sepals 4, free; petals 4, free, equal; stamens 6, tetradynamous. Fruit a silique, indehiscent, segments 2, lomentaceous and often breaking into 1-seeded units; seeds uniseriate.

A genus of 3 species; nearly cosmopolitan. [From the Greek, *raphanos*, radish.]
Selected reference: Warwick (2010c).

1. Fruit firm, linear, constricted between each seed ....**R. raphanistrum**
1. Fruit spongy, oblong-cylindric, slightly or not at all constricted between each seed .... **R. sativus**

Raphanus raphanistrum L. [Pre-Linnaean generic name.] WILD RADISH.

Raphanus raphanistrum Linnaeus, Sp. Pl. 669. 1753. *Rapistrum raphanistrum* (Linnaeus) Crantz, Cl.
  Crucif. Emend. 107. 1769. *Raphanus sylvestris* Lamarck, Fl. Franc. 2: 495. 1779 ("1778"), nom. illegit.
  Rapistrum arvense Allioni, Fl. Pedem. 1: 258. 1785, nom. illegit. *Raphanistrum lampsana* Gaertner,
  Fruct. Sem. Pl. 2: 300, t. 143(6). 1791. *Raphanistrum inocuum* Moench, Methodus 217. 1794, nom.
  illegit. *Raphanus infestus* Salisbury, Prodr. Stirp. Chap. Allerton 273. 1796, nom. illegit. *Durandea
  unilocularis* Delarbre, Fl. Auvergne, ed. 2. 365. 1800, nom. illegit. *Raphanus articulatus* Stokes, Bot.
  Mat. Med. 3: 482. 1812, nom. illegit. *Raphanistrum segetum* Baumgarten, Enum. Stirp. Transsilv.
  2: 280. 1816, nom. illegit. *Raphanistrum segetale* Henckel von Donnersmarck, Enum. Pl. Regio-
  mont. 172. 1817, nom. illegit. *Raphanistrum arvense* Meret de Vaumartoise, Nouv. Fl. Env. Paris,
  ed. 2. 2: 309, nom. illegit. *Raphanistrum vulgare* Gray, Nat. Arr. Brit. Pl. 2: 687. 1821, nom. illegit.

Raphanistrum sylvestre Ascherson, Fl. Brandenburg 1: 65. 1860, nom. illegit. *Raphanus stylosus* Dulac, Fl. Hautes-Pyrénées 194. 1867, nom. illegit. *Raphanus longistylus* Saint-Lager, Ann. Soc. Bot. Lyon 7: 65. 1880, nom. illegit. *Raphanus segetus* Clavaud, Actes Soc. Linn. Bordeaux 35: 288. 1881, nom. illegit. *Raphanistrum raphanistrum* (Linnaeus) H. Karsten, Deut. Fl. 673. 1882, nom. inadmiss. *Raphanus raphanistrum* Linnaeus var. *arvensis* Woerlein, Deutsche Bot. Monatsschr. 3: 50. 1885, nom. inadmiss. *Raphanus raphanistrum* Linnaeus forma *arvensis* Beck, Fl. Nieder-Oes-terreich 499. 1892, nom. inadmiss. *Raphanus raphanistrum* Linnaeus var. *typicus* Beck, Fl. Nieder-Oesterreich 499. 1892, nom. inadmiss. *Raphanistrum segetum* Baumgarten var. *arvense* Beckhaus, Fl. Westfalen 142. 1893. *Raphanus raphanistrum* Linnaeus subsp. *arvensis* (Woerlein) Schmalhau-sen, Fl. Ssredn. Jushn. Rossii 1: 81. 1895, nom. inadmiss. *Caulis raphanister* E.H.L. Krause, Bot. Centralb. 81: 207. 1900, nom. illegit. *Crucifera raphanistrum* (Linnaeus) E.H.L. Krause, in Sturm, Fl Deutschl., ed. 2. 6: 124. 1902. *Raphanus raphanistrum* Linnaeus subsp. *euraphanistrum* Briquet, Prodr. Fl. Corse 2(1): 97. 1913, nom. inadmiss. *Raphanus raphanistrum* Linnaeus subsp. *communis* Domin, Beih. Bot. Centrabl. 26(2): 255. 1914, nom. inadmiss. *Raphanus raphanistrum* Linnaeus subsp. *segetus* Bonnier ex Thellung, in Hegi, Ill. Fl. Mitt.-Eur. 4: 276. 1918, nom. inadmiss.

Annual herbs, to 8 dm; stem retrorsely hispid. Basal leaves with the blade oblong, obovate, or oblanceolate, lyrate, pinnatifid, or undivided, 3–15(22) cm long, 1–5 cm wide, the apex obtuse or acute, the base cuneate, the margin dentate, the lobes 1–4 on each side, oblong or ovate, to 4 cm long, smaller than the terminal, the upper and lower surfaces sparsely to densely pubes-cent, the petiole 1–6 cm long, the cauline leaves with the blade often undivided, the distal ones subsessile. Flowers in a raceme, the pedicel divaricate or ascending, 0.7–2.5 cm long in fruit, straight; sepals narrowly oblong, 7–11 mm long, sparsely pubescent; petals broadly obovate, 1.5–2.5 cm long, yellow or creamy white, the veins dark brown or purple, the claw to 1.5 cm long; stamens 9–14 mm long. Fruit cylindric or narrowly lanceolate, (1.5)2–11(14) cm long, 3–8(11) mm wide, strongly constricted between the seeds, strongly ribbed, the beak narrowly conic; seeds oblong or ovoid, 3–4 mm long, reddish brown or dark brown to black.

Disturbed sites. Frequent; nearly throughout. Nearly cosmopolitan. Native to Europe, Af-rica, and Asia. Spring.

Raphanus sativus L. [Cultivated, planted.] GARDEN RADISH.

Raphanus sativus Linnaeus, Sp. Pl. 2: 669. 1753. *Raphanus officinalis* Crantz, Cl. Crucif. Emend. 108. 1769, nom. illegit. *Raphanus sativus* Linnaeus var. *rapaceus* Bogenhard, Taschenb. Fl. Jena 159. 1850, nom. inadmiss. *Raphanus sativus* Linnaeus var. *vulgaris* Cosson & Germain de Saint-Pierre, Fl. Descr. Anal. Paris, ed. 2. 121. 1861, nom. inadmiss. *Raphanus raphanistrum* Linnaeus var. *sati-vus* (Linnaeus) Beck, Fl. Nieder-Oesterreich 500. 1892. *Caulis dubia* E.H.L. Krause var. *raphanus* E.H.L. Krause, Bot. Centralbl. 81: 207. 1900, nom. illegit. *Crucifera dubia* E.H.L. Krause var. *rapha-nus* E.H.L. Krause, in Sturm, Deutschl. Fl., ed. 2. 6: 168. 1902, nom. illegit. *Raphanus raphanistrum* Linnaeus subsp. *sativus* (Linnaeus) Domin, Beih. Bot. Centralbl. 26(Abt. 2): 255. 1910.

Annual or biennial herb, to 13 dm; stem sparsely scabrous or hispid. Basal leaves with the blade oblong, obovate, oblanceolate or spatulate, lyrate or pinnatisect or undivided, 2–60 cm long, 1–20 cm wide, the apex obtuse to acute, the base cuneate, the margin dentate, the lobes 1–12 on each side, oblong or ovate, to 10 cm long, the upper and lower surfaces sparsely scabrous or hispid, the petiole 1–30 cm long, the cauline leaves with the blade undivided, the distal ones subsessile. Flowers in a raceme, the pedicel spreading to ascending, 0.5–4 cm long in fruit;

sepals narrowly oblong, 6–10 mm long, glabrous or sparsely pubescent; petals obovate, 1.5–2.5 cm long, pink, purple, or white, the veins often darker, the claw to 1.4 mm long; stamens 5–12 mm long. Fruit fusiform to lanceolate, sometimes ovoid or cylindric, (1)3–15(25) cm long, (5)7–13(15) mm wide, smooth or rarely slightly constricted between the seeds, the beak conic to linear; seeds globose or ovoid, 3–4 mm long.

Disturbed sites. Rare; Lake, Polk, and Leon Counties. Escaped from cultivation. Nearly cosmopolitan. Native to Europe. Spring.

Rorippa Scop. 1760. YELLOWCRESS

Annual or perennial herbs. Leaves basal and cauline, alternate, simple, pinnate-veined, petiolate or epetiolate, estipulate. Flowers in racemes, pedicellate, bracteolate or ebracteate, actinomorphic, bisexual; sepals 4, free; petals absent or 4, free, equal; stamens 6, tetradynamous; intrastaminal nectar glands confluent. Fruit a silique, dehiscent, the replum rounded, the septum reduced to a rim or complete; seeds biseriate.

A genus of about 85 species; nearly cosmopolitan. [From the Saxon name, *rorippen*, used for some mustards.]

Neobeckia Greene (1896); *Radicula* Moench (1794).

Selected references: Al-Shehbaz (2010e); Nakayama et al. (2014).

1. Petals absent...**R. sessiliflora**
1. Petals present.
   2. Petals white; fruit septa produced to rims...**R. aquatica**
   2. Petals yellow; fruit septa complete.
      3. Leaves unlobed, or if lobed, then not divided to the midrib; fruit 2–5 mm long......**R. palustris**
      3. Leaves pinnately divided to the midrib; fruit 8–12 mm long.............................**R. teres**

Rorippa aquatica (Eaton) E. J. Palmer [Growing in water.] LAKECRESS.

Cochlearia armoracia Linnaeus var. *aquatica* Eaton, Man. Bot., ed. 3. 243. 1822. *Cochlearia aquatica* (Eaton) Eaton, Man. Bot., ed. 5. 181. 1829. *Neobeckia aquatica* (Eaton) Greene, Pittonia 3: 95. 1896. *Radicula aquatica* (Eaton) B. L. Robinson, Rhodora 10: 32. 1908. *Armoracia aquatica* (Eaton) Wiegand, Rhodora 27: 186. 1925; non Kosteletzky, 1836. *Rorippa aquatica* (Eaton) E. J. Palmer & Steyermark, Rhodora 40: 132. 1938.

Nasturtium natans de Candolle var. *americanum* A. Gray, Ann. Lyceum Nat. Hist. New York 3: 223. 1835. *Nasturtium lacustre* A. Gray, Gen. Amer. Bor. 1: 132. 1848. *Armoracia americana* (A. Gray) Hooker & Arnott, Brit. Fl., ed. 6. 28. 1850, nom. illegit. *Cardamine americana* (A. Gray) Kuntze, Revis. Gen. Pl. 1: 26. 1891, nom. illegit. *Rorippa americana* (A. Gray) Britton, Mem. Torrey Bot. Club 5: 169. 1894, nom. illegit. *Armoracia lacustris* (A. Gray) Al-Shehbaz & V. M. Bates, J. Arnold Arbor. 68: 357. 1987.

Aquatic perennial herbs, to 8(11) dm; stems glabrous. Leaves all cauline, the submerged ones 1- to 4-pinnatisect into filiform or capillary segments, the emergent ones lanceolate to oblong, the margin usually undivided, rarely lobed, (1.5)2–5.5(6.5) cm long, (5)7–15(20) mm wide, the margin entire or dentate, the submerged ones short-petiolate, the emergent sessile or the petiole to 2 mm. Flowers in a raceme, the pedicel divaricate to horizontal or slightly reflexed, straight or curved, (5)7–15 mm long in fruit; sepals oblong, 2–4 mm long; petals spatulate to obovate, 4–8 mm long, white; stamens 4–5 mm long. Fruit oblong to ellipsoidal, straight, 4–7 mm long, 2–3 mm wide, glabrous; seeds numerous, ovoid, ca. 1 mm long, brown to reddish brown, the surface reticulate.

Rivers and spring runs. Rare; Wakulla and Jackson Counties. Quebec south to Florida, west to Ontario, Minnesota, Iowa, Kansas, Oklahoma, and Texas. Spring.

Rorippa palustris (L.) Besser [Of swampy areas.] BOG YELLOWCRESS.

Sisymbrium amphibium Linnaeus var. *palustre* Linnaeus, Sp. Pl. 657. 1753. *Sisymbrium palustre* (Linnaeus) Pollich, Hist. Pl. Palat. 2: 230. 1777. *Myagrum palustre* (Linnaeus) Lamarck, Encycl. 1: 572. 1785. *Sisymbrium terrestre* Withering, Bot. Arr. Brit. Pl., ed. 2. 2: 692. 1787, nom. illegit. *Radicula palustris* (Linnaeus) Moench, Methodus 263. 1794. *Caroli-Gmelina palustris* (Linnaeus) P. Gaertner et al., Oekon. Fl. Watterau 2: 470. 1800. *Brachiolobos palustris* (Linnaeus) Clairville, Man. Herb. Suisse 218. 1811. *Nasturtium terrestre* W. T. Aiton, Hortus Kew. 4: 110. 1812, nom. illegit. *Nasturtium palustre* (Linnaeus) de Candolle, Syst. Nat. 2: 191. 1821. *Rorippa palustris* (Linnaeus) Besser, Enum. Pl. 27, 103. 1822. *Rorippa terrestris* Fuss, Fl. Transsilv. 47. 1866, nom. illegit. *Cardamine palustris* (Linnaeus) Kuntze, Revis. Gen. Pl. 1: 24. 1891. *Cardamine palustris* (Linnaeus) Kuntze var. *normalis* Kuntze, Revis. Gen. Pl. 1: 25. 1891, nom. inadmiss. *Cardamine palustris* (Linnaeus) Kuntze forma *communis* Kuntze, Revis. Gen. Pl. 1: 25. 1891, nom. inadmiss. *Crucifera palustris* (Linnaeus) E.H.L. Krause, in Sturm, Fl. Deutschl., ed. 2. 6: 90. 1902. *Radicula terrestris* Wooten & Standley, Contr. U.S. Herb. 19: 284. 1915, nom. illegit.

Nasturtium palustre (Linnaeus) de Candolle var. *glabrum* O. E. Schulz, in Urban, Symb. Antill. 3: 516. 1903. *Radicula glabra* (O. E. Schulz) Britton, Torreya 6: 30. 1906. *Rorippa palustris* (Linnaeus) Besser subsp. *glabra* (O. E. Schulz) Stuckey, Sida 4: 358. 1972. *Rorippa islandica* (Oeder ex Murray) Borbás var. *glabra* (O. E. Schulz) Welsh & Reveal, Great Basin Naturalist 37: 348. 1978 ("1977"). *Rorippa palustris* (Linnaeus) Besser var. *glabra* (O. E. Schulz) Roy L. Taylor & MacBryde, Canad. J. Bot. 56: 185. 1978.

Rorippa islandica (Oeder ex Murray) Borbás var. *fernaldiana* Butters & Abbe, Rhodora 42: 28. 1940. *Rorippa islandica* (Oeder ex Murray) Borbás subsp. *fernaldiana* (Butters & Abbe) Hultén, Ark. Bot., ser. 2. 7: 61. 1968. *Rorippa palustris* (Linnaeus) Besser subsp. *fernaldiana* (Butters & Abbe) Jonsell, Symb. Bot. Upsal. 19: 158. 1968. *Rorippa palustris* (Linnaeus) Besser var. *fernaldiana* (Butters & Abbe) Stuckey, Sida 4: 361. 1972.

Erect annual or perennial herb, to 10(14) dm; stem glabrous, rarely sparsely pubescent proximally. Basal leaves with the blade lyrate-pinnatisect, the cauline with the blade lyrate-pinnatisect, (1.5)2.5–10(18) cm long, (0.5)0.8–2.5(3) cm wide, the lateral lobes oblong or ovate when present, the lateral lobes smaller than the terminal, the apex rounded, the base auriculate or amplexicaul, the margin subentire or irregularly dentate, sinuate, serrate, or crenate, the upper surface glabrous, the lower surface glabrous or rarely sparsely pubescent proximally, petiolate or subsessile. Flowers in a terminal or axillary raceme, the pedicel divaricate or slightly to strongly reflexed, straight or curved, (2.5)3–10(14) in fruit; sepals oblong, ca. 2 mm long; petals spatulate, 2–3 mm long, pale yellow, stamens 1–3 mm long. Fruit oblong, ellipsoid, or oblong-ovoid, (3)4–10 mm long, 2–4 mm wide, straight or slightly curved, glabrous; seeds ovoid or subglobose, ca. 1 mm long, brown to yellowish brown.

Wet, disturbed sites. Occasional; nearly throughout. Nearly cosmopolitan. Native to North America, Europe, and Asia. Spring–fall.

Rorippa sessiliflora (Nutt.) Hitchc. [With stalkless flowers.] STALKLESS YELLOWCRESS.

Nasturtium sessiliflorum Nuttall, in Torrey & A. Gray, Fl. N. Amer. 1: 73. 1836. *Cardamine palustris* (Linnaeus) Kuntze forma *sessiliflora* (Nuttall) Kuntze, Revis. Gen. Pl. 1: 24. 1891. *Rorippa sessiliflora* (Nuttall) Hitchcock, Key Spring Fl. Manhattan 18: 1864. *Radicula sessiliflora* (Nuttall) Greene, Leafl. Bot. Observ. Crit. 1: 113. 1905.

Erect annual herb, to 4.5(6) dm; stem glabrous. Basal leaves with the blade margin dentate, the cauline with the blade oblong or oblanceolate to obovate, 1.5–7(13) cm long, (0.5)1–3(5) cm wide, the lateral lobes much smaller than the terminal, the apex obtuse or rounded, the base auriculate or not, the margin dentate, sinuate, denticulate, or entire, the upper and lower surfaces glabrous. Flowers in a terminal or axillary raceme, the pedicel divaricate-ascending to horizontal, straight, to 2(4) mm long in fruit; sepals ovate, 1–2 mm long, petals absent; stamens 1–2 mm long. Fruit oblong to oblong-linear, (4)6–9(12) mm long, 1–3 mm wide, glabrous; seeds cordiform, ca. 0.5 mm long, yellow-brown.

Along streams, rivers, and lakes. Rare; eastern and central panhandle. Massachusetts south to Florida, west to South Dakota, Nebraska, Kansas, Oklahoma, and Texas. Spring.

Rorippa teres (Michx.) Stuckey [Circular in transverse section, in reference to the fruit.] SOUTHERN MARSH YELLOWCRESS.

Cardamine teres Michaux, Fl. Bor.-Amer. 2: 29. 1803. *Sisymbrium teres* (Michaux) Torrey & A. Gray, Fl. N. Amer. 1: 93. 1838. *Rorippa teres* (Michaux) Stuckey, Sida 2: 409. 1966.

Nasturtium palustre Linnaeus var. *tanacetifolium* de Candolle, Syst. Nat. 2: 192. 1821. *Sisymbrium walteri* Elliott, Sketch Bot. S. Carolina 2: 146. 1824. *Erysimum walteri* (Elliott) Eaton, Man. Bot., ed. 5. 213. 1829. *Nasturtium tanacetifolium* (de Candolle) Hooker & Arnott, J. Bot. (Hooker) 1: 190. 1834, nom. illegit. *Nasturtium walteri* (Elliott) A. W. Wood, Class-Book Bot., ed. 1861. 228. 1861. *Cardamine indica* (Linnaeus) Kuntze var. *tanacetifolia* (de Candolle) Kuntze, Revis. Gen. Pl. 1: 24. 1891. *Rorippa tanacetifolia* (de Candolle) A. Heller, Contr. Herb. Franklin & Marshall Coll. 1: 40. 1895, nom. illegit. *Rorippa walteri* (Elliott) C. Mohr, Bull. Torrey Bot. Club 24: 23. 1897. *Radicula walteri* (Elliott) Greene, Leafl. Bot. Observ. Crit. 1: 114. 1905.

Prostrate, decumbent, or erect annual or perennial herb, to 4 dm; stem glabrous or pubescent. Basal leaves with the blade pinnatifid, petiolate, the cauline with the blade oblong, oblanceolate, obovate, or lyrate pinnatisect, (2)3.5–10(13) cm long, 1–4(5) cm wide, the lateral lobes oblong to ovate, the apex obtuse to rounded, the base auriculate or not, rarely 2-pinnatifid, the lateral margin dentate to crenate or sinuate, the upper surface glabrous or pubescent, the lower surface glabrous, short petiolate. Flowers in a terminal or axillary raceme, the pedicel ascending to horizontal, straight or curved-ascending, 1–5 mm long in fruit; sepals oblong, 1–2 mm long; petals spatulate, 2 mm long, yellow; stamens 1–2 mm long. Fruit linear to oblong-linear, 8–14(21) mm long, 1–3 mm wide, straight or curved, glabrous or pubescent; seeds cordiform, ca. 0.5 mm long, reddish brown.

Wet hammocks and disturbed sites. Frequent; nearly throughout. North Carolina south to Florida, west to Texas. Winter–spring.

EXCLUDED TAXON

Rorippa palustris (Linnaeus) Besser subsp. *hispida* (Desvaux) Jonsell—Reported for Florida by Fernald (1950), who misapplied the name to material of *R. palustris* subsp. *palustris.*

Sinapis L. 1753. MUSTARD

Annual herbs. Leaves basal and cauline, alternate, simple, pinnate-veined, petiolate, estipulate. Flowers in racemes, pedicellate, ebracteate, actinomorphic, bisexual; sepals 4, free; petals 4, free, equal; stamens 6, tetradynamous; intrastaminal nectar glands 4. Fruit a silique, segments 2, the terminal segment indehiscent, the basal dehiscent; seeds uniseriate.

A genus of 5 species; nearly cosmopolitan. [From the Greek *sinapi*, mustard, in reference to the flavor of the seeds.]

Selected reference: Warwick (2010d).

Sinapis arvensis L. [Of cultivated fields.] CHARLOCK MUSTARD.

Sinapis arvensis Linnaeus, Sp. Pl. 668. 1753. *Raphanus arvensis* (Linnaeus) Crantz, Cl. Crucif. Emend. 109. 1769. *Sinapis arvensis* Linnaeus var. *pinnatifida* Stokes, Bot. Mat. Med. 3: 478. 1812, nom. inadmiss. *Rhamphospermum arvense* (Linnaeus) Andrzejowski ex Besser, Enum. Pl. 83. 1822. *Napus agriasinapis* Schimper & Spenner, in Spenner, Fl. Friburg 3: 944. 1829, nom. illegit. *Sinapis arvensis* Linnaeus var. *leiocarpa* Gaudin, Syn. Fl. Helv. 570. 1836, nom. inadmiss. *Eruca arvensis* (Linnaeus) Noulet, Fl. Bass. Sous-Pyren. 34. 1837. *Sinapistrum arvense* (Linnaeus) Spach, Hist. Nat. Vég. 6: 345. 1838. *Sinapistrum arvense* (Linnaeus) Spach var. *vulgaris* Spach, Hist. Nat. Vég. 6: 345. 1838, nom. inadmiss. *Brassica arvensis* (Linnaeus) Rabenhorst, Fl. Lusat. 184. 1839; non Linnaeus, 1767. *Brassica sinapistrum* Boissier, Voy. Bot. Espagne 2: 39. 1839. *Sinapis arvensis* Linnaeus var. *vera* Babington, Prim. Fl. Sarnicae 8. 1839, nom. inadmiss. *Sinapis arvensis* Linnaeus var. *psilocarpa* Neilreich, Fl. Wien 496. 1846, nom. inadmiss. *Brassica sinapis* Visiani, Fl. Dalmat. 3: 136. 1850, nom. illegit.; non Noulet, 1837. *Sinapis arvensis* Linnaeus var. *vulgaris* Lange, Haandb. Danske Fl., ed. 3. 499. 1864, nom. inadmiss. *Sinapi arvense* (Linnaeus) Dulac, Fl. Hautes-Pyrénées 194, 1867. *Agrosinapis arvensis* (Linnaeus) Fourreau, Ann. Soc. Linn. Lyon, ser. 2. 16: 329. 1868. *Sinapis arvensis* Linnaeus var. *typica* Lindemann, Prodr. Fl. Cherson. 23. 1872, nom. inadmiss. *Brassica arvensis* (Linnaeus) Rabenhorst var. *normalis* Kuntze, Revis. Gen. Pl. 1: 19. 1891, nom. inadmiss. *Caulis sinapiaster* E.H.L. Krause, Bot. Centralbl. 81: 207. 1900, nom. illegit. *Crucifera sinapistra* E.H.L. Krause, in Sturm, Fl. Deutschl., ed. 2. 6: 134. 1902, nom. illegit. *Brassica sinapistrum* Boissier forma *leiocarpa* (Gaudin) Briquet, Prodr. Fl. Corse 2(1): 74. 1913, nom. inadmiss. *Brassica kaber* (de Candolle) L. C. Wheeler var. *pinnatifida* L. C. Wheeler, Rhodora 40: 308. 1938.

Sinapis kaber de Candolle, Syst. Nat. 2: 217. 1821. *Sinapistrum arvense* (Linnaeus) Spach var. *brevirostre* Spach, Hist Nat. Vég. 6: 346. 1838. *Sinapis arvensis* Linnaeus var. *brevirostris* (Spach) O. E. Schulz, in Engl. Pflanzenr. 4(Heft 70): 124. 1919. *Brassica kaber* (de Candolle) L. C. Wheeler, Rhodora 40: 306. 1938.

Erect annual herb, to 10(20) dm; stem hirsute, hispid, or glabrous. Basal leaves with the blade obovate, oblong, or lanceolate, (3)14–18(25) cm long, 1.5–5(7) cm wide, lyrate, pinnatifid, or undivided, the lobes 1–4 on each side, the margin coarsely toothed, the upper and lower surfaces sparsely pubescent, the petiole 1–4(7) cm long, the cauline leaves with the blade not divided, coarsely toothed, petiolate or subsessile. Flowers in terminal racemes, the pedicel ascending or erect, (2)3–7 in fruit; sepals narrowly oblong, 5–6 mm long, yellow or green; petals obovate, (8)9–12(17) mm long, bright yellow, the claw subequaling the sepals; stamens 4–6 mm long. Fruit linear, (1.5)2–4.5(5.5) cm long, 2–4 mm wide, glabrous or pubescent, the valvular segment terete, 2–4 cm long, (2)4- to 8(12)-seeded, the terminal segment conic or subulate, (0.7)1–1.6 cm long, seedless or 1-seeded; seeds globose, 1–2 mm long, reddish brown.

Disturbed sites. Rare; peninsula. Nearly cosmopolitan. Native to Europe and Asia. Spring.

EXCLUDED TAXON

Sinapis alba Linnaeus—Reported by Small (1933) and Al-Shehbaz (1985). No Florida specimens seen.

Sisymbrium L., nom. cons. 1753. HEDGEMUSTARD

Annual herbs. Leaves alternate, basal and cauline, simple, pinnate-veined, petiolate or epetiolate, stipulate. Flowers in racemes, pedicellate, ebracteate, actinomorphic, bisexual; sepals 4, free; petals 4, free, equal; stamens 6, tetradynamous; intrastaminal nectar glands confluent, subtending the base of the stamens, the median glands present. Fruit a silique, dehiscent, replum rounded, the septum complete; seeds uniseriate.

A genus of 41 species; North America, Central America, South America, Europe, Africa, Asia, Australia, and Pacific Islands. [From a Greek vernacular name used for several mustards.]
Norta Adans. (1763).
Selected reference: Al-Shehbaz (2010g).

1. Fruit subulate, 1–2 cm long; pedicel 2–3 mm long, closely appressed to the rachis..............**S. officinale**
1. Fruit linear, 2–3 cm long; pedicel 5–10 mm long, spreading or loosely ascending.
  2. Pedicel at maturity nearly as thick as the fruit...**S. altissimum**
  2. Pedicel at maturity obviously thinner than the fruit...**S. irio**

Sisymbrium altissimum L. [Most high.] TALL TUMBLEMUSTARD.

Sisymbrium altissimum Linnaeus, Sp. Pl. 659. 1753. *Hesperis altissima* (Linnaeus) Kuntze, Revis. Gen. Pl. 2: 936. 1861. *Nasturtium altissimum* (Linnaeus) E.H.L. Krause, Bot. Centralbl. 81: 231. 1900. *Crucifera altissima* (Linnaeus) E.H.L. Krause, in Sturm, Fl. Deutschl., ed 2. 6: 82. 1902. *Norta altissima* (Linnaeus) Britton, in Britton & A. Brown, Ill. Fl. N. U.S., ed. 2. 2: 174. 1913.

Erect annual herb, to 12(16) dm; stem sparsely to densely hirsute proximally, glabrous or glabrate distally. Basal leaves with the blade broadly oblanceolate, oblong, or lanceolate, (2)5–20(35) cm long, (1)2–8(10) cm wide, pinnatisect, pinnatifid, or runcinate, the lobes (3)4–6(8) on each side, oblong or lanceolate, smaller than the terminal lobe, the margin entire, dentate, or lobed, the upper and lower surfaces glabrous or pubescent, the petiole 1–10(15) cm long, the cauline leaves similar to the basal ones, the distalmost blade with linear or filiform lobes. Flowers in a terminal raceme, the pedicel divaricate or ascending, nearly as wide as the fruit,

(4)6–10(13) mm long in fruit; sepals oblong, 4–6 mm long; petals spatulate, (5)6–8(10) mm long, yellow, the claw 4–6 mm long; stamens 3–6 mm long. Fruit narrowly linear, usually straight, (4.5)6–9(12) cm long, 1–2 mm wide, smooth, glabrous; seeds oblong, ca. 1 mm long.

Disturbed sites. Rare; Glades County. Not recently collected. Nearly throughout North America; South America, Europe, Africa, Asia, and Pacific Islands. Native to Europe, Africa, and Asia. Spring.

Sisymbrium irio L. [Pre-Linnaean generic name for a mustard.] LONDONROCKET.

> *Sisymbrium irio* Linnaeus, Sp. Pl. 659. 1753. *Sisymbrium latifolium* S. Gray, Nat. Arr. Brit. Pl. 2: 679. 1821, nom. illegit. *Descurainia irio* (Linnaeus) Webb & Berthelot, Hist. Nat. Iles Canaries 3(2(1)): 73. 1836. *Sisymbrium filiforme* Dulac, Fl. Hautes-Pyrénées 197. 1867, nom. illegit. *Irio ruderalis* Fourreau, Ann. Soc. Linn. Lyon, ser. 2. 16: 331. 1868. *Erysimum irio* (Linnaeus) Ruprecht, Mém. Acad. Imp. Sci. Saint Petersbourg, ser. 7. 15: 89. 1869. *Phryne laxata* Bubani, Fl. Pyren. 3: 175. 1901, nom. illegit. *Crucifera irio* (Linnaeus) E.H.L. Krause, in Sturm, Fl. Deutschl., ed. 2. 6: 81. 1902. *Norta irio* (Linnaeus) Britton, in Britton & A. Brown, Ill. Fl. N. U.S., ed. 2. 2: 174. 1913.

Erect annual herb, to 6(7.5) dm; stem glabrous or sparsely pubescent, at least proximally. Basal leaves with the blade oblanceolate or oblong, (1.5)3–12(15) cm long, (0.5)1–6(9) cm wide, runcinate to pinnatisect, the lobes (1)2–6(8) on each side, oblong or lanceolate, smaller than the terminal lobe, the apex acute to obtuse, the base cuneate, the margin entire, dentate, or lobed, the petiole (0.5)1–4.5(6) cm long, the cauline leaves similar to the basal ones, the distalmost blade smaller, to 2 cm wide, the margin entire or 1- to 3-lobed, the upper and lower surfaces glabrous or sparsely pubescent. Flowers in a terminal raceme, the pedicel divaricate or ascending, narrower than the fruit, (0.5)0.7–1.2(2) cm long in fruit; sepals oblong, 2–3 mm long; petals oblong-oblanceolate, 3–4 mm long, yellow, the claw ca. 1 mm long; stamens 3–5 mm long. Fruit narrowly linear, (2.5)3–4(5) cm long, ca. 1 mm wide, straight or slightly inward curved, slightly torulose, glabrous; seeds oblong, ca. 1 mm long.

Disturbed sites. Rare; Miami-Dade County. Massachusetts, New York, Ohio, Michigan, South Carolina, and Florida, Wyoming and Utah south to Texas and California; South America; Europe, Africa, Asia, Australia, and Pacific Islands. Native to Europe, Africa, and Asia. Spring.

Sisymbrium officinale (L.) Scop. [Used in medicine.] HEDGEMUSTARD.

> *Erysimum officinale* Linnaeus, Sp. Pl. 660. 1753. *Sisymbrium officinale* (Linnaeus) Scopoli, Fl. Carniol., ed. 2. 2: 26. 1772. *Chamaepilum officinale* (Linnaeus) Wallroth, Sched. Crit. 377. 1822. *Velarum officinale* (Linnaeus) Reichenbach, in Mössler, Handb. Gewächsk., ed. 2. 2: 1165. 1828. *Valarum officinale* (Linnaeus) Schur, Enum. Pl. Transsilv. 54. 1866. *Erysimum vulgare* Ruprecht, Mém. Acad. Imp. Sci. Saint Petersbourg, ser. 7. 15(2): 91. 1869, nom. illegit.
>
> *Hesperis officinalis* (Linnaeus) Kuntze, Revis. Gen. Pl. 2: 935. 1891. *Nasturtium sisymbrium* E.H.L. Krause, Bot. Centralbl. 81: 231. 1900. *Phryne vulgaris* Bubani, Fl. Pyren. 173. 1901, nom. illegit. *Crucifera sisymbrium* (E.H.L. Krause) E.H.L. Krause, in Sturm, Fl. Deutschl., ed. 2. 6: 84. 1902, nom. illegit. *Kibera officinalis* (Linnaeus) Celestani, Nuovo Giorn. Bot. Ital., ser. 2. 15: 379. 1908. *Sisymbrium officinale* (Linnaeus) Scopoli, var. *genuinum* Briquet, Prodr. Fl. Corse 2(1): 22. 1913, nom. inadmiss.

Sisymbrium officinale (Linnaeus) Scopoli var. *leiocarpum* de Candolle, Syst. Nat. 2: 460. 1821. *Sisymbrium leiocarpum* (de Candolle) Jordan, Ann. Soc. Linn. Lyon, ser. 2. 7: 507. 1860. *Erysimum officinale* Linnaeus var. *leiocarpum* (de Candolle) Farwell, Pap. Michigan Acad. Sci. 3: 97. 1924.

Erect annual herb, to 8 dm; stem hirsute, rarely glabrate distally. Basal leaves with the blade, broadly oblanceolate or oblong-obovate, (2)3–10(15) cm long, (1)2–5(8) cm wide, lyrate-pinnatifid, pinnatisect, or runcinate, the lobes (2)3–4(5) on each side, oblong or lanceolate, smaller than the terminal lobe, the margin entire, dentate or lobed, the upper and lower surfaces glabrous or pubescent, the terminal lobe suborbiculate dentate, the margin dentate, the petiole (1)2–7(10) cm long, the cauline leaves similar to the basal, the blade with the lobe margins dentate or subentire. Flowers in a terminal raceme, the pedicel erect, appressed to the rachis, narrower than the fruit, 2–3(4) mm long; sepals oblong-ovate, 2–3 mm long; petals spatulate, 3–4 mm long, yellow, the claw 1–2 mm long; stamens 2–3 mm long. Fruit subulate, (0.7)1–1.4(1.8) cm long, 1–2 mm wide, straight, slightly torulose or smooth, glabrous or pubescent; seeds oblong, ca. 1 mm long.

Disturbed sites. Rare; Alachua, Leon, and Calhoun Counties. Nearly throughout North America; West Indies, Central America, and South America; Europe, Africa, Asia, Australia, and Pacific Islands. Native to Europe, Africa, and Asia. Spring.

Thlaspi L. 1753. PENNYCRESS

Annual or biennial herbs. Leaves alternate, basal and cauline, simple, pinnate-veined, petiolate or epetiolate, estipulate. Flowers in racemes, pedicellate, ebracteate, actinomorphic, bisexual; sepals 4, free; petals 4, free, equal; stamens 6, tetradynamous; intrastaminal nectar glands 2 or 4, lateral, usually 1 on each side of the lateral stamen, the median gland absent. Fruit a silique, strongly compressed, winged, the replum rounded, the septum complete; seeds uniseriate.

A genus of 6 species; North America, South America, Europe, Africa, Asia, Australia, and Pacific Islands. [From the Greek *thalos, thals-*, to compress, in reference to the flattened fruits.]

Selected reference: Al-Shehbaz (2010i).

Thlaspi arvense L. [Of cultivated land.] FIELD PENNYCRESS.

Thlaspi arvense Linnaeus, Sp. Pl. 646. 1753. *Thlaspidea arvensis* (Linnaeus) Opiz, Seznam 96. 1852. *Thlaspidium arvense* (Linnaeus) Bubani, Fl. Pyren. 3: 214. 1901. *Crucifera thlaspi* E.H.L. Krause, in Sturm, Fl. Deutschl., ed. 2. 6: 150. 1902, nom. illegit. *Teruncias arvensis* (Linnaeus) Lunell, Amer. Midl. Naturalist 4: 364. 1916.

Erect annual or biennial herb, to 5(8) dm; stem glabrous, sometimes glaucous. Basal leaves with the blade oblanceolate, spatulate, or obovate, 1–5 cm long, 4–23 mm wide, the apex rounded, the base attenuate or cuneate, the margin entire, repand, or coarsely dentate, the upper and lower surfaces glabrous, sometimes glaucous, the petiole 0.5–3 cm long, the cauline with the blade oblong, (0.5)1.5–4(8) cm long, (2)5–15(25) mm wide, the apex rounded, obtuse, or subacute, the base auriculate, sessile. Flowers in a terminal and axillary raceme, the pedicel straight or slightly upcurved, (5)9–13(15) mm in fruit; sepals ovate or oblong, 2–3 mm long;

petals spatulate, 2–5 mm long, white, narrowed to a clawlike base ca. 1 mm long, the apex obtuse or emarginate; stamens 1–2 mm long. Fruit obovate to suborbicular, (0.6)0.9–2 cm long, (0.5)0.7–2 cm wide, the base obtuse or rounded, the apex deeply emarginate, the notch ca. 5 mm deep, the wings 1–5 mm wide; seeds ovoid, ca. 2 mm long, concentrically striate.

Disturbed sites. Rare; northern counties, central peninsula. Nearly throughout North America; South America; Europe, Africa, Asia, Australia, and Pacific Islands. Native to Europe and Asia. Spring.

Warea Nutt. 1834. PINELANDCRESS

Annual herbs. Leaves alternate, simple, pinnate-veined, petiolate or epetiolate, estipulate. Flowers in corymbose racemes, pedicellate, ebracteate, actinomorphic, bisexual; sepals 4, free; petals 4, free, equal, clawed; stamens 6, tetradynamous; intrastaminal nectar glands usually 6, confluent, subtending the stamen bases, median glands present. Fruit a silique, dehiscent, the replum rounded, the septum complete; seeds uniseriate.

A genus of 4 species; North America. [Commemorates Nathaniel A. Ware (1789–1854), teacher, lawyer, and plant collector in the southeastern United States.]

Selected reference: Al-Shehbaz (2010h).

1. Leaves sessile, the blade rounded or round-auriculate at the base.
    2. Leaves conspicuously auriculate, the auricles adnate, clasping; central peninsula ....
    .... **W. amplexifolia**
    2. Leaves inconspicuously or not at all auriculate, the auricles (if present) free, not clasping; panhandle .... **W. sessilifolia**
1. Leaves subpetiolate or petiolate, the blade cuneate at the base.
    3. Petal claws conspicuously pubescent; gynophore shorter than the pedicel; central and southern peninsula .... **W. carteri**
    3. Petal claws smooth or merely papillose; gynophore longer than the pedicel; central panhandle......
    .... **W. cuneifolia**

Warea amplexifolia (Nutt.) Nutt. [With the leaves clasping the stem.] WIDELEAF PINELANDCRESS; CLASPING WAREA.

Standleya amplexifolia Nuttall, Amer. J. Sci. Arts 5: 297. 1822. *Warea amplexifolia* (Nuttall) Nuttall, J. Acad. Nat. Sci. Philadelphia 7: 83. 1834. TYPE: FLORIDA: "East Florida," s.d., *Ware s.n.* (lectotype: BM). Lectotypified by Channell and James (1964: 23).
Warea auriculata Shinners, Shinners, Sida 1: 106. 1962. TYPE: FLORIDA: Osceola Co.: Lake Wilson Road, Loughman, Kissimmee, 25 Sep 1937, *Singletary s.n.* (holotype: DUKE).

Erect annual herb, to 7(8) dm; stem glabrous, somewhat glaucous. Leaves with the blade ovate, oblong, to lanceolate, (1)1.5–4 cm long, 4–14(16) mm wide, the apex acute to obtuse, the base amplexicaul to auriculate, the auricles ovate, 3–9 mm long, the margin entire, the upper and lower surfaces glabrous, somewhat glaucous, sessile. Flowers in a raceme 1–5(8) cm long in fruit, the pedicel 8–15 mm long in fruit; sepals linear-oblanceolate, 5–8 mm long, white or pinkish, spreading or deflexed; petals broadly obovate to orbicular, 7–10 mm long, the blade

3–5 mm long, the claw 4–5 mm long, minutely papillate, the margin entire; stamens 12–15 mm long; gynophore (8)10–15 mm long. Fruit linear, (3)4–7 cm long, 1–2 mm wide, glabrous; seeds 1–2 mm long, the surface concentrically striate.

Sandhills and scrub. Rare; Marion, Lake, Orange, Polk and Osceola Counties. Endemic. Spring–summer.

Warea amplexifolia is listed as endangered in Florida (Florida Administrative Code, Chapter 5B-40) and as endangered in the United States (U.S. Fish and Wildlife Service, 50 CFR 23).

Warea carteri Small [Commemorates Joel Jackson Carter (1843–1912) from Pennsylvania, who often accompanied John K. Small on his collecting trips in Florida.] CARTER'S PINELANDCRESS; CARTER'S MUSTARD.

Warea carteri Small, Bull. Torrey Bot. Club 36: 159. 1909. TYPE: FLORIDA: Miami-Dade Co.: between Cutler and Black Point, 13–16 Nov 1903, *Small & Carter 831* (holotype: NY).

Erect annual herb, to 14 dm; stem glabrous, usually glaucous. Leaves with the blade linear or linear-oblanceolate to oblanceolate or narrowly oblong, 1–4.5 cm long, 1–6(10) mm wide, the apex acute to obtuse, the base cuneate to attenuate, the margin entire, the upper and lower surfaces glabrous, somewhat glaucous, the petiole 1–8 mm long. Flowers in a raceme 0.4–3(4) cm long in fruit, the pedicel 4–10 mm long in fruit; sepals linear-oblanceolate, 3–5 mm long, white, spreading or reflexed; petals broadly obovate to suborbicular, 4–6 mm long, the blade 2–3 mm long, the claw 2–3 mm long, coarsely papillate to pubescent, the margin crisped; stamens 5–7 mm long; gynophore 3–6(7) mm long. Fruit linear, 3–5(6) cm long, 1–2 mm wide, glabrous; seeds 1–2 mm long, the surface concentrically striate.

Sandhills and scrub. Occasional; central and southern peninsula. Endemic. All year.

Warea carteri is listed as endangered in Florida (Florida Administrative Code, Chapter 5B-40) and as endangered in the United States (U.S. Fish and Wildlife Service, 50 CFR 23).

Warea cuneifolia (Muhl. ex Nutt.) Nutt. [Leaf base cuneate.] CAROLINA PINELANDCRESS.

Cleome cuneifolia Muhlenberg ex Nuttall, Gen. N. Amer. Pl. 2: 73. 1818. *Warea cuneifolia* (Muhlenberg ex Nuttall) Nuttall, J. Acad. Nat. Sci. Philadelphia 7: 84. 1834.

Erect annual herb, to 6.5(8) dm; stems glabrous, somewhat glaucous. Leaves with the blade, linear to linear-oblanceolate or oblanceolate, (0.7)1–3(4) cm long, 2–6(8) mm wide, the apex rounded to obtuse, the base cuneate, the margin entire, the upper and lower surfaces glabrous, somewhat glaucous, the petiole 1–3 cm long, obsolete distally. Flowers in a raceme 0.3–2(3) cm in fruit, the pedicel (4)5–9(11) mm long; sepals linear-oblanceolate, 3–5(7) mm long, white or purplish, spreading or reflexed; petals broadly obovate to spatulate, 4–9 mm long, white or pink, the blade 2–5 mm long, the claw 2–4 mm long, smooth or obscurely papillate, the margin entire; stamens 6–8(10) mm long; gynophore (5)7–11 mm long. Fruit linear, 2–4(5) cm long, ca. 1 mm wide, glabrous; seeds ca. 1 mm long, the surface concentrically striate.

Sandhills. Rare; Jackson, Gadsden, and Liberty Counties. North Carolina south to Florida, west to Alabama. Summer.

Warea sessilifolia Nash [With sessile leaves.] SESSILELEAF PINELANDCRESS.

Warea sessilifolia Nash, Bull. Torrey Bot. Club 23: 101. 1890. TYPE: FLORIDA: Leon Co.: Bellair, 3 Sep 1895, *Nash 2544* (holotype: NY; isotype: NY).

Erect annual herb, to 6.5(8) dm; stem glabrous, somewhat glaucous. Leaves with the blade ovate to lanceolate, (0.8)1–2.5(4) cm long, 3–15(30) mm wide, the apex acute to obtuse, the base obtuse or minutely auriculate, the auricles ca. 2 mm long, the margin entire, the upper and lower surfaces glabrous, somewhat glaucous, sessile. Flowers in a raceme 1–3 cm long in fruit, the pedicel 9–12 mm long in fruit; sepals linear-oblanceolate, 6–7 mm long, white or purplish, strongly reflexed; petals broadly obovate to suborbicular, 7–11 mm long, purple to pink, the blade 2–5 mm long, the claw 4–6 mm long, minutely papillate, the margin entire; stamens 9–15 mm long; gynophore 10–16 mm long. Fruit linear, 2.5–4.5 cm long, 1–2 mm wide, glabrous; seeds ca. 1 mm long, the surface concentrically striate.

Sandhills. Occasional; central and western panhandle. Georgia, Alabama, and Florida. Summer.

EXCLUDED GENERA

Arabidopsis thaliana (Linnaeus) Heynhold—Reported for all southeastern states, which of course includes Florida, by Al-Shehbaz (1988) and followed by Wunderlin and Hansen (2003). Al-Shehbaz (pers. comm.) now indicates that no Florida specimens are known.

Barbarea orthoceras Ledebour—Reported for Florida by Small (1903, 1913a, both as *B. stricta* Andrzejowski, misapplied; 1933, as *Campe stricta* (Andrzejowski) W. Wight ex Piper, misapplied). No Florida specimens are known and this northern taxon is unlikely in Florida.

Barbarea verna (Miller) Ascherson—Reported for Florida by Small (1903, 1913a, both as *B. praecox* (Smith) W. T. Aiton; 1933, as *Campe verna* (Miller) A. Heller). No Florida specimens are known.

Barbarea vulgaris R. Brown—Reported for Florida by Al-Shehbaz (1988). No Florida specimens are known.

Eruca vesicaria (Linnaeus) Cavanilles subsp. *sativa* (Miller) Thellung—Reported by Small (1933, as *Eruca eruca* (Linnaeus) Ascherson & Graebner), as "nearly throughout Cult. N.A.," which would presumably include Florida. This plant is not known from the southeastern United States.

XIMENIACEAE Horan 1834. TALLOW WOOD FAMILY

Trees or shrubs. Leaves alternate, simple, pinnate-veined, petiolate, estipulate. Flowers in axillary cymes, bracteate, actinomorphic, bisexual; sepals 4(5), free; petals 4(5), free; stamens 8(10), free, the anthers 2-loculate, dehiscing by longitudinal slits; ovary superior, 4-carpellate, 4-loculate. Fruits 1-seeded drupes.

A family of 4 genera and 12 species; North America, West Indies, Mexico, Central America, South America, Africa, Asia, Australia, and Pacific Islands.

Selected reference: Robertson (1982).

Ximenia L. 1753. TALLOW WOOD

Trees or shrubs; branches armed with axillary spines. Leaves alternate, simple, pinnate-veined, petiolate, estipulate. Flowers in axillary cymes, pedicellate; sepals 4(5), free; petals 4(5), free; stamens 8(10), free; nectariferous disk surrounding the ovary base; style 1, the stigma 1. Fruit a 1-seeded drupe.

A genus of 9 species; North America, West Indies, Mexico, Central America, South America, Africa, Asia, Australia, and Pacific Islands. [Commemorates Francisco Ximénez (d. 1612), Dominican monk and naturalist in Mexico.]

Ximenia is usually placed in the Olacaceae by authors but recent evidence supports its being placed in the Ximeniaceae with four closely related monotypic genera (Malécot and Nickrent, 2008).

Selected reference: DeFilips (1968).

Ximenia americana L. [Of America.] TALLOW WOOD; HOG PLUM.

Ximenia americana Linnaeus, Sp. Pl. 1193. 1753.
Pimecaria odorata Rafinesque, Alsogr. Amer. 65. 1838. TYPE: FLORIDA: "R. St. Juan" [Saint Johns River].

Shrub or small tree, to 10 m; bark smooth, reddish, the branches armed with axillary spines. Leaves with the blade oblong or elliptic to oblanceolate or suborbicular, 3–7 cm long, 2–6 cm wide, the apex rounded or obtuse, mucronate or retuse, the base rounded or obtuse, the margin entire, the upper and lower surfaces glabrous, the petiole 5–10 mm long. Flowers 2–8(10) in an umbellate, subumbellate, or racemose cyme, short-pedunculate, the pedicel 4–12 cm long; bracts minute; sepals cupulate, 0.5–1.5 mm long, 4- to 5-lobed, ciliate; petals linear, 6–10 mm long, yellow or whitish yellow, reflexed, the inner surface densely pubescent; stamens 2.5–4 mm long; ovary glabrous, partly emersed in the nectariferous disk, the style 2.5–4 mm long. Fruit ellipsoidal, ovoid, or globose, 1.5–2.5 cm long, yellow, the endocarp crustaceous, light reddish brown; seed 1–2 cm long, white, the surface smooth.

Hammocks, scrub, and flatwoods. Frequent; peninsula. Florida; West Indies, Mexico, Central America, and South America; Africa, Asia, and Australia. Spring–fall.

SCHOEPFIACEAE Blume 1850. GRAYTWIG FAMILY

Trees or shrubs. Leaves alternate, simple, pinnate-veined, petiolate, estipulate. Flowers in cymes, bracteate, actinomorphic, bisexual; sepals 4–5, connate; petals 4–5, basally connate; stamens 4–5, basally connate, adnate to the corolla tube, the anthers 2-loculate, dehiscing by longitudinal slits; ovary inferior, 4- to 5-carpellate, 2-loculate in 4-merous flowers and 3-loculate in 5-merous ones. Fruits 1-seeded drupes.

A family of 3 genera with about 55 species; North America, West Indies, Mexico, Central America, South America, Africa, and Asia.

Selected reference: Robertson (1982).

Schoepfia Schreb. 1789. GRAYTWIG

Trees or shrubs. Leaves alternate, simple, pinnate-veined, petiolate, estipulate. Flowers in axillary cymes, pedicellate; bracts and bracteoles fused to form a 2-lobed calyx-like structure; sepals 4–5, basally connate; petals 4–5, basally connate; stamens 4–5, adnate to the corolla tube; ovary inferior, 4- to 5-carpellate, 2- to 3-loculate, topped with a thick nectariferous disk, the style 1, stigma 2- to 3-lobed. Fruit a 1-seeded drupe, surmounted by the calyx ring.

A genus of about 25 species; North America, West Indies, Mexico, Central America, South America, Africa, and Asia. [Commemorates Johann David Schöpf (1752–1800), German botanist and physician who traveled in North America and the West Indies.]

Schoepfia is usually placed in the Olacaceae by authors but recent evidence supports its being placed in the Schoepfiaceae with two other genera (Malécot and Nickrent, 2008; Der and Nickrent, 2008).

Schoepfia chrysophylloides (A. Rich.) Planch. [Resembling *Chrysophyllum* (Sapotaceae).] GRAYTWIG.

Diplocalyx chrysophylloides A. Richard, in Sagra, Hist. Fis. Nat. Cuba 11: 81. 1850. *Schoepfia chrysophylloides* (A. Richard) Planchon, Ann. Sci. Nat., Bot., ser. 4. 2: 261. 1854.

Shrub or small tree, to 8 m; bark deeply fissured, the branches glabrous. Leaves with the blade ovate to lanceolate, 3–8 cm long, 2–3 cm wide, somewhat plicate-falcate, the apex obtuse to bluntly acuminate, the base cuneate, the margin entire or obscurely toothed, the upper and lower surface glabrous, the petiole 3–8 mm long, marginate. Flowers 1–3 in an axillary cyme, the pedicel 1–4 mm long, reddish; calyx cupulate, the rim entire or slightly 4- or 5-lobed; corolla tube narrowly campanulate, 2–3 mm long, the lobes 1–2 mm long, triangular, the apex acute, recurved; stamens borne at the base of the corolla lobes, the anthers sessile. Fruit ovoid to subglobose, 8–10 mm long, bright red, nearly enclosed by the accrescent calyx.

Tropical hammocks. Occasional; Volusia County south along the east coast, southern Florida. Florida; West Indies. All year.

EXCLUDED TAXA

Schoepfia schreberi J. F. Gmelin—This West Indian species was reported by Chapman (1897, as *S. arborescens* (Vahl) Schultes), Small (1903, 1913a), and Long and Lakela (1971), the name misapplied to material of *S. chrysophylloides*.

SANTALACEAE R. Br. 1810. SANDALWOOD FAMILY

Shrubs. Leaves opposite, simple, pinnate-veined, petiolate, estipulate. Flowers in terminal and axillary paniculate or umbellate cymes, bracteate, actinomorphic, bisexual; sepals 4, free; petals 4, free; stamens 4, free, the anthers 2-loculate, basifixed, dehiscing by longitudinal slits; nectariferous disk present; ovary inferior, 3- to 4-carpellate, 1-loculate. Fruit a drupe.

A family of about 37 genera and about 480 species. North America, South America, Europe, Africa, Asia, Australia, and Pacific Islands.

Santalum L. 1753. SANDALWOOD

Shrubs. Leaves opposite, simple, pinnate-veined, petiolate, estipulate. Flowers in terminal and axillary paniculate or umbellate cymes, bracteate; sepals 4, free; petals 4, free; receptacle nectariferous disk with 4 petaloid lobes alternating with the stamens; ovary 1-carpellate, the stigma 1, the style 3- to 4-lobed. Fruit a drupe; seed 1.

A genus of about 20 species; North America, Asia, Australia, and Pacific Islands. [From the Greek *santalon*, the sandalwood tree.]

Santalum album L. [White, in reference to the color of the young wood.] SANDALWOOD.

Santalum album Linnaeus, Sp. Pl. 349. 1753.

Shrub, to 3 m; stem erect or sometimes scandent, slightly angular-striate. Leaves with the blade ovate to broadly lanceolate-elliptic, 2.5–7 cm long, 1.5–4 cm wide, the apex acute, the base cuneate, the margin flat or slightly revolute, the upper and lower surfaces smooth, the lower surface pale green, the petiole 5–15 mm long, 2-ribbed. Flowers 1–6 in a terminal or axillary paniculate or umbellate cyme, the peduncle 4–11 mm long, the pedicel to 1 mm long; sepal lobes minute; petals triangular to ovate, ca. 2 mm long, red, the apex obtuse, with tufts of long, coarse trichomes; receptacle nectariferous disk ca. 2 mm long, with 4 prominent, petaloid, ovate lobes; style ca. 2 mm long, the stigma 3- to 4-lobed. Fruit ellipsoid, 7–8 mm long, sessile or subsessile, yellow to black, with a small apical collar, apiculate with the persistent style base, the endocarp smooth.

Disturbed sites. Rare; Miami-Dade County. Escaped from cultivation. Florida; Asia and Australia. Native to Asia and Australia. All year.

VISCACEAE Batsch 1802. MISTLETOE FAMILY

Shrubs, parasitic on the branches of other woody, dicot angiosperms. Leaves opposite, simple, pinnate-veined, petiolate, estipulate. Flowers in axillary spikes, actinomorphic, unisexual (plants dioecious). Staminate flowers with the perianth segments 4, free, arranged around a central disk; stamens 4, adnate to the perianth, the anthers 2-loculate, opening by a longitudinal slit. Carpellate flowers with the perianth segments 3, the carpels 3, the ovary 1-loculate, inferior. Fruit a drupe.

A family of 7 genera and about 520 species; nearly cosmopolitan.

Selected reference: Kuijt (1982).

Phoradendron Nutt. 1848. MISTLETOE

Shrubs. Leaves simple, pinnipalmate-veined, petiolate, estipulate. Flowers in axillary spikes, unisexual (plants dioecious). Staminate flowers with a sessile anther attached to the middle of each perianth segment. Carpellate flowers with the stigma sessile, scarcely differentiated. Fruit a 1-seeded drupe with a glutinous, mucilaginous mesocarp, the perianth persistent.

A genus of about 235 species; North America, West Indies, Mexico, Central America, and South America. [From the Greek *phor*, a thief, and *dendron*, tree, referring to the parasitic habit.]

Phoradendron is sometimes placed in the Loranthaceae or the Santalaceae s.l., but we prefer maintaining the separate family Viscaceae (Der and Nickrent, 2008; Su et al., 2015).

1. Branchlets terete; fruit white or yellowish white .... **P. leucarpum**
1. Branchlets 4-angled; fruit lemon-yellow to orange ....**P. rubrum**

Phoradendron leucarpum (Raf.) Reveal & M. C. Johnst. [White-fruited.] OAK MISTLETOE.

> *Viscum leucarpum* Rafinesque, Fl. Ludov. 79. 1817. *Phoradendron leucarpum* (Rafinesque) Reveal & M. C. Johnston, Taxon 38: 107. 1989.
> *Viscum serotinum* Rafinesque, Ann. Gen. Sci. Phys. 5: 348. 1820. *Phoradendron serotinum* (Rafinesque) M. C. Johnston, Southw. Naturalist 2: 45. 1957.
> *Viscum oblongifolium* Rafinesque, New Fl. 3: 23. 1838 ("1836"). TYPE: FLORIDA.
> *Viscum ochroleucum* Rafinesque, New Fl. 3: 23. 1838 ("1836"). TYPE: "From New Jersey to Florida."
> *Phoradendron eatonii* Trelease, in Small, Shrubs Florida 121, 133. 1913. TYPE: FLORIDA: Collier Co.: Delp [Deep] Lake, 1905, *Eaton 1310* (holotype, MO; isotype, NY).
> *Phoradendron macrotomum* Trelease, in Small, Shrubs Florida 121, 133. 1913. *Phoradendron flavescens* (Pursh) Nuttall var. *macrotomum* (Trelease) Fosberg, Lloydia 4: 279. 1941. *Phoradendron serotinum* (Rafinesque) M. C. Johnston var. *macrotomum* (Trelease) M. C. Johnston, Southw. Naturalist 2: 47. 1957. TYPE: FLORIDA: Duval Co.: Jacksonville, 1894, *Curtiss 4569* (holotype, MO; isotypes, F, GH, K, NY, US).

Shrub, to 1 m; stems erect or somewhat pendent, terete, thick, brittle, diffusely branched, glabrate or glabrous, the nodes swollen and articulated, the internodes 14–20 mm long. Leaves with the blade oblanceolate to obovate-elliptic or suborbicular, 1.5–6 cm long, 1–4 cm wide, coriaceous, the apex rounded, the base cuneate, the margin entire, short-petiolate. Flowers 1–3 at a node in an axillary spike, the rachis glabrous or sometimes puberulent, the bracts often ciliate. Fruit subglobose, 4–6 mm long, white or yellowish white, glabrous.

Parasitic on branches of various dicot tree species. Frequent; nearly throughout. New York south to Florida, west to Kansas and New Mexico. Fall–winter.

Phoradendron rubrum (L.) Griseb. [Red, in reference to the fruit color.] MAHOGANY MISTLETOE.

> *Viscum rubrum* Linnaeus, Sp. Pl. 1023. 1753. *Phoradendron rubrum* (Linnaeus) Grisebach, Fl. Brit. W.I. 314. 1860.

Shrub, to 0.5 m; stems 4-angled, pseudodichotomous, internodes short, 4-angled, upwardly somewhat enlarged and becoming 2-edged. Leaves with the blades linear-lanceolate to oblanceolate or oblong spatulate to broadly or narrowly elliptic, 2–9 cm long, 1–4 cm wide, the apex obtuse to rounded, the base cuneate, the margin entire, the petiole 3–8 mm long. Flowers in an axillary spike to ca. 3 cm long, 2- to 3-jointed, each joint bearing ca. 12 flowers. Fruit subglobose, ca. 4 mm in diameter, smooth, lemon yellow to orange or red.

Tropical hammocks; parasitic on branches of various dicot tree species. Rare; Miami-Dade County, Monroe County keys. Florida; West Indies. Fall–winter.

Phoradendron rubrum is listed as endangered in Florida (Florida Administrative Code, Chapter 5B-40).

EXCLUDED TAXA

Phoradendron flavens (Swartz) Grisebach—Reported by Chapman (1860, 1883, 1897, as *P. flavescens* (Pursh) Nuttall) and Small (1903, 1913a, 1913d, 1933, as *P. flavescens*), who did not realize that the illegitimate name *P. flavescens* was automatically typified by the type of *Viscum flaven* Swartz. They therefore misidentified material of *P. leucarpum* (see Johnston, 1957; Reveal and Johnston, 1989).

Phoradendron trinervium (Lamarck) Grisebach—Long & Lakela (1971) give a second-hand report of this West Indian species as occurring in Monroe County. No Florida specimens seen.

TAMARICACEAE Link 1821. TAMARISK FAMILY

Shrubs or trees. Leaves alternate, simple, pinnate-veined, epetiolate, estipulate. Flowers in simple or compound racemes, actinomorphic, bisexual, bracteate; sepals 5, basally connate; petals 5, free; stamens 5, free; nectariferous disk present; ovary 1, 3-carpellate and -loculate, the styles and stigmas 3. Fruit a capsule; seeds numerous.

A family of 4 genera and about 78 species; North America, Mexico, South America, Europe, Africa, Asia, Australia, and Pacific Islands.

Selected references: Crins (1989); Gaskin (2015).

Tamarix L. 1753. TAMARISK

Shrubs or trees. Leaves scalelike, epetiolate, estipulate. Flowers in simple or compound racemes; bract 1; sepals 5, basally connate, dimorphic; petals 5, free; stamens 5, free, inserted onto a nectariferous disk (aborted inner whorl of stamens). Fruit a loculicidal capsule; seeds numerous.

A genus of about 55 species; North America, Mexico, South America, Europe, Africa, Asia, Australia, and Pacific Islands. [Of the Tamaris River in Spain.]

Selected references: Baum (1967, 1978); Crins (1989).

Tamarix canariensis Willd. [Of the Canary Islands.] CANARY ISLAND TAMARISK.

Tamarix canariensis Willdenow, Abh. Königl. Akad. Wiss. Berlin 1812–1813: 79. 1816. *Tamarix gallica* Linnaeus var. *canariensis* (Willdenow) Ehrenberg, Linnaea 2: 268. 1827.

Shrub or tree, to 3 m; stem reddish brown, the branchlets papillose. Leaves scalelike, the blade lanceolate, 1.5–2.5 mm long, the apex acuminate, the base cuneate, the margin entire, the upper and lower surfaces glabrous. Flowers in a simple or compound raceme; bract linear-triangular, scalelike and boat-shaped, longer than the pedicel, the apex long-acuminate, the upper surface often papillate; pedicel 1–2 mm long, subequaling the calyx; sepals triangular-ovate, 1–2 mm long, the apex acuminate, the margin erose-denticulate; petals obovate to narrowly obovate, 1–2 mm long, the margin entire; stamens inserted onto the lobes of a nectariferous disk. Fruit a loculicidal capsule; seeds 0.5–0.7 mm long, elongate-obovoid, straight, comose at one end, otherwise glabrous.

Beaches. Rare; Duval and Franklin Counties. Escaped from cultivation. North Carolina south to Florida, also Louisiana and Arizona; Mexico; Europe and Africa. Native to Europe. Summer.

EXCLUDED TAXA

Tamarix gallica Linnaeus—Reported by Small (1933), Radford et al. (1964, 1968), Long & Lakela (1971), and Clewell (1985). All Florida material seen is of *T. canariensis*.

Tamarix parviflora de Candolle—Reported by Wunderlin (1998) and Wunderlin and Hansen (2003), based on a report by Crins (1989). All Florida material seen is of *T. canariensis*.

PLUMBAGINACEAE Juss. 1789. LEADWORT FAMILY

Perennial herbs or shrubs. Leaves alternate, simple, pinnate-veined, petiolate or epetiolate, estipulate. Flowers in terminal or axillary racemes or panicles, actinomorphic, bisexual, bracteate; sepals 5, basally connate; petals 5, basally connate; stamens 5, epipetalous or free, the anthers 2-celled, dorsifixed, introrse, longitudinally dehiscent; ovary 1, superior, 5-carpellate, 1-loculate, the style(s) 1 or 5. Fruit a utricle or a capsule; seed 1.

A family of 24 genera and about 775 species; nearly cosmopolitan.

Armeriaceae Horan. (1834).

Selected references: Luteyn (1990).

1. Subscapose herbs, the principal leaves basal, the upper leaves scalelike; corolla tubular-funnelform, the petals free nearly to the base.. **Limonium**
1. Subscandent shrubs or suffrutescent herbs, the principal leaves cauline; corolla salverform, the petals connate for most of their length.. **Plumbago**

Limonium Mill. 1754. SEALAVENDER

Perennial herbs. Leaves in basal rosettes, alternate, simple, pinnate-veined, petiolate or epetiolate, estipulate. Flowers in secund panicles, each subtended by 3–4 bracts; sepals 5, the calyx tubular-funnelform, 5-ribbed and -lobed; petals 5, free nearly to the base; stamens epipetalous at the corolla base; styles 5, the stigmas linear-clavate. Fruit a utricle.

A genus of about 300 species; nearly cosmopolitan. (From the Greek *leimonion*, a meadow, in reference to the frequent occurrence of some species in salt meadows.)

Selected references: Luteyn (1976); Smith (2005a).

Limonium carolinianum (Walter) Britton [Of the Carolinas.] CAROLINA SEALAVENDER.

Statice caroliniana Walter, Fl. Carol. 118. 1788. *Statice limonium* Linnaeus var. *caroliniana* (Walter) A. Gray, Manual, ed. 2. 270. 1856. *Limonium carolinianum* (Walter) Britton, Mem. Torrey Bot. Club 5: 225. 1894.

Statice brasiliensis Boissier var. *angustata* A. Gray, Syn. Fl. N. Amer. 2(1): 54. 1878. *Limonium angustatum* (A. Gray) Small, Bull. Torrey Bot. Club 24: 488. 1897. *Statice angustata* (A. Gray) Wangerin, Z. Naturwiss. 82: 439. 1911. *Limonium carolinianum* (Walter) Britton var. *angustatum* (A. Gray) S. F. Blake, Rhodora 25: 56. 1913. *Limonium nashii* Small var. *angustatum* (A. Gray) Ahles, J. Elisha Mitchell Sci. Soc. 80: 173. 1964. TYPE: FLORIDA: Monroe Co.: Pine Key, s.d., *Blodgett s.n.* (holotype: GH; isotype: NY).

Limonium nashii Small, Bull. Torrey Bot. Club 24: 491. 1897. *Statice nashii* (Small) Wangerin, Z. Naturwiss. 82: 440. 1911. *Limonium carolinianum* (Walter) Britton var. *nashii* (Small) B. Boivin, Naturaliste Canad. 93: 643. 1966. TYPE: FLORIDA: Brevard Co.: Titusville, 31 Jul 1895, *Nash 2305* (lectotype: NY; isolectotypes: GH, MO, US). Lectotypified by Luteyn (1976: 311).

Limonium obtusilobum S. F. Blake, Rhodora 18: 63. 1916. *Limonium carolinianum* (Walter) Britton var. *obtusilobum* (S. F. Blake) Ahles, J. Elisha Mitchell Sci. Soc. 80: 173. 1964. TYPE: FLORIDA: s.d., *Chapman s.n.* (holotype: GH).

Statice tracyi Gandoger, Bull. Soc. Bot. France 66: 221. 1919. TYPE: FLORIDA: Franklin Co.: St. Vincent Island, 4 Sep 1899, *Tracy 6475* (lectotype: LY?; isolectotypes: GH, MO, NY). Lectotypified by Luteyn (1976: 311).

Scapose herb, to 25(40) cm. Basal leaves 20(25) in a dense rosette, the blade elliptic, obovate, oblanceolate to spatulate (rarely linear), 5–15(30) cm long, 0.5–5(7.5) cm wide, coriaceous, the apex rounded or acute to retuse, mucronate, the base narrowly cuneate, the margin undulate, the upper and lower surfaces glabrous, epetiolate or the petiole to 20 cm long; upper leaves scalelike, to 6 mm long. Flowers in a loose, secund panicle 10–60(90) cm long, 1–3(5) per spikelet; bracts 2–6 mm long, the apex obtuse, sheathing, glabrous; sepals obconic, 4–6 mm long, whitish, the tube 3–5 mm long, the lobes ca. 2 mm long, erect, alternating with 2 short, ovate to lanceolate teeth, glabrous or pilose along the midrib; petals 6–8 mm long, blue or lavender (rarely white), long-clawed, the lobes spatulate, the apex obtuse or retuse. Fruit oblanceolate, 3–5 mm long, 5-angled, the apex with the 5 persistent style bases; seed oblong-ovate, smooth and shiny.

Brackish marshes and salt flats. Frequent; nearly throughout. Labrador south to Florida, west to Texas; West Indies and Mexico. All year.

EXCLUDED TAXON

Limonium brasiliensis (Boissier) Small—Reported by Small (1903, 1913a, 1913b, 1913d), who misapplied the name to material of *L. carolinianum*.

Plumbago L. 1753. LEADWORT

Perennial herbs or shrubs. Leaves cauline, alternate, simple, pinnate-veined, petiolate or epetiolate. Flowers in spiciform racemes or panicles, each subtended by a bract and 2 bracteoles;

sepals 5, the calyx tubular, 5-ribbed and -lobed; petals 5, the corolla salverform; stamens free; style 1, the stigmas 5, linear-clavate. Fruit a 5-valved capsule.

A genus of 12 species; nearly cosmopolitan. (*Plumbago*, a lead-like ore, alluding to the historical use as a cure for lead poisoning.)

Selected reference: Smith (2005b).

1. Calyx puberulent, the ribs bearing stalked glands only on the distal portion; petiole auriculate............
.... **P. auriculata**
1. Calyx glabrous, the ribs bearing stalked glands their entire length; petiole not auriculate.....................
.... **P. scandens**

Plumbago auriculata Lam. [Eared, in reference to the projections at the petiole base.] CAPE LEADWORT.

Plumbago auriculata Lamarck, Encycl. 270. 1786. *Plumbagidium auriculatum* (Lamarck) Spach, Hist. Nat. Vég. 10: 339. 1841.
Plumbago capensis Thunberg, Prodr. Pl. Cap. 33. 1794.

Erect or clambering shrub, to 3 m; stem diffusely branched, glabrous or puberulous on the youngest branchlets. Leaves with the blade elliptic, oblanceolate, or spatulate, 3–9 cm long, 1.5–2 cm wide, the apex acute or obtuse, apiculate, the base long-attenuate, sometimes auriculate, the margin entire, the upper surface glabrous, the lower surface pale-lepidote, epetiolate or sometimes short petiolate. Flowers in a spiciform raceme or panicle, the rachis puberulent; bracts ovate or lanceolate, 3–9 mm long, mucronate; sepals 10–13 mm long, the tube short-pilose and with capitate, gland-like protuberances ca. 1 mm long along the ribs in the distal ½–⅔; petals 3.5–5 cm long, pale blue, the tube 2.5–4 cm long, the lobes broadly obovate, 1–1.5 cm long, 6–15 mm wide. Fruit linear-oblong, ca. 8 mm long, long-beaked; seed ca. mm long, brown.

Disturbed sites. Rare; Brevard, Hillsborough, Martin, and Lee Counties. Escaped from cultivation. Florida, Louisiana, and California; West Indies, Mexico, Central America, and South America; Africa and Pacific Islands. Native to Africa. All year.

Plumbago zeylanica L. [Of Ceylon (Sri Lanka).] DOCTORBUSH.

Plumbago zeylanica Linnaeus, Sp. Pl. 1: 151. 1753. *Plumbago flaccida* Moench, Methodus 429. 1794, nom. illegit. *Plumbago lactea* Salisbury, Prodr. Stirp. Chap. allerton 122. 1796, nom. illegit.
Plumbago scandens Linnaeus, Sp. Pl., ed. 2. 215. 1762. *Plumbago Americana* Weigel, Diss. Hort. Cryph. 12. 1782, nom. illegit. *Molubda scandens* (Linnaeus) Rafinesque, Sylva Tellur. 123. 1838. *Plumbagidium scandens* (Linnaeus) Spach, Hist. Nat. Vég. 10: 339. 1841. *Plumbago scandens* Linnaeus var. *normalis* Kuntze, Revis. Gen. Pl. 2: 396. 1891, nom. inadmiss.
Plumbago floridana Nuttall, Amer. J. Sci. Arts 5: 290. 1822. TYPE: FLORIDA: s.d., *Ware s.n.* (holotype: PH?).

Prostrate, climbing, or erect suffrutescent herb, to 1 m; stem branched, glabrous. Leaves with the blade ovate, elliptic-lanceolate, or spatulate to oblanceolate, (3)5–9(15) cm long, (1)2.5–4(7) cm wide, the apex acute, acuminate or obtuse, the base attenuate, the margin entire, the upper and lower surfaces glabrous, epetiolate or the petiole to 1.5 cm long. Flowers in a spiciform raceme or panicle 0.5–1.5(3) cm long, the rachis glandular; bracts lanceolate, 3–7 mm long; sepals 7–11(13) mm long, the tube with stalked, capitate, gland-like protuberances along the ribs, otherwise glabrous; petals 17–33 mm long, white, the tube 12–25 mm long, the lobes obovate, 5–12 mm long, 3–4 mm wide. Fruit linear-oblong, ca. 8 mm long, long-beaked; seed reddish brown to dark brown, 5–6 mm long.

Coastal hammocks. Occasional; central and southern peninsula. Florida, Texas, and Arizona; West Indies, Mexico, Central America, and South America; Africa, Asia, Australia, and Pacific Islands. All year.

POLYGONACEAE Juss. 1789. BUCKWHEAT FAMILY

Annual or perennial herbs, shrubs, trees, or vines; stems and branches with or without swollen nodes; tendrils present or absent. Leaves alternate or rarely whorled or opposite, the blade simple, pinnate- or palmate-veined, petiolate or epetiolate, the stipules (ocrea) fused into a sheath surrounding the stem or absent. Flowers terminal or terminal and axillary, cymose, composed of simple or branched fascicles and spiciform, racemiform, paniculiform, or corymbiform, each fascicle subtended by a bract and each flower by a sheath (ocreola) and occasionally by 2 bracteoles, rarely fascicles subtended by an involucre and the flowers subtended only by bracts (*Eriogonum*), actinomorphic, bisexual or unisexual (plants dioecious, monoecious, or variously polygamous); perianth segments 4–6, petaloid or sepaloid, basally connate or free, sometimes developing keels or wings; stamens (3)5–8(9), often in 2 whorls and the filaments of 2 lengths, the anthers 2-loculate, longitudinally dehiscent, versatile or basifixed; ovary (2)3-carpellate, 1-loculate, the styles 2–3. Fruit an achene.

A family of about 50 genera and about 1,200 species; nearly cosmopolitan.

Generic limits within the Polygonaceae, especially circumscription of *Polygonum*, have been variously set and no consensus has been achieved. For the purposes of this flora, we slightly differ in our treatment from that in the *Flora of North America* by recognizing *Reynoutria* as distinct from *Fallopia*. Schuster et al. (2015) merge *Emex* with *Rumex*, but here it is kept distinct as in the *Flora of North America*, pending further study. *Polygonella*, though quite distinct in our flora, appears closely related to the extralimital *Polygonum* section *Duravia* (see Ronse Decraene et al., 2004) but is also kept distinct following *Flora of North* America, again pending further study.

Selected references: Freeman and Reveal (2005); Galasso et al. (2009); Graham and Wood (1965); Ronse Decraene et al. (2004); Sanchez et al. (2011); Schuster et al. (2015).

1. Vines with tendrils.

    2. Perianth segments green to greenish white; floral tube enlarging in fruit to ca. 3 cm in length; pedicel 3-winged; leaf base truncate or subcordate to broadly cuneate; ocrea persistent.............................
    .... **Brunnichia**

    2. Perianth segments pink (rarely white); floral tube not enlarging in fruit; pedicel not winged; leaf base deeply cordate; ocrea deciduous .... **Antigonon**

1. Trees, shrubs, or herbs, if scandent, then lacking tendrils.
   3. Flowers in an involucrate head (bracts sometimes small); stipules (ocrea) absent......... **Eriogonum**
   3. Flowers not in a head, or if in a head, then not involucrate; stipules (ocrea) present.
      4. Trees or shrubs.
         5. Floral tube becoming chartaceous, the outer 3 perianth segments enlarged into conspicuous oblanceolate wings.. **Triplaris**
         5. Floral tube becoming fleshy and berrylike, the perianth segments not enlarged ....
             ....**Coccoloba**
      4. Herbs, subshrubs, or vines.
         6. Perianth segments 6, the outer 3 with a stout apical spine..**Emex**
         6. Perianth segments 3, 4, or 5, the outer ones lacking an apical spine.
            7. Inner 3 perianth segments larger than the outer 3 in fruit .... **Rumex**
            7. Inner 3 perianth segments subequal or smaller than the outer 3 in fruit.
               8. Branches partially adnate to the stem and appearing to arise internodally; leaf venation appearing parallel with the secondary veins not conspicuous ....**Polygonella**
               8. Branches arising from the nodes; leaf venation evidently palmate or pinnate with the secondary veins conspicuous.
                  9. Outer perianth segments winged or keeled.
                     10. Stem twining, scandent, or sprawling; stigmas capitate ....**Fallopia**
                     10. Stem erect or prostrate; stigmas fimbriate..**Reynoutria**
                  9. Outer perianth segments not winged or keeled.
                     11. Perianth segments free .... **Fagopyrum**
                     11. Perianth segments connate to ⅔ their length.
                        12. Perianth segments with 1 primary vein..**Polygonum**
                        12. Perianth segments with 3 primary veins...**Persicaria**

Antigonon Endl. 1837. LOVECHAIN

Perennial vines; lateral branches leafy or modified as tendrils. Leaves alternate, pinnate-veined, petiolate; stipules (ocrea) present. Flowers in terminal and axillary racemiform cymes bearing lateral and terminal tendrils, each subtended by an ocreola, actinomorphic, bisexual; perianth segments 5, basally connate, petaloid, dimorphic, the outer 3 larger than the inner 2; stamens (7)8(9), the filaments alternating and united with small, triangular, nectariferous teeth to form a broad cupulate staminal tube adnate to the perianth tube; styles 3, free. Fruit a trigonous achene, enclosed by the accrescent perianth.

A genus of 6 species; Nearly cosmopolitan. [From the Greek *anti*, against or opposite, and *gony*, joint or knee, perhaps in reference to the geniculate stems.]

Corculum Stuntz (1913).

Selected reference: Freeman (2005a).

Antigonon leptopus Hook. & Arn. [From the Greek *leptos*, slender or small, and *pous*, a foot, referring to the petiole base.] CORAL VINE; QUEEN'S JEWELS.

Antigonon leptopus Hooker & Arnott, Bot. Beech. Voy. 308, t. 69. 1838. *Corculum leptopus* (Hooker & Arnott) Stuntz, Bull. Bur. Pl. Industr. U.S.D.A. 282: 86. 1913.

Herbaceous or woody vine, to 15 m; stems glabrous or sparsely to densely brownish or red-dish pubescent, with terminal or axillary tendrils. Leaves with the blade ovate, 5–14 cm long, (2)4–10 cm wide, the apex acute to acuminate, the base cordate, the margin entire, ciliate, the upper and lower surfaces glabrous or pubescent, especially on the veins, the petiole 1–2.5(5) cm long, often winged distally, glabrous or pubescent; stipules (ocrea) to 2 mm long, decidu-ous. Flowers pendulous in an axillary, branched, racemiform inflorescence 4–20 cm long, the axis puberulent to pilose, the peduncle angular, 1–5 cm long, the pedicel 3–5(10) mm long, articulated proximally; perianth segments ovate to elliptic, the outer 3 8–20 mm long, 4–15 mm wide, the inner 2 4–8 mm long, 2–6 mm wide, pink or rarely white, glabrous, the apex acute, the margins entire. Fruit ovate to bluntly trigonous, 8–12 mm long, brown, shiny, glabrous or pubescent, subtended by persistent perianth segments.

Hammock margins and disturbed sites. Frequent; peninsula, central and western panhan-dle. Escaped from cultivation. Nearly cosmopolitan. Native to Mexico and Central America. Spring–fall.

Antigonon leptopus is listed as a Category II invasive species in Florida by the Florida Exotic Pest Plant Council (FLEPPC, 2015).

Brunnichia Banks ex Gaertn. 1788. BUCKWHEATVINE

Perennial vines; stems with lateral branches modified as tendrils. Leaves alternate, pinnate-veined, petiolate; stipules (ocrea) present. Flowers in terminal and axillary paniculiform or racemiform cymes with a few terminal and lateral tendrils, each subtended by an ocreola, actinomorphic, bisexual; perianth segments 5, basally connate, petaloid, slightly dimorphic, the outer 3 larger than the inner 2; stamens 8, the filaments free, adnate to the perianth tube; styles 3, free. Fruit a trigonous achene enclosed by the accrescent perianth.

A genus of 4 species; North America, West Indies and Africa. [Commemorates Morten Thrane Brünnich (1737–1827), Danish naturalist.]

Selected reference: Holmes (2005).

Brunnichia ovata (Walter) Shinners [Egg-shaped, in reference to the leaves.] AMERICAN BUCKWHEATVINE.

Rajania ovata Walter, Fl. Carol. 247. 1788. *Brunnichia ovata* (Walter) Shinners, Sida 3: 115. 1967. *Brunnichia cirrhosa* Gaertner, Fruct. Sem. Pl. 1: 214, t. 45. 1788.

Herbaceous or soft-woody vine; stem angled, climbing by means of branched tendrils termi-nating some of the short, lateral branches and inflorescences, glabrous or with a band of tri-chomes at the top edge of the internode, the bark of the woody stems reddish with gray stripes and lenticels. Leaves with the blade ovate to ovate-lanceolate, 3–15 cm long, 1.5–8 cm wide, the apex acute to acuminate, the base truncate to subcordate, the margin entire, the upper surface glabrous, the lower usually glabrous to slightly pubescent, the petiole 7–25 mm long, dilated at the base, glabrous or puberulent; stipules (ocrea) obscure, to 1 mm long, fringed with red-dish brown trichomes. Flowers in an inflorescence 5–26 cm long, the pedicel 5–10 mm long, articulate below the middle, 3-winged, 1 wing larger, dilated distally; perianth segments linear-oblong to oblong, subequal, the apex acute, the margin entire. Fruit narrowly ovoid, 8–10 mm long, brown, sublustrous.

Floodplain forests and moist hammocks. Frequent; panhandle. Virginia south to Florida, west to Missouri, Oklahoma, and Texas; West Indies. Summer.

Coccoloba P. Browne 1756. SEAGRAPE; PIGEON PLUM

Trees or shrubs. Leaves alternate, pinnate-veined, petiolate; stipules (ocrea) present. Flowers in axillary or subterminal, racemiform cymes, 1–several flowers subtended by an ocreola, actinomorphic, unisexual (plants dioecious). Staminate flowers in fascicles of 2–5 at a node, each subtended by an ocreola that encloses the adjacent flower and its ocreola; perianth segments 5, subequal, basally connate, monomorphic; stamens 8, the filaments dilated at the base and connate into an annulus, exserted, the aborted gynoecium present, the styles included. Carpellate flowers solitary at the nodes, subtended by an ocreola; perianth segments as in staminate; sterile stamens present; styles exserted. Fruit a subtrigonous achene, the floral tube persistent, fleshy, surrounding the achene and berrylike.

A genus of about 120 species; North America, West Indies, Mexico, Central America, and South America. [From the Greek *coccos*, seed or berry, and *lobos*, capsule or pod, in reference to the fleshy perianth tube surrounding the fruit.]

Selected reference: Freeman (2005b).

1. Leaf blades ovate to lanceolate or elliptic, usually 2–3 times longer than wide, the base cuneate to rounded .... **C. diversifolia**
1. Leaf blades orbicular to subreniform, the length equaling or less than the width, the base cordate.......
.... **C. uvifera**

Coccoloba diversifolia Jacq. [Diverse-leaved, those of different shapes borne on the same plant.] TIETONGUE; PIGEON PLUM.

Coccoloba diversifolia Jacquin, Enum. Syst. Pl. 19. 1760.
Coccoloba floridana Meisner, in A. de Candolle, Prodr. 14: 165. 1856. TYPE: FLORIDA: Monroe Co.: Key West, Mar 1846, *Rugel 140* (holotype: NY).

Shrub or tree, to 10(18) m; bark thin, gray with brown mottling, the branchlets grayish green, becoming gray in age, glabrous. Leaves with the blade ovate to lanceolate or elliptic, (3)5–10(13) cm long, (1)3–5(7) cm wide, coriaceous, the apex acute to obtuse, the base cuneate to rounded, the margin entire, the upper surface glabrous, shiny, minutely punctate, the lower glabrous, dull, the petiole 5–15 mm long, glabrous or puberulent; stipules (ocrea) 3–5 mm long, coriaceous and persistent proximally, membranaceous and deciduous distally, the margin oblique, tan or brown, glabrous or puberulent. Flowers in a raceme 5–15 cm long, glabrous, the peduncle 1–6 cm long, the pedicel 2–4 mm long. Staminate flowers in a fascicle of 1–3; perianth segments round to broadly elliptic, ca. 5 mm long and wide, greenish, the apex obtuse, the margin entire. Carpellate flowers solitary, the perianth segments as in the staminate. Fruit (fleshy floral tube) subglobose to obpyriform, 9–14 mm long, red-purple, the included achene 6–10 mm long, shiny.

Tropical coastal hammocks. Occasional; Brevard and Lee Counties southward. Florida; West Indies, Mexico, and Central America. Spring–fall.

Coccoloba uvifera (L.) L. [Grape-bearing, in reference to the fruits resembling grapes.] SEAGRAPE.

Polygonum uvifera Linnaeus Sp. Pl. 365. 1753. *Coccoloba uvifera* (Linnaeus) Linnaeus, Syst. Nat., ed. 10. 1007. 1759. *Guaiabara uvifera* (Linnaeus) House, Amer. Midl. Naturalist 8: 64. 1922.

Shrub or tree, to 15 m; bark gray with light brown mottling, the branchlets green, becoming gray in age, finely pubescent, soon becoming glabrous. Leaves with the blade orbicular to sub-reniform, (6)10–20(27) cm long, 6–20(27) cm wide, coriaceous, the apex rounded or slightly emarginate, the base cordate, the margin entire, the upper surface glabrous, shiny or dull, minutely punctate, the lower surface glabrous, dull, the petiole to 5–15 mm long, puberulent or pilose; stipules (ocrea) 3–8 mm long, coriaceous and persistent proximally, membranaceous and deciduous distally, the margin oblique, brown or reddish brown, glabrous or puberulent. Flowers in a raceme 10–30 cm long, glabrous or puberulent, the peduncle 1–5 cm long, the pedicel 1–3 mm long. Staminate flowers in a fascicle of 1–7; perianth segments round to broadly elliptic, ca. 6 mm long and wide, whitish, the apex obtuse, the margin entire. Carpellate flowers solitary, the perianth segments as in the staminate. Fruit (fleshy floral tube) subglobose to obpyriform, 12–20 mm long, red-purple, the included achene 8–11 mm long, shiny.

Coastal hammocks and beach strands. Frequent; central and southern peninsula. Florida and Mississippi; West Indies, Mexico, Central America, and South America; Pacific Islands. Native to North America and tropical America. Spring–fall.

EXCLUDED TAXON

Coccoloba laurifolia Jacquin—Small (1903, 1913a, 1913b, 1913c, 1913d, 1933) misapplied the name of this West Indian species to Florida material of *C. diversifolia*.

Emex Campderá, nom. cons. 1819. THREECORNERJACK.

Annual herbs. Leaves alternate, pinnate-veined, petiolate; stipules (ocrea) present. Flowers in terminal and axillary racemiform cymes, actinomorphic, unisexual (plants monoecious). Staminate flowers in fascicles of 1–8 at a node, ocreolate; perianth segments 6, free, subequal; stamens 4–6, the filaments free. Carpellate flowers in fascicles of 2–7 at a node, perianth segments 6, basally connate, dimorphic, the outer 3 with the apex ending in a stout spine, the inner 3 erect, tuberculate; styles 3, free. Fruit an achene enclosed in the accrescent perianth.

A genus of 2 species; North America; West Indies and South America; Europe, Africa, Asia, Australia, and Pacific Islands. [*Ex*, from, and *Rumex*, in reference to its segregation from that genus.]

Emex is nested in *Rumex* (Schuster et al., 2015), but is morphologically distinct from that genus in Florida by its number of perianth segments, the outer 3 with a stout spine. Further study is needed.

Selected reference: Freeman (2005c).

Emex spinosa (L.) Campd. [Spiny.] SPINY THREECORNERJACK.

> *Rumex spinosus* Linnaeus, Sp. Pl. 337. 1753. *Emex spinosa* (Linnaeus) Campderá, Monogr. Rumex 58. 1819.

Ascending or erect annual herb, to 60(80) cm; stem often much branched at the base, reddish. Leaves with the blade ovate to ovate-oblong or triangular, 3–13 cm long, 1–12 cm wide, the apex obtuse to acute, the base truncate or subcordate, the margin crenulate to dentate or entire, the upper and lower surfaces glabrous, the petiole 2–29 cm long; stipules (ocrea) chartaceous, loose, oblique, glabrous, deciduous. Staminate flowers in a terminal or axillary racemiform fascicle of 1–8, pedunculate; perianth segments narrowly oblong to oblanceolate, ca. 2 mm long. Carpellate flowers in an axillary, sessile whorl of 2–7; outer perianth segments ovate to oblong, 4–6 mm long, the apex acute, the inner ones linear-lanceolate, 5–6 mm long, the apex acute; styles erect, the stigmas penicillate. Fruit enclosed in accrescent perianth, the outer 3 perianth segments ribbed and foveolate, ending in stout, spreading spines, the inner 3 erect, 3-ribbed, tuberculate, the achene 4–5 mm long, trigonous, glabrous, shiny.

Disturbed sites. Rare; Escambia County. Not collected since 1901. Massachusetts, New Jersey, Florida, Texas, and California; South America; Europe, Africa, Asia, Australia, and Pacific Islands. Native to Europe, Africa, and Asia. Summer–fall.

Eriogonum Michx. 1803. WILD BUCKWHEAT

Perennial herbs. Leaves alternate, opposite, or whorled, pinnate-veined, petiolate or epetiolate; stipules (ocrea) absent. Flowers in terminal paniculiform cymes, involucrate, actinomorphic, bisexual; bracts (3–5), foliaceous, basally connate; perianth segments 6, basally connate, petaloid, monomorphic or dimorphic; stamens 9, two opposite each outer sepal, 1 opposite each inner perianth segment; styles 3. Fruit trigonous, enclosed by the accrescent perianth.

A genus of about 250 species; North America; Mexico. [From the Greek *erion*, wool, and *gony*, knee, in reference to the woolly swollen nodes of the species first described, *E. tomentosum.*]

Selected reference: Reveal (2005).

1. Leaves of the flowering stem alternate; perianth segments monomorphic, the outer surface white- or silvery-tomentose; filaments glabrous..**E. longifolium**
1. Leaves of the flowering stem opposite or whorled; perianth segments dimorphic, the outer surface tannish to rusty-lanate; filaments pilose .... **E. tomentosum**

Eriogonum longifolium Nutt. var. **gnaphalifolium** Gand. [Long-leaved; with leaves like *Gnaphalium* (Asteraceae).] LONGLEAF WILD BUCKWHEAT; SCRUB BUCKWHEAT.

Eriogonum longifolium Nuttall var. *gnaphalifolium* Gandoger, Bull. Soc. Bot. Belg. 42: 190. 1906. TYPE: FLORIDA: Lake Co.: near Eustis, Jun–Jul 1894, *Hitchcock s.n.* (holotype: LY; isotypes: F, KSC, MIN).

Eriogonum floridanum Small, Fl. S.E. U.S. 367, 1329. 1903. TYPE: FLORIDA: Lake Co.: vicinity of Eustis, 1–15 May 1894, *Nash 704* (holotype: NY; isotypes: DAO, E, F, G, GH, K, MICH, MIN, MO, MSC, NEB, P, UC, US, W).

Eriogonum longifolium Nuttall var. *floridanum* Gandoger, Bull. Soc. Bot. Belg. 42: 190. 1906. TYPE: FLORIDA: Seminole Co.: Sanford [miscited as Tampa], s.d., *Williamson s.n.* (holotype: LY; isotype: UC).

Eriogonum longifolium Nuttall var. *longidens* Gandoger, Bull. Soc. Bot. Belg. 42: 19. 1906. TYPE: FLORIDA: Polk Co.: 7 Jul 1894, *Ohlinger 302* (holotype: LY; isotypes: F, ILL, MO).

Erect perennial herb, to 1.3 m. Basal leaves with the blade narrowly oblanceolate, (3)8–20 cm long, 3–10 mm wide, the apex obtuse, the base cuneate, the margin entire, the upper surface glabrous, the lower surface white-lanate, the petiole 0.5–20 cm long, the cauline leaves alternate, progressively reduced upward, sessile. Flowers in an open paniculiform cyme, the peduncle 1–3 cm long; bracts 3, scalelike, triangular, to 5 mm long; involucre turbinate, 6–7 mm long, 5-toothed; perianth segments yellow, the outer surface white- to silvery-tomentose, monomorphic, lanceolate to narrowly elliptic, 4–8 mm long, spreading, the stipe-like base 2–4(7); stamens 2–3 mm long, the filaments glabrous. Fruit ovoid to ellipsoidal, 5–6 mm long, brown, white-tomentose.

Sandhills and scrub. Occasional; Putnam County, central peninsula. Endemic. Spring–fall.

Eriogonum longifolium var. *gnaphalifolium* (as *E. floridanum*) is listed as endangered in Florida (Florida Administrative Code, Chapter 5B-40) and as threatened in the United States (U.S. Fish and Wildlife Service, 50 CFR 23).

Eriogonum tomentosum Michx. [With dense, matted, soft trichomes.]
DOGTONGUE WILD BUCKWHEAT.

Eriogonum tomentosum Michaux, Fl. Bor.-Amer. 1: 246. 1803.

Erect perennial herb, to 1.2 m. Basal leaves with the blade elliptic to oblong, 4–15 cm long, 1.5–4 cm wide, the apex rounded, the base cuneate, the apex obtuse to acute, the margin entire, the upper surface glabrous or subglabrous, the lower surface white- to rusty-tomentose, the petiole 2–7 cm long, the cauline leaves in whorls of 3–4, progressive reduced upward, sessile. Flowers in an open paniculiform cyme, lacking a peduncle; bracts 3–5, leaflike, the proximal ones ovate to elliptic, 1–6 cm long, 0.5–2.5 cm wide, the distal ones elliptic, 4–30 mm long, 1–2 mm wide; involucre turbinate-campanulate to campanulate, 3–5 mm long, 5-toothed; perianth segments cream to light tan, the outer surface tan to rusty brown, dimorphic, those of the outer whorl of 3 broadly lanceolate, 3–5 mm long, spreading, those of the inner whorl of 3 obovate, to suborbicular, 3–6 mm long, 3–4 mm wide, erect, the stipe-like base 3–4 mm long; stamens 4–5 mm long, the filaments pilose. Achene ovoid to ellipsoidal, 4–5 mm long, brown, glabrous.

Sandhills. Frequent; northern counties, central peninsula. North Carolina, South Carolina, Georgia, Florida, and Alabama. Summer–fall.

EXCLUDED TAXON

Eriogonum longifolium Nuttall—Because infraspecific categories were not recognized, the typical variety was reported by implication by Chapman (1860, 1883, 1897). All Florida material is of var. *gnaphalifolium*.

Fagopyrum Mill. 1754. BUCKWHEAT

Annual herbs. Leaves alternate, palmate-veined, petiolate or epetiolate; stipules (ocrea) present. Flowers in terminal and axillary paniculiform cymes, ocreolate, bracteate, actinomorphic, bisexual; perianth segments 5, basally connate, petaloid, dimorphic, the outer 3 smaller than the inner 2; stamens 8, the filaments free; styles 3, free. Fruit a trigonous achene.

A genus of 16 species; North America, Central America, South America, Europe, Africa, Asia, Australia, and Pacific Islands. [*Fagus*, beech, and the Greek *pyros*, wheat, in reference to the resemblance of the fruit of the beechnut.]

Selected reference: Hinds and Freeman (2005a).

Fagopyrum esculentum Moench [*Esculen*, edible.] BUCKWHEAT.

Polygonum fagopyrum Linnaeus, Sp. Pl. 364. 1753. *Fagopyrum esculentum* Moench, Methodus 290. 1794. *Polygonum cereale* Salisbury, Prodr. Stirp. Chap. Allerton 259. 1796, nom. illegit. *Fagopyrum sarracenicum* Dumortier, Fl. Belg. 18. 1827, nom. illegit. *Fagopyrum vulgare* T. Nees von Esenbeck, Gen. Fl. Germ. 1: pl. 53. 1834, nom. illegit. *Fagopyrum cereale* Rafinesque, Fl. Tellur. 3: 10. 1837 ("1836"), nom. illegit. *Phegopyrum esculentum* (Moench) Petermann, Fl. Bienitz 92. 1841, nom. illegit. *Fagopyrum esculentum* Moench var. *vulgare* Alefeld, Landw. Fl. 286. 1866, nom. inadmiss. *Fagopyrum fagopyrum* (Linnaeus) H. Karsten, Deut. Fl. 522. 1881, nom. inadmiss. *Helxine fagopyrum* (Linnaeus) Kuntze, Revis. Gen. Pl. 2: 553. 1891. *Fagopyrum polygonum* MacCloskie, Torreya 5: 198. 1905, nom. illegit.

Annual herb, to 9 dm; stem glabrous or sometimes pubescent at the nodes. Leaves with the blade ovate or triangular, the apex acute to acuminate, the base truncate or cordate to sagittate, the margin entire or slightly 3-lobed, ciliolate, basal primary veins 7–9, the upper surface glabrous, the lower surface glabrous or puberulent, the lower leaves with a petiole 1–6(9) cm long, the upper ones sessile; stipules (ocrea) funnelform, 2–8 mm long, sheathing but open on one side, brownish hyaline, the margin truncate, eciliolate, glabrous to puberulent proximally. Flowers in a paniculiform cyme 1–4 cm long, the peduncle 0.5–4 cm long, puberulent in lines, the pedicel 2–4 mm long; bracts ovate, the apex acute, glabrous; perianth segments elliptic to obovate, 3–5 mm long, the outer ones smaller than the inner, creamy white to pale pink, the apex obtuse to acute, the margin entire; stamens 4–5 mm long; styles ca. 2 mm long. Fruit sharply trigonous, 4–6 mm long, dark brown or black, the surface smooth, lustrous, strongly exserted.

Disturbed sites. Rare; peninsula, Franklin County. Escaped from cultivation. Nearly throughout North America; Central America and South America; Europe, Africa, Asia, Australia, and Pacific Islands. Native to Asia. Spring–summer.

Fallopia Adans., nom. cons. 1763. FALSE BUCKWHEAT

Annual or perennial vines. Leaves alternate, the blade pinnipalmate-veined, petiolate; stipules (ocrea) present. Flowers in axillary spiciform cymes, actinomorphic, bisexual; perianth segments 5, basally connate, petaloid, dimorphic, the outer 3 winged or keeled, larger than the inner ones; stamens 8, free; styles 3, basally connate, the stigmas capitate. Fruit a trigonous achene, enclosed by the accrescent perianth.

A genus of about 10 species; North America, Mexico, South America, Europe, Africa, and Asia. [Commemorates Gabriele Falloppio (1532–1562), Italian physician and anatomist.

Bilderdykia Dumort. (1827); *Tiniaria* (Meisner) Rchb. (1837).

Selected reference: Freeman and Hinds (2005).

1. Fruiting perianth narrowly keeled; achene dull...**F. convolvulus**
1. Fruiting perianth broadly winged; achne lustrous .... **F. scandens**

Fallopia convolvulus L. [Old generic name for twining plants.] BLACK BINDWEED.

> *Polygonum convolvulus* Linnaeus, Sp. Pl. 364. 1753. *Fagopyrum carinatum* Moench, Methodus 290. 1794, nom. illegit. *Bilderdykia convolvulus* (Linnaeus) Dumortier, Fl. Belg. 18. 1827. *Tiniaria convolvulus* (Linnaeus) Webb & Moquin-Tandon ex Webb & Berthelot, Hist. Nat. Iles Canaries 3(2(3)): 221. 1846. *Fagopyrum convolvulus* (Linnaeus) H. Gross, Bull. Geogr. Bot. 23: 21. 1913. *Reynoutria convolvulus* (Linnaeus) Shinners, Sida 3: 117. 1967. *Fallopia convolvulus* (Linnaeus) Á. Löve, Taxon 19: 300. 1970.

Annual vine, to 1 m; stem branched proximally, puberulent. Leaves with the blade cordate-ovate, cordate hastate, or sagittate, 2–6(15) cm long, 2–5(10) cm wide, pinnipalmate-veined, the apex acuminate, the base cordate, the margin undulate, the upper surface glabrous or puberulent on nerves and near the margin, minutely punctate, the lower surface minute puberulent, at least on the nerves, minutely punctate, often mealy, the petiole 0.5–5 cm long, puberulent; stipules (ocrea) 2–4 mm long, chartaceous, the margin oblique, eciliate, reddish-brown, the surface glabrous or scabrid on the nerves, with reflexed trichomes and slender bristles proximally. Flowers in an axillary spiciform inflorescence, 3–6 per ocreolate fascicle, bisexual, the peduncle to 10 cm long or absent, the pedicel 1–3 mm long; perianth segments 5, elliptic to obovate, 3–5 mm long, greenish white or purple-blotched, the apex acute or obtuse, the outer 3 obscurely keeled; stamens 8. Fruit obovoid, 3–5 mm long, trigonous, black, the surface minutely granular, dull.

Disturbed clearings in swamps and hydric hammocks. Rare; Duval, Alachua, Manatee, and Santa Rosa Counties. Nearly throughout North America; South America; Europe, Africa, Asia, Australia, and Pacific Islands. Native to Europe, Africa, and Asia. Spring–fall.

Fallopia scandens (L.) Holub [Climbing.] CLIMBING FALSE BUCKWHEAT.

> *Polygonum scandens* Linnaeus, Sp. Pl. 364. 1753. *Anredera scandens* (Linnaeus) Smith, in Rees, Cycl. 39(Addend.): Anredera #1. 1818. *Polygonum dumetorum* Linnaeus var. *scandens* (Linnaeus) A. Gray, Manual, ed. 5. 418. 1867. *Polygonum convolvulus* Linnaeus var. *scandens* (Linnaeus) Kuntze, Revis. Gen. Pl. 2: 555. 1891. *Tiniaria scandens* (Linnaeus) Small, Fl. S.E. U.S. 382, 1330. 1903.

Bilderdykia scandens (Linnaeus) Greene, Leafl. Bot. Obs. Crit. 1: 23. 1904. *Fagopyrum scandens* (Linnaeus) H. Gross, Bull. Geogr. Bot. 23: 22. 1913. *Reynoutria scandens* (Linnaeus) Shinners, Sida 3: 118. 1967. *Fallopia scandens* (Linnaeus) Holub, Folia Geobot. Phytotax. 6: 176. 1971.

Polygonum dumetorum Linnaeus, Sp. Pl., ed. 2. 52. 1762. *Fagopyrum dumetorum* (Linnaeus) Schreber, Spic. Fl. Lips. 42. 1771. *Bilderdykia dumetorum* (Linnaeus) Dumortier, Fl. Belg. 18. 1827. *Helxine dumetorum* (Linnaeus) Rafinesque, Fl. Tellur. 3: 10. 1837 ("1836"). *Tiniaria dumetorum* (Linnaeus) Opiz, Seznam 98. 1852. *Polygonum convolvulus* Linnaeus var. *dumetorum* (Linnaeus) Kuntze, Revis. Gen. Pl. 2: 555. 1891. *Polygonum scandens* Linnaeus var. *dumetorum* (Linnaeus) Gleason, Phytologia 4: 23. 1952. *Reynoutria scandens* (Linnaeus) Shinners var. *dumetorum* (Linnaeus) Shinners, Sida 3: 118. 1967. *Fallopia dumetorum* (Linnaeus) Holub, Folia Geobot. Phytotax. 6: 176. 1971.

Polygonum cristatum Engelmann & A. Gray, Boston J. Nat. Hist. 5: 259. 1845. *Polygonum convolvulus* Linnaeus var. *cristatum* (Engelmann & A. Gray) Kuntze, Revis. Gen. Pl. 2: 556. 1891. *Tiniaria cristata* (Engelmann & A. Gray) Small, Fl. S.E. U.S. 382, 1330. 1903. *Bilderdykia cristata* (Engelmann & A. Gray) Greene, Leafl. Bot. Obs. Crit. 1: 23. 1904. *Polygonum dumetorum* Linnaeus forma *cristatum* (Engelmann & A. Gray) B. L. Robinson, Rhodora 10: 31. 1908. *Polygonum scandens* Linnaeus var. *cristatum* (Engelmann & A. Gray) Gleason, Phytologia 4: 23. 1952. *Reynoutria scandens* (Linnaeus) Shinners var. *cristata* (Engelmann & A. Gray) Shinners, Sida 3: 118. 1967. *Fallopia cristata* (Engelmann & A. Gray) Holub, Folia Geobot. Phytotax. 6: 176. 1971. *Bilderdykia scandens* (Linnaeus) Greene var. *cristata* (Engelmann & A. Gray) C. F. Reed, Phytologia 50: 461. 1982.

Scandent annual or perennial herb, to 5 m; stem branched, obscurely angled to terete, glabrous or puberulent. Leaves with the blade cordate, deltate, or hastate, pinnipalmate-veined, the apex acuminate, the base cordate, the margin entire, the upper and lower surfaces glabrous or papillose to scabrid, at least on the veins and the margin, rarely minutely glandular-punctate, the petiole 0.3–10 cm long, glabrous or scabrid in lines; stipules (ocrea) 1–6 mm long, tan or brown, the margin oblique, eciliate, the surface glabrous or scabrid, usually deciduous. Flowers axillary in a racemiform inflorescence, 2–6 per ocreate fascicle, bisexual, the peduncle to 7 cm long or absent, scabrid, the pedicel 4–8 mm long, glabrous; perianth segments 5, elliptic to obovate, 4–8 mm long, green to white or pinkish, the apex obtuse to acute, the outer 3 winged, decurrent on the floral stipe; stamens 8. Fruit ovate, 2–6 mm long, trigonous, dark brown to black, the surface smooth, lustrous; wings of the fruiting perianth to 2 mm wide or more, undulate, incised, or flat, truncate or decurrent on the stipe nearly to the articulation.

Swamps, wet hammocks, and disturbed sites. Occasional; northern peninsula south to Brevard County, central panhandle, Escambia County. Quebec south to Florida, west to Alberta, Wyoming, Colorado, and Texas; Europe and Asia. Spring–fall.

Fallopia scandens has a complex nomenclatural history resulting in a variety of treatments. Characters used to separate *F. scandens* and *F. cristatum* from *F. dumetorum* are highly variable and intergrade. We have not seen convincing evidence to treat the complex other than as consisting of a single polymorphic taxon.

Persicaria (L.) Mill. 1754. KNOTWEED; SMARTWEED; TEARTHUMB.

Annual or perennial herbs. Leaves alternate, pinnate- or pinnipalmate-veined, petiolate or epetiolate; stipules (ocrea) present. Flowers in terminal or axillary spiciform, paniculiform, or capitate cymes, actinomorphic, bisexual or functionally unisexual (*P. hydropiperoides*);

perianth segments 4 or 5, basally connate, dimorphic, the outer 3 larger than the inner 2; stamens 5–8, free or basally connate; styles 2 or 3, free or basally connate. Fruit a lenticular or trigonous achene, often enclosed by the accrescent perianth.

A genus of about 100 species; nearly cosmopolitan. [*Persea*, peach, and *-aria*, to resemble, in reference to the resemblance of the leaves to some species of *Persea* (Rosaceae).]

Antenoron Raf. (1817); *Tovara* Adans., nom. rej. (1763); *Tracaulon* Raf. (1837).

Selected reference: Hinds and Freeman (2005b).

1. Style persistent on the fruit as 2 hooked beaks .... **P. virginiana**
1. Style deciduous, falling before the fruit matures.
  2. Stem retrorsely barbed with small prickles.
    3. Leaf blades broadly hastate to hastate-cordate or triangular; perianth segments 4 ......**P. arifolia**
    3. Leaf blades linear-lanceolate to oblong; perianth segments 5.
      4. Leaves petiolate, the blade base sagittate to cordate; peduncle not stipitate-glandular............
      .... **P. sagittata**
      4. Leaves sessile or subsessile, the blade base cordate to truncate or cuneate; peduncle stipitate-
      glandular, at least distally...**P. meisneriana**
  2. Stem not retrorsely barbed.
    5. Flowers in capitate inflorescences; leaves with amplexicaul basal auricles....................**P. capitata**
    5. Flowers in spiciform inflorescences; leaves lacking amplexicaul basal auricles.
      6. Ocrea eciliate or the cilia less than 0.5 mm long.
        7. Peduncles stipitate-glandular and/or densely setose just below the inflorescence..............
        .... **P. pensylvanica**
        7. Peduncles not stipitate-glandular, either glabrous to sparsely setose just below the
        inflorescence.
          8. Perianth segments with prominent distally bifurcate anchor-shaped veins; fruit con-
          cave on one or more sides; plant annual..**P. lapathifolia**
          8. Perianth segments without anchor-shaped veins; fruit biconvex; plant a rhizomatous
          perennial... **P. glabra**
      6. Ocrea ciliate with bristles 1–12 mm long.
        9. Peduncles and the upper stem stipitate-glandular..**P. careyi**
        9. Peduncles and the upper stem other than stipitate-glandular.
          10. Stem hirsute at least near the inflorescence.
            11. Leaves broadly ovate, the petiole more than 1 cm long; ocrea with a broad herba-
            ceous flange... **P. orientalis**
            11. Leaves linear-lanceolate, the petiole less than 3 mm long; ocrea without a broad
            herbaceous flange...**P. hirsuta**
          10. Stem glabrous or appressed pubescent.
            12. Perianth glandular-dotted.. **P. punctata**
            12. Periath eglandular.
              13. Ocreola strongly overlapping.
                14. Cilia of the ocreola margin as long as the ocreola tube ....**P. cespitosa**
                14. Cilia of the ocreola margin less than 1 mm long and much shorter than the
                ocreola tube.. **P. maculosa**
              13. Ocreola separated or merely reaching the base of the one above.

15. Leaves 1.5–4 cm wide; ocrea cilia 5–15 mm long; trichomes on the ocrea tube flexuous, spreading; flowers white ....**P. setacea**

15. Leaves 0.5–1.5(2) cm wide; ocrea 1–7 mm long; trichomes on the ocrea tube stiff, ascending-appressed; flowers pink or greenish ......**P. hydropiperoides**

Persicaria arifolia (L.) Haraldson [Leaves resembling *Arum* (Araceae).] HALBERDLEAF TEARTHUMB.

Polygonum arifolium Linnaeus, Sp. Pl. 364. 1753. *Tracaulon arifolium* (Linnaeus) Rafinesque, Fl. Tellur. 3: 13. 1837 "1836." *Truellum arifolium* (Linnaeus) Soják, Preslia 46: 145. 1974. *Persicaria arifolia* (Linnaeus) Haraldson, Symb. Bot. Upsal. 22: 72. 1978.

Scandent annual herb, to 1.5 m; stem branching proximally, ribbed, retrorsely barbed. Leaves with the blade broadly hastate to hastate-cordate or triangular, (2)6.5–13(18) cm long, (1)6–11(16) cm wide, the apex acuminate, the base truncate to truncate-cordate, the margin entire, ciliate, sometimes also retrorsely barbed, the upper surface appressed-pubescent, rarely glabrous, the lower surface stellate-pubescent, rarely glabrous, often retrorsely barbed on the midrib, the petiole 1–7 cm long, retrorsely barbed; stipules (ocrea) 8–15 mm long, chartaceous, the margin oblique, ciliate with bristles 0.5–2.5 mm long, the surface glabrous or appressed-spreading pubescent. Flowers in a terminal or axillary capitate or paniculiform cluster, bisexual, the peduncle 1–8 cm long, retrorsely barbed, proximally, stellate-pubescent and stipitate-glandular distally, the glands red or pink, the pedicel 2–3 mm long; perianth segments 4, broadly elliptic, 5–6 mm long, pink or red, often greenish white proximally, the apex acute to obtuse; stamens (6)7. Fruit ovate, 4–6 mm long, lenticular, brown to black, the surface smooth lustrous.

Swamps and marshes along rivers. Rare; Nassau County. Quebec south to Florida, west to Ontario, Minnesota, Missouri, and Georgia, also Louisiana and Washington. Summer–fall.

Persicaria capitata (Burch.-Ham. ex D. Don) H. Gross [With a knob-like head, in reference to the inflorescence.] NEPALESE SMARTWEED.

Polygonum capitatum Buchanan-Hamilton ex D. Don, Prodr. Fl. Nepal. 73. 1825. *Persicaria capitata* (Buchanan-Hamilton ex D. Don) H. Gross, Bot. Jahrb. Syst. 49: 277. 1913. *Cephalophilon capitatum* (Buchanan-Hamilton ex D. Don) Tzvelev, Novosti Sist. Vyss. Rast. 24: 76. 1987.

Procumbent or ascending annual herb, to 0.5 m; stem often rooting at the proximal nodes, glabrous except for fleshy, retrorse, white trichomes at the nodes. Leaves with the blade ovate-deltate, 2–5 cm long, 1–4 cm wide, pinnate-veined, the apex acute, the base truncate to rounded, the margin entire, glabrous or scabrous, the upper surface glabrous, the lower surface pilose and glandular-punctate, the petiole 1–30 mm long, winged at the base; stipules (ocrea) chartaceous, 4–10 mm long, brownish, the margin oblique, eciliate, the surface glabrous or with bristlelike trichomes proximally. Flowers in a terminal or axillary head subtended by clasping leafy bracts, 1–2 per ocreolate fascicle, the peduncle 2–20 mm long, stipitate-glandular distally, the pedicel to 1 mm long; perianth segments (4)5, oblong to elliptic, ca. 3 mm long, white to pink, the apex acute to obtuse; stamens (5)8. Fruit ovate, ca. 2 mm long, lenticular, dark brown to black, the surface striate and minutely punctate, dull.

Disturbed sites. Rare; Alachua and Hillsborough Counties. Escaped from cultivation. Florida, Louisiana, Oregon, and California; Asia, Australia, and Pacific Islands. Native to Asia. Summer.

Persicaria careyi (Olney) Greene [Commemorates John Carey (1797–1880), British trader and botanist who traveled and collected in the United States.] CAREY'S KNOTWEED.

> *Polygonum careyi* Olney, Proc. Providence Franklin Soc. 1: 29. 1847. *Persicaria careyi* (Olney) Greene, Leafl. Bot. Obs. Crit. 1: 24. 1904.

Erect annual herb, to 1.5 m; stem branching distally, ribbed distally, hirsute proximally, stipitate-glandular distally. Leaves with the blade lanceolate, 6–18 cm long, 1–3.5 cm wide, pinnately veined, the apex acuminate, the base cuneate, the margin entire, the upper and lower surfaces sparsely hirsute, hirsute and sometimes stipitate-glandular on the veins, the petiole 0.5–1.5 cm long or the leaves sometimes sessile, hirsute; stipules (ocrea) 8–20 mm long, chartaceous, the margin truncate, ciliate with bristles 2–7 mm long, the surface spreading-hirsute, rarely stipitate-glandular. Flowers in terminal and axillary loosely branched, drooping, spiciform inflorescence, 2–8 per ocreolate fascicle, bisexual, the peduncle stipitate-glandular, the pedicels 1–4 mm long; perianth segments 5, obovate, the apex obtuse, 2.5–3 mm long, pink or rose; stamens 5 (or 8). Fruit ovoid or obovoid, ca. 2 mm long, dark brown to black, thickly biconvex, the surface smooth, lustrous.

Disturbed sites. Rare; Leon County. Not recently collected. Quebec south to West Virginia, west to Minnesota and Illinois, also Florida. Native to the north of our area. Summer–fall.

Persicaria cespitosa (Blume) Nakai var. **longiseta** (Bruijn) C. F. Reed [Long-bristled, in reference to the ocrea.] ORIENTAL LADY'S-THUMB.

> *Polygonum longisetum* Bruijn, in Miquel, Pl. Jungh. 307. 1854. *Polygonum donii* Meisner var. *longisetum* (Bruijn) Miquel, Fl. Ned. Ind. 1: 1000. 1855. *Polygonum blumei* Meisner var. *longisetum* (Bruijn) Meisner, in Miquel, Ann. Mus. Bot. Lugduno-Batavi 2: 58. 1865. *Polygonum cespitosum* Blume var. *longisetum* (Bruijn) Steward, Contr. Gray Herb. 88: 67. 1930. *Persicaria longiseta* (Bruijn) Kitagawa, Rep. Inst. Sci. Res. Manchoukuo 1: 322. 1937. *Persicaria cespitosa* (Blume) Nakai var. *longiseta* (Bruijn) C. F. Reed, Phytologia 63: 410. 1987.

Decumbent to ascending annual herb, to 80 cm; stem branching, not noticeably ribbed, glabrous. Leaves with the blade ovate-lanceolate to linear-lanceolate, 2–8 cm long, 1–3 cm wide, pinnate-veined, the apex acute to accuminate, the base cuneate, the margin entire, antrorsely strigose, the upper surface glabrous or sparsely strigose along the veins and the margin, the lower surface glabrous or sparsely strigose along the veins, the petiole 1–3(6) mm long, glabrous, the leaves sometimes sessile; stipules (ocrea) 5–12 mm long, chartaceous, the margin truncate, ciliate with bristles 4–12 mm long, the surface glabrous or strigulose. Flowers in a terminal or sometimes axillary spiciform inflorescence, 1–5 per ocreolae fascicle, bisexual, the peduncle 1–5 cm long, the pedicel 1–2 mm long; perianth segments 5, obovate, 2–3 mm, dull rose-purple with paler lobe margins, the apex obtuse to rounded; stamens 5.

Fruit ovate, ca. 2 mm long, dark brown to black, trigonous, the surface smooth, lustrous. Moist, disturbed sites. Rare; central and western panhandle. Quebec south to Florida, west to Minnesota, Nebraska, Kansas, and New Mexico, also Alberta and British Columbia; Europe and Asia. Native to Asia. Spring–fall.

Persicaria glabra (Willd.) M. Gómez [Glabrous.] DENSEFLOWER KNOTWEED.

Polygonum glabrum Willdenow, Sp. Pl. 447. 1799. *Persicaria glabra* (Willdenow) M. Gómez de la Maza y Jiménez, Anales Inst. Segunda Enseñ. 2: 278. 1896.

Polygonum densiflorum Meisner, in Martius, Fl. Bras. 5(1): 13. 1855. *Persicaria densiflora* (Meisner) Moldenke, Torreya 34: 7. 1934.

Polygonum densiflorum Meisner var. *imberbe* Meisner, in A. de Candolle, Prodr. 14: 121. 1856. *Polygonum densiflorum* Meisner forma *imberbe* (Meisner) Fassett, Caldasia 4: 226. 1946.

Polygonum pensylvanicum Linnaeus var. *densiflorum* A. W. Wood, Amer. Bot. Fl. 283. 1873.

Polygonum portoricense Bertero ex Small, Mem. Dept. Bot. Columbia Coll. 1: 46. 1895. *Persicaria portoricensis* (Bertero ex Small) Small, Fl. S.E. U.S. 377, 1330. 1903.

Decumbent to erect perennial herb, to 1.5 m; stem usually branched distally, not noticeably ribbed, rooting at the proximal nodes, reddish brown to purple glabrous or rarely pubescent distally, sometimes glandular-punctate. Leaves with the blade lanceolate to oblong lanceolate, (10)15–30 cm long, 2–5 cm wide, pinnate-veined, the apex acute to acuminate, the base cuneate, the margin entire, glabrous or antrorsely strigose, the upper and lower surfaces glabrous or scabrous along the midvein, sometimes glandular-punctate, the petiole 0.5–2 cm long, scabrous; stipules (ocrea) 12–23 mm long, chartaceous, the margin truncate, eciliate, the surface glabrous, usually obscurely glandular punctate. Flowers in a terminal or sometimes an axillary, erect to slightly nodding spiciform inflorescence, (1)3–8 per ocreolate fascicle, the peduncle glabrous or scabrid, glandular-punctate, the pedicel 2–5 mm long; perianth segments 5, obovate, 3–4 mm long, greenish white to white or pink, glandular dotted, the apex obtuse or rounded; stamens 5–7. Fruit broadly oval to orbicular, ca. 2 mm long, lenticular, biconvex, dark brown to nearly black, the surface smooth, lustrous.

Swamps, marshes, and shallow water of streams. Frequent; nearly throughout. New Jersey and Pennsylvania south to Florida, west to Kansas and Texas; West Indies, Central America, and South America; Africa, Asia, Australia, and Pacific Islands. Native to North America, West Indies, Central America, South America, Africa, Asia, and Pacific Islands. Spring–fall.

Persicaria hirsuta (Walter) Small [Coarsely pubescent.] HAIRY SMARTWEED.

Polygonum hirsutum Walter, Fl. Carol. 132. 1788. *Persicaria hirsuta* (Walter) Small, Fl. S.E. U.S. 379, 1330. 1903.

Polygonum hirsutum Walter var. *dasyphyllum* Meisner, in A. de Candolle, Prodr. 14: 103. 1856.

Polygonum hirsutum Walter var. *glabrescens* Meisner, in A. de Candolle, Prodr. 14: 103. 1856. *Persicaria hirsuta* (Walter) Small var. *glabrescens* (Meisner) Small, Fl. S.E. U.S. 379, 1330. 1903. TYPE: FLORIDA: 1843–1845, *Rugel 86* (holotype: ?).

Decumbent to ascending perennial herb, to 9 dm; stem branched, rooting at the proximal nodes, not noticeably ribbed, brownish hirsute at the nodes. Leaves with the blade ovate to lanceolate, (2)4–8 cm long, (0.5)1–2.5 cm wide, pinnate-veined, the apex acute to acuminate, the base truncate to slightly cordate, the margin entire, the upper and lower surfaces strigose to hirsute, the petiole 1–3 mm long or the blade sessile; stipules (ocrea) 6–12 mm long, chartaceous, the margin truncate, ciliate with bristles 4–8 mm long, the surface hirsute. Flowers in a terminal, erect, spiciform inflorescence, 1–3 per ocreolate fascicle, the peduncle 3–6 cm long, hirsute, at least proximally, the pedicel 1–2 mm long; perianth segments 5, obovate, ca. 2 mm long, pink, occasionally whitish, the apex obtuse or rounded; stamens 5. Fruit broadly ovate, ca. 2 mm long, trigonous, dark brown to brownish black, the surface smooth, lustrous.

Swamps, lake shores, and pond margins. Occasional; northern and central peninsula west to central panhandle. North Carolina south to Florida, west to Mississippi. Spring–fall.

Persicaria hydropiperoides (Michx.) Small [Resembling *Polygonum hydropiper*.] MILD WATERPEPPER; SWAMP SMARTWEED.

Polygonum hydropiperoides Michaux, Fl. Bor.-Amer. 1: 239. 1803. *Polygonum mite* Persoon, Syn. Pl. 1: 440. 1805, nom. illegit. *Persicaria hydropiperoides* (Michaux) Small, Fl. S.E. U.S. 378, 1330. 1903.

Polygonum virgatum Chamisso & Schlechtendal, Linnaea 3: 45: 1828; non Loiseleur Deslongchamps, 1827. *Polygonum hydropiperoides* Michaux var. *virgatum* Meisner, in A. de Candolle, Prodr. 14: 103. 1856.

Polygonum opelousanum Riddell, New Orleans Med. Surg. J. 9: 611. 1853. *Persicaria opelousana* (Riddell) Small, Fl. S.E. U.S. 378, 1330. 1903. *Polygonum hydropiperoides* Michaux subsp. *opelousanum* (Riddell) W. Stone, Pl. S. New Jersey 422. 1912 ("1911").

Polygonum hydropiperoides Michaux var. *macerum* Stanford, Rhodora 28: 26. 1926. TYPE: FLORIDA: Pinellas Co.: near St. Petersburg, 28 Oct 1907, *Deam 2987* (holotype: GH).

Polygonum hydropiperoides Michaux var. *sanibelense* Stanford, Rhodora 28: 27. 1926. TYPE: FLORIDA: Lee Co.: Sanibel Island, 17 May 1901, *Tracy 7547* (holotype: GH).

Decumbent to ascending perennial herb, to 1 m; stem usually branched, not noticeably ribbed, glabrous or obscurely strigose distally, rooting at the nodes proximally. Leaves with the blade broadly lanceolate to linear-lanceolate, 5–25 cm long, 0.5–3.5 cm wide, pinnate-veined, the apex acuminate, the base cuneate, the margin entire, the upper and lower surfaces strigose or glabrous, sometimes punctate on the lower surface, the petiole 2–20 mm long, glabrous or strigose; stipules (ocrea) 5–23 mm long, chartaceous, brown, the margin truncate, ciliate with bristles (2)4–10 cm long, the surface glabrous or strigose. Flowers in a terminal or sometimes an axillary spiciform, erect inflorescence, 2–6 per ocreolate fascicle, bisexual or occasionally some unisexual and staminate, the peduncle 1–3 cm long, glabrous or strigose, the pedicel 1–2 mm long; perianth segments 5, obovate, 3–4 mm long (2–3 mm in staminate flowers), roseate proximally, roseate, white, or greenish white distally, the apex obtuse or rounded; stamens 8. Fruit ovate, 2–3 mm long, trigonous, brown to brownish-black, the surface smooth, lustrous.

Marshes, swamps, cypress domes, lake shores, and pond margins. Common; nearly throughout. Nearly throughout North America; West Indies, Mexico, Central America, and South America. Spring–fall.

Persicaria lapathifolia (L.) Delarbre [Leaves like *Lapathum* (=*Rumex*).] CURLYTOP KNOTWEED; PALE SMARTWEED.

Polygonum lapathifolium Linnaeus, Sp. Pl. 360. 1753. *Persicaria lapathifolia* (Linnaeus) Delarbre, Fl. Auvergyne, ed. 2. 519. 1800. *Polygonum persicaria* Linnaeus var. *lapathifolium* (Linnaeus) Meisner, Monogr. Polyg. 69. 1826. *Discolenta lapathifolia* (Linnaeus) Rafinesque, Fl. Tellur. 3: 15. 1837 ("1838"). *Persicaria lapathifolia* (Linnaeus) Gray subsp. *genuina* H. Goss, Bot. Centralbl. 37(2): 113. 1919, nom. inadmiss.

Polygonum incarnatum Elliott, Sketch Bot. S. Carolina 1: 456. 1817. *Polygonum nodosum* Persoon var. *incarnatum* (Elliott) A. Gray, Manual, ed. 2. 372. 1856. *Polygonum lapathifolium* Linnaeus var. *incarnatum* (Elliott) S. Watson, in A. Gray, Manual, ed. 6. 440. 1890. *Persicaria incarnata* (Elliott) Small, Fl. S.E. U.S. 377, 1330. 1903.

Persicaria paludicola Small, Man. S.E. Fl. 457, 1504. 1933. TYPE: FLORIDA: Miami-Dade Co.: Everglades, Camp Jackson to Camp Longview, 21–23 Feb 1911, *Small et al. 3494* (holotype: NY).

Ascending to erect annual herb, to 1.5 m; stem simple or branched, sometimes rooting at the proximal nodes, slightly ribbed, glabrous or rarely strigose distally, sometimes glandular-punctate or stipitate-glandular distally. Leaves with the blade linear-lanceolate to oblong-ovate, 4–12(22) cm long, 0.5–4(6) cm wide, pinnate-veined, the apex acuminate, the base cuneate, the margin entire, antrorsely scabrous, the upper and lower surfaces strigose on the main nerves, especially on the lower surface, sometimes glandular-punctate on the lower surface, the petiole 1–16 mm long, strigose or glabrous; stipules (ocrea) 4–24(35) mm long, chartaceous, brownish, the margin truncate, eciliate or ciliate with bristles to 1 mm long, the surface glabrous, rarely strigose. Flowers in a terminal or sometimes axillary, usually nodding, spiciform inflorescence, 4–14 per fascicle, bisexual, the peduncle 2–25 mm long, glabrous or stipitate-glandular, the pedicel 0.5–2 mm long; perianth segments 4(5), obovate to elliptic, 2–3 mm long, greenish white or pink, the outer ones with prominent bifurcate, anchor-shaped veins, the apex obtuse to rounded; stamens 5–6. Fruit ovate, 2–3 mm long, lenticular or trigonous, concave on 1 or more sides, dark brown to black, the surface smooth, lustrous or dull.

Marshes, wet clearings, and stream margins. Occasional; peninsula, central and western panhandle. Nearly throughout North America; West Indies, Mexico, Central America, and South America; Europe, Africa, Asia, Australia, and Pacific Islands. Native to North America. Summer–fall.

Persicaria maculosa Gray [Spotted, in reference to the purplish spot near the center of the leaf blade.] SPOTTED LADY'S-THUMB.

Polygonum persicaria Linnaeus, Sp. Pl. 361. 1753. *Polygonum ruderale* Salisbury, Prodr. Stirp. Chap. Allerton 259. 1796, nom. illegit. *Persicaria mitis* Delarbre, Fl. Auvergne, ed. 2. 518. 1800, nom. rej. *Persicaria maculosa* Gray, Nat. Arr. Brit. Pl. 2: 269. 1821, nom. cons. *Polygonum persicaria* Linnaeus var. *vulgare* Meisner, Monogr. Polyg. 68. 1826, nom. inadmiss. *Persicaria vulgaris* Webb & Moquin-Tandon ex Webb & Berthelot, Hist. Nat. Iles Canaries 3(2(3)): 219. 1846, nom. illegit. *Persicaria persicaria* (Linnaeus) Small, Fl. S.E. U.S. 378, 1330. 1903, nom. inadmiss. *Persicaria ruderalis* C. F. Reed, Phytologia 50: 461. 1982, nom. illegit.

Procumbent or ascending to erect annual herb, to 1 m; stem simple or branched, often rooting at the proximal nodes, glabrous or strigose. Leaves with the blade lanceolate to linear-lanceolate,

3–14 cm long, 4–15 mm wide, pinnate-veined, often with a prominent triangular- or crescent-shaped purplish spot near the middle, apex acute to acuminate, the base cuneate, the margin entire, the upper and lower surface glabrous or strigose, especially along the midvein, sometimes the lower surface glandular-punctate, the margin antrorsely strigose; stipules (ocrea) 4–10(15) mm long, chartaceous, light brown, the margin truncate, ciliate with bristles 1–5 mm long, the surface glabrous or strigose. Flowers in a terminal or axillary erect spiciform inflorescence, 4–14 per ocreolate fascicle, bisexual, the peduncle 1–5 cm long, glabrous or sometimes pubescent, the pedicel 1–2 mm long; perianth segments 4–5, obovate, 2–4 mm long, pink to greenish pink, the apex obtuse to rounded, glabrous or sometimes with a few superficial resinous glands; stamens 4–8. Fruit ovate, 2–3 mm long, lenticular or 3-angled, the surface smooth, lustrous.

Disturbed sites. Rare; Duval, St. Johns, Lee, and Miami-Dade Counties. Not recently collected. Nearly throughout North America; Europe, Africa, Asia, and Pacific Islands. Native to Europe and Asia. Spring–fall.

Persicaria meisneriana (Cham. & Schldl.) M. Gómez var. **beyrichiana** (Cham. & Schldl.) C. C. Freeman. [Commemorates Carl Daniel Friedrich Meisner (1800–1874), Swiss botanist; Heinrich Karl Beyrich (1796–1834), German botanist who collected in the Americas.] BRANCHED TEARTHUMB; MEXICAN TEARTHUMB.

Polygonum beyrichianum Chamisso & Schlechtendal, Linnaea 3: 42. 1828. *Polygonum meisnerianum* Chamisso & Schlechtendal var. *beyrichianum* (Chamisso & Schlechtendal) Meisner, in Martius, Fl. Bras. 5(1): 19. 1855. *Tracaulon beyrichianum* (Chamisso & Schlechtendal) Small, Fl. S.E. U.S. 380, 1330. 1903. *Persicaria meisneriana* (Chamisso & Schlechtendal) M. Gómez de la Maza y Jiménez var. *beyrichiana* (Chamisso & Schlechtendal) C. C. Freeman, Sida 21: 291. 2004.

Erect or scandent annual herb, to 1 m; stem ribbed, glabrous or pubescent, retrorsely barbed and with stipitate-glandular trichomes. Leaves with the blade linear-lanceolate to lanceolate, 6–14 cm long, 0.8–1.5 cm wide, pinnate-veined, the apex acuminate, the base cordate to truncate or cuneate, the margin entire, ciliate, the upper and lower surfaces glabrous, rarely strigose with eglandular or with glandular trichomes, retrorsely barbed on the main veins, sessile or the petiole 1–5 mm long; stipules (ocrea) 8–20 mm long, chartaceous, brown, the margin oblique, eciliate or ciliate with bristles to 0.5 mm long, the surface glabrous or rarely glandular-pubescent, sometimes with prickles. Flowers in a terminal or axillary subcapitate or racemiform inflorescence, 2–3 per ocreolate fascicle, bisexual, the peduncle 1–8 cm long, stipitate-glandular-pubescent (at least distally), often with retrorse prickles proximally, 2–3 per fascicle, the pedicel 2–4 cm long; perianth segments 5, elliptic to ovate, 2–3 mm long, white to pink, the apex obtuse; stamens 5. Fruit ovate, 3–4 mm long, 3-gonous, dark brown, the surface smooth, lustrous.

Swamps, wet hammocks, and lake margins. Rare; Alachua, Jefferson, and Leon Counties. South Carolina, Georgia, Florida, Louisiana, and Texas; West Indies, Mexico, Central America, and South America; Africa. Summer–fall.

Persicaria meisneriana var. *beyrichiana* (as *Polygonum meisnerianum*) is listed as endangered in Florida (Florida Administrative Code, Chapter 5B-40).

Persicaria orientalis (L.) Spach [Of the Orient.] KISS-ME-OVER-THE-GARDEN-GATE; PRINCESS-FEATHER.

Polygonum orientale Linnaeus, Sp. Pl. 362. 1753. *Polygonum altissimum* Moench, Methodus 630. 1794, nom. illegit. *Polygonum speciosum* Salisbury., Prodr. Stirp. Chap. Allerton 259. 1796, nom. illegit. *Heptarina orientalis* (Linnaeus) Rafinesque, Fl. Tellur. 3: 15. 1837 ("1836"). *Persicaria orientalis* (Linnaeus) Spach, Hist. Nat. Vég. 10: 537. 1841. *Lagunea orientalis* (Linnaeus) Nakai, J. Jap. Bot. 18: 112. 1842.

Erect annual herb, to 2.5 m; stem simple or branched distally, terete or ribbed, usually ribbed, pilose to hispid distally, strigose or glabrescent proximally. Leaves with the blade ovate, 6–25(30) cm long, 3–17 cm wide, pinnate-veined, the apex acuminate, the base cordate to truncate, the margin entire, scabrous to ciliate, the upper and lower surfaces minutely strigose to densely tomentose, especially along the veins on the lower surface, the petiole 1–8.5(14) cm long, densely pilose to hispid; stipules (ocrea) 1–2 cm long, chartaceous and brown proximally, green and foliaceous distally, the margin truncate, ciliate with bristles 1–3 mm long, the surface densely strigose to hispid. Flowers in a terminal or axillary nodding or erect spiciform inflorescence, (1)2–5 per ocreolate fascicle, bisexual, the peduncle 2–10 cm long, hirsute, the pedicel 1–4 mm long; perianth segments 5, 3–5 mm long, pink to red; stamens 6–8. Fruit suborbicular, 3–4 mm long, lenticular, dark brown to black, the surface smooth or minutely granulate, shiny or dull.

Disturbed sites. Rare; Hernando County. Escaped from cultivation. Quebec south to Florida, west to Minnesota, Nebraska, Kansas, Oklahoma, and Texas, also Washington, Oregon, and California; Europe, Asia, and Australia. Native to Asia. Summer–fall.

Persicaria pensylvanica (L.) M. Gómez [Of Pennsylvania.] PENNSYLVANIA SMARTWEED.

Polygonum pensylvanicum Linnaeus, Sp. Pl. 362. 1753. *Polygonum scabrum* Moench, Suppl. Meth. 267. 1802, nom. illegit; non Moench, 1794. *Persicaria pensylvanica* (Linnaeus) M. Gómez de la Maza y Jiménez, Anales Inst. Segunda Enseñ. 2: 278. 1896. *Polygonum pensylvanicum* Linnaeus var. *genuinum* Fernald, Rhodora 19: 72. 1917, nom. inadmiss.

Polygonum bicorne Rafinesque, Fl. Ludov. 29. 1817. *Persicaria bicornis* (Rafinesque) Nieuwland, Amer. Midl. Naturalist 3: 201. 1914.

Polygonum longistylum Small, Bull. Torrey Bot. Club 21: 169. 1894. *Persicaria longistyla* (Small) Small, Fl. S.E. U.S. 377, 1330. 1903.

Polygonum pensylvanicum Linnaeus var. *durum* Stanford, Rhodora 27: 180. 1925. *Persicaria pensylvanica* (Linnaeus) M. Gómez de la Maza y Jiménez var. *dura* (Stanford) C. F. Reed, Phytologia 50: 461. 1982. TYPE: FLORIDA: Gadsden Co.: river bottom at Chattahoochee Landing, 12 Sep 1884, *Curtiss s.n.* (holotype: GH).

Ascending to erect or rarely decumbent annual herb, to 2 m; stem simple or branched, sometimes rooting at the proximal nodes, ribbed, glabrous or strigose to pubescent distally, sometimes stipitate-glandular distally. Leaves with the blade lanceolate, 3–20 cm long, 0.5–5 cm wide, pinnate-veined, the apex acute to acuminate, the base cuneate, the margin entire, the upper surface glabrous, sometimes glandular-punctate, the lower surface glabrous or strigose, sometimes glandular-punctate, the petiole 1–2(3) cm long, glabrous or strigose; stipules (ocrea)

5–20 mm long, chartaceous, brownish, the margin truncate, eciliate or with bristles to 1 mm long, the surface glabrous or strigose. Flowers in a terminal or axillary racemiform inflorescence, 2–14 per ocreolate fascicle, bisexual, the peduncle 1–6 cm long, glabrous or pubescent, usually stipitate-glandular, the pedicel 2–5 mm long; perianth segments 5, obovate to elliptic, 3–5 mm long, pink or occasionally greenish, the apex obtuse or rounded; stamens 6–8, sometimes heterostylic and either the stamens or styles conspicuously exserted from the calyx. Fruit broadly oval to subreniform, 2–3 mm long, lenticular or rarely 3-gonous, brown to black, the surface smooth, lustrous.

Floodplain forests, marshes, and stream banks. Occasional; nearly throughout. Nearly throughout North America; South America; Europe. Native to North America. Summer–fall.

Persicaria bicorne is recognized by some authors (e.g., Hinds and Freeman, 2005b) based on its heterostylic flowers, the achenes usually with a prominent hump in the center of one side of the fruit, and narrower leaves. We have not found this to be convincing evidence for recognition as a distinct species.

Persicaria punctata (Elliott) Small [Dotted.] DOTTED SMARTWEED.

Polygonum punctatum Elliott, Sketch Bot. S. Carolina 1: 455. 1817. *Peutalis punctata* (Elliott) Rafinesque, New Fl. 4: 49. 1938. *Discolenta punctata* (Elliott) Rafinesque, New Fl. 4: 49. 1838, nom. Alt. *Persicaria punctata* (Elliott) Small, Fl. S.E. U.S. 379, 1330. 1903. *Polygonum punctatum* Elliott var. *typicum* Fassett, Brittonia 6: 371. 1949, nom. inadmiss.

Polygonum acre Kunth, in Humboldt et al., Nov. Gen. Sp. 2: 179. 1817; non Lamarck, 1779. *Persicaria acris* M. Gómez de la Maza y Jiménez, Anales Inst. Segunda Enseñ. 2: 277. 1896.

Polygonum antihaemorrhoidale Martius forma *aquatile* Martius, Reise Bras. 2: 550. 1828. *Polygonum antihaemorrhoidale* Martius var. *aquatile* (Martius) Martius, Linnaea 5(Litt.-Ber.): 41. 1830. *Polygonum acre* Kunth var. *aquatile* (Martius) Meisner, in Martius, Fl. Bras. 5(1): 18. 1855. *Polygonum acre* Kunth var. *leptostachyum* Meisner, in A. de Candolle, Prodr. 14: 108. 1856, nom. illegit. *Polygonum punctatum* Elliott var. *leptostachyum* Small, Bull. Torrey Bot. Club 19: 356. 1892, nom. illegit. *Persicaria punctata* (Elliott) Small var. *leptostachya* Small, Fl. SE. U.S. 379, 1330. 1903, nom. illegit. *Polygonum punctatum* Ell. var. *aquatile* (Martius) Fassett, Caldasia 4: 221. 1946.

Polygonum acre Kunth var. *brachystachyum* Meisner, in A. de Candolle, Prodr. 14: 108. 1856. TYPE: FLORIDA: Gadsden Co.: Aspalaga, s.d., *Rugel s.n.* (holotype: ?).

Polygonum acre Kunth var. *confertiflorum* Meisner, in A. de Candolle, Prodr. 14: 108. 1856. *Persicaria acris* M. Gómez de la Maza y Jiménez var. *confertiflora* (Meisner) H. Gross ex Loesener, Repert. Spec. Nov. Regni Veg. 12: 218. 1913. *Polygonum punctatum* Elliott var. *confertiflorum* (Meisner) Fassett, Brittonia 6: 377. 1949. *Persicaria punctata* (Elliott) Small var. *confertiflora* (Meisner) D. Löve & Bernard, Svensk Bot. Tidskr. 53: 391. 1959.

Polygonum acre Kunth var. *majus* Meisner, in A. de Candolle, Prodr. 14: 108. 1856. *Polygonum punctatum* Elliott var. *majus* (Meisner) Fassett, Brittonia 6: 373. 1949. TYPE: FLORIDA: Wakulla Co.: St. Marks River, s.d., *Rugel 615* (holotype: NY).

Polygonum punctatum Elliott var. *robustius* Small, Bull. Torrey Bot. Club 21: 477. 1894. *Persicaria punctata* (Elliott) Small var. *robustior* (Small) Small, Fl. S.E. U.S. 379, 1330. 1903. *Persicaria robustior* (Small) E. P. Bicknell, Bull. Torrey Bot. Club 36: 455. 1909. *Polygonum punctatum* Elliott subsp. *robustius* (Small) W. Stone, Pl. S. New Jersey 423. 1912 ("1911"). *Polygonum acre* Kunth var. *robustius* (Small) Farwell, Rep. (Annual) Michigan Acad. Sci. 20: 174. 1919. *Polygonum robustius* (Small) Fernald, Rhodora 23: 147. 1921.

Ascending to erect annual or perennial herb, to 2 m; stem branched, rooting at the proximal nodes, inconspicuously ribbed, glabrous, glandular-punctate. Leaves with the blade lanceolate to lanceolate-ovate or subrhombic, 4–20 cm long, 0.6–2.4(4.5) cm wide, pinnate-veined, the apex acute to acuminate, the base cuneate, the margin entire, the upper and lower surfaces glabrous or scabrous on the main veins, glandular-punctate, the margin antrorsely strigose, the petiole 2–20 mm long; stipules (ocrea) (4)9–18 mm long, chartaceous, the margin truncate, ciliate with bristles 2–12 mm long, the surface glabrous or strigose, glandular punctate. Flowers in a terminal or axillary, erect, spiciform inflorescence, 2–6 per ocreate fascicle, bisexual, the peduncle 3–6 cm long, the pedicel 1–5 mm long; perianth segments 5, obovate, 3–4 mm long, white to greenish white, rarely tinged with pink, glandular-punctate, the apex obtuse to rounded; stamens 6–8. Fruit ovate, 2–3 mm long, 3-gonous or lenticular, dark brown to brownish black, the surface smooth, lustrous.

Freshwater and brackish marshes, swamps, and wet ditches. Common; nearly throughout. Nearly throughout North America; West Indies, Mexico, Central America, and South America; Pacific Islands. Spring–fall.

Fassett (1949) subdivided the species into 12 varieties and two forms which most recent authors have not recognized. Most recently, Hinds and Freeman (2005b) recognized *Persicaria punctata* with no varieties, but retained *Persicaria robustior* as a distinct species. Since there is great overlap in the characters used by Hinds and Freeman, we do not support the latter, even at the varietal level.

Persicaria sagittata (L.) H. Gross ex Nakai [Arrowhead-shaped.] ARROWLEAF TEARTHUMB.

Polygonum sagittatum Linnaeus, Sp. Pl. 363. 1753. *Helxine sagittatum* (Linnaeus) Rafinesque, Fl. Tellur. 3: 10. 1837 ("1836"). *Polygonum sagittatum* Linnaeus var. *typicum* R. Keller, Bull. Soc. Roy. Bot. Belgique 30(2): 46. 1891, nom. inadmiss. *Tracaulon sagittatum* (Linnaeus) Small, Fl. S.E. U.S. 381, 1330. 1903. *Persicaria sagittata* (Linnaeus) H. Gross ex Nakai, Fl. Saishu & Kwan Isl. 41. 1914. *Truellum sagittatum* (Linnaeus) Soják, Preslia 46: 149. 1974.

Polygonum sagittatum Linnaeus var. *boreale* Meisner, Monogr. Polyg. 65. 1826.

Polygonum sagittatum Linnaeus var. *americanum* Meisner, in A. de Candolle, Prodr. 14: 132. 1856. *Persicaria americana* (Meisner) Ohki, Bot. Mag. (Tokyo) 40: 55. 1926. *Persicaria sagittata* (Linnaeus) H. Gross ex Nakai var. *americana* (Meisner) Miyabe, J. Fac. Agric. Hokkaido Univ. 26: 520. 1934.

Scandent annual herb, to 2 m; stem ribbed, retrorsely barbed. Leaves with the blade broadly lanceolate to oblong, 2–8 cm long, 1–3 cm wide, pinnate-veined, the apex obtuse to acute, the base sagittate to cordate, the margin entire, ciliate or eciliate, the upper surface glabrous or strigose, the lower surface glabrous or strigose, usually retrorsely barbed on the midrib, the petiole 0.5–5 cm long, glabrous or strigose, retrorsely barbed; stipules (ocrea) (3)5–13 mm long, chartaceous, the margin oblique, eciliate or ciliate with bristles to 1 mm long, the surface glabrous. Flowers in a terminal or axillary capitate cluster, 2–3 per ocreate fascicle, bisexual, the peduncle 1–8 cm long, glabrous, sometimes with retrorse prickles proximally, the pedicel 1–2 mm long; perianth segments 5, elliptic, 3–5 mm long, white, greenish white, or pink, the apex obtuse; stamens 8. Fruit ovate, 3–4 mm long, 3-gonous, brown to black, the surface smooth or minutely punctate, dull or lustrous.

Marshes, wet hammocks, and stream banks. Rare; Nassau, Liberty, and Franklin Counties. Newfoundland to Florida, west to Saskatchewan, North Dakota, South Dakota, Nebraska, Colorado, Oklahoma, and Texas, also Oregon; Asia. Summer–fall.

Persicaria setacea (Baldwin) Small [Bristly.] BOG SMARTWEED.

Polygonum setaceum Baldwin, in Elliott, Sketch Bot. S. Carolina 1: 455. 1817. *Persicaria setacea* (Baldwin) Small, Fl. S.E. U.S. 379, 1330. 1903. *Polygonum hydropiperoides* Michaux var. *setaceum* (Baldwin) Gleason, Phytologia 4: 23. 1952.

Ascending to erect perennial herb, to 1.5 m; stem branched distally, rooting at the proximal nodes, slightly ribbed, glabrous or loosely appressed- to spreading-hirsute near the nodes. Leaves with the blade lanceolate, (3)6–15(18) cm long, (1.5)2–3(4.5) cm wide, pinnate-veined, the apex acute to acuminate, the base cuneate to truncate, the margin entire, the upper and lower surfaces hirsute, the petiole 1–5 mm long, hirsute, the leaves sometimes sessile; stipules (ocrea) 1–2 cm long, chartaceous, brown, the margin truncate, ciliate with bristles 6–12 mm long, the surface strigose and with loosely antrorse to spreading trichomes (at least proximally). Flowers terminal or sometimes axillary in an erect, spiciform inflorescence, 1–5 per ocreate fascicle, bisexual, the peduncle 1–7 cm long strigose, the pedicel 1–3 mm long; perianth segments 5, obovate, 2–3 mm long, the apex obtuse to rounded, green proximally, cream to tan distally, occasionally pink-tinged; stamens 5. Fruit ovate, 2–3 mm long, trigonous, brown to black, the surface smooth, lustrous.

Floodplain forests, lake margins, and marshes. Frequent; nearly throughout. New York and Massachusetts south to Florida, west to Michigan, Illinois, Kansas, Oklahoma, and Texas, also Washington. Spring–fall.

Persicaria virginiana (L.) Gaertn. [Of Virginia.] JUMPSEED.

Polygonum virginianum Linnaeus, Sp. Pl. 360. 1753. *Persicaria virginiana* (Linnaeus) Gaertner, Fruct. Sem. Pl. 2: 180. 1790. *Polygonum muticum* Moench, Suppl. Meth. 266. 1802, nom. illegit. *Tovara virginiana* (Linnaeus) Rafinesque, Fl. Tellur. 3: 12. 1837 ("1836"). *Sunania virginiana* (Linnaeus) H. Hara, J. Jap. Bot. 37: 330. 1962. *Antenoron virginianum* (Linnaeus) Roberty & Vautier, Boissiera 10: 35. 1964.

Tovara virginiana (Linnaeus) Rafinesque var. *glaberrima* Fernald, Rhodora 39: 404. 1937. *Polygonum virginianum* Linnaeus var. *glaberrimum* (Fernald) Steyermark, Rhodora 62: 130. 1960. *Sunania virginiana* (Linnaeus) H. Hara var. *glaberrima* (Fernald) H. Hara, J. Jap. Bot. 37: 330. 1962. *Antenoron virginianum* (Linnaeus) Roberty & Vautier var. *glaberrimum* (Fernald) H. Hara, J. Jap. Bot. 40: 32. 1965.

Erect perennial herb, to 6(13) dm; stem rhizomatous, simple or sparingly branched distally, ribbed, glabrous or strigose. Leaves with the blade broadly lanceolate to ovate, 5–16 cm long, 2–10 cm wide, pinnate-veined, the apex acute to acuminate, the base acute to rounded, the margin entire, the upper surface strigose and scabrous, the lower surface pubescent, at least on the main veins; petiole 1–20 mm long or the leaf sometimes sessile; stipules (ocrea) 1–2 cm long, chartaceous, brownish, the margin truncate, ciliate with bristles 0.5–4 mm long, the surface strigose to tomentose. Flowers terminal or axillary in a spiciform inflorescence, 1–3 per ocreate fascicle, bisexual, the peduncle 1–7 cm long, pubescent or glabrous distally, the pedicel ca. 1 mm long; perianth segments 4, elliptic to obovate, 3–4 mm long, greenish white, white, or sometimes pink, the apex acute to acuminate; stamens 5. Fruit ovate, ca. 4 mm long, lenticular, brown, the styles persistent, bifid, reflexed distally, the surface smooth to rugose, dull or lustrous.

Floodplain forests, moist hammocks, and seepage areas. Occasional; northern peninsula south to Hernando County, central panhandle. Quebec south to Florida, west to Minnesota, Nebraska, Kansas, Oklahoma, and Texas; Mexico. Summer–fall.

The persistent hooked style bifid to the base and the 4-lobed perianth make this species very distinctive, resulting in its sometimes being placed in the segregate genus *Tovara* Adanson or *Antenoron* Raf. However, Haraldson (1978) and Ronse Decraene and Akeroyd (1988) demonstrate that it is closely allied with *Polygonum* section *Persicaria*. *Persicaria* is treated at the generic level with *Tovara* as a section by these authors and Hinds and Freeman (2005b).

EXCLUDED TAXA

Persicaria amphibia (L.) Delarbre—Reported by Chapman (1897, as *Polygonum muhlenbergii* S. Watson) and Small (1903, 1913, both as *Persicaria emersa* (Michaux)Small; 1933, as *Persicaria muhlenbergii* S. Watson) as "throughout N.A.," which would presumably include Florida. Also reported by Correll & Johnston (1970, as *Persicaria coccinea* (Muhlenberg ex Willdenow) Greene) as "throughout N.A.," which would presumably include Florida and Godfrey and Wooten (1981, as *Polygonum amphibium* L. var. *emersum* Michaux). No Florida specimens known. Excluded from Florida by Clewell (1985), Wunderlin (1998, 2003), and Hinds and Freeman (2005b).
Persicaria hydropiper (Linnaeus) Spach—Reported from Florida by Small (1933), Correll and Johnston (1970) as "throughout N.A.," which would presumably include Florida, and Hinds and Freeman (2005b). No Florida specimens known. Florida material is probably based on a misidentification of *P. punctata*, with which it is frequently confused.
Persicaria opelousana Riddell—Reported for Florida by Radford et al. (1964, 1968, as "*Polygonum hydropiperoides* Michaux var. *opelousanum* (Riddell) W. Stone"). However, the combination by Stone for this taxon was made at the subspecies level and is a heterotypic synonym of *P. hydropiperoides*.
Persicaria nepalensis (Meisner) Miyabe—Reported for Florida by Wunderlin (1998, as *Polygonum nepalense* Meisner), Wunderlin and Hansen (2003, 2011, both as *Polygonum nepalense* Meisner), and Hinds and Freeman (2005b), but reports are based on misidentification of material of *Persicaria capitata*.

Polygonella Michx. 1803. JOINTWEED

Herbaceous annuals or suffrutescent to woody perennials; branches adnate to the stem and appearing to arise internodally. Leaves alternate, palmate-veined, the venation appearing parallel and the secondary veins inconspicuous, epetiolate; stipules (ocrea) present. Flowers in terminal, racemiform cymes, actinomorphic, bisexual or functionally unisexual (plants dioecious, gynodioecious, or gynomonoecious), 1 per ocreolate fascicle; perianth segments 5, basally connate, petaloid, dimorphic; stamens 8, in 2 series with 5 outer and 3 inner, free; styles (2)3, free. Fruit an achene, (2)3-gonous.

A genus of 11 species; North America. [The genus name *Polygonum*, and *ella*, diminutive.]

Polygonella is nested within *Polygonum* (Sanchez et al., 2011; Schuster et al., 2015). However, it is morphologically distinct from that genus by having its branches adnate to the stem and appearing to arise internodally. Further study is needed.

Delopyrum Small (1913); *Dentoceras* Small (1924); *Gonopyrum* Fisch. & C. A. Mey. (1840); *Thysanella* A. Gray (1845).

Selected reference: Freeman (2005d).

1. Ocrea ciliate.
   2. Inner perianth segments fimbriate.
      3. Leaves with hyaline margins, these frequently clustered on short basal branches; stems below the inflorescence smooth or sparsely scabrous along the angles .... **P. robusta**
      3. Leaves without hyaline margins, these usually distributed uniformly over the whole plant; stems below the inflorescence minutely but densely scabrous .... **P. fimbriata**
   2. Inner perianth segments entire or erose, but not fimbriate.
      4. Plant with one main stem, the lowest secondary stem branching well above ground level .... **P. ciliata**
      4. Plant branching into several erect stems near ground level .... **P. basiramia**
1. Ocrea eciliate.
   5. Leaves early deciduous, the stem appearing leafless at anthesis .... **P. gracilis**
   5. Leaves persistent, the stem leafy at anthesis.
      6. Leaves 8–20 mm wide .... **P. macrophylla**
      6. Leaves 6 mm wide or less.
         7. Plant prostrate; stigma and style 0.5–0.8 mm long; outer perianth segments appressed .... **P. myriophylla**
         7. Plant erect or only basally decumbent; stigma and style 0.1–0.2 mm long; outer perianth segments reflexed .... **P. polygama**

Polygonella basiramia (Small) Nesom & V. M. Bates [Basally branching.] FLORIDA JOINTWEED; TUFTED WIREWEED.

Delopyrum basiramia Small, Bull. Torrey Bot. Club 51: 380. 1924. *Polygonella ciliata* Meisner var. *basiramia* (Small) Horton, Brittonia 15: 195. 1963. *Polygonella basiramia* (Small) G. L. Nesom & V. M. Bates, Brittonia 36: 40. 1984. TYPE: FLORIDA: Highlands Co.: sand dunes N of Kuhlman, 23 May 1921, *Small s.n.* (holotype: NY).

Erect subherbaceous perennial, to 80 cm; stem branching at the base, strongly fluted, glabrous to densely but minutely stellate-pubescent. Leaves with the blade narrowly elliptic, (3.5)7–19(28) mm long, ca. 0.5 mm wide, caducous, the apex acute, the base cuneate, the margin entire, the upper and lower surfaces glabrous; stipule (ocrea) margin ciliate. Flowers in a racemiform inflorescence 1–2.5(3) cm long, the pedicel ca. 0.5 mm long, the ocreola with the apex acuminate, the margin scarious, bisexual or carpellate (plants gynodioecious); outer perianth segments narrowly oblong, ca. 1 mm long, white to pinkish, loosely appressed, inconspicuously keeled, the margin entire, the inner ones narrowly oblong, ca. 1 mm long, white to pinkish, loosely appressed, the margin entire; filaments dimorphic, pubescent at the base, the anthers dark red. Fruit trigonous, ca. 2.5 mm long, the surface rich brown or light yellow-brown, smooth, shining.

Scrub. Rare; Polk and Highlands Counties. Endemic. Summer–fall.

Polygonella basiramia is listed as endangered in Florida (Florida Administrative Code, Chapter 5B-40) and in the United States (U.S. Fish and Wildlife Service, 50 CFR 23).

Polygonella ciliata Meisner [Ciliate.] HAIRY JOINTWEED.

Polygonella ciliata Meisner, in A. de Candolle, Prodr. 14: 81. 1856. *Delopyrum ciliatum* (Meisner) Small, Fl. Miami 65, 200. 1913. TYPE: FLORIDA: Manatee Co.: near the Manatee River, Jul (?) 1845, *Rugel 429* (holotype: BM).

Erect subherbaceous annual, to 1 m; stem solitary, few branched well above the ground, strongly fluted, glabrous or minutely stellate-pubescent. Leaves with the blade linear, (0.5)1–3(5) cm long, 0.5–1(3) mm wide, the apex acute, the base cuneate, the margin entire, the upper and lower surfaces glabrous, caducous except on the lower lateral branches; stipule (ocrea) margin ciliate. Flowers in a racemiform inflorescence (1)2–3(4.5) cm long, the pedicel 0.5–1 mm long, the ocreola acuminate, the margin scarious, bisexual or carpellate (plants gynodioecious); outer perianth segments narrowly oblong, ca. 1 mm long, white, appressed, inconspicuously keeled, the margin entire, the inner ones narrowly oblong, ca. 1 mm long, white, appressed; filaments dimorphic, glabrous, the anthers dark red. Fruit trigonous, 2–3 mm long, the surface reddish brown or yellow-brown, smooth, shining.

Sandhills and scrub. Occasional; central and southern peninsula. Endemic. Summer–fall.

Polygonella fimbriata (Elliott) Horton [Fimbriated.] SANDHILL JOINTWEED.

Polygonum fimbriatum Elliott, Sketch Bot. S. Carolina 1: 583. 1824. *Thysanella fimbriata* (Elliott) A. Gray, Boston J. Nat. Hist. 5: 232. 1845. *Polygonella fimbriata* (Elliott) Horton, Brittonia 15: 190. 1963.

Erect herbaceous annual, to 6 cm; stem solitary, minutely but densely scabrous. Leaves with the blade linear, (1)2–3.5(5) cm long, 1–1.5 (3) mm wide, the apex acute, the base cuneate, the margin entire, the upper and lower surfaces minutely scabrous; stipule (ocrea) margin ciliate. Flowers in a racemiform inflorescence (0.5) 1–2.5(3) cm long, the pedicel less than 1 mm long, the ocreola with the apex acuminate, the margin scarious, bisexual or carpellate (plants gyno-monoecious); outer perianth segments, ovate, ca. 2 mm long, pink with a white margin, loosely appressed, the margin erose, the inner ones oblong, ca. 2 mm long, pink with a white margin, the margin fimbriate; filaments monomorphic, the anthers pink. Fruit trigonous, (1.5)2(2.5) mm long, the surface yellow-brown, smooth and glossy proximally, minutely pebbled and dull distally.

Sandhills. Rare; Holmes County. Georgia, Alabama, and Florida. Summer–fall.

Polygonella gracilis Meisner [Slender.] TALL JOINTWEED.

Polygonum gracile Nuttall, Gen. N. Amer. Pl. 1: 255. 1818; non Salisbury, 1796; nec R. Brown, 1810. *Polygonella gracilis* Meisner, in A. de Candolle, Prodr. 14: 80. 1856. *Delopyrum gracile* (Meisner) Small, Fl. Miami 65, 200. 1913.
Delopyrum filiforme Small, Man. S.E. Fl. 451, 1504. 1933.

Erect herbaceous annual, to 1.5 m; stem solitary, simple or branched, strongly fluted, glabrous, often glaucous. Leaves with the blade filiform to broadly spathulate, (1)2–4(6.5) cm long, 1–5(8) mm wide, the apex obtuse, the base cuneate, the margin entire, the upper and lower surfaces glabrous, often glaucous; stipule (ocrea) margin eciliate. Flowers in a raceme-like inflorescence (1)2–4(7) cm long, the pedicel to 0.5 mm long, the ocreola abruptly flared upward, the apex acute, the margin scarious, unisexual (plants dioecious). Staminate flowers with the outer perianth segments elliptic, 1–2 mm long, white, loosely appressed, the margin entire, the inner ones elliptic, 1–2 mm long, white, loosely appressed, the margin entire; filaments dimorphic, the anthers dark red. Carpellate flowers with the outer perianth segments oblong to elliptic, ca. 1 cm long, white to pink, loosely appressed, the margin entire or obscurely erose, the inner ones broadly elliptic to ovate, 2–3 mm long, white to pink, loosely appressed, the margin entire or obscurely erose. Fruit (2)3-gonous, 2–3 mm long, the surface dark red-brown to yellow-brown, smooth, glossy.

Sandhills, dunes, and scrub. Frequent; nearly throughout. North Carolina south to Florida, west to Mississippi. Summer–fall.

Polygonella macrophylla Small [Large-leaved.] LARGELEAF JOINTWEED.

Polygonella macrophylla Small, Bull. Torrey Bot. Club 23: 407. 1896. TYPE: FLORIDA: Franklin Co.: Carrabelle, Oct 1896, *Chapman 5541.* (holotype: NY; isotypes: NY).

Erect suffrutescent perennial, to 1.1 m; stem solitary, simple or branched distally, strongly fluted, glabrous. Leaves with the blade oblanceolate to obovate or spatulate, (1)2.5–5.5(6.5) cm long, 1–2(2.5) cm wide, somewhat fleshy, the apex rounded or obtuse, the base cuneate to obtuse, the margin entire, hyaline, the upper and lower surfaces glabrous; stipule (ocrea) margin eciliate. Flowers in a racemiform inflorescence (1.5)3–5(7) cm long, the pedicel 1–2 mm long, the ocreola widely flared, the apex acute, the margin scabrous, bisexual or carpellate (plants gynodioecious); outer perianth segments obovate, 1–2 mm long, white, pink, or red, loosely appressed, the margin entire, the inner ones obovate, 2–3 mm long, white, red, or pink, loosely appressed, the margin entire; filaments dimorphic, the anthers white. Fruit trigonous, 3–4 mm long, the surface yellow-brown, smooth, glossy.

Coastal dunes and scrub. Rare; central and western panhandle. Florida and Alabama. Summer–fall.

Polygonella macrophylla is listed as threatened in Florida (Florida Administrative Code, Chapter 5B-40).

Polygonella myriophylla (Small) Horton [Many-leaved.] SMALL'S JOINTWEED; WOODY WIREWEED; SANDLACE.

> *Dentoceras myriophyllum* Small, Bull. Torrey Bot. Club 51: 389. 1924. *Polygonella myriophylla* (Small) Horton, Brittonia 15: 196. 1963. TYPE: FLORIDA: Polk Co.: S of Frostproof, 18 Apr 1920, *Small & DeWinkeler 9603* (holotype: NY; isotype: NY).

Prostrate woody perennial, to 0.5 m; stem dark brown. Leaves with the blade clavate to spatulate, 2–8 mm long, ca. 0.5 mm wide, the apex obtuse, the base cuneate, the margin entire, the upper and lower surfaces glabrous; stipule (ocrea) margin eciliate. Flowers in a racemiform inflorescence 0.5–1.3 cm long, the pedicel 1–2 mm long, the ocreola widely flared, the apex acute, the margin scarious, bisexual; outer perianth segments elliptic to oblong, 1.5–2 mm long, white, pink, or yellow, loosely appressed, the margin entire, the inner ones elliptic to suborbicular, 1.5–2.5 mm long, white, pink, or yellow, the margin entire; filaments dimorphic, the anthers pink or yellow. Fruit trigonous, 2–3 mm long, the surface reddish brown, slightly pebbled, glossy.

Scrub. Occasional; Orange, Polk, Osceola, and Highlands Counties. Endemic. Summer–fall.

Polygonella myriophylla is listed as endangered in Florida (Florida Administrative Code, Chapter 5B-40) and in the United States (U.S. Fish and Wildlife Service, 50 CFR 23).

Polygonella polygama (Vent.) Engelm. & A. Gray [Polygamous, alluding to the dioecious condition.] OCTOBER FLOWER.

Erect or basally decumbent suffrutescent perennial, to 70 cm; stem usually branched, glabrous or minutely pubescent. Leaves with the blade linear to narrowly clavate or spatulate, (3)4–16(36) mm long, 0.5–6 mm wide, the apex obtuse, the base cuneate, the margin hyaline at least above the middle, the upper and lower surfaces glabrous; stipule (ocrea) margin eciliate. Flowers in a racemiform inflorescence 0.5–2(3) cm long, the pedicel 0.5–1 mm long, ocreola with the apex acute to acuminate, the margin scarious, unisexual (plants dioecious). Staminate flowers with the outer perianth segments broadly elliptic, 1–2 mm long, white, loosely appressed, the margin entire, the inner ones elliptic, 1–2 mm long, white, appressed, the margin entire; filaments dimorphic, glabrous, the anthers pink, orange, or yellow. Carpellate flowers with the outer perianth segments broadly elliptic or ovate, 0.5–1 mm long, white to pink, loosely appressed, the margin entire, the inner ones broadly elliptic to ovate, 0.5–1.5 mm long, white to pink, loosely appressed, the margin entire. Fruit trigonous, 1.5–2 mm long, the surface brown or yellow-brown, smooth, glossy.

1. Leaves (2)3–6 mm wide ....var. **polygama**
1. Leaves 0.5–1 mm wide ....var. **brachystachya**

Polygonella polygama (Vent.) Engelm. & A. Gray var. **polygama** OCTOBER FLOWER.

> *Polygonum polygamum* Ventenat, Descr. Pl. Nouv. pl. 65. 1802. *Polygonella polygama* (Ventenat) Engelmann & A. Gray, Boston J. Nat. Hist. 5: 231. 1845.
> *Polygonella parvifolia* Michaux, Fl. Bor.-Amer. 2: 240. 1803. *Stopinaca parvifolia* (Michaux) Rafinesque, Fl. Tellur. 3: 11. 1837 ("1836").

Polygonella parvifolia Michaux var. *subenervis* Meisner, in A. de Candolle, Prodr. 14: 80. 1856. TYPE: FLORIDA: Manatee Co.: along the Manatee River, s.d., *Rugel 455* (holotype: ?).

Polygonella brachystachya Meisner var. *laminigera* Fernald, Rhodora 39: 406. 1937. TYPE: FLORIDA: Brevard Co.(?): Indian River, Sep, *Curtiss 2433* (holotype: GH).

Leaves with the blade (2)3–6 mm wide.

Sandhills, flatwoods, and scrub. Frequent; nearly throughout. Virginia south to Florida, west to Texas. Summer–fall.

Polygonella polygama (Vent.) Engelm. & A. Gray var. **brachystachya** (Meisn.) Wunderlin [Short-spiked.] OCTOBER FLOWER.

Polygonella brachystachya Meisner, in A. de Candolle, Prodr. 14: 80. 1856. *Polygonella polygama* (Ventenat) Engelmann & A. Gray var. *brachystachya* (Meisner) Wunderlin, Florida Sci. 44: 79. 1981. TYPE: FLORIDA: Miami-Dade Co.: near Miami, 1846, *Rugel s.n.* (holotype: NY).

Leaves with the blade 0.5–1 mm wide, otherwise plants as in var. *polygama*.

Flatwoods. Occasional; central and southern peninsula. Endemic. Summer–fall.

Polygonella robusta (Small) G. L. Nesom & V. M. Bates [Robust.] LARGEFLOWER JOINTWEED; SANDHILL WIREWEED.

Thysanella robusta Small, Bull. Torrey Bot. Club 36: 159. 1909. *Polygonella fimbriata* (Elliott) Horton var. *robusta* (Small) Horton, Brittonia 15: 191. 1963. *Polygonella robusta* (Small) G. L. Nesom & V. M. Bates, Brittonia 36: 43. 1984. TYPE: FLORIDA: Manatee Co.: Braidentown [Bradenton], 28 Nov 1901, *Tracy 7638* (holotype: NY; isotype: NY).

Erect herbaceous annual or perennial to 1 m; stem simple or branched at the base, glabrous or minutely scabrous along the angles. Leaves with the blade linear, (1)2.5–4(7) cm long, (1)1.5–2.5(3) mm wide, the apex acute, the base cuneate, the margin entire, hyaline, the upper and lower surfaces minutely scabrous; stipule (ocrea) margin ciliate. Flowers in a racemiform inflorescence (1) 2.5–4.5 (6.5) cm long, the pedicel to 1 mm long, the ocreola with the apex acuminate, the margin scarious, bisexual or a few usually distal ones carpellate (plants gynomonoecious); outer perianth segments, ovate, 2–3(4) mm long, pink to white, loosely appressed, the margin erose, the inner ones oblong, 2–4 mm long, pink to white, loosely appressed, the margin fimbriate; filaments monomorphic, glabrous, the anthers pink. Fruit trigonous, 2–2.5 mm long, the surface yellow-brown, smooth and shiny proximally, minutely pebbled and dull distally.

Sandhills and scrub. Common; northern and central peninsula, west to central panhandle. Endemic. Summer–fall.

EXCLUDED TAXA

Polygonella articulata (Linnaeus) Meisner—Reported by Small (1903, 1913, both as *Gonopyrum articulatum* (Linnaeus) Small; 1933, as *Delopyrum articulatum* (Linnaeus) Small ex Rydberg). No Florida specimens known. Most likely based on imperfect locality data on a coastal Georgia specimen.

Polygonella polygama (Ventenat) Engelmann & A. Gray var. *croomii* (Chapman) Fernald]—Reported by Small (1933, as *Polygonella croomii* Chapman) for north Florida. A specimen at NY from the

Chapman Herbarium said to be collected by Croom and from "south Florida" is of questionable provenance and was annotated as the holotype by Vernon Bates in 1985. Wunderlin (1981) and Freeman (2005d) report it only from North Carolina and South Carolina.

Polygonum L. 1753. KNOTWEED

Annual or perennial herbs or vines. Leaves alternate, pinnate- or pinnipalmate-veined, petiolate or epetiolate; stipules (ocreae) present. Flowers in axillary clusters or in terminal and axillary spiciform, racemiform, paniculiform, or capitate cymes, actinomorphic, bisexual or unisexual (plants dioecious or polygamous); perianth segments 4–5, basally connate, monomorphic or dimorphic, petaloid; stamens 5–8, free or basally connate; styles 2–3, free or connate. Fruit a lenticular or trigonous achene, often enclosed by the persistent perianth.

A genus of about 65 species; nearly cosmopolitan. [From the Greek *poly*, many, and *gony*, knee, referring to the often thickened joints of the stem.]

Selected references: Costea et al. (2005); Mertens and Raven (1965).

1. Stem erect; flowers mostly terminal in loose, leafy spikelike inflorescences; leaves in the distal part of the inflorescence reduced, shorter than or equaling the flowers ....**P. argyrocoleon**
1. Stem trailing; flowers entirely axillary; leaves in the distal part of the inflorescence longer than the flowers.
   2. Persistent perianth spreading in fruit; achene smooth, lustrous .... **P. glaucum**
   2. Persistent perianth appressed in fruit; achene tuberculate-striate, dull .... **P. aviculare**

Polygonum argyrocoleon Steud. ex Kunze [From the Greek *argyros*, silver, and *koleos*, sheath.] SILVERSHEATH KNOTWEED.

Polygonum argyrocoleon Steudel ex Kunze, Linnaea 20: 17. 1847.

Erect or ascending annual herb, to 1 m; stem branched primarily proximally, striate, somewhat glaucous. Leaves with the blade lanceolate to linear-lanceolate, 1.5–5 cm long, 2–8 mm wide, pinnate-veined, the secondary veins inconspicuous, the apex acute, the base cuneate, the margin entire, the upper and lower surfaces glabrous, somewhat glaucous, the petiole to 1.5 mm long; stipules (ocrea) 4–8 mm long, green or dark-brown at the base, silvery-scarious above, 2-lobed, the margin eciliate, soon disintegrating into persistent fibers distally. Flowers in an axillary or terminal spiciform inflorescence, 4–6 in an ocreolate fascicle, bisexual, the distal leaves reduced and shorter than or equaling the floral clusters, the pedicel 1–2 mm long; perianth segments 5, oblong to obovate, ca. 2 mm long, green or white, sometimes with pink, not keeled; stamens 7–8. Fruit ovate, ca. 2 mm long, brown, trigonous, the surface smooth, lustrous.

Disturbed sites. Rare; Orange and Miami-Dade Counties. Nearly throughout North America; Europe and Asia. Native to Europe and Asia. Spring–fall.

Polygonum aviculare L. [Pertaining to birds, which are said to eat the young leaves and seeds.] PROSTRATE KNOTWEED.

Polygonum aviculare Linnaeus, Sp. Pl. 362. 1753. *Polygonum centinodium* Lamarck, Fl. Franc. 3: 237. 1779 ("1778"), nom. illegit. *Polygonum aviculare* Linnaeus var. *procumbens* Meisner, Monogr. Polyg. 87. 1826, nom. inadmiss. *Centinodium axillare* Friche-Joset & Montandon, Syn. Fl. Jura 270. 1856, nom. illegit. *Centinodium aviculare* (Linnaeus) Fourreau, Ann. Soc. Linn. Lyon, ser. 2. 17: 146. 1869. *Polygonum heterophyllum* Lindman, Svensk Bot. Tidskr. 6: 690. 1912, nom. illegit. *Polygonum aviculare* Linnaeus subsp. *heterophyllum* Ascherson & Graebner, Syn. Mitteleur. Fl. 4: 848. 1913, nom. inadmiss. *Polygonum aviculare* Linnaeus var. *heterophyllum* Munshi & Javeid, J. Econ. Taxon. Bot., Addit. Ser. 2: 55. 1986, nom. inadmiss.

Polygonum neglectum Besser, Enum. Pl. 45. 1822. *Polygonum aviculare* Linnaeus subsp. *neglectum* (Besser) Arcangeli, Comp. Fl. Ital. 583. 1882.

Polygonum aviculare Linnaeus var. *depressum* Meisner, in A. de Candolle, Prodr. 14: 98. 1856. *Polygonum aviculare* Linnaeus subsp. *depressum* (Meisner) Arcangeli, Comp. Fl. Ital. 583. 1882.

Procumbent to ascending annual herb, to 1 m; stem branching proximally, terete or somewhat grooved, somewhat glaucous. Leaves with the blade elliptic to oblanceolate, the lower ones 2.5–6 cm long, 0.5–2 mm wide, pinnate-veined, the secondary veins inconspicuous, the apex acute or obtuse, the base cuneate, the margin entire, the upper and lower surfaces somewhat glaucous, the petiole 1–6 mm long; stipules (ocrea) 3–10 mm long, silvery distally, hyaline, 2-lobed, the margin eciliate, soon disintegrating into persistent fibers. Flowers in an axillary inflorescence, 1–6 in an ocreolate fascicle, bisexual, the pedicel 1–5 mm long; perianth segments 5, oblong to obovate, 2.5–3 mm long, green with a white or pink margin; stamens 5–7. Fruit ovate, ca. 3 mm long, brown, (2)3-angled, the surface coarsely striate-tubercled, dull.

Disturbed sites. Rare; Putnam County, central and western panhandle. Nearly throughout North America; Europe, Asia, Australia, and Pacific Islands. Native to Europe and Asia. Spring–fall.

Polygonum aviculare is taxonomically complex and controversial. Small (1933) recognized *P. aviculare* and *P. depressum* for Florida. More recently, Costea et al. (2005) recognized six subspecies in North America, with two occurring in Florida (subsp. *aviculare* and subsp. *neglectum*). We have been unable to satisfactorily separate our plants into discrete entities and prefer to treat *P. aviculare* as a single polymorphic taxon.

Polygonum glaucum Nutt. [Glaucous.] GLAUCOUS KNOTWEED.

Polygonum glaucum Nuttall, Gen. N. Amer. Pl. 1: 254. 1818. *Polygonum aviculare* Linnaeus var. *glaucum* (Nuttall) Torrey, Ann. Lyceum Nat. Hist. New York 2: 240. 1827.

Prostrate or ascending annual herb, to 7 dm; stem branched from the base, glaucous. Leaves with the blade lanceolate or narrowly elliptic, 1–3 cm long, 2–8 mm wide, pinnate-veined, the secondary veins inconspicuous, the apex obtuse or acute, the base cuneate, the margin entire, revolute, the upper and lower surfaces glabrous, glaucous, the petiole 0.5–3 mm long, glaucous; stipules (ocrea) 7–15 mm long, 2-lobed, silvery-scarious distally, the lower half becoming brown, the apex entire or lacerate. Flowers axillary, 1–3 per ocreolate fascicle, the pedicel 3–5 mm long; perianth segments 5, oblong-obovate to spatulate, 2–4 mm long, white with a white or pink margin, the apex obtuse or rounded; stamens 8. Fruit ovate, 3–4 mm long, trigonous, reddish brown to dark brown, the surface smooth, lustrous.

Dunes. Rare; Duval, Franklin, and Gulf Counties. Not recently collected. Massachusetts south to Florida. Spring–fall.

EXCLUDED TAXA

Polygonum maritimum Linnaeus—Reported by Small (1903, 1913), possibly misapplied to *P. glaucum*.

Reynoutria Houtt. 1777. KNOTWEED

Perennial herbs. Leaves alternate, pinnipalmate-veined, petiolate; stipules (ocrea) present. Flowers in terminal and axillary paniculiform or racemiform cymes, actinomorphic, bisexual or carpellate; perianth segments 5, dimorphic, the outer 3 winged and larger than the keeled inner 2; stamens 8, free; styles 3, basally connate, the stigma fimbriate. Fruit a trigonous achene, enclosed by the accrescent perianth.

A genus of about 6 species; North America, South America, Europe, Asia, Australia, and Pacific Islands. [Commemorates Baron van Reynoutre (fl. 16th century), Dutch botanist.

Pleuropterus Turcz. (1848).

Selected reference: Freeman and Hinds (2005).

Reynoutria japonica Houtt. [Of Japan.] JAPANESE KNOTWEED.

Polygonum cusidatum Siebold & Zuccarini, Abh. Math.-Phys. Cl. Königl. Bayer. Akad. Wiss. 4: 208. 1846. *Polygonum zuccarinii* Small, Mem. Dept. Bot. Columbia Coll. 1: 158. 1895, nom. illegit. *Polygonum sieboldii* de Vriese, in L. H. Bailey, Cycl. Amer. Hort. 3: 1393. 1901, nom. illegit.; nom. Meisner, 1856. *Pleuropterus zuccarinii* Small, in Britton & A. Brown, Ill. Fl. N. U.S., ed. 2. 1: 676. 1913, nom. illegit. *Pleuropterus cuspidatus* (Siebold & Zuccarini) H. Gross, Beih. Bot. Centralbl. 37(2): 114. 1919.

Reynoutria japonica Houttuyn, Nat. Hist. 2(8): 640, t. 51(1). 1777. *Tiniaria japonica* (Houttuyn) Hedberg, Svensk Bot. Tidskr. 40: 299. 1946. *Fallopia japonica* (Houttuyn) Ronse Decraene, Bot. J. Linn. Soc. 98: 369. 1988. *Polygonum japonicum* (Houttuyn) S. L. Welsh, Utah Fl., ed. 4. 594. 2008; non Meisner, 1856.

Erect perennial herb, to 3 m; stem branched, glabrous, glaucous. Leaves with the blade ovate, 5–15 cm long, 2–10 cm wide, pinnate-veined, the apex abruptly cuspidate, the base truncate to narrowly cuneate, the margin glabrous or scabrous to ciliate, the upper surface glabrous, the lower surface minutely dotted with trichomes along the veins, glaucous, the petiole 1–3 cm long, glabrous; stipules (ocrea) usually deciduous, 4–6(10) mm long, brownish, with reflexed trichomes and slender bristles at the base, otherwise glabrous or puberulent, the margin eciliate. Flowers in a terminal or axillary paniculiform or racemiform inflorescence 4–12 cm long, 3–8(15) per ocreate fascicle, bisexual or carpellate, the axis puberulent, the peduncle to 2.5 cm long or absent, puberulent, the pedicel 3–5 mm long, glabrous, the perianth segments obovate to elliptic, 4–6 mm long, greenish or white to pink, keeled; stamens 8. Fruit elliptic, 2–4 mm long, trigonous, the wings 1–2 mm wide, these decurrent on the stipe-like base nearly to the articulation, glabrous.

Disturbed sites. Rare; Santa Rosa County. Nearly throughout North America; South America; Europe, Asia, Australia, and Pacific Islands. Native to Asia. Summer–fall.

Rumex L. 1753. DOCK

Annual or perennial herbs. Leaves alternate, simple, pinnate-veined, petiolate; stipules (ocrea) present. Flowers terminal and axillary in spiciform or paniculiform cymes, actinomorphic, bisexual or unisexual (plants moneocious or dioecious, rarely polygamomonoecious); perianth segments 6, basally connate, dimorphic, the outer 3 remaining small, the inner 3 enlarging and in some species developing thickened callosities (tubercles) on 1–3 segments; stamens 6, free; styles 3. Fruit a trigonous achene, enclosed by the accrescent perianth.

A genus of about 200 species; nearly cosmopolitan. [Classical Latin name for sorrel, probably derived from *rumo*, to suck, in reference to the practice among the Romans of sucking the leaves to relieve thirst.]

Rumex obtusifolius, *R. obovatus*, *R. paraguayensis*, and *R. pulcher* are difficult to identify without mature and complete reproductive and vegetative material.

Acetosella (Meisn.) Fourr. (1869).

Selected reference: Mosyakin (2005).

1. Leaves (at least some) sagittate or hastate; flowers unisexual (plants dioecious); inner perianth segments etuberculate.
    2. Inner perianth segments subequaling and enclosing the achene; pedicel joint just below the base of the perianth segments .... **R. acetosella**
    2. Inner perianth segments expanding into broad reticulate wings; pedicel joint below the middle of the perianth segments .... **R. hastatulus**
1. Leaves not sagittate or hastate; flowers bisexual or if unisexual, then plant monoecious or polygamo-monoecious; at least 1 inner perianth segment with a tubercle.
    3. Inner perianth segments with subulate teeth.
        4. Tubercle finely alveolate-reticulate.
            5. Inner perianth segments 4–5 mm long, 3–4 mm wide, mostly with 4–5 marginal teeth on each side ....**R. obovatus**
            5. Inner perianth segments ca. 3 mm long, ca. 2 mm wide, mostly with 2–3 marginal teeth on each side .... **R. paraguayensis**
        4. Tubercle not alveolate-reticulate.
            6. Pedicel little if any longer than the fruiting perianth; achene minutely pebbled, lustrous .... **R. pulcher**
            6. Pedicel considerably longer than the fruiting perianth; achene smooth, dull.......................... .... **R. obtusifolius**
    3. Inner perianth segments entire or crenulate.
        7. Inner perianth segments narrow, cuneate or truncate at the base; fruit elliptic to pandurate...... ....**R. cuneifolius**
        7. Inner perianth segments broad, truncate or rounded at the base; fruit ovate, deltoid, or somewhat hastate.
            8. Floral stipe mostly 10 mm long or longer .... **R. verticillatus**
            8. Floral stipe less than 10 mm long.

    9. Tubercle 1(2–3), smooth ....**R. patientia**
    9. Tubercles 3, reticulate-rugose.
        10. Verticels of flowers usually contiguous; floral stipe 3 mm long or longer ... **R. crispus**
        10. Verticels of flowers usually separate; floral stipe usually 3 mm or shorter ....
        ....**R. chrysocarpus**

Rumex acetosella L. [Old generic name meaning "little sorrel."] SHEEP SORREL.

Rumex acetosella Linnaeus, Sp. Pl. 338. 1753. *Acetosa acetosella* (Linnaeus) Miller, Gard. Dict., ed. 8. 1768. *Lapathum arvense* Lamarck, Fl. Franc. 3: 8. 1779 ("1778"), nom. illegit. *Acetosa hastata* Moench, Methodus 357. 1794, nom. illegit. *Rumex infestus* Salisbury, Prodr. Stirp. Chap. Allerton 258. 1796, nom. illegit. *Rumex acetosella* Linnaeus var. *vulgaris* W.D.J. Koch, Syn. Fl. Germ. Helv. 616. 1837, nom. inadmiss. *Acetosella vulgaris* Fourreau, Ann. Soc. Linn. Lyon, ser. 2. 17: 145. 1869. *Acetosella acetosella* (Linnaeus) Small, Man. S.E. Fl. 446, 1504. 1933, nom. inadmiss.

Erect or ascending perennial herb, to 4 dm; stem rhizomatous, branched in the distal half, glabrous. Leaves with the blade obovate-oblong to lanceolate-elliptic, 2–6 cm long, 0.3–2 cm wide, chartaceous, the apex acute or obtuse, the base hastate or broadly cuneate, the margin entire, the upper and lower surfaces glabrous, the lower and median leaves with a petiole usually about as long as the blade; stipules (ocrea) thin-hyaline, loose, friable. Flowers in a terminal, spiciform or paniculiform inflorescence, interrupted, (3)5–8(10) per ocreolate fascicle, dioecious or rarely polygamomonoecious, the pedicel articulate near the base of the perianth segments; outer perianth segments lanceolate, ca. 1 mm long, erect, the inner ones broadly ovate, 1–2 mm long, ca. 1 mm wide, the apex obtuse or subacute, the base cuneate, etuberculate. Fruit ovate, 1–1.5 mm long, brown, the surface smooth, lustrous.

    Disturbed sites. Rare; Gulf and Escambia Counties. Not recently collected. Nearly cosmopolitan. Native to Europe, Africa, and Asia. Spring–summer.

Rumex chrysocarpus Moris [Golden-fruited.] AMAMASTLA.

Rumex chrysocarpus Moris, Mem. Reale Accad. Sci. Torino 38: 46, t. 2. 1835.

Erect or ascending perennial herb, to 6(8) dm; stem often with short lateral branches below the inflorescence, glabrous. Leaves with the blade linear-lanceolate to lanceolate, 5–12 cm long, 2–4 cm wide, coriaceous, the apex obtuse, the base cuneate or rounded, the margin entire or crenulate or crisped, the upper and lower surfaces glabrous, the lower and median leaves with a petiole usually about as long as the blade; stipules (ocrea) membranous, longitudinally striate, friable. Flowers in a terminal or axillary, spiciform or paniculiform inflorescence, interrupted, 5–15 per ocreolate fascicle, bisexual, the pedicel articulated in the proximal ⅓ to near the middle; outer perianth segments lanceolate, 1–2 mm long, erect, the inner ones suborbicular or ovate-deltoid, 4–5 mm long, 3–4 mm long, the apex acute, the base truncate or subacute, the margin entire, tubercles 3, smooth. Fruit ovate, 2–3 mm long, brown or dark reddish brown, the surface smooth, lustrous.

    Swamps and disturbed, seasonally wet sites. Rare; Franklin, Santa Rosa, and Escambia Counties. Florida west to Texas; Mexico. Native to Texas and Mexico. Summer–fall.

Rumex crispus L. [Curled, in reference to the leaf margins.] CURLY DOCK.

Rumex crispus Linnaeus, Sp. Pl. 335. 1753. *Lapathum crispum* (Linnaeus) Scopoli, Fl. Carniol., ed. 2. 1: 261. 1771. *Rumex patientia* Linnaeus var. *crispus* (Linnaeus) Kuntze, Revis. Gen. 2: 560. 1891.

Erect perennial herb, to 1(1.5) m; stem branched above the middle, glabrous. Leaves with the blade lanceolate to linear-lanceolate, 15–30(35) cm long, 2–6 cm wide, chartaceous, the apex acute, the base cuneate to subcordate, the margin conspicuously crisped and undulate, the upper and lower surfaces glabrous, the lower leaves with a petiole shorter than the blade; stipules (ocrea) membranous, longitudinally striate, friable. Flowers in a terminal, elongate, paniculiform inflorescence, contiguous or interrupted, 10–25 per ocreolate fascicle, bisexual, the pedicel (3)4–8 mm long, articulated in the proximal ⅓; outer perianth segments lanceolate, 1–2 mm long, erect or spreading, the inner ones suborbiculate-ovate or ovate-deltoid, 4–6 mm long, the apex obtuse to subacute, the base truncate to subcordate, the margin entire or slightly erose, tubercles (1–2)3, 1 usually larger, smooth. Fruit ovate, 2–3 mm long, reddish brown, the surface smooth, lustrous.

Disturbed sites. Occasional; nearly throughout. Nearly throughout North America; West Indies, Mexico, Central America, and South America; Europe, Africa, Asia, Australia, and Pacific Islands. Native to Europe, Africa, and Asia. Spring–summer.

Rumex cuneifolius Campd. [With wedge-shaped leaves.] WEDGELEAF DOCK.

Rumex cuneifolius Campderá, Monogr. Rumex 66, 95. 1819.

Ascending or erect perennial herb, to 3(4) dm; stem branching in the distal ⅔, rhizomatous and/or stoloniferous, glabrous. Leaves with the blade obovate to obovate-elliptic, 5–8(12) cm long, (2)3–5(7) cm wide, coriaceous, the apex obtuse to rounded, the base rounded or cuneate, the margin entire or crisped, the upper and lower surfaces glabrous, the lower leaves with a petiole shorter than the blade; stipules (ocrea) membranous, longitudinally striate, friable. Flowers in a terminal or axillary, spiciform or paniculiform inflorescence, contiguous at least in the distal part, 5–20 per ocreolate fascicle, the pedicel 3–5 mm long, articulated near the middle or the proximal ⅓; outer perianth segments lanceolate, 1–2 mm long, erect, the inner ones ovate-triangular, 4–5 mm long, the apex acute to subacute, the base cuneate, the margin entire, tubercles 3, finely punctate. Fruit elliptic, ca. 3 mm long, dark reddish brown, the surface smooth, glossy.

Disturbed sites. Rare; Escambia County. Not recently collected. Florida, Alabama, Oregon, and California; South America; Europe and Australia. Native to South America. Summer–fall.

Mohr (1901) lists this species as firmly established on ballast in Pensacola, Wilhelm (1984) reports that it is no longer extant there.

Rumex hastatulus Baldwin [Slightly halberd-shaped.] HEARTWING DOCK; HASTATELEAF DOCK.

Rumex hastatulus Baldwin, in Elliott, Sketch Bot. S. Carolina 1: 416. 1817. *Acetosa hastatula* (Baldwin) Á. Löve, Univ. Inst. Applied Sci. Iceland, Dept. Agric. Rep. ser. B. 3: 107. 1948.
Rumex engelmannii Meisner, in A. de Candolle, Prodr. 14: 64. 1856.

Erect or ascending annual or short-lived perennial herb, to 4.5 dm; stem solitary or few-branched from the base, branched distally, glabrous. Leaves with the blade obovate-oblong, ovate-lanceolate, oblong-lanceolate, or lanceolate, 1–6(10) cm long, 0.5–2 cm wide, charta-ceous, the apex obtuse or subacute, the base hastate, auriculate, or sometimes without evident lobes, the margin entire, the upper and lower surfaces glabrous, the lower leaves with a petiole usually about as long as the blade; stipules (ocrea) thin, hyaline, friable. Flowers in a terminal, narrowly paniculiform inflorescence, interrupted, 3–6(8) per ocreolate fascicle, the pedicel 2–3 mm long, articulated proximally, unisexual (plants dioecious); outer perianth segments lanceolate, ca. 1 mm long, reflexed, the inner ones suborbicular, or broadly ovate, ca. 3 mm long and wide, the apex obtuse to subacute, the base cordate or rounded, etuberculate. Fruit ovate, ca. 1 mm long, brown, the surface with a translucent dark line at each angle, lustrous.

Disturbed sites. Common; northern counties, central peninsula. New York and Massachu-setts south to Florida, west to Kansas, Oklahoma, and New Mexico. Spring.

Rumex obovatus Danser [Obovate, in reference to the leaf shape.] TROPICAL DOCK.

Rumex obovatus Danser, Ned. Kruidk. Arch. 1920: 241. 1921.

Erect annual herb, to 7 dm; stem branched in the distal ⅔, glabrous. Leaves with the blade ob-ovate 4–7(11) cm long, 2–5 cm wide, coriaceous, the apex obtuse or rounded, the base cuneate to truncate, the margin undulate or crisped, the upper and lower surfaces glabrous, the lower leaves with a petiole usually as long as the blade; stipules (ocrea) hyaline, friable. Flowers in a terminal, paniculiform inflorescence, contiguous or interrupted, at least in the proximal ½, 10–20 per ocreolate fascicle, bisexual, the pedicel 3–5 mm long, articulated in the proximal ⅓; outer perianth segments obovate, 1–2 mm long, the inner ones ovate-triangular or triangular-deltoid, 4–5 mm long, the apex acute to subacute, the base truncate, the margin dentate with 3–5 subulate teeth on each side, tubercles usually 3, distinctly reticulate-rugose. Fruit ovate, 2–3 mm long, brown, the surface smooth, lustrous.

River banks and pond margins. Occasional; northern and central peninsula, central and western panhandle. Florida and Louisiana; South America; Europe. Native to South America. Summer–fall.

Rumex obtusifolius L. [Blunt-leaved.] BITTER DOCK.

Rumex obtusifolius Linnaeus, Sp. Pl. 335. 1753. *Lapathum sylvestre* Lamarck, Fl. Franc. 3: 4. 1779 ("1778"), nom. illegit. *Lapathum obtusifolium* (Linnaeus) Moench, Methodus 356. 1794. *Rumex sylvestris* Campderá, Monogr. Rumex 146. 1819, nom. illegit. *Acetosa obtusifolia* (Linnaeus) M. Gómez de la Maza y Jiménez, Anales Inst. Segunda Enseñ. 2: 277. 1896.

Erect perennial herb, to 1.5 m; stem branched distally to middle or occasionally in distal ⅔, glabrous. Leaves with the blade oblong to ovate-oblong or ovate, 20–40 cm long, 10–15 cm wide, chartaceous, the apex obtuse to subacute, the base cordate or rounded, rarely truncate, the margin undulate or rarely crisped, the upper and lower surfaces glabrous, the lower leaves with a petiole about as long as the blade; stipules (ocrea) membranous, longitudinally striate, friable. Flowers in a terminal, elongate, paniculiform inflorescence, interrupted, 10–25 per ocreolate fascicle, bisexual, the pedicel 3–8(10) mm long, articulated in the proximal ⅓ or near the middle; outer perianth segments oblong, ca. 2 mm long, spreading or deflexed, the inner ones triangular ovate, 3–6 mm long, the apex obtuse to subacute, the base truncate, the margin dentate with 2–5 subulate teeth, tubercle 1, sometimes 3, and then 1 distinctly larger, smooth. Fruit ovate, 2–3 mm long, brown to reddish brown, the surface smooth, dull.

Disturbed sites. Occasional; central and southern peninsula, central and western panhandle. Nearly throughout North America; Europe, Africa, Asia, and Pacific Islands. Native to Europe, Africa, and Asia. Summer–fall.

Rumex paraguayensis Parodi [Of Paraguay.] PARAGUAYAN DOCK.

Rumex paraguayensis Parodi, Anales Soc. Ci. Argent. 5: 160. 1878.

Erect annual herb, to 4(7) dm; stem branched above the middle or in the distal ⅔, glabrous. Leaves with the blade obovate, 4–6(7) cm long, 2–4(5) cm wide, chartaceous, the apex obtuse, the base cuneate or truncate, the margin entire or crisped, the upper and lower surfaces glabrous, the lower leaves with a petiole as long as the blade or sometimes shorter; stipules (ocrea) hyaline, friable. Flowers in a terminal paniculiform inflorescence, contiguous or interrupted, 10–20 per ocreolate fascicle, the pedicel 3–6 mm long, articulated in the proximal ⅓; outer perianth segments, lanceolate, 1–2 mm long, the inner ones obovate-triangular to deltoid, 3–4 mm long, the apex acute, the base truncate, the margin with 2–3(4) triangular or subulate teeth on each side, tubercles usually 3, smooth or minutely punctate. Fruit broadly elliptic, ca. 2 mm long, reddish brown, the surface smooth, lustrous.

Wet, disturbed sites. Rare; central peninsula, central panhandle. Florida, Louisiana, and Texas; South America. Native to South America. Spring–summer.

Rumex patientia L. [The old colloquial name.] PATIENCE DOCK.

Rumex patientia Linnaeus, Sp. Pl. 333. 1753. *Lapathum hortense* Lamarck, Fl. Franc. 3: 3. 1779 ("1778"), nom. illegit. *Acetosa patientia* (Linnaeus) M. Gómez de la Maza y Jiménez, Anales Inst. Segunda Enseñ. 2: 277. 1896.

Erect perennial herb, to 2 m; stem branched from above the middle, inflorescence, glabrous. Leaves with the blade ovate-oblong to oblong-lanceolate, 30–40(50) cm long, 10–15 cm wide, chartaceous, the apex acute or subacute, the base truncate, cuneate, or cordate, the margin entire or undulate, the upper and lower surfaces glabrous, the lower leaves with a petiole as long as the blade or sometimes shorter; stipules (ocrea) hyaline, friable. Flowers in a terminal, paniculiform inflorescence, contiguous or interrupted, 10–20(25) per ocreolate fascicle, bisexual, the pedicel 5–10(17) mm long, articulated in the proximal ⅓; outer perianth segments 1–2 mm long, spreading or deflexed, the inner ones suborbicular, 5–8(10) mm long, the apex obtuse to subacute, the base cordate, the margin entire or weakly erose, tubules 1(3), smooth. Fruit ovate, ca. 3 mm long, brown, the surface smooth, lustrous.

Disturbed sites. Rare; Gadsden and Wakulla Counties. Nearly throughout North America; Europe and Asia. Native to Europe and Asia. Spring–summer.

Rumex pulcher L. [Beautiful.] FIDDLE DOCK.

Rumex pulcher Linnaeus, Sp. Pl. 336. 1753. *Lapathum sinuatum* Lamarck, Fl. Franc. 3: 5. 1779 ("1778"), nom. illegit. *Lapathum pulchrum* (Linnaeus) Moench, Suppl. Meth. 121. 1802.

Erect perennial herb, to 7 dm; stem branched in the distal ⅔, glabrous. Leaves with the blade oblong-ovate, sometimes lanceolate or panduriform, 4–10(15) cm long, (2)3–5 cm wide, chartaceous, the apex obtuse to subacute, the base truncate or subcordate, the margin undulate or somewhat crisped, the upper surface glabrous, the lower surface glabrous or pubescent on the veins, the lower leaves with a petiole usually shorter than the blade; stipules (ocrea) thin, membranous, friable. Flowers in a terminal, paniculiform inflorescence, interrupted, 10–20 per ocreolate fascicle, bisexual, the pedicel 2–5 mm long, articulated in the proximal ⅓ or occasionally near the middle; outer perianth segments lanceolate, 2–3 mm long, spreading, the inner ones ovate-triangular, 3–6 mm long, 2–3 mm wide, the margin with 2–5(9) teeth on each side, tubercles (1)3. Fruit ovate, 2–3 mm long, dark reddish brown to nearly black, the surface minutely pebbled, lustrous.

Disturbed areas. Occasional; peninsula, central and western panhandle. Massachusetts south to Florida, west to New Mexico, also Oregon and California; West Indies, Mexico, and South America; Europe, Africa, Asia, and Australia. Native to Europe, Africa, and Asia. Spring–fall.

Rumex verticillatus L. [Whorled, in reference to inflorescence with axillary fascicles.] SWAMP DOCK.

Rumex verticillatus Linnaeus, Sp. Pl. 334. 1753. *Lapathum verticillatum* (Linnaeus) Nieuwland, Amer. Midl. Naturalist 3: 237. 1914.

Rumex floridanus Meisner, in A. de Candolle, Prodr. 14: 46. 1856. *Rumex verticillatus* Linnaeus subsp. *floridanus* (Meisner) Á. Löve, Taxon 35: 613. 1986. TYPE: FLORIDA: Manatee Co.: Manatee [River], 1845, *Rugel s.n.* (holotype: NY).

Rumex hydrolapathum Hudson var. *floridanus* A. W. Wood, Amer. Bot. Fl. 281. 1870. TYPE: FLORIDA.

Rumex fascicularis Small, Bull. Torrey Bot. Club 22: 367, pl. 246. 1895. *Rumex verticillatus* Linnaeus subsp. *fascicularis* (Small) A. Löve, Taxon 35: 613. 1986. TYPE: FLORIDA: Lake Co.: marshy shore of Lake Harris, near Eldorado, 1–15 Jun 1884, *Nash 898* (holotype: NY).

Erect or ascending perennial herb, to 1.5 m; stem simple or few-branched, often rooting at the lower nodes, glabrous. Leaves with the blade lanceolate or elliptic-lanceolate, 5–30(40) cm long, 2–5(10) cm wide, chartaceous, the apex acute or acuminate, the base cuneate, the margin entire or slightly undulate, the upper and lower surfaces glabrous, the lower leaves with a petiole somewhat shorter than the blade; stipules (ocrea) thin, membranous, longitudinally striate, friable. Flowers in a terminal or axillary, paniculiform inflorescence, interrupted at least in the proximal part, 10–15(25) per ocreolate fascicle, bisexual or unisexual (plants polygamomonoecious); the pedicel 10–15 mm long, articulated in the proximal ½; outer perianth segments oblanceolate, 1–2 mm long, spreading, the inner ones triangular-ovate, 4–5 mm long, 3–4 mm wide, the apex acute or subacute, the margin entire or undulate, tubercles 3, minutely punctate or transversely rugose proximally. Fruit ovate, 2–3 mm long, brown, the surface smooth, lustrous.

Swamps, marshes, and pond and river margins. Frequent; nearly throughout. Quebec south to Florida, west to Minnesota, Nebraska, Kansas, Oklahoma, and Texas. Spring–summer.

EXCLUDED TAXA

Rumex altissimus A. W. Wood—Reported for Florida by Mohr (1901) and Horton (1972). According to Wilhelm (1984), all Florida specimens determined as this taxon are referable to other species. Recently reported by Mosyakin (2005). No Florida specimens seen.

Rumex conglomeratus Murray—Reported for Leon County by Moldenke (1944), most probably based on misidentification. No Florida specimens seen.

Rumex frutescens Thouars—Reported for Florida by Mohr (1901), Wilhelm (1984), Wunderlin (1998), and Wunderlin and Hansen (2003), misapplied to our material of *R. cuneifolius*.

Rumex salicifolius J. A. Weinmann—This western species was reported from Florida by Small (1903, 1913a), probably based on a misidentification. No Florida specimens seen.

Triplaris Loefl. ex L. 1759. LONG JOHN

Trees. Leaves alternate, pinnate-veined, petiolate, stipules (ocreae) present. Flowers axillary or terminal in spiciform racemiform cymes, 1-several flowers subtended by an ocreola, unisexual (plant dioecious). Staminate flowers in fascicles subtended by an ocreola; perianth segments 6 in 2 whorls of 3 each, basally connate; stamens 9, the filaments adnate to the perianth tube in their basal ½; gynoecium absent. Carpellate flowers solitary, subtended by an ocreola; perianth segments 6, in 2 strongly unequal whorls of 3, the outer connate into a floral tube, the inner 3 smaller, free or partly connate basally into the floral tube; sterile stamens present or absent; styles 3. Fruit a trigonous achene, the floral tube persistent, surrounding the achene, the outer perianth lobes enlarged into wings.

A genus of 17 species; North America, Mexico, Central America, and South America. [*Triplus*, three, and *aris*, provided with, in reference to the flowers and fruits.]

Selected reference: Brandbyge (1986).

Triplaris melaenodendron (Bertol.) Standl. & Steyerm. [From the Greek *melaeno*, dark, apparently in reference to the dark inflorescences.] LONG JOHN.

Vellasquezia melaenodendron Bertoloni, Novi Comment. Acad. Sci. Inst. Bononiensis 4: 440, t. 46. 1840. *Triplaris melaenodendron* (Bertoloni) Standley & Steyermark, Publ. Field Mus. Nat. Hist., Bot. Ser. 23: 5: 1943.

Tree, to 12(20) m; bark whitish brown to brown mottled, the branchlets glabrous to subglabrous. Leaves with the blade ovate to ovate-elliptic, 15–35 cm long, 8–18(22) cm wide, chartaceous, the apex acute or abruptly short-acuminate, the base rounded to slightly cordate, the margin entire, the upper surface pilose to glabrous, the lower pilose to puberulous, the midrib strigose, the petiole 1–2 cm long strigose; stipules (ocrea) to 10 cm long, chartaceous, sparsely whitish pubescent, caducous and leaving a ringlike scar. Flowers in a spiciform raceme 2–25 cm long, gray-yellow to pale brown pilose-tomentose to pilose-strigose, the peduncle 1–4 cm long, the pedicel to 4 mm long. Staminate flowers 3–7 in a fascicle; perianth segments elliptic,

2–3 mm long and wide, gray-yellow to pale brown, the apex obtuse, the margin entire. Carpellate flowers solitary in the raceme; perianth segments as in the staminate. Fruit 4–5(6) cm long, reddish sericeous-pilose, the floral tube narrowly oblong-campanulate, 1.5–2 cm long, the persistent outer perianth segments oblanceolate, 2.5–4 cm long, the apex rounded to subacute; achene 10–11(17) mm long, dark yellowish brown, the surface glabrous to subglabrous.

Disturbed sites. Rare; Collier County. Escaped from cultivation. Florida; Mexico, Central America, and South America. Native to Mexico, Central America, and South America. Spring.

DROSERACEAE Salisb. 1808. SUNDEW FAMILY

Annual or perennial herbs. Leaves alternate, the blade infolded or circinate in vernation, modified as active traps (*Dionaea*) or with mucilage-tipped, sensitive trichomes (*Drosera*), petiolate or the blade not distinguished from the petiole, stipulate or estipulate. Flowers in scapose umbelliform cymes or 1-sided circinate racemes, bracteate, actinomorphic, bisexual; sepals 5, free or basally connate; petals 5, free or basally connate; stamens 5 or 10–20, the anthers versatile, 4-loculate; ovary superior, 3- to 5-carpellate, 1-loculate, the styles 1 or 3. Fruit a 3-valved or an irregularly dehiscent capsule; seeds numerous.

A family of 4 genera and about 85 species; nearly cosmopolitan.

Dionaeaceae Raf. (1837).

Selected references: Mellichamp (2015); Wood (1960).

1. Leaves (at least some) with a trap mechanism consisting of the blade hinged lengthwise along the midrib and folding inward, the margins with interlocking bristles ....**Dionaea**
1. Leaves with a trap mechanism consisting of numerous gland-tipped projections ....**Drosera**

Dionaea Sol. ex J. Ellis 1768. VENUS FLYTRAP

Perennial herbs. Leaves with the blade medially hinged and modified to form sensitive, active traps, petiolate, estipulate. Flowers in scapose, umbelliform cymes; sepals 5, free; petals 5, free; stamens (10)15(20); ovary 5-carpellate, 1-loculate, the style 1, the stigma capitate. Fruit an irregularly dehiscent capsule; seeds numerous.

A genus of 1 species; North America and West Indies. [From the Greek *Dione*, goddess, of Greek mythology.]

Dionaea muscipula Ellis [*Musc*, fly.] VENUS FLYTRAP.

Dionaea muscipula J. Ellis, St. James's Chron. Brit. Eve. Post 1172: [4]. 1768. *Dionaea sensitiva* Salisbury, Prodr. Stirp. Chap. Allerton 321. 1796, nom. illegit. *Dionaea corymbosa* Rafinesque, Med. Fl. 2: 217. 1830, nom. illegit.

Scapose herb. Leaves rosulate, the bases forming a bulb-like structure around a short, horizontal stem, the blade ca. 3 cm long, hinged along the midrib, with 3 sensitive bristles along the midrib, each blade half elongate-subreniform, the margin with conspicuous spinelike bristles that interlock when the leaf is closed, the petiole ca. 6 cm long, usually winged and distally expanded. Flowers in a scape 10–30 cm long, the basal bract triangular, 3–5 mm long; sepals triangular, 4–5 mm long; petals spatulate, 10–12 mm long; stigma with numerous elongate papillae. Fruit ovoid, 4–5 mm long; seeds obovoid, 1–2 mm long, black, lustrous.

Roadside seepage areas. Rare; Liberty and Franklin Counties. New Jersey, North Carolina, South Carolina, and Florida; West Indies. Native to North and South Carolina. Spring.

Drosera L. 1753. SUNDEW

Annual or perennial herbs. Leaves circinate, the margins and upper surface of the blades with sensitive, gland-tipped trichomes, petiolate or without a distinction between the blade and the petiole, stipulate or estipulate. Flowers in 1-sided circinate racemes; sepals 5, basally connate; petals 5, free or basally connate; stamens 5, free; ovary 3-carpellate, 1-loculate, the styles 3, bifid. Fruit a 3-valved capsule; seeds numerous.

A genus of about 190 species; nearly cosmopolitan. [From the Greek *droseros*, dewy, from the glistening, mucilaginous droplets secreted by the glandular-tipped trichomes.]

Selected references: Shinners (1962a); Wynne (1944).

1. Leaves filiform, without a distinct blade and petiole.
    2. Petals 0.7–1 cm long; leaves 8–25 cm long, less than 1 mm wide, glandular trichomes purple, drying dark brown..**D. filiformis**
    2. Petals 1–3 cm long; leaves 25–40 cm long, more than 1 mm wide, glandular trichomes green, drying light brown... **D. tracyi**
1. Leaves with a distinct blade and petiole.
    3. Scape glandular-pubescent except at the base; stipules absent or vestigial................... **D. brevifolia**
    3. Scape glabrous; stipules prominent.
        4. Petiole glabrous, filiform; petals white; seeds uniformly papillose.........................**D. intermedia**
        4. Petiole with few to many nonglandular trichomes; petals pink; seeds short-papillose-ridged ....
        ....**D. capillaris**

Drosera brevifolia Pursh [Short-leaved.] DWARF SUNDEW.

Drosera brevifolia Pursh, Fl. Amer. Sept. 211. 1814.
Drosera uniflora Rafinesque, Atl. J. 148. 1832. TYPE: FLORIDA.
Drosera leucantha Shinners, Sida 1: 57. 1962.

Scapose herb, the basal rosette to 3.5 cm wide. Leaves with the blade suborbicular to spatulate, 4–10 mm long, the margin and upper surface with sensitive gland-tipped trichomes, the petiole 5–10 mm long and wide, glabrous; stipules absent or with only a few vestigial setaceous segments. Flowers 1–8 in a scape 4–8 cm long, glandular-stipitate, the pedicel glandular-stipitate, the basal bract linear-lanceolate, 3–4 mm long; sepals oblong-ovate, glandular-pubescent, the apex obtuse; petals obovate, 4–5 mm long, 2–3 mm wide, white to pink or rose-purple, the apex rounded, the margin slightly erose. Fruit ovoid, 3–4 mm long; seeds oblong to obovate, caudate at the base, ca. 0.4 mm long, black, with crateriform markings in 10–12 rows.

Flatwoods, savannas, bogs, and pond margins. Frequent; nearly throughout. Virginia south to Florida, west to Kansas, Oklahoma, and Texas; West Indies, Mexico, Central America, and South America. Spring.

Drosera capillaris Poir. [Hairlike, in reference to the flowering scape.] PINK SUNDEW.

Drosera capillaris Poiret, in Lamarck, Encycl. 6: 299. 1804. *Drosera rotundifolia* Linnaeus var. *capillaris* (Poiret) Eaton & J. Wright, Man. Bot., ed. 8. 230. 1840.
Drosera sessilifolia Rafinesque, Atl. J. 148. 1832. TYPE: FLORIDA.

Scapose herb, the basal rosette to 10(12) cm wide. Leaves with the blade obovate to spatulate, 5–10 mm long, 4–9 mm wide, the margin and upper surface with sensitive gland-tipped trichomes, the petiole 1–4 cm long, with few to many gland-tipped trichomes; stipules free or slightly basally adnate, fimbriate, the segments 3–5 mm long. Flowers 2–20 in a scape 4–20 cm long, glabrous, the pedicel glabrous, the basal bract linear-lanceolate, 3–4 mm long; sepals oblong-elliptic, 3–4 mm long, 1–2 mm wide, glabrous, the apex obtuse; petals obovate, 6–7 mm long, 2–3 mm wide, pink, the apex rounded, the margin slightly erose. Fruit ellipsoid-ovoid, 4–5 mm long; seeds ovate-oblong to elliptic, ca. 0.5 mm long, brown, coarsely papillose-corrugated in 14–16 ridges.

Wet flatwoods and bogs. Frequent; nearly throughout. Maryland and Delaware south to Florida, west to Texas; West Indies, Mexico, Central America, and South America. Spring.

Drosera filiformis Raf. [Threadlike, in reference to the leaves.] THREADLEAF SUNDEW.

Drosera filiformis Rafinesque, Med. Repos., ser. 2. 5: 360. 1808. *Drosera filiformis* Rafinesque var. *typica* Wynne, Bull. Torrey Bot. Club 71: 171. 1944, nom. inadmiss.

Scapose herbs. Leaves erect, filiform, without a distinction between the blade and the petiole, 8–25 cm long, to 1 mm wide, with purple glandular trichomes; stipules basally adnate, long-fimbriate, forming a wool-like mat among the close-set leaf bases. Flowers 4–16 in a scape 6–22 cm long, glabrous, the pedicel glandular-pubescent, the basal bract linear-lanceolate, 2–3 mm long; sepals oblong to elliptic, 4–7 mm long, 2–3 mm wide, glandular-stipitate, the apex obtuse; petals obovate, 7–15 mm long, 5–8 mm wide, purple, the apex rounded, the margin erose. Fruit obovoid, 5–6 mm long; seeds ellipsoidal, abruptly caudate at both ends, ca. 0.5 mm long, black, coarsely crateriform with the pits in 16–20 rows.

Exposed lake bottoms. Rare; Washington and Bay Counties. Nova Scotia, Massachusetts, and New York south to Maryland, also North Carolina and Florida. Spring.

Drosera filiformis is listed as endangered in Florida (Florida Administrative Code, Chapter 5B-40).

Drosera intermedia Hayne [Intermediate between two species.] WATER SUNDEW; SPOONLEAF SUNDEW.

Drosera intermedia Hayne, J. Bot. (Schrader) 1800(1): 37. 1801. *Rorella intermedia* (Hayne) Nieuwland, Amer. Midl. Naturalist 4: 56. 1915.

Scapose herb. Leaves in a rosette or along a solitary or branching stem to 10 cm long, the blade oblong-spatulate to obovate, 8–20 mm long, 3–5 mm wide, often very gradually tapering into the petiole, the margin and upper surface with sensitive gland-tipped trichomes, the petiole to 5 cm long, glabrous; stipules basally adnate, fimbriate, the segments to 5 mm long. Flowers 9–20 in a scape 9–20 cm long, the pedicel glabrous or sparsely pubescent, the basal bract linear-lanceolate, 3–4 mm long; sepals oblong, 3–4 mm long, 1–1.5 mm wide, glabrous or sparsely pubescent, the apex obtuse; petals obovate, 5–8 mm long, 3–5 mm wide, white or pinkish, the apex obtuse, the margin slightly undulate. Fruit ellipsoid, 4–5 mm long; seeds oblong, ca. 1 mm long, reddish brown to black, densely elongate-papillose.

Pond margins. Occasional; northern counties, central peninsula. Labrador south to Florida, west to Nunavut, Ontario, Minnesota, Illinois, Arkansas, and Texas, also Idaho; West Indies and South America; Europe. Spring.

Drosera intermedia is listed as threatened in Florida (Florida Administrative Code, Chapter 5B-40).

Drosera tracyi (Diels) Macfarl. [Commemorates Samuel Mills Tracy (1847–1920), Mississippi botanist and agronomist who collected extensively in the southeastern United States.] TRACY'S SUNDEW.

> *Drosera filiformis* Rafinesque var. *tracyi* Diels, in Engler, Pflanzenr. 4(Heft 26): 92. 1906. *Drosera tracyi* (Diels) Macfarlane, in L. H. Bailey, Stand. Cycl. Hort. 2: 1077. 1914.

Scapose herb. Leaves erect, the blade filiform, 25–35 cm long, to 1 mm wide. Leaves filiform, to 40 cm long, 1–2 mm wide, with green glandular-stipitate trichomes, without a distinction between the blade and the petiole; stipules basally adnate, long-fimbriate, forming a wool-like mat among the close-set leaf bases. Flowers 14–16 in a scape 25–45 cm long, this glabrous, the pedicel glandular-pubescent, the basal bract linear-lanceolate, 2–3 mm long; sepals oblong to elliptic, 4–7 mm long, 2–3 mm wide, glandular-pubescent, the apex obtuse; petals obovate, 12–15 mm long, 12–15 mm wide, purple-rose, the apex rounded, the margin erose. Fruit obovoid, 5–6 mm long; seeds ellipsoidal, abruptly caudate at both ends, ca. 0.5 mm long, black, coarsely crateriform with the pits in 16–20 rows.

Bogs and flatwoods. Frequent; central and western panhandle. Georgia south to Florida, west to Louisiana, also California. Summer.

EXCLUDED TAXA

> *Drosera longifolia* Linnaeus—Reported by Chapman (1860, 1883), the name misapplied to material of *D. intermedia*.
> *Drosera rotundifolia* Linnaeus—Reported by Chapman (1860, 1883) and Small (1903, 1913a, 1933). No Florida specimens known.

CARYOPHYLLACEAE Juss., nom. cons. 1789. PINK FAMILY

Annual or perennial herbs; stems commonly with swollen nodes. Leaves opposite, sometimes with axillary leaf fascicles (short leafy lateral branches) and appearing whorled, the blades simple, pinnate-veined, petiolate or epetiolate, stipulate or estipulate. Flowers in dichasial cymes or solitary, axillary or terminal, actinomorphic, bisexual or rarely unisexual, bracteate; sepals (3–4)5, free or basally connate; petals (3–4)5, free, sometimes reduced or absent; stamens (1–4)5 or 10, the filaments free or basally adnate to the petals; ovary superior, 2- to 5-carpellate, 1-loculate, the styles 1–5, free or united. Fruits a few- to many-seeded capsule dehiscing longitudinally or by apical teeth or a 1-seeded indehiscent utricle.

A family of about 90 genera and about 2,200 species; nearly cosmopolitan.

Alsinaceae Bartl., nom. cons. (1825); *Corrigiolaceae* (Dumort.) Dumort. (1829); *Scleranthaceae* J. Presl & C. Presl (1822).

Selected reference: Rabeler and Thieret (1988).

1. Fruit 1-seeded, indehiscent.
  2. Stipules present.. **Paronychia**
  2. Stipules absent ....**Scleranthus**
1. Fruit few- to many-seeded, dehiscent.
  3. Sepals united for most of their length to form a campanulate, cylindric, or often inflated calyx tube.
    4. Styles 3–5.
      5. Calyx lobes longer than the tube.. **Agrostemma**
      5. Calyx lobes shorter than the tube .... **Silene**
    4. Styles 2.
      6. Calyx subtended by 1–3 pairs of bracts .... **Dianthus**
      6. Calyx ebracteate.
        7. Flowers to 0.5 cm long .... **Gypsophila**
        7. Flowers 2 cm long or more.
          8. Calyx ovoid, angle-winged, 5-nerved... **Vaccaria**
          8. Calyx tubular, not angle-winged, 20-nerved ....**Saponaria**
  3. Sepals separate or basally united.
    9. Floral bracts and sepals scarious ....**Polycarpaea**
    9. Floral bracts and sepals with hyaline margins, but not scarious.
      10. Styles partly united.
        11. Stem leaves scalelike... **Stipulicida**
        11. Stem leaves suborbicular.
          12. Leaves always 2 at a node..**Drymaria**
          12. Leaves often appearing as 4 at a node due to the presence of axillary leafy fascicles (short leafy lateral branches)... **Polycarpon**
      10. Styles separate.
        13. Stipules present.
          14. Leaves appearing whorled due to the presence of axillary leafy fascicles (short leafy lateral branches)..**Spergula**
          14. Leaves always 2 at a node ....**Spergularia**
        13. Stipules absent.
          15. Petals deeply 2-lobed.
            16. Capsule ovoid or ellipsoidal, dehiscing nearly the entire length .... **Stellaria**
            16. Capsule long-cylindric, dehiscing apically ....**Cerastium**
          15. Petals entire or absent.
            17. Styles 4 or 5 .... **Sagina**
            17. Styles 3.

18. Valves of the capsule 6 .... **Arenaria**
18. Valves of the capsule 3.
    19. Leaves acicular, 0.3–1 cm long, densely congested on the lower stem; peren-
        nial herb.. **Sabulina**
    19. Leaves linear, 1–3.5 cm long, disposed evenly along the stem; perennial
        herb .... **Mononeuria**

Agrostemma L. 1753. CORNCOCKLE

Annual herbs. Leaves simple, opposite, pinnate-veined, epetiolate, estipulate. Flowers few in terminal or subterminal cymes or solitary, actinomorphic, bisexual; sepals 5, basally connate and forming a tube; petals 5, free; stamens 10; styles (4)5. Fruit a capsule, dehiscing by (4)5 apical teeth; seeds numerous.

A genus of 2 species; North America, West Indies, South America, Europe, Africa, Asia, Australia, and Pacific Islands. [From the Greek *agros*, field, and *stemma*, crown, in reference to the crown-like calyx lobes and the plant growing as a weed in fields.]

Selected reference: Thieret (2005).

Agrostemma githago L. [A pre-Linnaean name.] COMMON CORNCOCKLE.

Agrostemma githago Linnaeus, Sp. Pl. 435. 1753. *Lychnis githago* (Linnaeus) Scopoli, Fl. Carniol., ed. 2. 1: 310. 1771. *Lychnis segetum* Lamarck, Fl. Franc. 3: 50. 1779 ("1778"), nom. illegit. *Githago segetum* Link, Diss. Bot. 62. 1795. *Agrostemma hirsuta* Stokes, Bot. Mat. Med. 2: 559. 1812, nom. illegit. *Lychnis agrostemma* Ledebour, Fl. Altaic. 2: 184. 1830, nom. illegit.

Erect annual, to 10 dm; stem simple to freely branched, often prominently 4-angled, thinly white-silky pilose. Leaves with the blade linear to linear-lanceolate, 4–15 cm long, 3–10 mm wide, the apex acute, the base cuneate, the margin entire, the upper and lower surfaces thinly white-silky pilose. Flowers solitary or several in a cyme, borne on a slender terminal or sub-terminal axillary pedicel to 20 cm long; calyx tube elliptic-ovoid, 12–18 mm long, strongly 10-ribbed, becoming inflated and hardened in fruit, the lobes linear-lanceolate, 2–4 cm long, the apex acute or acuminate; petals oblanceolate, 2–4 cm long, the apex shallowly 2-lobed, red or reddish purple. Fruit 15–20 mm long; seeds reniform, ca. 3 mm long, the surface tuberculate, black.

Disturbed sites. Rare; northern peninsula. Nearly throughout North America; West Indies and South America; Europe, Africa, Asia, Australia, and Pacific Islands. Native to Europe, Africa, and Asia. Summer–fall.

Arenaria L. 1753. SANDWORT

Annual or perennial herbs. Leaves opposite, simple, pinnate-veined, epetiolate, estipulate. Flowers in a terminal cyme, actinomorphic, bisexual; sepals 5, free or slightly basally connate; petals 5, free; stamens 10, inserted with the petals at the base of a nectariferous disk; styles 3. Fruit a capsule, dehiscent by 6 apical valves; seeds numerous.

A genus of about 210 species; nearly cosmopolitan. [Growing on sand.]
Selected reference: Hartman et al. (2005).

1. Leaves elliptic-lanceolate, 7–32 mm long; plant perennial...**A. lanuginosa**
1. Leaves ovate, 2.5–5 mm long; plant annual ....**A. serpyllifolia**

Arenaria lanuginosa (Michx.) Rohrb. [Woolly, in reference to the fine pubescence.] SPREADING SANDWORT.

Spergulastrum lanuginosum Michaux, Fl. Bor.-Amer. 1: 275. 1803. *Micropetalon lanuginosum* (Michaux) Persoon, Syn. Pl. 1: 509. 1805. *Stellaria lanuginosa* (Michaux) Torrey & A. Gray, Fl. N. Amer. 1: 187. 1838. *Arenaria lanuginosa* (Michaux) Rohrbach, in Martius, Fl. Bras. 14(2): 274. 1872. *Arenaria lanuginosa* (Michaux) Rohrbach var. *genuina* Rohrbach, Linnaea 37: 260. 1872, nom. inadmiss.

Arenaria diffusa Elliott, Sketch Bot. S. Carolina 1: 519. 1821. *Arenaria lanuginosa* (Michaux) Rohrbach var. *diffusa* (Elliott) Macloskie, Rep. Princeton Univ. Exp. Patagonia, Botany 1: 394. 1905; non Rohrbach, 1872.

Perennial herb, to 5 dm; stem retrorsely puberulent. Leaves with the blade elliptic to lanceolate or oblanceolate, 15–30 mm long, 2–8 mm wide, 1-veined, the apex acute, the base cuneate to rounded, the margin entire, ciliolate at the base, the upper and lower surfaces somewhat pustulate, sessile. Flowers axillary, solitary, the pedicel 1.5–4 cm long, ascending to reflexed, puberulent; sepals lanceolate, 3–4 mm, the apex acute, 1- to 3-nerved, glabrous, often pustulate, sometimes ciliolate at the base; petals minute or absent. Fruit ovoid, 3–6 mm long, loosely to tightly enclosed by the persistent calyx; seeds globose-reniform, somewhat keeled dorsally, the surface smooth, black or dark reddish black, lustrous.

Shaded stream banks and swamp margins. Frequent; northern counties, central peninsula. Virginia south to Florida, west to Texas; West Indies, Mexico, Central America, and South America. Spring–fall.

Arenaria serpyllifolia L. [With leaves of *Serpyllum* (= *Thymus*) (Lamiaceae).] THYMELEAF SANDWORT.

Annual herb, to 3 dm; stem puberulent. Leaves with the blade ovate, 3–8 mm long, 3–5 mm wide, 3- to 5-veined, the apex acute, the base cuneate to rounded, the margin entire, the upper and lower surfaces sparsely scabrid-puberulent, sessile. Flowers axillary, solitary, the pedicel 4–8 mm long; sepals lance-ovate, 2.5–4 mm long, the apex acuminate, 3- to 5-nerved, somewhat keeled, the margin scarious, scabrid-puberulent or glandular-puberulent; petals usually shorter than the sepals. Fruit ovoid or ovoid-conical, distinctly swollen at the base, less than twice as long as wide, exceeding the persistent sepals; seeds globose-reniform, the surface tesselate-tuberculate, gray-black or reddish brown.

1. Fruit ovoid or ovoid-conical, distinctly swollen at the base, less than 2 times as long as wide ....
.... subsp. **serpyllifolia**
1. Fruit cylindric to cylindric-obovoid, not or scarcely swollen at the base, usually 2 times as long as wide .... subsp. **leptoclados**

Arenaria serpyllifolia L. subsp. **serpyllifolia** THYMELEAF SANDWORT.

> *Arenaria serpyllifolia* Linnaeus, Sp. Pl. 423. 1753. *Alsine serpyllifolia* (Linnaeus) Crantz, Inst. Rei Herb. 2: 406. 1766. *Stellaria serpyllifolia* (Linnaeus) Scopoli, Fl. Carniol., ed. 2. 1: 319. 1771. *Alsinanthus serpyllifolius* (Linnaeus) Desvaux, J. Bot. (Desvaux) 3: 222. 1816. *Alsinella serpyllifolia* (Linnaeus) Gray, Nat. Arr. Brit. Pl. 2: 665. 1821. *Euthalia serpyllifolia* (Linnaeus) Ruprecht, Fl. Caucasi 220. 1869. *Arenaria serpyllifolia* Linnaeus subsp. *euserpyllifolia* Briquet, Prodr. Fl. Corse 1: 536. 1910, nom. inadmiss.

Fruit ovoid or ovoid-conical, distinctly swollen at the base, less than twice as long as wide.

Disturbed sites. Occasional; northern counties south to Hernando County. Nearly throughout North America; West Indies and South America; Europe, Africa, Asia, Australia, and Pacific Islands. Native to Europe, Africa, and Asia. Spring.

Arenaria serpyllifolia L. subsp. **leptoclados** (Rchb.) Nyman [From the Greek *leptos*, fine or slender, and *clados*, branch.] THYMELEAF SANDWORT.

> *Arenaria serpyllifolia* Linnaeus var. *leptoclados* Reichenbach, Icon. Fl. Germ. Helv. 5: 32, pl. 216(4941B). 1842. *Arenaria leptoclados* (Reichenbach) Gussone, Fl. Sicul. Syn. 2: 824. 1844. *Arenaria serpyllifolia* Linnaeus subsp. *leptoclados* (Reichenbach) Nyman, Consp. Fl. Eur. 115. 1878.
>
> *Arenaria serpyllifolia* Linnaeus var. *tenuior* Mertens & W.D.J. Koch, in Röhling, Deutschl. Fl., ed. 3. 3: 266. 1831. *Arenaria serpyllifolia* Linnaeus subsp. *tenuior* (Mertens & W.D.J. Koch) Arcangeli, Comp. Fl. Ital. 101. 1882. *Arenaria tenuior* (Mertens & W.D.J. Koch) Gürke, in K. Richter, Pl. Eur. 2: 273. 1899.

Fruit cylindric to cylindric-obovoid, not or scarcely swollen at the base, usually twice as long as wide.

Disturbed sites. Rare; Alachua and Levy Counties, panhandle. Pennsylvania south to Florida, west to Missouri, Arkansas, and Louisiana; West Indies; Europe, Africa, Asia, and Australia. Native to Europe, Africa, and Asia. Spring.

Cerastium L. 1753. CHICKWEED

Annual or perennial herbs. Leaves opposite, simple, pinnate-veined, epetiolate, estipulate. Flowers in terminal cymes, actinomorphic, bisexual; sepals 5, free; petals 5, free, 2-lobed; stamens (5)10, free. Fruit a capsule, dehiscing by apical teeth; seeds numerous.

A genus of about 100 species; nearly cosmopolitan. [From the Greek *kerastes*, horned, in reference to the shape of the slender and often recurved capsule apices.]

Selected reference: Morton (2005a).

1. Flowers arranged in dense glomerules, the pedicel usually 1–3 mm long......................**C. glomeratum**
1. Flowers in an open inflorescence, the pedicel usually 4 mm long or more.
   2. Plant with long, silvery trichomes giving it a gray appearance; sepals and bracts wholly herbaceous ....**C. brachypetalum**
   2. Plant lacking long, silvery trichomes, not appearing gray; sepals and bracts with a scarious margin.
      3. Plant perennial, often with short, nonflowering branches in the leaf axil ....**C. fontanum**
      3. Plant annual, never with short, nonflowering branches in the leaf axil ....**C. semidecandrum**

Cerastium brachypetalum Pers. [From the Greek, *brachy*, short, and the Latin, *petalum*, petal.] GRAY CHICKWEED.

Cerastium brachypetalum Persoon, Syn. Pl. 1: 520. 1805.

Annual herb, to 3 dm; stem erect, simple or branched at the base, with long, spreading or ascending, silvery trichomes. Lower leaves with the blade oblanceolate to spatulate, the apex obtuse, the main cauline leaves lanceolate to elliptic, 4–15(20) mm long, 1.5–5(7) mm wide, the apex acute, the base cuneate to rounded, the margin entire, the upper and lower surfaces with long, glandular or eglandular, silvery trichomes. Flowers 3–30 in a lax dichasium, the pedicel longer than the calyx; bracts herbaceous, lanceolate, with long, ascending, glandular or eglandular, silvery trichomes; sepals lanceolate, 4–5 mm long, foliaceous, wholly herbaceous, with long, ascending, glandular or eglandular, silvery trichomes exceeding the tips; petals oblanceolate, 2–3 mm long, the apex 2-fid, sparsely ciliate proximally. Fruit 5–7 mm long, slightly curved distally; seeds obovate-reniform, the surface tuberculate, pale brown.

Disturbed sites. Rare; panhandle. New York south to Florida, west to Kansas, Oklahoma, and Texas, also Idaho and Oregon; Europe and Asia. Native to Europe and Asia. Spring.

Cerastium fontanum Baumg. subsp. **vulgare** (Hartm.) Greuter & Burdet [*Fontanus*, spring of water, in reference to the plant's habitat of wet or damp places; *vulgaris*, common.] BIG CHICKWEED.

Cerastium vulgare Hartman, Handb. Skand. Fl. 182. 1820. *Cerastium holosteoides* Fries var. *vulgare* (Hartman) Hylander, Uppsala Univ. Arsskr. 1945(7): 151. 1945. *Cerastium fontanum* Baumgarten subsp. *vulgare* (Hartman) Greuter & Burdet, Willdenowia 12: 37. 1985.

Perennial herb, to 3.5 dm; stem erect to decumbent, few- to many-branched, with lateral branches or short leafy fascicles at the lower nodes, hirsute, glandular-pubescent above or throughout. Lower leaves with the blade obovate to spatulate or elliptic, the apex obtuse to rounded, the main cauline leaves ovate to elliptic-ovate or oblong, 5–30 mm long, 2–10 mm wide, the apex obtuse to acute, the base cuneate, the margin entire, the upper and lower surfaces with spreading, eglandular trichomes. Flowers in a spreading dichasium, the pedicel 1–3 times as long as the calyx; bracts often purplish-tipped, the uppermost slightly to prominently scarious-margined and tipped; sepals lanceolate, 4–6 mm long, usually purplish-tipped, hirsute, often glandular; petals shorter than to slightly exceeding the sepals. Fruit 7–11 mm long, straight or curved; seeds obovate-reniform, the surface papillate-tuberculate, reddish brown.

Disturbed sites. Rare; Jackson and Walton Counties. Nearly throughout North America; West Indies, Mexico, and South America; Europe, Africa, Asia, Australia, and Pacific Islands. Native to Europe, Africa, and Asia. Spring.

Cerastium glomeratum Thuill. [Close together in a head, in reference to the head-like inflorescence.] MOUSE-EAR CHICKWEED; STICKY CHICKWEED

Cerastium glomeratum Thuillier, Fl. Env. Paris, ed. 2. 226. 1799. *Cerastium vulgatum* Linnaeus var. *glomeratum* (Thuillier) Edgeworth & Hooker f, in Hooker f., Fl. Brit. Ind. 1: 228. 1874.

Annual herb, to 3 dm; stem erect or decumbent, solitary or few-branched from the base, finely hirsute throughout and usually glandular-pubescent above. Leaves with the blade obovate to spatulate below, oval to ovate or elliptic-ovate above, 0.5–2(3) cm long, 3–9(15) mm wide, the apex obtuse to rounded, the base cuneate, the margin entire, the upper and lower surfaces hirsute. Flowers in a dense cluster at the tip of dichotomous branches, the pedicel shorter than the calyx; bracts herbaceous, hirsute and glandular-pubescent; sepals lanceolate, 3–5 mm long, hirsute and glandular-pubescent, with some eglandular trichomes projecting beyond the tip, usually purple-spotted at the apex, the margin scarious; petals shorter than to slightly exceeding the sepals, the claw glabrous. Fruit 5–8 mm long, straight or curved; seeds obovate-reniform, ca. 0.5 mm long, the surface papillate-tuberculate, light brown.

Disturbed sites. Frequent; northern counties, central peninsula. Nearly cosmopolitan. Native to Europe, Africa, and Asia. Spring.

Cerastium semidecandrum L. [From the Latin *semi*, half, and the Greek *deca*, ten, and *andros*, male or stamen, in reference to the five fertile stamens.] FIVESTAMEN CHICKWEED.

Cerastium semidecandrum Linnaeus, Sp. Pl. 438. 1753. *Centunculus semidecander* (Linnaeus) Scopoli, Fl. Carniol., ed. 2. 1: 321. 1771. *Myosotis semidecandrum* (Linnaeus) Moench, Methodus 225. 1794. *Cerastium semidecandrum* Linnaeus var. *genuinum* Rouy & Foucaud, Fl. France 3: 219. 1896, nom. inadmiss.

Annual herb, to 2 dm; stem glandular-pubescent. Lower leaves with the blade oblanceolate, the cauline ones ovate to broadly elliptic, 0.5–1(1.5) cm long, 2–5 mm wide, the apex acute to obtuse, the base cuneate, the margin entire, the upper and lower surfaces white-pubescent. Flowers in a moderately open cyme, the pedicel usually subequaling or a little longer than the sepals; bracts evidently scarious-margined and -tipped; sepals lanceolate, 3–5 mm long, the apex acute, stipitate-glandular, usually with a few eglandular trichomes, the margin scarious; petals shorter than the sepals, shallowly notched, the claw glabrous; stamens 5 or sometimes 10. Fruit 4–7 mm long; seeds obovate-reniform, the surface smooth or nearly so, light brown.

Disturbed sites. Rare; Okaloosa, Santa Rosa, and Escambia Counties. Nova Scotia and New Brunswick, Massachusetts and New York south to Florida, west to Ontario, Wisconsin, Nebraska, Kansas, Arkansas, and Louisiana, also British Columbia, Washington, and Oregon; Europe, Africa, and Asia. Native to Europe, Africa, and Asia. Spring.

EXCLUDED TAXA

Cerastium nutans Rafinesque—Reported by several authors, including Correll and Johnston (1970) and Small (1903, 1913a, 1933, all as *C. longepedunculatum* Muhlenberg ex W.P.C. Barton). No Florida specimens known.

Cerastium viscosum Linnaeus—This name of confused application, now a nomen rejiciendum (see Taxon 46: 775–778. 1997; 49: 262. 2000), was used for *C. glomeratum* in Florida by several authors, including Chapman (1860, 1883) and Small (1903, 1913a, 1933).

Cerastium vulgatum Linnaeus—This name of confused application, now a nomen rejiciendum (see Taxon 46: 775–778. 1997; 49: 262. 2000), was used for *C. fontanum* subsp. *vulgare* in Florida by several authors, including Chapman (1860, 1883) and Small (1903, 1913a, 1933).

Dianthus L. 1753. PINK

Annual or biennial herbs. Leaves opposite, simple, pinnate-veined, epetiolate, estipulate. Flowers in terminal, head-like cymes or occasionally solitary, bisexual; sepals 5, basally connate into a tube; petals 5, free; stamens 10, free; styles 2. Fruit a capsule dehiscing by 4-teeth; seeds numerous.

A genus of about 300 species; North America, South America, Europe, Africa, Asia, Australia, and Pacific Islands. [From the Greek *Dios*, god, and *anthos*, flower.]

Selected reference: Rabeler and Hartman (2005b).

Dianthus armeria L. [Roman name of the sea-pink, *Armeria* (Plumbaginaceae), with which this species was once associated.] DEPTFORD PINK.

Dianthus armeria Linnaeus, Sp. Pl. 410. 1753. *Caryophyllus armeria* (Linnaeus) Moench, Methodus 59. 1794. *Cylichnanthus maculatus* Dulac, Fl. Hautes-Pyrénées 261. 1867, nom. illegit. *Diosanthos armeria* (Linnaeus) Saint-Lager, Ann. Soc. Bot. Lyon 7: 87, 124. 1880.

Erect annual or biennial herb, to 7 dm; stem dichotomously branched above, glabrous or crisp-pubescent at the nodes or in the inflorescence. Leaves with the blade linear, 3–10 cm long, 1–5 mm wide, the basal ones sometimes oblanceolate, to 8 mm wide, the apex attenuate or blunt, the base connate with the opposite leaf base for 2–4 mm, the margin entire, the upper and lower surfaces strigillose. Flowers 2–several in a head-like cyme or occasionally solitary, sessile or nearly so, each subtended by 1–3 pairs of linear-attenuate bracts; calyx tube 1–1.5 cm long, 20- to 25-nerved, strigillose, the lobes 3–6 mm long; petals long-clawed, the blade rhombic-obovate, 4–5 mm long, dentate, pink or rose, dotted with white. Fruit subequaling the persistent calyx; seeds subglobose, the surface minutely tuberculate, dark brown.

Disturbed sites. Rare; Bay County. Escaped from cultivation. Nearly throughout North America; South America, Europe, Asia, Australia, and Pacific Islands. Native to Europe. Spring.

Drymaria Willd. ex Schult. 1819. DRYMARY

Sprawling annual or perennial herbs. Leaves opposite, simple, pinnate-veined, petiolate, stipulate. Flowers in terminal or axillary cymes, actinomorphic, bisexual; sepals 5, free; petals 3–5, free; stamens 2–5, the filaments flattened, slightly connate; styles 3, slightly basally connate. Fruit a capsule dehiscing into 3 entire valves; seeds few.

A genus of about 48 species; North America, West Indies, Mexico, Central America, South America, Africa, Asia, Australia, and Pacific Islands. [From the Greek *drym*, pertaining to the forest.]

Selected reference: Rabeler and Hartman (2005a).

Drymaria cordata (L.) Willd. ex Schult. [Heart-shaped, in reference to the leaves.] DRYMARY; WEST INDIAN CHICKWEED.

Holosteum cordatum Linnaeus, Sp. Pl. 88. 1753. *Drymaria cordata* (Linnaeus) Willdenow ex Schultes, in Roemer & Schultes, Syst. Veg. 5: 406. 1819.

Decumbent annual or perennial herb; stem 2–6 dm long, slender, much branched, often rooting at the nodes, glabrate or glandular-puberulent. Leaves with the blade orbicular to reniform, 0.5–2.5 mm long, 0.5–3 mm wide, the apex rounded to mucronulate, the base rounded to cordate, the margin entire, the upper and lower surfaces glabrate to glandular-puberulent, the petiole 2–5(15) mm long; stipules to 2 mm long, lacerate, scarious. Flowers in a terminal or axillary, lax cyme, the pedicel with a band of glandular trichomes on the lower part; sepals lanceolate to ovate, 2–4 mm long, glabrous or with a patch of short, stiff trichomes in the middle; petals 2–3 mm long, deeply bifid, white; stamens 2–5; styles bifid or trifid. Fruit 1.2–2.5 mm long; seeds 2–8, subglobose, the surface tuberculate in lines, dark reddish brown.

Moist to wet hammocks and disturbed sites. Frequent; nearly throughout. Georgia, Florida, Mississippi, and Louisiana; West Indies, Mexico, Central America, and South America; Africa, Asia, Australia, and Pacific Islands. Native to Mexico, Central America, and tropical America. All year.

Gypsophila L. 1753. BABY'S-BREATH

Perennial herbs. Leaves opposite, simple, pinnate-compound, epetiolate, estipulate. Flowers in terminal paniculiform cymes, actinomorphic, bisexual; sepals 5, basally connate; petals 5, free; stamens 10, free; styles 2. Fruit a capsule dehiscing by 4 broad teeth; seeds few to numerous.

A genus of about 150 species; North America, South America, Europe, Africa, Asia, Australia, and Pacific Islands. [From the Greek *gypsos*, gypsum, and *philein*, to love, in reference to the habitat of some species.]

Selected reference: Pringle (2005).

Gypsophila paniculata L. [Flowers in panicles.] BABY'S-BREATH.

Gypsophila paniculata Linnaeus, Sp. Pl. 407. 1753. *Gypsophila parviflora* Moench, Methodus 60. 1794, nom. illegit. *Arrostia paniculata* (Linnaeus) Rafinesque, Fl. Tellur. 2: 54. 1837 ("1836").

Erect perennial herb, to 10 dm; stem diffusely branched, glabrous or slightly glandular-puberulent near the base, glaucous. Leaves with the blade linear to lanceolate, 1.5–6(9) cm long, 2–8(10) cm wide, reduced above, the apex acute, the base cuneate, the margin entire, the upper and lower surfaces slightly glandular-puberulent. Flowers in a diffuse paniculiform cyme, the pedicel 3–6 mm long; bracts with scarious margins, the midvein green or purplish; calyx campanulate to turbinate, 1.5–3 mm long, purple or purplish green on the nerves, white-scarious between the nerves, incised to about the middle or slightly below, the lobes broadly rounded and white-scarious around the margin; petals 2–4 mm long, the apex truncate to emarginate, white or pinkish. Fruit globose, subequaling or exceeding the calyx; seeds subreniform, 1–2 mm long, the surface coarsely tuberculate.

Disturbed sites. Rare; Jackson County. Escaped from cultivation. Quebec south to Pennsylvania, west to British Columbia, Washington, Oregon, and California, also Florida; South America; Europe, Asia, Australia, and Pacific Islands. Native to Europe and Asia. Spring.

Mononeuria Rchb. 1841.

Annual herbs. Leaves opposite, simple, pinnate-veined, epetiolate, estipulate. Flowers in terminal cymes, actinomorphic, bisexual; sepals 5, free; petals 5, free; stamens 10, free; ovary 3-carpellate, 1-loculate, the styles 3. Fruit a capsule, 3-valved; seeds numerous.

A genus of 9 species; North America. [One-nerved, in reference to the sepals of some species.]

Mononeuria is sometimes placed in the genus *Arenaria* (e.g., Gleason and Cronquist, 1991; Larsen, 1986; Radford et al., 1964, 1968) or *Minuartia* (e.g., McNeill, 1980; Rabeler et al., 2005). A recent molecular study (Dillenberger and Kadereit, 2014) shows *Minuartia* to be polyphyletic and it was subsequently divided into 11 genera, including *Mononeuria* and *Sabulina* of Florida.

Mononeuria paludicola (Fernald & B. G. Schub.) Dillenb. & Kadereit [*Palud*, pertaining to marshes, and *cola*, dweller.] GODFREY'S STITCHWORT; GODFREY'S SANDWORT.

> *Stellaria paludicola* Fernald & B. G. Schubert, Rhodora 50: 197, t. 1104. 1948. *Arenaria godfreyi* Shinners, Sida 1: 51. 1962. *Minuartia godfreyi* (Shinners) McNeill, Rhodora 82: 498. 1980. *Mononeuria paludicola* (Fernald & B. G. Schubert) Dillenberger & Kadereit, Taxon 63: 84. 2014.

Sprawling and weakly ascending annual herb; stem to ca. 4 dm long, diffusely branched, prostrate below, weakly ascending above, glabrous or occasionally stipitate-glandular at the nodes and near the base. Leaves linear or linear-lanceolate, 1–3.5 cm long, the apex acute, the base broadly cuneate, the margin entire, sometimes glandular-pubescent along the proximal margin, the upper and lower surfaces glabrous. Flowers 3–5(7) in diffuse, open, leafy cymes, the pedicel filiform, 1–3 cm long, glandular pubescent; bracts linear to lanceolate; sepals lance-ovate to oblong-elliptic, 3–5 mm long, the apex acute, 3-nerved, glandular-pubescent; petals ca. 2 times as long as sepals, white, the apex notched. Fruit ovate, subequaling the calyx, glabrous; seeds suborbicular, the surface papillose, dark brown, dull.

Wet woods and seepage areas. Rare; Volusia and Taylor Counties. North Carolina south to Florida, west to Arkansas. Spring.

Mononeuria paludicola (as *Minuartia godfreyi*) is listed as endangered in Florida (Florida Administrative Code, Chapter 5B-40).

EXCLUDED TAXON

> *Mononeuria uniflora* (Walter) Dillenb. & Kadereit—Reported by Chapman (1897, as *Stellaria uniflora* Walter) and by Small (1903, 1913a, both as *Alsinopsis uniflora* (Walter) Small; 1913a; 1933, as *Sabulina uniflora* (Walter) Small), the name misapplied to material of *M. paludicola*.

Paronychia Mill. 1754. NAILWORT

Annual, biennial, or perennial herbs. Leaves opposite, simple, pinnate-veined, petiolate or epetiolate, stipulate. Flowers in terminal or axillary cymes, actinomorphic, bisexual or rarely unisexual (plants polygamodioecious); sepals (3–4)5, basally connate and forming a tube; petals (3–4)5, subulate-filiform and staminodelike (pseudostaminodes), inserted on the hypanthium rim, or absent; stamens (4)5, inserted on the hypanthium rim; styles 2, united or free, the ovary 2 to 3-carpellate, 1-loculate. Fruit a utricle enclosed by the persistent calyx, indehiscent; seed 1.

A genus of about 50 species; nearly cosmopolitan. [Greek for whitlow, a disease of the nails and for a plant with whitish scaly parts used to cure it.]

We have taken a rather broad interpretation of taxa for this treatment and accordingly do not recognize some entities given specific or infraspecific rank by other authors.

Anychia Michx. (1803); *Anychiastrum* Small (1903); *Gastronychia* Small (1933); *Gibbesia* Small (1898); *Nyachia* Small (1925); *Odontonychia* Small (1903); *Siphonychia* Torrey & A. Gray (1838).

Selected references: Chaudhri (1968); Core (1939, 1941); DeLaney (2010); Hartman, Thieret, and Rabeler (2005); Shinners (1962b); Ward (1977).

1. Leaves strongly hirsute, ending abruptly in a short, apical spine or mucro; sepals inconspicuously nerved.. **P. herniarioides**
1. Leaves with various kinds of trichomes, but not strongly hirsute or ending abruptly in a short, apical spine or mucro; sepals strongly 3-nerved.
  2. Sepals straight or only slightly hooded at the apex.
    3. Flowering branches diffuse .... **P. patula**
    3. Flowering branches corymbiform.
      4. Sepals acuminate, with an evident subterminal cusp ....**P. rugelii**
      4. Sepals rounded or obtuse, lacking an evident subterminal cusp................................**P. erecta**
  2. Sepals distinctly hooded at the apex.
    5. Leaves scalelike, revolute ....**P. chartacea**
    5. Leaves not scalelike or revolute.
      6. Sepals with a distinct dorsal hyaline cusp; hypanthium lacking uncinate trichomes ....
      ....**P. baldwinii**
      6. Sepals lacking a distinct dorsal hyaline cusp; hypanthium with uncinate trichomes.
        7. Plant many-stemmed; flowers 1.5–2 mm long; sepal lobes oblong, about ⅔ the petal length, constricted near the middle, the margin thickened and raised proximally into conspicuous, usually reddish tubercles; plant perennial .... **P. discoveryi**
        7. Plant single-stemmed; flowers 1–1.5 mm long; sepal lobes obovate, about ½ the petal length, constricted near the base, the margin not thickened into tubercles; plant annual ....**P. americana**

Paronychia americana (Nutt.) Fenzl ex Walp. [Of America.] AMERICAN NAILWORT.

Herniaria americana Nuttall, Amer. J. Sci. Arts 5: 291. 1822. *Siphonychia americana* (Nuttall) Torrey & A. Gray, Fl. N. Amer. 1: 173. 1838. *Paronychia americana* (Nuttall) Fenzl ex Walpers, Repert. Bot. Syst. 1: 262. 1842. *Buinalis americana* (Nuttall) Kuntze, Revis. Gen. Pl. 2: 534. 1891. TYPE: FLORIDA: *Ware s.n.* (Holotype: PH?).

Buinalis floridana Rafinesque, New Fl. 4: 40. 1838 ("1836." TYPE: FLORIDA. This name is tentatively placed in synonymy here based on Rafinesque citing a specimen labeled as "*Herniaria americana*" from the Collins Herbarium and one labeled as "*Anychia floridana* Baldw." in the Baldwin herbarium. Kuntze (1891) places the Rafinesque name as a synonym of *Buinalis americana*. Similarly, Merrill (1949) places the Rafinesque name as a synonym of *Siphonychia americana*. However, other workers have not dealt with the name and there is some uncertainty regarding its identity. Until the name is typified, its placement here is tentative.

Siphonychia pauciflora Small, Fl. S.E. U.S. 402, 1330. 1903. *Paronychia americana* (Nuttall) Fenzl ex Walp. subsp. *pauciflora* (Small) Chaudhri, Revis. Paronych. 110. 1968.

Prostrate, diffusely spreading annual or occasional biennial herb; stem to 6 dm long, short-retrorsely pubescent or minutely puberulous to glabrate. Leaves with the blade spatulate to oblanceolate, 5–20 mm long, 2–4 mm wide, the apex obtuse, the base cuneate, the margin serrulate, ciliolate, somewhat papillose, the upper and lower surfaces minutely pubescent, glabrescent in age, sessile or the petiole to 2 mm long; stipules ovate-lanceolate, 2–5(7) mm long. Flowers terminal on a short lateral branch, numerous, compact, forming a subspheroid glomerule 3–6 mm in diameter or sometimes few on a lax branch, 1–2 mm long, urceolate, mostly reddish brown, subsessile, bisexual; bracts foliaceous or sometimes reduced; hypanthium ca. 1 mm long, the base densely covered with uncinate trichomes; sepals obovate, 0.5–1 mm long, with a broad, rounded, incurved apex, rarely with a minute apical cusp; petals ca. 0.5 mm long; stamens 5, the anthers yellow; style shortly 2-cleft. Fruit ovoid-ellipsoid, ca. 1 mm long.

Scrub and coastal dunes. Occasional; peninsula, west to central panhandle. South Carolina, Georgia, Florida, and Alabama. Spring–fall.

Chaudhri (1968) recognizes subsp. *pauciflora*, but we agree with Shinners (1962b), Ward (1977), and Hartman, Thieret, and Rabeler (2005) who reduce it to synonymy.

Paronychia baldwinii (Torr. & A. Gray) Fenzl ex Walp. [Commemorates William Baldwin (1779–1819), U.S. physician and botanist who collected in the southeastern United States.] BALDWIN'S NAILWORT.

Anychia baldwinii Torrey & A. Gray, Fl. N. Amer. 1: 172. 1838. *Paronychia baldwinii* (Torrey & A. Gray) Fenzl ex Walpers, Repert. Bot. Syst. 1: 262. 1842. *Anychiastrum baldwinii* (Torrey & A. Gray) Small, Fl. S.E. U.S. 401, 1330. 1903. *Plagidia baldwinii* (Torrey & A. Gray) Nieuwland, Amer. Midl. Naturalist 3: 154. 1913. TYPE: FLORIDA.

Paronychia riparia Chapman, Fl. South. U.S., ed. 2. 607. 1883. *Anychiastrum riparium* (Chapman) Small, Fl. S.E. U.S. 401, 1330. 1903. *Plagidia riparia* (Chapman) Nieuwland, Amer. Midl. Naturalist 3: 154. 1913. *Paronychia baldwinii* (Torrey & A. Gray) Fenzl ex Walpers subsp. *riparia* (Chapman) Chaudhri, Revis. Paronych. 128. 1968.

Paronychia baldwinii (Torrey & A. Gray) Fenzl ex Walpers var. *ciliata* Chaudhri, Revis. Paronych. 128. 1968. TYPE: FLORIDA: Leon Co.: *Godfrey 55630* (holotype: WVA).

Prostrate annual, biennial, or occasional perennial herb; stem to 1.2 m long, branched at the base, sometimes filiform with numerous lateral branches, minutely pubescent to glabrous. Leaves with the blade ovate-lanceolate, oblong, or linear-elliptic, 5–20 mm long, 2–4 mm wide, the apex acute, the base cuneate, the margin ciliolate, the upper and lower surfaces minutely pubescent, sessile or the petiole to 2 mm long; stipules linear-lanceolate, 2–6 mm long. Flowers in a diffuse cyme, cylindric-oblong, 1–2 mm long, subsessile, bisexual; bracts minute; hypanthium shallow; sepals ovate to oblong, ca. 1 mm long, the apex slightly hooded, minutely cuspidate, conspicuously 3-ribbed, the surface minutely puberulent or glabrate, the margin ciliate, sometimes the cilia few or lacking, white with brown margins; petals ca. 0.5 mm long; stamens 5, the anthers orange-yellow, often reddish brown at apex; style 2-cleft near the top or to the middle. Fruit obovoid, ca. 1 mm long.

Hammocks, riverbanks, and dunes. Frequent; northern counties, central peninsula. Virginia south to Florida, west to Alabama. Summer.

A perennial form with sepals nearly glabrous and with marginal cilia few or lacking and the stem glabrous or pubescent only in longitudinal lines has been recognized by Core (1941) as *P. riparia* and by Ward (1977) and Chaudhri (1968) as subsp. *riparia*. This taxon intergrades with typical *P. baldwinii* and is thus not recognized here.

Paronychia chartacea Fernald [Papery, in reference to the bracts.] PAPER NAILWORT; PAPERY WHITLOW-WORT.

> *Nyachia pulvinata* Small, Torreya 25: 12. 1925. *Paronychia pulvinata* (Small) Pax & K. Hoffmann, in Engler & Prantl, Nat. Pflanzenfam., ed. 2. 16(c): 300. 1934; non A. Gray, 1864. *Paronychia chartacea* Fernald, Rhodora 38: 418. 1936. TYPE: FLORIDA: Highlands Co.: between Avon Park and Sebring, 13 Dec 1920, *Small 9782* (holotype: NY).
> *Paronychia chartacea* Fernald subsp. *minima* L. C. Anderson, Sida 14: 436. 1991. *Paronychia chartacea* Fernald var. *minima* (L. C. Anderson) R. L. Hartman, Sida 21: 754. 2004. TYPE: FLORIDA: Washington Co.: upper shoreline of Crystal Lake, 11 air mi S of Vernon, T1N, R15W, Sec. 35, E ½ of NE ¼, 1 Oct 1990, *Anderson 13301* (holotype: NY; isotypes: AUA, FLAS, FSU, GA, GH, MO, SMU, UNC, US, USF, VDB).

Low-growing annual herb; stem to 15(24) cm long, sparsely to much branched at the base, often minutely purple-spotted, sparsely to densely retrorsely pubescent, usually only on 1 side. Leaves with the blade oblong-lanceolate to triangular-ovate, 1.5–5 mm long, 1–2.5 mm wide, the apex acute, the base truncate to subauriculate, the margin strongly revolute, the upper and lower surfaces sparsely pubescent, sessile; stipules lanceolate, 0.5–2.5 mm long, membranaceous, the margin fimbriate. Flowers in a much-branched, open cyme, in small clusters 1.5–4 mm wide, tubular-oblong, 0.5–1 mm long, bisexual or unisexual (plants polygamodioecious); bracts shorter than to slightly longer than the flowers; hypanthium shallow, sparsely pubescent with straight to somewhat uncinate trichomes; sepals (3)4–5, oblong, 0.5–1 mm long, the apex prominently hooded, obtuse, usually with a subapical acute to obtuse mucro, greenish or yellow-green to brownish, sometimes whitish-margined; petals rudimentary or absent; stamens 4–5, the anthers orange; styles 2–3, free. Fruit ovoid to ellipsoidal, ca. 0.5 mm long, nearly smooth.

Scrub. Occasional; central peninsula, Washington and Bay Counties. Endemic. Summer.

The predominantly staminate plants are more openly branched while the predominantly bisexual or carpellate plants are more densely matted with more numerous, shorter branches. Sexual dimorphism in this species is believed to be unique in the genus.

The differences between the panhandle plants and those of the central ridge region are indeed a matter of degree as stated by Anderson (1991). The quantitative characters of the width of the caudex, leaf, and flower cluster overlap. The qualitative differences as given by Anderson are in general strong tendencies, but exceptions can be found in all cases. In the absence of definitive characters that can be used to distinguish between the two taxa with certainty, other than geography, and in consideration of the considerable intrapopulation variability, a single taxon is recognized here.

Paronychia chartacea is listed as endangered in Florida (Florida Administrative Code, Chapter 5B-40) and as threatened in the United States (U.S. Fish and Wildlife Service, 50 CFR 23).

Paronychia discoveryi DeLaney [Commemorates the Space Shuttle Discovery.] FLORIDA PERENNIAL NAILWORT.

> *Paronychia discoveryi* DeLaney, Bot Explor. 4: 71, f. 2, 4, 5, 7, 9–11. 2010. TYPE: FLORIDA: St. Johns Co.: along powerline right-of-way ca. 1.8 mi. W of Crescent Beach, SW corner of the junction of FL 206 and Cypress Point Drive, 7 Oct 2009, *DeLaney 5741* (holotype: USF; isotypes: FLAS, FSU, FTG, MO, NY, US, USF).

Prostrate, diffusely spreading perennial herb; stem to 2.5(4) dm long, short retrorsely pubescent, glabrous distally. Leaves with the blade linear-spatulate to linear-oblanceolate, 4–6(12) mm long, 0.8–1.2 mm wide, the apex obtuse to subacute, the base cuneate, the margin antrorsely ciliate, the upper surface evenly strigillose, the lower surface strigose to glabrate; the stipules ovate-lanceolate, 2–2.5 mm long, sessile or subsessile. Flowers terminal on a short lateral branch, numerous, compact, forming subspherical glomerules 3–6 mm in diameter or sometimes few in lax branches, 1.5–2 mm long, broadly cuneate, mostly reddish-brown, subsessile, bisexual; bracts foliaceous; hypanthium ca. 1 mm long, the base covered with uncinate trichomes; sepals oblong, 0.5–1 mm long, tapering to the hypanthium, widest near the base, the apex strongly inflexed and forming a truncate hood, usually with a minute apical cusp, the margin thickened and raised proximally into conspicuously, usually reddish tubercles; petals ca. 0.5 mm long; stamens (4)5, the anthers yellow; style 2(3)-cleft. Fruit ovoid-subglobose, ca. 1 mm long.

Scrub and scrubby flatwoods. Rare; St. Johns County south to Brevard and Hardee Counties. Endemic. Summer–fall.

Paronychia erecta (Chapm.) Shinners [Erect.] SQUAREFLOWER.

> *Siphonychia erecta* Chapman, Fl. South. U.S. 47. 1860. *Buinalis erecta* (Chapman) Kuntze, Revis. Gen. Pl. 2: 534. 1891. *Odontonychia erecta* (Chapman) Small, Fl. S.E. U.S. 401, 1330. 1903. *Paronychia erecta* (Chapman) Shinners, Sida 1: 102. 1962. TYPE: FLORIDA: s.d., *Chapman s.n.* (holotype: NY; isotype: NY).
>
> *Siphonychia corymbosa* Small, Bull. Torrey Bot. Club 24: 337. 1897. *Odontonychia corymbosa* (Small) Small, Fl. S.E. U.S. 402, 1330. 1903. *Paronychia erecta* (Chapman) Shinners var. *corymbosa* (Small) Chaudhri, Revis. Paronych. 112. 1968.

Erect cespitose perennial or occasionally biennial herb, to 4 dm long; stem simple below, much branched above, glabrous and glaucous or minutely gray-pubescent. Leaves with the blade spatulate-oblanceolate to linear, 0.5–3 cm long, 1–3 mm wide, the apex obtuse, the base attenuate, ascending to erect, the margin ciliate to subentire, the upper and lower surfaces glabrous or minutely appressed pubescent, sessile or the petiole to 2 mm long; stipules ovate-lanceolate, 1 cm long, silvery. Flowers in a compound terminal corymbiform inflorescence, tubular, 2–3 mm long, sessile, bisexual; bracts foliaceous, shorter than the flowers; hypanthium ca. 1 mm long, distinctly ribbed, glabrous or sometimes short-hirtellous; sepals oblong to oblong-lanceolate, ca. 2 mm long, the apex slightly narrowed, obtuse to acute or obscurely mucronate, slightly incurved and forming a small hood, the cusp indistinct, glabrous, pale brown with a white margin; petals absent; stamens 5, the anthers dark-brown; style shortly bilobed at the apex. Fruit ovoid-oblong, ca. 1 mm long.

Coastal dunes and open, dry, sandy sites. Occasional; central and western panhandle. Florida, Alabama, Mississippi, and Louisiana. Spring–fall.

Chaudhri (1968) and Ward (1977) recognize var. *corymbosa* as distinct. However, we agree with Shinners (1962b) that it is not separable from the typical and thus recognize only a single variable taxon.

Paronychia herniarioides (Michx.) Nutt. [Resembling the genus *Herniaria* (Caryophyllaceae).] COASTALPLAIN NAILWORT.

Anychia herniarioides Michaux, Fl. Bor.-Amer. 1: 113. 1803. *Paronychia herniarioides* (Michaux) Nuttall, Gen. N. Amer. Pl. 1: 159. 1818. *Anychiastrum herniarioides* (Michaux) Small, Fl. S.E. U.S. 401, 1330. 1903. *Plagidia herniarioides* (Michaux) Nieuwland, Amer. Midl. Naturalist 3: 154. 1913. *Gastronychia herniarioides* (Michaux) Small, Man. S.E. Fl. 480, 1504. 1933. *Plagidia rufa* Rafinesque, New. Fl. 4: 43. 1838 ("1836"). TYPE: FLORIDA.

Prostrate annual herb; stem to 2 dm long, much branched at the base, repeatedly forked toward the top, scabrous-pubescent. Leaves with the blade oblong to oblong-elliptic to spatulate, 3–12(15) mm long, 2–4 mm wide, the apex obtuse and short-mucronate, spinose-cuspidate, the base cuneate, the upper and lower surfaces scabrous-pubescent, the margin ciliate, sessile; stipules ovate-lanceolate, 7 mm long. Flowers 3–7 in a cluster in a compact, subterminal cyme, urceolate, ca. 2 mm long, sessile, bisexual; bracts foliaceous; hypanthium globose, ca. 1 mm long, scabrous-pubescent; sepals lanceolate-subulate, ca. 1 mm long, strigose, the margin narrow-membranaceous, slightly hooded, conspicuously awned; petals ca. 5 mm long, white with a reddish tip; stamens 5, the anthers orange-brown; style shortly bilobate. Fruit subglobose, ca. 0.7 mm long.

Sandhills and scrub. Occasional; northern and central peninsula, Liberty County. South Carolina, Georgia, Alabama, and Florida. Summer.

Paronychia patula Shinners [Spreading.] PINELAND NAILWORT.

Siphonychia diffusa Chapman, Fl. South. U.S. 47. 1860. *Buinalis diffusa* (Chapman) Kuntze, Revis. Gen. Pl. 2: 534. 1891. *Paronychia patula* Shinners, Sida 1: 102. 1962. TYPE: FLORIDA: s.d., *Chapman s.n.* (holotype: NY; isotype: NY).

Prostrate annual herb; stem to 5 dm long, much branched, strigose. Leaves with the blade oblanceolate to linear-oblong, to 15 mm long, to 2 mm wide, the apex obtuse to subacute, the base cuneate, the margin entire, the upper and lower surfaces strigose, sessile; stipules lanceolate,

3–4 mm long. Flowers in a subspheroid cluster on the end of a short lateral branch, 1.5–2 mm long, narrowly tubular, white with the lower half reddish brown, subsessile, bisexual; bracts foliaceous; hypanthium reddish brown, with short-uncinate trichomes; sepals linear to oblong, 1.5–2 mm long, the apex concave and slightly hooded with a short but distinct cusp, white; petals ca. 5 mm long, white with a reddish filiform apex; stamens 5, the anthers dark brown; style shortly bilobate. Fruit ovoid, ca. 1 mm long.

Sandhills and scrub. Occasional; northern counties, central peninsula. Georgia, Alabama, and Florida. Summer.

Paronychia rugelii (Chapm.) Shuttlew. ex Chapm. [Commemorates Ferdinand Ignatius Xavier Rugel (1806–1878), German-born American botanical explorer, pharmacist, and surgeon, who collected for R. J. Shuttleworth in the southeastern United States.] RUGEL'S NAILWORT.

> *Siphonychia rugelii* Chapman, Fl. South. U.S. 47. 1860. *Buinalis rugelii* (Chapman) Kuntze, Revis. Gen. Pl. 2: 534. 1891. *Paronychia rugelii* (Chapman) Shuttleworth ex Chapman, Fl. South. U.S., ed. 3. 397. 1897. *Forcipella rugelii* (Chapman) Small, Bull. Torrey Bot. Club 25: 150. 1898. *Gibbesia rugelii* (Chapman) Small, Bull. Torrey Bot. Club 25: 621. 1898. TYPE: FLORIDA: "ad fluv. Whittlecouchy, prope Camp Island," Jul 1848, *Rugel 54* (holotype: NY; isotype: GH).
>
> *Odontonychia interior* Small, Man. S.E. Fl. 483, 1504. 1933. *Siphonychia interior* (Small) Core, J. Elisha Mitchell Sci. Soc. 55: 344. 1939. *Paronychia rugelii* (Chapman) Shuttleworth ex Chapman var. *interior* (Small) Chaudhri, Revis. Paronych. 113. 1968. TYPE: FLORIDA: Dixie Co.: along the Suwannee River E of Old Town, 14 Jul 1924, *Small et al. 11465* (holotype: NY).

Erect annual herb; stem to 4 dm long, branched somewhat above the base, much branched above, densely retrorse-pubescent. Leaves with the blade linear-oblanceolate to linear-oblong, 1–2 cm long, 2–3 mm wide, the apex acute or subacute, the base cuneate, the margin entire, the upper and lower surfaces strigose, sessile; stipules ovate-lanceolate, to 5 mm long, often deeply split. Flowers in a corymbose cyme, these compact or sometimes lax, narrowly tubular, 2–3 mm long, light brown, sessile, bisexual; bracts foliaceous, ca. 2 mm long; hypanthium ca. 1 mm long, dark-brown, pubescent; sepals linear to linear-lanceolate, ca. 2 mm long, slightly hooded but distinctly awned, nearly white; petals ca. 5 mm long, white with a reddish tip; stamens 5, the anthers purple-black; style filiform, the apex shortly bilobate. Fruit subglobose, ca. 1 mm long.

Sandhills and flatwoods. Occasional; northern and central peninsula, eastern and central panhandle. Georgia and Florida. Summer.

Chaudhri (1968) recognized var. *interior* on the basis of more lax inflorescences. We follow Shinners (1962b), Ward (1977), and Hartman, Thieret, and Rabeler (2005) in recognizing only a single variable species.

EXCLUDED TAXA

> *Paronychia canadensis* (Linnaeus) A. W. Wood—Reported by Small (1903, 1913a, both as *Anyachia dichotoma* Michaux) and Core (1941), who does not cite any Florida specimens in the exsiccatae, the nearest being from Madison County, Alabama, and Floyd County, Georgia.

Paronychia fastigiata (Rafinesque) Fernald—Reported by Small (1933, as *Anychia polygonoides* Rafinesque), Correll and Johnston (1970), and Core (1941), who does not cite any Florida specimens in the exsiccatae, the nearest being from Baldwin County, Alabama, and Floyd County, Georgia. Excluded by Ward (1977) on the basis of an apparent misidentification of our material as *P. baldwinii*.

Polycarpaea Lam., nom. cons. 1792.

Annual herbs. Leaves opposite, simple, pinnate-veined, petiolate or epetiolate, stipulate. Flowers in terminal corymbiform cymes, bisexual; sepals 5, basally connate and forming a tube; petals 5, adnate with the stamens and basally inserted on the hypanthium rim; stamens 5; ovary 3-carpellate, 1-loculate, stipitate, the stigmas 3. Fruit a capsule, valves 3; seeds few.

A genus of about 50 species; North America, Central America, South America, Africa, Asia, and Australia. [From the Greek *poly*, many, and *carpos*, fruit, in reference to the many capsules.]

Selected reference: Thieret and Rabeler (2005a).

Polycarpaea corymbosa (L.) Lam. [Flowers in a corymb.] OLD-MAN'S CAP.

Achyranthes corymbosa Linnaeus, Sp. Pl. 205. 1753. *Polycarpaea corymbosa* (Linnaeus) Lamarck, Tabl. Encycl. 2: 129. 1797. *Lahaya corymbosa* (Linnaeus) Schultes, in Roemer & Schultes, Syst. Veg. 5: 405. 1819. *Mollia corymbosa* (Linnaeus) Willdenow ex Sprengel, Syst Veg. 1: 795. 1824 ("1925"). *Celosia corymbosa* (Linnaeus) Willdenow ex Roxburgh, Fl. Ind., ed. 1832. 1: 681. 1832. *Polycarpaea corymbosa* (Linnaeus) Lamarck var. *genuina* Pax, Bot. Jahrb. Syst. 17: 590. 1893, nom. inadmiss.

Polycarpaea nebulosa Lakela, Rhodora 65: 35. 1963. TYPE: FLORIDA: Hillsborough Co.: Temple Terrace, south-facing slope of sandhill W of 56th Street, E of Overlook Drive, 7 Nov 1962, *Lakela 25565* (holotype: USF; isotypes: F, GH, SMU, US).

Erect annual herb, to 18 cm; stem simple or few branched below, much branched above, pilose or glabrate. Leaves with the blade linear, 1–2.5 cm long, 1–1.5 mm wide, fleshy, the apex acute, the base cuneate, the margin slightly revolute, 1-nerved, the midvein terminating in a bristle ca. 1 mm long, the upper and lower surfaces glabrous, sessile or subsessile; stipules scarious, apically cleft, 3–5 mm long. Flowers in a leafy compound cyme ca. 3 mm long, white, lustrous; bracts scarious; hypanthium short, cup-shaped; sepals ovate or lanceolate, 2–3 mm long, glabrous, scarious, with a reddish orange median basal spot, the apex acute; petals elliptic to ovate, ca. 1 mm long, the margin entire or erose toward the tip, pink, fading to orange; filaments subulate, curved inward, the anthers white. Fruit ovate, ca. 1.5 mm long; seeds subreniform, ca. 7, the surface with obscure transverse lines, pale brown.

Disturbed sites. Occasional; Hernando, Pasco, Hillsborough, Polk, and Highlands Counties. Florida; Central America and South America; Africa, Asia, and Australia. Native to Africa, Asia, and Australia. Spring.

Polycarpon Loefl. ex L. 1759. MANYSEED

Annual, biennial, or perennial herbs. Leaves opposite, often appearing whorled with 4 at a node due to the presence of axillary leafy fascicles (short leafy lateral branches), simple, pinnate-veined, petiolate, stipulate. Flowers in terminal cymes, bisexual; sepals 5, free; petals

5, free; stamens (1)3–5, the filaments basally connate; ovary 3-carpellate, 1oculate, the styles 3-cleft. Fruit a capsule, valves 3; seeds numerous.

A genus of about 16 species; nearly cosmopolitan. [From the Greek *poly*, many, and *carpos*, fruit, in reference to the many capsules.]

Selected reference: Thieret and Rabeler (2005b).

Polycarpon tetraphyllum (L.) L. [From the Greek *tetra*, four, and *phyllos*, leaf, in reference to the frequent condition of leaves appearing as though in a whorl of 4.] FOURLEAF MANYSEED.

> *Mollugo tetraphylla* Linnaeus, Sp. Pl. 89. 1753. *Polycarpon tetraphyllum* (Linnaeus) Linnaeus, Syst. Nat., ed. 10. 881. 1759. *Alsine polycarpon* Cranz, Inst. Rei Herb. 2: 405. 1766, nom. illegit.

Ascending to prostrate annual, biennial, or perennial herb; stem to 15 cm long, much branched at the base, glabrous. Leaves with the blade broadly ovate to obovate, to 15 mm long, to 5 mm wide, the apex obtuse or rounded, the base broadly cuneate, the margin entire, the upper and lower surfaces glabrous, short-petiolate; stipules lanceolate, scarious. Flowers numerous in a cyme; bracts lanceolate, scarious; sepals lanceolate, 2–3 mm long, the prominent keel obscurely serrulate, the margin scarious; petals oblanceolate, shorter than the sepals, slightly emarginate, white, membranaceous; style short, 3-cleft. Fruit ovoid; seeds ovoid, lenticular, ca. 0.5 mm long, punctulate.

Moist, shaded, disturbed sites. Rare; Duval, Alachua, Pinellas, and Escambia Counties. Nearly cosmopolitan. Native to Europe, Africa, and Asia. Spring–summer.

Sabulina Rchb. 1832. STITCHWORT

Perennial herbs. Leaves opposite, simple, pinnate-veined, epetiolate, estipulate. Flowers in terminal cymes, actinomorphic, bisexual; sepals 5, free; petals 5, free; stamens 10; ovary 3-carpellate, 1-loculate, the styles 3. Fruit a capsule, 3-valved; seeds numerous.

A genus of about 65 species; North America, South America, Europe, Africa, and Asia. [*Sabulo*, of sandy places.]

Sabulina is sometimes placed in the genus *Arenaria* (e.g., Gleason and Cronquist, 1991; Larsen, 1986; Radford et al., 1964, 1968) or *Minuartia* (e.g., McNeill, 1980; Rabeler et al., 2005). A recent molecular study (Dillenberger and Kadereit, 2014) shows *Minuartia* to be polyphyletic and it was subsequently divided into 11 genera, including *Mononeuria* and *Sabulina* in Florida.

Alsinopsis Small (1903).

Sabulina caroliniana (Walter) Small [Of Carolina.] PINEBARREN STITCHWORT.

> *Arenaria caroliniana* Walter, Fl. Carol. 141. 1788. *Alsinopsis caroliniana* (Walter) Small, Fl. S.E. U.S. 421, 1330. 1903. *Minuartia caroliniana* (Walter) Mattfeld, Bot. Jahrb. Syst. 57(Beibl. 126): 28. 1921. *Sabulina caroliniana* (Walter) Small, Man. S.E. Fl. 499, 1504. 1933. *Minuopsis caroliniana* (Walter) W. A. Weber, Phytologia 67: 427. 1989.
>
> *Arenaria squarrosa* Michaux, Fl. Bor.-Amer. 1: 273. 1803. *Alsine squarrosa* (Michaux) Fenzl ex A. Gray, Manual, ed. 2. 57. 1856.

Low-growing perennial herb, to 3 dm; stem much branched at the base, forming a dense basal cushion, the flowering stem sparingly forked near the top, nearly leafless, glandular-pubescent. Leaves with the blade subulate, 3–12 mm long, 0.5–1 mm wide, the apex acute, the base cuneate, the margin entire, scarious in the proximal ½, the upper and lower surfaces glabrous. Flowers in a terminal cyme, the pedicel filiform, 1–5 cm long, glandular-puberulent; sepals ovate, 2–4.5 mm long, the apex acute to obtuse, the margin scarious; petals oblanceolate, 5–12 mm long, the apex rounded. Fruit triangular-ovoid, 4–6 mm long; seeds subreniform, ca. 0.5 mm long, the surface coarsely tuberculate.

Pond margins and sandhill depressions. Occasional; central and western panhandle. New York south to Florida. Spring–summer.

Sagina L. 1753. PEARLWORT

Annual herbs. Leaves opposite, simple, pinnate-veined, epetiolate, estipulate. Flowers solitary, terminal or axillary, actinomorphic, bisexual; sepals (4)5, free; petals (4)5, free; stamens 10 or fewer, in 2 whorls, the outer whorl nectariferous at the base and always present; ovary 3-carpellate, 1-loculate, the styles (4)5. Fruit a capsule, (4)5-valved to the middle; seeds numerous.

A genus of about 15 species; nearly cosmopolitan. [*Sagina*, fattening, previously applied to *Spergula*, which is planted in Europe as a forage.]

Selected references: Crow (1978, 2005).

Sagina decumbens (Elliott) Torr. & A. Gray [Depressed, but with ascending tips, in reference to the habit.] TRAILING PEARLWORT.

> *Spergula decumbens* Elliott, Sketch Bot. S. Carolina 1: 523. 1821. *Sagina decumbens* (Elliott) Torrey & A. Gray, N. Amer. Fl. 1: 177. 1838. *Sagina elliottii* Fenzl ex A. Gray, Manual, ed. 2. 61. 1856, nom. illegit.

Ascending or decumbent annual herb; stem to 10(15) cm long, often with short, sterile, axillary fascicles, frequently purple tinged, glabrous. Leaves with the blade linear, 4–22 mm long, ca. 1 mm wide, the upper cauline ones becoming subulate, 1–5 mm long, the apex apiculate, the base cuneate, the margin entire, hyaline, the upper and lower surfaces glabrous. Flowers solitary, axillary or terminal, the pedicels filiform, 10–15 mm long, glabrous or glandular pubescent; sepals ovate, 1.5–3 mm long, glandular-pubescent or glabrous toward the base, the margin and apex frequently purple-tinged; petals elliptic, 1–2 mm long, slightly exceeding the sepals at anthesis, white; stamens 1–2 mm long. Fruit ellipsoid, 2–3 mm long, the valves thin, the sepals remaining appressed to the capsule; seeds obliquely triangular, flattened, with a dorsal groove, the surface smooth or pebbled to strongly tuberculate with a delicate reticulate pattern, light tan.

Disturbed sites. Frequent; northern counties, central peninsula. Quebec south to Florida, west to Kansas, Oklahoma, and Texas; West Indies. Native to North America. Spring.

Saponaria L. 1753. SOAPWORT

Perennial herbs. Leaves opposite, simple, pinnate-veined, petiolate or epetiolate, estipulate. Flowers in terminal or subterminal cymes, actinomorphic, bisexual; sepals 5, basally connate to form a 5-lobed tube; petals 5, basally appendaged, adnate to the carpophore; stamens 10, the filaments adnate to the carpophore; ovary 3-carpellate, 1-loculate, the styles 2(3). Fruit a capsule, dehiscing by 4 valves; seeds numerous.

A genus of about 40 species; North America, Mexico, South America, Europe, Africa, Asia, Australia, and Pacific Islands. [*Sapo*, soap, the mucilaginous juice from *S. officinalis* forming a lather in water.]

Selected reference: Thieret and Rabeler (2005e).

Saponaria officinalis L. [Of the shops, a plant used for medicinal purposes.] BOUNCINGBET; SOAPWORT.

> *Saponaria officinalis* Linnaeus, Sp. Pl. 408. 1753. *Bootia vulgaris* Necker, Delic. Gallo-Belg. 193. 1768, nom. illegit. *Bootia saponaria* Necker, Hist. & Commentat. Acad. Elect. Sci. Theod.-Palat. 2: 484. 1770, nom. illegit. *Lychnis officinalis* (Linnaeus) Scopoli, Fl. Carniol., ed. 2. 1: 303. 1771. *Silene saponaria* Fries ex Willkomm & Lange, Bot. Not. 10: 172. 1842. *Lychnis saponaria* Jessen, Deut. Excurs.-Fl. 279. 1879, nom. illegit.

Erect perennial herb, to 8 dm; stem simple, glabrous. Leaves with the blade elliptic to elliptic-ovate or oblanceolate, 4–10 cm long, 1–4 cm wide, prominently 3-nerved, the apex acute, the base cuneate, the margin entire, the upper and lower surfaces glabrous, short-petiolate or sessile. Flowers in a congested and subcapitate to open and oblong-pyramidal inflorescence to 15 cm long; primary bracts foliaceous, the ultimate ones scarious; calyx tubular, 1.5–2.5 cm long, membranaceous, 20-nerved, the lobes triangular-attenuate, the tube often becoming deeply bilobed; petals with a distinct claw and blade, the blade oblong to oblong-ovate, 8–15 mm long, the apex retuse or emarginate, with a pair of subulate appendages at the juncture between the claw and blade, white or pinkish; stamens exserted, basally connate with the petals around a short carpophore. Fruit membranaceous, included in the calyx, dehiscent by 4 short valves; seeds suborbicular, ca. 1.5 mm long, the surface finely reticulate-papillose in concentric rows.

Disturbed sites. Occasional; panhandle. Nearly throughout North America; Mexico and South America; Europe, Africa, Asia, Australia, and Pacific Islands. Native to Europe and Asia. Summer–fall.

Scleranthus L. 1753. KNOTGRASS

Annual or biennial herbs. Leaves opposite, simple, pinnate-veined, epetiolate, estipulate. Flowers in compact, terminal and axillary cymose clusters, actinomorphic, bisexual; sepals 5, basally connate; petals absent; stamens 5–10 or rarely fewer; ovary 3-carpellate, 1-loculate, the styles 2, free. Fruit a utricle surrounded by the hypanthium and the persistent sepals; seed 1.

A genus of about 10 species; North America, Mexico, Central America, South America, Europe, Africa, Asia, Australia, and Pacific Islands. [From the Greek *scleros*, hard, and *anthos*, flower, in reference to the hardened calyx-tube.]

Selected reference: Thieret and Rabeler (2005c).

Scleranthus annuus L. [Annual.] GERMAN KNOTGRASS.

Scleranthus annuus Linnaeus, Sp. Pl. 406. 1753. *Scleranthus annuus* Linnaeus var. *cymosus* Fenzl, in Ledebour, Fl. Ross. 2: 157. 1843, nom. inadmiss.

Prostrate or ascending annual or perennial herb; stem to 15 dm long, branched at the base, glabrous or puberulent. Leaves with the blade linear, 5–25 mm long, ca. 1 mm wide, connate at the base with a membranaceous flange, this often with a puberulent-ciliolate margin, the apex subulate-tipped, the base cuneate, the margin entire, the upper and lower surfaces glabrous or puberulent. Flowers in a compact, terminal or axillary cymose cluster 2.5–4 mm long, sessile or subsessile; sepals basally connate ⅓–½ their length, 10-ribbed, the lobes lanceolate, the apex acute, with a narrow scarious border near the apex, 1-nerved; filaments broadened at the base and attached to a membranaceous disk lining the rim of the hypanthium. Fruit membranaceous, surrounded by the hardened hypanthium and the persistent sepals; seed obovoid to globose, ca. 1 mm long, beaked at the micropylar end, pale yellow.

Disturbed sites. Rare; Leon, Santa Rosa, and Escambia Counties. Quebec south to Florida, west to Minnesota, Nebraska, Kansas, and Oklahoma, also Saskatchewan west to British Columbia, south to California; Mexico, Central America, and South America; Europe, Africa, Asia, and Pacific Islands. Native to Europe, Africa, and Asia. Summer.

Silene L. 1753, nom. cons. CATCHFLY

Annual or perennial herbs. Leaves opposite, simple, pinnate-veined, petiolate or epetiolate, estipulate. Flowers few to many in terminal, bracteate cymes, actinomorphic, bisexual or unisexual (plants dioecious); sepals 5, connate into a 5-lobed tube, (8–9)10- or 20-nerved; petals 5 or sometimes absent, narrow-clawed, often with a pair of auricles at the junction between the blade and the claw, usually with a pair of appendages on the ventral surface at the juncture; stamens 10, adnate with the petals near the base to form a short tube around the carpophore; ovary borne on a short carpophore, the styles 3(4–5). Fruit a capsule, the valves the same as or twice the number of the styles, enclosed by the accrescent calyx; seeds numerous.

A genus of about 700 species; nearly cosmopolitan. [From the mythological Silenus, the intoxicated foster-father of the Greek god Bacchus, who was described as being covered with foam, apparently in reference to the viscid excretions of many *Silene* species.]

Lychnis L., nom. rej. (1753).

Selected references: Hitchcock and Maguire (1947); Morton (2005c).

1. Calyx 6–9 mm long.
    2. Stem and calyx glabrous; leaves linear or slightly spatulate ....**S. antirrhina**
    2. Stem and calyx long-pilose; leaves oblanceolate...**S. gallica**
1. Calyx 15–30 mm long.
    3. Petals bright red.
        4. Petals 2-lobed; cauline leaves 2–4 pairs... **S. virginica**
        4. Petals entire or irregularly dentate; cauline leaves 10–20 pairs ....**S. regia**
    3. Petals pink, purplish, or white.

5. Petals deeply lacerate... **S. catesbaei**
5. Petals 2-lobed.
   6. Calyx glabrous.. **S. armeria**
   6. Calyx hirsute or glandular-pubescent.
      7. Calyx lobes with the apex obtuse to rounded; plant to 2 dm tall.................. **S. caroliniana**
      7. Calyx lobes with the apex acute; plant 6–10 dm tall.
         8. Calyx lobes broadly ovate to lanceolate, to 6 mm long; flowers (at least some) unisexual (plants dioecious, the carpellate flowers (4)5-stylous, diurnal..........**S. latifolia**
         8. Calyx lobes linear-lanceolate, 5–12 mm long; flowers all bisexual, 3-stylous, nocturnal ....**S. noctiflora**

Silene antirrhina L. [From *Antirrhinum* (Veronicaceae), in reference to the similarity of the leaves.] SLEEPY CATCHFLY.

Silene antirrhina Linnaeus, Sp. Pl. 419. 1753.
Silene antirrhina Linnaeus var. *linaria* A. W. Wood, Amer. Bot. Fl. 53. 1870. TYPE: "Ga. & Fla."
Silene antirrhina Linnaeus forma *apetala* Farwell, Pap. Michigan Acad. Sci. 3: 97. 1924.

Erect annual herb, to 8 dm; stem simple or few branched, usually retrorse-puberulent near the base, the internodes (at least the upper ones) often with a broad, dark, glutinous band. Leaves with the blade linear to oblanceolate or spatulate, 2–6 cm long, 1–15(20) mm wide, the apex acute, the base cuneate, the margin entire, the upper and lower surfaces glabrous, ciliate near the base, sessile. Flowers few to many in a compact to open, forked cyme, bisexual, the pedicel stiffly erect to ascending, 4–25 mm long; sepals connate into a tube 4–10 mm long, this 10-nerved, often constricted at the mouth, glabrous, the lobes short-triangular, 1–2 mm long, purple tipped; petals (sometimes absent) slightly exceeding the calyx, 2-lobed, white to pink or purplish, the appendages minute or absent; stamens included; styles 3, included. Fruit subequaling the calyx, the valves 6; seeds rotund-reniform, ca. 0.5 mm long, the surface papillate, brownish or grayish black.

Open, often disturbed sites. Occasional; northern counties, central peninsula. Nearly throughout North America; Mexico and South America; Europe. Native to North America and Mexico. Spring–summer.

Silene armeria L. [Roman name of the sea-pink, Armeria (Plumbaginaceae), with which this species was once associated.] SWEET WILLIAM

Silene armeria Linnaeus, Sp. Pl. 420. 1753. *Lychnis armoraria* Scopoli, Fl. Carniol., ed. 2. 1: 310. 1771, nom. illegit. *Cucubalus fasciculatus* Lamarck, Fl. Franc. 3: 27. 1779 ("1778"), nom. illegit. *Silene glauca* Salisbury, Prodr. Stirp. Chap. Allerton 302. 1796, nom. illegit.

Erect annual herb, to 4 dm; stem simple, glabrous, glaucous, sometimes glutinous distally. Leaves with the blade elliptic or ovate-lanceolate, 1–6 cm long, 0.5–2.5 cm wide, the apex acute, the base cuneate, the margin entire, the upper and lower surfaces glabrous, glaucous, the lower leaves short petiolate, the cauline ones sessile. Flowers in a terminal, compact, corymbiform cyme, bisexual, the pedicel 1–5 mm long; sepals connate into a clavate tube, this prominently 10-nerved, usually purple-tinged, glabrous, the lobes ovate triangular, ca. 1 mm long; petals pink, rose, or rarely white, the blade obovate, ca. 5 mm long, unlobed, the claw 6–8 mm long, the appendages linear to lanceolate, 2–3 mm long; stamens included; styles 3(4), exserted. Fruit oblong, 7–10 mm long, the valves 6(8), glabrous, the carpophore 7–8 mm long; seeds rotund-reniform, 0.5–1 mm long, the surface rugose, dark brown.

Disturbed sites. Rare; Leon County. Escaped from cultivation. Quebec south to Florida, west to Minnesota, Iowa, Missouri, Arkansas, and Louisiana, also Montana and Utah, Alaska south to California; West Indies and South America; Europe, Africa, Asia, Australia, and Pacific Islands. Native to Europe and Asia. Spring.

Silene caroliniana Walter [Of Carolina.] CAROLINA CATCHFLY.

Silene caroliniana Walter, Fl. Carol. 142. 1788. *Silene caroliniana* Walter subsp. *typica* R. T. Clausen, Rhodora 41: 578. 1939, nom. inadmiss.

Erect or ascending perennial herb, to 20 cm; few-branched, puberulent to hirsute, stipitate-glandular distally. Leaves mostly basal, the blade oblong-spatulate to oblanceolate, 3–12 cm long, 1.5–3 cm wide, the apex obtuse, the base cuneate, the margin entire, the upper and lower surfaces glandular-pubescent, the petiole broadly winged, the cauline leaves in 2–4 pairs, linear-oblong, 3–5 cm long, 5–8 mm wide, the apex acute, the base cuneate, the margin entire, the upper and lower surfaces pilose or stipitate-glandular, sessile. Flowers in a dense corymb, bisexual, the pedicel 2–8 mm long; sepals connate into a tube 15–20 mm long, this prominently 8- to 10-nerved, densely glandular-pubescent, the lobes rounded, 1–3 mm long, usually purple-tinged; petals pink, the blade obovate, 8–13 mm long, the apex slightly notched or entire, the appendages oblong, ca. 2 mm long, the claw slightly exceeding the calyx, ciliate; stamens equaling the claw; styles 3(4), slightly exceeding the claw. Fruit ellipsoid to obovoid, 8(10) mm long, the valves 6(8), the carpophore 5–8 mm long; seeds reniform-rotund, ca. 1.5 mm long, the surface coarsely and evenly papillate, dark brown.

Dry, often rocky hammocks. Rare; Okaloosa County. New Hampshire south to Florida, west to Ohio, Missouri, and Alabama. Spring.

Silene caroliniana is listed as endangered in Florida (Florida administrative Code, Chapter 5B-40).

Silene catesbaei Walter [Commemorates Mark Catesby (1682–1749), English naturalist.] EASTERN FRINGED CATCHFLY; FRINGED PINK.

Silene catesbaei Walter, Fl. Carol. 141. 1788.
Cucubalus polypetalus Walter, Fl. Carol. 141. 1788. *Silene polypetala* (Walter) Fernald & B. G. Schubert, Rhodora 50: 198. 1948.
Silene baldwynii Nuttall, Gen. N. Amer. Pl. 1: 288. 1818. *Silene fimbriata* Baldwin, in Elliott, Sketch Bot. S. Carolina 1: 515. 1821, nom. illegit.; non Sims, 1806. *Melandrium baldwynii* (Nuttall) Rohrbach, Monogr. Silene 231. 1869.

Decumbent to ascending perennial herb; stem to 4 dm long, with slender stolon-like rhizomes and leafy offshoots terminating in overwintering rosettes, several-branched from the base, simple or sparingly branched above, pilose. Leaves with the blade spatulate, 3–9 cm long, 1–2 cm wide, the apex rounded, short-apiculate, the base long attenuate to a short, winged petiole, reduced upward and becoming elliptic to oblong or lanceolate, the base rounded to clasping and sessile, the margin entire, ciliolate, the upper and lower surfaces sparsely pilose. Flowers in a few-flowered cyme or sometimes solitary, bisexual; bracts foliose; sepals connate into a tube 2–3 cm long, this prominently 10-nerved, glandular-pilose, the lobes triangular-lanceolate, 7–9 mm long; petals pink or white, the blade obdeltoid, 2–3.5 cm long, the apex lacerate, without appendages, the claw subequaling the calyx; stamens slightly exserted; styles 3(4), equaling the calyx. Fruit ovoid, 7–9 mm long, the valves 6(8), the carpophore 7–10 mm long; seeds broadly reniform, ca. 1 mm long, the surface rugose on the sides, papillate on the margins, dark reddish brown.

Bluff forests. Rare; Gadsden and Jackson Counties. Georgia and Florida. Spring.

Silene catesbaei (as *S. polypetala*) is listed as endangered in Florida (Florida Administrative Code, Chapter 5B-40) and in the United States (U.S. Fish and Wildlife Service, 50 CFR 23).

Silene gallica L. [Of France.] COMMON CATCHFLY.

Silene gallica Linnaeus, Sp. Pl. 417. 1753, nom. cons. *Cucubalus sylvestris* Lamarck, Fl. Franc. 3: 28. 1779 ("1778"), nom. illegit. *Oncerum gallicum* (Linnaeus) Dulac, Fl. Hautes-Pyrénées 256. 1867. *Silene sylvestris* Gaterau, Descr. Pl. Montauban 32. 1789, nom. illegit. *Corone gallica* (Linnaeus) Fourreau, Ann. Soc. Linn. Lyon, ser. 2. 16: 344. 1868. *Silene gallica* Linnaeus var. *genuina* Hallier et al., in W.D.J. Koch, Syn. Deut. Schweiz. Fl., ed. 3. 370. 1891, nom. inadmiss.

Erect annual herb, to 4 dm; stem much branched at the base, conspicuously hirsute proximally, hirsute and glandular-puberulent distally. Leaves with the blade oblanceolate to spatulate or linear-oblong, 1.5–3(4) cm long, 1–8(15) mm wide, the apex rounded, apiculate, the base clasping, ciliate, the margin entire, the upper and lower surfaces scabrous-pubescent. Flowers in a racemose, secund cyme, bisexual, pedicellate; bracts foliose; sepals connate into a tube 6–9 mm long, this 10-nerved, hirsute and glandular-puberulent, the lobes 2–3 mm long; petals white to pink, the blade 8–12 mm long, the apex entire or sometimes slightly toothed, the appendages linear, entire, the claw equaling the calyx; stamens about equaling the calyx; styles 3, about equaling the calyx. Fruit ovoid, 6–8 mm long, the valves 6, the carpophore to 1 mm long; seeds angular with concave faces, 0.5–1 mm long, the surface finely corrugate-rugose, dark reddish brown or blackish.

Disturbed sites. Rare; St. Johns County, central and western panhandle. Nearly throughout North America; West Indies, Central America, and South America; Europe, Africa, Asia, Australia, and Pacific Islands. Native to Europe, Africa, and Asia. Spring–summer.

Silene latifolia Poir. BLADDER CAMPION.

Silene latifolia Poiret, Voy. Barbarie 2: 165. 1789. *Melandrium latifolium* (Poiret) Maire, Bull. Soc. Hist. Nat. Afrique N. 27: 211. 1936.

Erect annual or short-lived perennial herb, to 1 m; stem few-branched from the base, finely hirsute, glandular-pubescent distally. Leaves with the blade oblong-lanceolate to elliptic, 3–12 cm long, 0.6–3 cm wide, the apex acute, the base cuneate, the margin entire, the upper and lower surfaces hirsute, the basal ones short-petiolate (usually absent at flowering time), the cauline sessile. Flowers several to many in an open cyme, unisexual (plants dioecious), the pedicel erect to ascending, 1–5 cm long; bracts lanceolate, much reduced; sepals connate into a tube, 1–2(2.4) cm long, 8–15 mm wide, 10-nerved in the staminate flowers, 20-veined in the carpellate, hirsute and glandular-pubescent, the lobes broadly ovate, to 6 mm long, the apex acute; petals white, the blade broadly ovate, ca. 2 times the calyx length, unlobed or 2-lobed; stamens equaling or slightly longer than the calyx; styles (4)5, slightly longer than the calyx. Fruit ovoid, equaling the calyx in length, the valves opening by 4(5) bifid teeth, the carpophore 1–2 mm long; seeds reniform-rotund, 1–2 mm long, dark gray-brown, coarsely reticulate.

Disturbed sites. Rare; Leon County. Nearly throughout North America; Europe, Africa, and Asia. Native to Europe, Africa, and Asia.

Silene noctiflora L. [Night-flowering.] NIGHTFLOWERING CATCHFLY.

Silene noctiflora Linnaeus, Sp. Pl. 419. 1753. *Lychnis noctiflora* (Linnaeus) Schreber, Spicil. Fl. Lips. 31. 1771. *Silene viscida* Moench, Methodus 708. 1794, nom. illegit. *Melandrium noctiflorum* (Linnaeus) Fries, Bot. Not. 1842: 178. 1842. *Elisanthe moctiflora* (Linnaeus) Ruprecht, Fl. Ingr. 161. 1860.

Erect to decumbent annual herb, to 6(8) dm; stem simple or few-branched from the base, often branched above, hirsute below with multicellular trichomes, these becoming intermixed with short, glandular-viscid trichomes above, especially in the inflorescence. Leaves with the blade spatulate to oblanceolate or ovate to lanceolate, 3–12 cm long, 0.5–5 cm wide, the apex acute, the base cuneate, the margin entire, the upper and lower surfaces hirsute to pubescent, sessile or short-petiolate. Flowers few to many in an open, loosely branched cyme, bisexual, the pedicel erect to ascending, 3–40 mm long; bracts lanceolate, reduced; sepals connate basally into a tube, 1.5–2.5 cm long, this prominently 10-nerved, the nerves interconnected by a few anastomosing veins, hirsute and glandular-pubescent on the nerves, the lobes linear-lanceolate, 5–12 mm long, the apex acute; petals white to pinkish, the blade 7–10 mm long, 2-lobed to about the middle, inrolled during the day, opening at night, with auricles 1–1.5 mm long, the appendages ca. 1 mm long, broad, entire or erose, the claw about equaling the blade; stamens about equaling the petal claws; styles 3, shorter than the petals. Fruit ovoid, 2–3 cm long, the valves 6, the carpophore 1–3 mm long; seeds rotund-reniform, ca. 1 mm long, the surface strongly tuberculate and reticulate, dark brown to black with a grayish coating.

Disturbed sites. Rare; Volusia County. Not recently collected. Nearly throughout North America; Europe, Asia, and Pacific Islands. Native to Europe and Asia. Summer.

Silene regia Sims [Royal.] ROYAL CATCHFLY.

Silene regia Sims, Bot. Mag. 41: t. 1724. 1815. *Melandrium illinoense* Rohrbach, Linnaea 36: 250. 1869, nom. illegit.

Erect perennial herb, to 16 dm; stem usually unbranched to the inflorescence, glabrous or puberulent below to glandular-puberulent above, especially in the inflorescence. Leaves with the blade broadly ovate to lanceolate, 3.5–12 cm long, 1.5–7 cm wide, the apex acute to acuminate, the base rounded to cordate-clasping, the margin entire, the upper and lower surfaces glabrous to densely puberulent, sessile. Flowers numerous in a narrow compound cyme, bisexual, the pedicel 5–20 mm long, densely glandular-pubescent; bracts ovate, much reduced; sepals connate into a tube 17–27 mm long, this prominently 10-nerved, constricted around the carpophore, glandular-pubescent, the lobes triangular, 2–4 mm long, the margin often red-tinged, scarious; petals bright red, the blade (9)12–20 mm long, the apex rounded, entire or irregularly toothed, the auricles rounded, 1–2 mm long, entire or erose, the appendages tubular, 2–4 mm long, the claw about equaling the calyx; stamens exserted; styles 3(5), exserted. Fruit ovoid-ellipsoid, the valves (4)6(10), the carpophore 3–5 mm long; seeds subreniform, somewhat angled, 1.5–2 mm long, the surface shallowly tuberculate, dark reddish brown, lustrous.

Open hammocks. Rare; Jackson County. Ohio south to Florida, west to Kansas and Oklahoma. Spring.

Silene regia is listed as endangered in Florida (Florida Administrative Code, Chapter 5B-40).

Silene virginica L. [Of Virginia.] FIRE PINK.

Silene virginica Linnaeus, Sp. Pl. 419. 1753. *Silene coccinea* Moench, Suppl. Meth. 305. 1802, nom. illegit.

Erect perennial herb, to 8 dm; stem simple, glandular-pubescent. Basal leaves with the blade oblanceolate or spatulate, 4–10 cm long, 8–18 mm wide, the apex acute to obtuse, the base cuneate, the margin entire, the upper and lower surfaces glabrous or rarely puberulent, the petiole ciliate, the cauline in 2–4 pairs, sessile or broadly petiolate, the blade oblanceolate to narrowly elliptic, 1–15(30) cm long, 4–15(30) mm wide, the apex acute or acuminate, the base cuneate, the margin entire, ciliate, the upper and lower surfaces glabrous. Flowers in an open cyme, the pedicel ascending, often elongate; bracts foliaceous; sepals united into a clavate tube 8–22 mm long, this 10-nerved, glandular-pubescent, the lobes lanceolate to oblong, 3–4 mm long; petals red, 2 times longer than the calyx, the blade obtriangular to oblong, 15–22 mm long, 2-lobed with 2 small lateral lobes, the appendages tubular, ca. 3 mm long, the claw longer than the calyx; stamens exserted, shorter than the petals; styles 3(4), equaling the stamens. Fruit ovoid, the valves 3(6), the carpophore 2–3(4) mm long, the pedicel sharply deflexed at the base, 7–12 mm long; seeds reniform, 1–2 mm long, the surface with large inflated papillae, ash-gray.

Open hammocks. Rare; Bay County. New York and Ontario south to Florida, west to Minnesota, Iowa, Kansas, Oklahoma, and Louisiana. Spring–summer.

Silene virginica is listed as endangered in Florida (Florida Administrative Code, Chapter 5B-40).

Spergula L. 1753. SPURRY

Annual herbs. Leaves opposite, appearing whorled or with 2 opposite sets of 6–8 clustered at each node due to the presence of axillary leafy fascicles (short leafy lateral branches), simple, epetiolate, stipulate. Flowers in terminal, compound, leafy bracteate cymes, bisexual; sepals 5, free; petals 5, free; stamens 10 or rarely 5; ovary 5-carpellate, 1-loculate, the styles 5, free to the base. Fruit a capsule, the valves 5; seeds numerous.

A genus of 5 species; nearly cosmopolitan. [*Spargere*, to scatter, from the sowing of seeds to produce a quick-growing early forage.]

Selected reference: Hartman and Rabeler (2005a).

Spergula arvensis L. [Of cultivated fields.] CORN SPURRY.

Spergula arvensis Linnaeus, Sp. Pl. 440. 1753. *Alsine arvensis* (Linnaeus) Crantz, Inst. Rei Herb. 2: 408. 1766. *Stellaria arvensis* (Linnaeus) Scopoli, Fl. Carniol., ed. 2. 318. 1771. *Arenaria arvensis* (Linnaeus) Wallroth, Sched. Crit. 200. 1822. *Spergula sativa* Boenninghausen, Prodr. Fl. Monast. Westphal. 135. 1824, nom. illegit. *Spergularia arvensis* (Linnaeus) Cambessedes, in A. Saint-Hilaire, Fl. Bras. Merid. 2: 179. 1829. *Spergula arvensis* Linnaeus var. *sativa* Alefeld, Landw. Fl. 108. 1866, nom. indamiss. *Spergula arvensis* Linnaeus subsp. *euarvensis* Briquet, Prodr. Fl. Corse 1: 493. 1910, nom. inadmiss.

Erect or ascending annual herb, to 40 cm; stem simple or few-branched from the base, sparingly puberulent to glandular-puberulent or glabrous, sometimes glaucous. Leaves with the blade linear-filiform, 1–5 cm long, ca. 1 mm wide, the apex acute, the base cuneate, the margin entire, revolute, the upper and lower surfaces glabrous, sometimes glaucous; stipules ca. 1 mm long, scarious. Flowers in a diffuse, compound cyme, the pedicel erect to spreading or reflexed, 4–25(40) mm long, glandular-puberulent; sepals ovate, 3–5 mm long, glandular-puberulent, the margin scarious; petals white, obovate, subequaling the sepals. Fruit broadly ovoid, 3–5 mm long; seeds subrotund, convex, 1–2 mm long, the surface smooth or papillate, black with a white or tan winged margin.

Disturbed sites. Occasional; northern and central peninsula, central and western panhandle. Nearly cosmopolitan. Native to Europe, Africa, and Asia. Spring.

Spergularia (Pers.) J. Presl & C. Presl 1819, nom. cons. SANDSPURRY

Annual herbs. Leaves opposite, simple, pinnate-veined, epetiolate, stipulate. Flowers in leafy, bracteate, racemiform cymes, bisexual; sepals 5, free; petals 5, free; stamens 2–5, free; ovary 3-carpellate, 1-loculate, the styles 3, free to the base. Fruit a capsule, the valves 3; seeds numerous.

A genus of about 60 species; nearly cosmopolitan. [A derivative of *Spergula*, in reference to its general resemblance to that genus.]

Tissa Adans., nom. rej. (1763).

Selected references: Hartman and Rabeler (2005b); Rossbach (1940).

Spergularia marina (L.) Griseb. [Marine.] SALT SANDSPURRY.

Arenaria rubra Linnaeus var. *marina* Linnaeus, Sp. Pl. 423. 1753. *Arenaria marina* (Linnaeus) Allioni, Fl. Pedem. 2: 114. 1785. *Stipularia marina* (Linnaeus) Haworth, Syn. Pl. Succ. 104. 1812. *Leptigonum marinum* (Linnaeus) Wahlberg, Fl. Gothob. 1: 47. 1820. *Alsine marina* (Linnaeus) Mertens & W.D.J. Koch, in Röhling, Deutschl. Fl., ed. 3. 2: 293. 1826. *Alsinella media* Hornemann, Nomencl. Fl. Danic. 32. 1827, nom. illegit. *Buda marina* (Linnaeus) Dumortier, Fl. Belg. 110. 1827. *Spergularia marina* (Linnaeus) Grisebach, Spic. Fl. Rumel. 1: 213. 1843. *Alsine heterosperma* Gussone, Fl. Sicul. Syn. 1: 501. 1843, nom. illegit. *Spergularia rubra* (Linnaeus) J. Presl & C. Presl var. *marina*

(Linnaeus) A. Gray, Manual 64. 1848. *Tissa marina* (Linnaeus) Britton, Bull. Torrey Bot. Club 16: 126. 1889. *Corion marinum* (Linnaeus) N. E. Brown, in Smith, Engl. Bot., ed. 3B, suppl. 48. 1892. *Alsine maritima* Pallas ex House, Amer. Midl. Naturalist 7: 134. 1921, nom. illegit.

Spergularia salina J. Presl & C. Presl, Fl. Cech. 95. 1819. *Arenaria salina* (J. Presl & C. Presl) Seringe, in de Candolle, Prodr. 1: 401. 1824. *Lepigonum salinum* (J. Presl & K. Presl) Fries, Novit. Fl. Suec. Mant., Alt. 3, 34. 1839. *Spergula salina* (J. Presl & C. Presl) D. Dietrich, Syn. Pl. 2: 1598. 1840. *Alsine marina* (Linnaeus) Mertens & W.D.J. Koch var. *minor* Heynhold, Nom. Bot. Hort. 38. 1840. *Spergularia neglecta* (Kindberg) Syme var. *salina* (J. Presl & C. Presl) Syme, in Smith, Engl. Bot., ed. 3B. 2: 130. 1864. *Spergularia canina* (Leffler) Hartman var. *salina* (J. Presl & C. Presl) Hartman, Handb. Skand. Fl., ed. 11. 248. 1879. *Tissa salina* (J. Presl & C. Presl) Britton, Bull. Torrey Bot. Club 16: 127. 1889. *Spergularia salina* J. Presl & C. Presl var. *genuina* Briquet, Prodr. Fl. Corse 1: 492. 1910, nom. inadmiss.

Erect or decumbent annual herb; stem to 20 cm, simple or branched from the base, glandular-puberulent. Leaves with the blade linear, 8–25(40) mm long, 1–2 mm wide, the apex obtuse or mucronulate, the base cuneate, the margin entire, the upper and lower surfaces glandular-puberulent; stipules broadly deltoid, 2–4 mm long, the apex short-acuminate, the margin entire or lacerate, often basally connate, glandular-puberulent. Flowers in a lax cyme, the pedicel 1–10 mm long, often reflexed; bracts foliose, 1–3 mm long; sepals ovate, 3–5 mm long, the apex obtuse, the margin scarious; petals pink or white, ovate to oblong, ½ to nearly as long as sepals. Fruit ovoid, 3–6 mm long; seeds obliquely obovate, ca. 1 mm long, compressed, the surface glandular-papillate, brown.

Beaches and saline flats. Occasional; northern and central peninsula, central and western panhandle. Nearly throughout North America; Mexico; Europe, Africa, Asia, Australia, and Pacific Islands. Native to North America, Mexico, Europe, and Africa. Spring.

EXCLUDED TAXON

Spergularia rubra (Linnaeus) J. Presl & C. Presl—Reported by Chapman (1860, 1883), who misapplied the name to material of *S. marina*.

Stellaria L. 1753. CHICKWEED; STARWORT

Annual or perennial herbs. Leaves opposite, simple, pinnate-veined, petiolate or epetiolate, estipulate. Flowers in terminal or axillary, bracteate or ebracteate and solitary, bisexual; sepals 5, free; petals 5, free, or absent; stamens 1–8 or 10, sometimes absent (*S. pallida*), arising from a nectariferous disk; ovary 3- or 4-carpellate, 1-loculate, the styles 3(4), free to the base. Fruit a capsule, the valves 6(8); seeds numerous.

A genus of about 120 species; nearly cosmopolitan. [*Stella*, star, in allusion to the star-shaped flowers.]

Alsine L. (1753).

Selected reference: Morton (2005b).

1. Midstem and lower leaves sessile or subsessile.
    2. Leaves linear-lanceolate to narrowly lanceolate; flowers in terminal branched cymes ....
    .... **S. graminea**

2. Leaves elliptic to broadly obovate; flowers solitary in the axil of the distal leaves .... **S. parva**
1. Midstem and lower leaves petiolate.
    3. Petals shorter than to equaling the sepals or absent.
        4. Petals usually present; sepals 4.5 mm long or longer; stamens 3–5(8); seeds 0.9–1.3 mm in diameter .... **S. media**
        4. Petals usually absent; sepals 4 mm long or less; stamens 1–3 or absent; seeds 0.5–0.9 mm in diameter .... **S. pallida**
    3. Petals longer than the sepals.
        5. Leaf blade long attenuate at the base .... **S. prostrata**
        5. Leaf blade sessile or subsessile .... **S. puber**

Stellaria graminea L. GRASS-LIKE STARWORT.

Stellaria graminea Linnaeus, Sp. Pl. 422. 1753. *Alsine graminea* (Linnaeus) Britton, Mem. Torrey Bot. Club 5: 150. 1894.

Decumbent or ascending perennial herb; stem to 9 dm long, rooting at the nodes, glabrous. Leaves with the blade linear-lanceolate to narrowly lanceolate, 1.5–4 cm long, 1–6 mm wide, the apex acute, the base rounded to cuneate, the margin entire, the upper and lower surfaces glabrous, often ciliate along the margin proximally, sessile. Flowers 5–many in a terminal, open, branched cyme, glabrous; bracts narrowly lanceolate, 1–5 mm long, scarious; pedicel 1–3 cm long, glabrous; sepals narrowly lanceolate, 3–7 mm long, the apex acute, the outer surface glabrous; petals equaling or longer than the sepals, 2-lobed; stamens 10; styles 3. Fruit narrowly ovoid, 5–7 mm long, longer than the sepals; seeds ca. 1 mm long, rugose in concentric rings, reddish brown.

Disturbed sites. Rare; Leon County. Nearly throughout North America; South America; Europe, Asia, Australia, and Pacific Islands. Native to Europe and Asia. Spring–fall.

Stellaria media (L.) Vill. [Intermediate between two related species.] COMMON CHICKWEED.

Alsine media Linnaeus, Sp. Pl. 272. 1753. *Cerastium medium* (Linnaeus) Crantz, Inst. Rei Herb. 2: 401. 1766. *Stellaria media* (Linnaeus) Villars, Hist. Pl. Dauphiné 3: 615. 1789. *Alsine vulgaris* Moench, Methodus 228. 1794, nom. illegit. *Stellaria media* (Linnaeus) Villars var. *oligandra* Fenzl, in Ledebour, Fl. Ross. 1: 377. 1842, nom. inadmiss. *Stellaria media* (Linnaeus) Villars var. *typica* Beck, Fl. Nieder-Oesterreich 364. 1890, nom. inadmiss. *Stellaria media* (Linnaeus) Villars var. *genuina* Rouy, Fl. France 3: 228. 1896, nom. inadmiss. *Stellaria media* (Linnaeus) Villars subsp. *typica* Beguinot, Nuovo Giorn. Bot. Ital. 17: 351. 1910, nom. inadmiss. *Stellaria oligandra* Hassler, Pl. Hassler., Addend. 6. 1917, nom. illegit.

Decumbent annual or short-lived perennial herb; stem to 5 dm long, rooting at the nodes, much branched from the base, glabrous below, becoming finely hirsute in 1–2 longitudinal lines distally. Leaves with the blade elliptic or rotund-ovate, 1–3 cm long, 3–15 mm wide, the apex acute to short-acuminate, the base rounded to cuneate, the margin entire, the upper and lower surfaces glabrous or sometimes ciliate on the margin at the base, sessile or the middle and lower ones usually with a winged, ciliate petiole as long as the blade. Flowers 5–many in a cyme, the pedicel slender, 3–40 mm long, ascending or reflexed in age, usually with a line of trichomes; bracts ovate to lanceolate, 1–40 mm long, foliaceous; sepals oblong-lanceolate, 4–6 mm long, the apex acute to obtuse, the margin scarious, the outer surface hirsute and somewhat glandular; petals shorter than to equaling the sepals, 2-lobed nearly to the base or absent; stamens 3–5(8); styles 3. Fruit ovoid-oblong, 3–5 mm long, somewhat longer than the sepals; seeds subrotund, ca. 1 mm long, tuberculate, reddish brown.

Disturbed sites. Frequent; nearly throughout. Nearly cosmopolitan. Native to Europe, Africa, and Asia. Winter–spring.

Stellaria pallida (Dumort.) Crépin [Pale, in reference to its yellowish green color.] LESSER CHICKWEED.

Alsine pallida Dumortier, Fl. Belg. 109. 1827. *Stellaria pallida* (Dumortier) Crépin, Man. Fl. Belgique, ed. 2. 19. 1866. *Stellaria media* (Linnaeus) Villars subsp. *pallida* (Dumortier) Ascherson & Graebner, Fl. Nordostdeut. Flachl. 310. 1898.

Prostrate annual herb; stem to 4 dm long, much branched, with a single line of trichomes along each internode, otherwise glabrous. Leaves with the blade obovate to elliptic, 0.3–1.5 mm long, 1–7 mm wide, the apex short acuminate, the base rounded to cuneate, the margin entire, sometimes with a few short cilia, the upper surface glabrous, the lower surface glabrous or with a few cilia on the midrib, sessile distally, petiolate proximally. Flowers 3–35 in a cyme, the pedicel slender, 1–10 mm long, spreading or reflexed in age, pubescent; bracts lanceolate, 2–10 mm long, foliaceous; sepals lanceolate, 3–4 mm long, the apex acute, the margin entire, the outer surface pubescent; petals usually shorter than or equaling the sepals or absent; stamens 1–3 or absent; styles 3. Fruit ovoid, 2–4(5) mm long, equaling or slightly longer than the sepals; seeds ca. 1 mm long, tuberculate, pale yellow brown.

Disturbed sites. Rare; Volusia and Gadsden Counties. Ontario and Pennsylvania south to Florida, west to Nebraska, Kansas, and Texas, also Washington and California; Mexico; Europe. Native to Europe. Spring–fall.

Stellaria parva Pederson [Small.] PYGMY STARWORT.

Stellaria parva Pederson, Bot. Tidsskr. 57: 44, f. 4. 1961.

Prostrate annual or perennial herb; stem to 3 dm long, much branched, rooting at the lower nodes, glabrous or sparsely pubescent on 2 sides. Leaves with the blade elliptic to obovate, 0.3–1.5 cm long, 1–6 mm wide, the apex obtuse to acute, the base cuneate, the margin entire, the upper surface glabrous, the lower surface glabrous or pubescent basally, sessile or subsessile. Flowers solitary in the distal nodes, the pedicel 2–10 mm long, spreading, reflexed in age, glandular-pubescent; bracts absent; sepals ovate-lanceolate, ca. 3 mm long, the apex obtuse, the margin membranaceous, the outer surface glandular-pubescent; petals equaling the sepals; stamens 4–5; styles 3–4. Fruit ovoid, 4–5 mm long, longer than the sepals; sepals ca. 1 mm long, papillate, brown.

Disturbed sites near hydric hammocks and swamp margins. Rare; Polk and Hillsborough Counties. Florida, Louisiana, and Texas; South America. Native to South America. Spring–fall.

Stellaria prostrata Baldwin [Lying down.] PROSTRATE STARWORT.

Stellaria prostrata Baldwin, in Elliott, Sketch Bot. S. Carolina 1: 518. 1821. *Alsine prostrata* (Baldwin) A. Heller, Cat. N. Amer. Pl. 3. 1898; non Forsskäl, 1775. *Alsine baldwinii* Small, Fl. S.E. U.S. 422, 1330. 1903. *Stellaria cuspidata* Willdenow ex Schlechtendal subsp. *prostrata* (Baldwin) J. K. Morton, Sida 21: 888. 2004. TYPE: FLORIDA: Duval Co.: Fort George Island, s.d. *Baldwin s.n.* (Holotype: NY).

Decumbent annual herb; stem to 7 dm long, much branched, glandular-puberulent. Leaves with the blade broadly ovate to deltoid, 1–2(5) cm long, 6–28 mm wide, the apex acute to acuminate, the base cordate or truncate, the margin entire, the upper and lower surfaces glandular-pubescent, the petiole of the proximal leaves subequaling or longer than the blade, the distal ones sessile. Flowers (3)5–35 in an open, leafy cyme, the pedicel slender, 1–2(3) cm long, ascending or erect, deflexed in age, glandular-pubescent; bracts lanceolate to ovate, 3–30 mm long, foliaceous; sepals ovate-lanceolate, 3–4 mm long, the apex obtuse to subacute, the margin usually ciliate, the outer surface pubescent on the midrib; petals equaling or shorter than the sepals; stamens 3–8; styles 3. Fruit ovoid, subequaling the sepals; seeds ovoid, ca. 0.8 mm long, the surface with conspicuous and peg-like papillae, reddish brown.

Moist sites. Rare; northern peninsula south to Hernando County, Franklin County. Virginia, South Carolina, Alabama, Florida, Louisiana, and Texas; West Indies and Mexico. Spring.

Stellaria puber Michx. [Minutely downy.] GREAT CHICKWEED.

Stellaria puber Michaux, Fl. Bor.-Amer. 1: 273. 1803. *Alsine puber* (Michaux) Britton, Mem. Torrey Bot. Club 4: 107. 1893. *Stellaria puber* Michaux subsp. *typica* Beguinot, Nuovo Giorn. Bot. Ital. 17: 384. 1910, nom. inadmiss.

Erect or ascending perennial herb, to 4 dm; stem rhizomatous, often with sterile or few-flowered shoots larger than the earlier mainly fertile stem, puberulent above, often in 2 lines. Leaves with the blade elliptic to lanceolate or oblanceolate, 2–9 cm long, 1–3 cm wide, the apex acute, the base cuneate, the margin entire, the upper surface glabrous or sparsely puberulent, the lower surface glabrous, sessile or subsessile. Flowers 3–70 in a leafy cyme, the pedicel 1–4 cm long, ascending to divergent, puberulent; bracts elliptic to lanceolate, 0.7–6.5 cm long, foliaceous; sepals ovate to oblong-lanceolate, 4–6 mm long, the apex obtuse or acute, puberulent, often inconspicuously ciliate; petals slightly exceeding the sepals, cleft half their length or nearly to the base; stamens 10; styles 3. Fruit globose, 4–6 mm long, equaling or shorter than the sepals; seeds asymmetrically reniform, ca. 2 mm long, coarsely sulcate-papillate, brown.

Dry hammocks. Rare; Gadsden and Liberty Counties. Vermont south to Florida, west to Minnesota, Illinois, Nebraska, and Louisiana. Spring.

EXCLUDED TAXA

Stellaria alsine Grimm—Reported by Rabeler and Thieret (1988), but no specimens cited from the state, the nearest is from Augusta-Richmond County, Georgia.

Stellaria cuspidata Willdenow ex Schlechtendal—A species of Texas and Mexico reported from Florida by Long and Lakela (1971). No Florida specimens known.

Stipulicida Michx. 1803. PINELAND SCALYPINK

Annual or short-lived perennial herbs. Leaves opposite, simple, pinnate-veined, petiolate or epetiolate, stipulate. Flowers in condensed, minutely bracteate, terminal fasciculate cymes, bisexual; sepals 5, free or basally connate, dimorphic; petals 5, free; stamens 3–5; ovary 3-carpellate, 1-loculate, the styles 1, the stigmas 3. Fruit a capsule, the valves 3; seeds few.

A genus of 1 species; North America and West Indies. [In reference to bristlelike segments of the stipules.]

Selected references: Judd (1983); Poindexter et al. (2014); Swanson and Rabeler (2005).

Stipulicida setacea Michaux [*Seta*, bristle, in reference to the thin, bristlelike stems.] PINELAND SCALYPINK.

Erect annual or short-lived perennial herb, 20 cm; stem, wiry, solitary or many-branched from the base, rigid, subdichotomously branched above. Basal leaves with the blade 0.5–2 cm long, 1.5–4 mm wide, the apex rounded, the base cuneate, the margin entire, the upper and lower surfaces glabrous, the petiole to 5 mm long or epetiolate on flowering stems, slender, the cauline leaves scalelike, 1–2 mm long, sessile, the bases sometimes connate; stipules basally connate, consisting of a few fringing bristles. Flowers few, forming a sessile or short-pedicellate, subcapitate cyme; bracts minute; sepals 1–2 mm long, the outer 2 elliptic-ovate, the inner 3 obovate, all with expanded, white-hyaline margins, these subentire or distinctly lacerate-fimbriate; petals white, slightly exceeding the sepals, the apex subentire. Fruit subglobose, 1–2 mm long, slightly longer than the perianth; seeds obovate, ca. 1 mm long, the surface minutely alveolate.

1. Sepal margin lacerate-fimbriate .... var. **lacerata**
1. Sepal margin entire or frayed, but not lacerate-fimbriate .... var. **setacea**

Stipulicida setacea Michx. var. **setacea** PINELAND SCALYPINK.

> *Stipulicida setacea* Michaux, Fl. Bor.-Amer. 1: 26, t, 6. 1803. *Polycarpon stipulifidum* Persoon, Syn. Pl. 1: 111. 1805, nom. illegit.
>
> *Stipulicida filiformis* Nash, Bull. Torrey Bot. Club 22: 148. 1895. *Stipulicida setacea* Michaux var. *filiformis* (Nash) D. B. Ward, Novon 11: 361. 2001. TYPE: FLORIDA: Lake Co.: vicinity of Eustis, 12–31 Mar 1894, *Nash 14* (holotype: NY).

Sepal margins entire or frayed, the outer sepal apex acute to obtuse.

Flatwoods and scrub. Frequent; nearly throughout. Virginia south to Florida, west to Louisiana. Spring–fall.

Apparent intermediates with var. *lacerata* are sometimes encountered.

Stipulicida setacea Michx. var. **lacerata** C. W. James PINELAND SCALYPINK.

> *Stipulicida setacea* Michaux var. *lacerata* C. W. James, Rhodora 59: 98. 1957. *Stipulicida lacerata* (C. W. James) D. B. Poindexter et al. J. Bot. Res. Inst. Texas 8: 426. 2014. TYPE: FLORIDA: Pinellas Co.: Dunedin, 14 Apr 1900, *Tracy 6828* (holotype: GH; isotypes: MSC, NY, US).

Sepal margins lacerate-fimbriate, the outer sepal apex mucronate.

Flatwoods and scrub. Frequent; peninsula. Florida; West Indies. Spring–fall.

Vaccaria Wolf 1776. SOAPWORT

Annual herbs. Leaves opposite, simple, pinnate-veined, petiolate or epetiolate, estipulate. Flowers in paniculiform cymes, bisexual; sepals 5, connate into a tube; petals 5, free; stamens 10; Ovary 2-carpellate, 1-loculate, the styles 2. Fruit a capsule, the valves 4, enclosed by the inflated, persistent calyx; seeds numerous.

A genus of 1 species; North America, South America, Europe, Africa, Asia, Australia, and Pacific Islands. [*Vacca*, cow, and *aria*, pertaining to, in reference to a plant's being eaten by cattle.]

Selected reference: Thieret and Rabeler (2005d).

Vaccaria hispanica (Mill.) Rauschert [Of Spain.] COW SOAPWORT; COW HERB.

Saponaria hispanica Miller, Gard. Dict., ed 8. errat. 1768. *Vaccaria hispanica* (Miller) Rauschert, Feddes Repert. 73: 52. 1966.

Saponaria vaccaria Linnaeus, Sp. Pl. 409. 1753. *Saponaria segetalis* Necker, Delic. Gallo-Belg. 1: 194. 1768, nom. illegit. *Lychnis vaccaria* (Linnaeus) Scopoli, Fl. Carniol., ed. 2. 1: 303. 1771. *Saponaria rubra* Lamarck, Fl. Franc. 2: 541. 1779 ("1778"), nom. Illegit. *Vaccaria pyramidata* Medikus, Philos. Bot. 1: 96. 1789. *Vaccaria parviflora* Moench, Methodus 63. 1794, nom. illegit. *Gypsophila vaccaria* (Linnaeus) Sibthorp & Smith, Fl. Graec. Prodr. 1: 279. 1809. *Vaccaria vulgaris* Host, Fl. Austriac. 1: 518. 1827, nom. illegit. *Vaccaria arvensis* Link, Handbuch 2: 240. 1829, nom. illegit. *Vaccaria sessilifolia* Sweet, Hort. Brit., ed. 2. 51. 1830, nom. illegit. *Vaccaria segetalis* Garke ex Ascherson, Fl. Brandenburg 1: 84. 1860, nom. illegit. *Vaccaria vaccaria* (L.) Huth, Helios 11: 136. 1893, nom. inadmiss. *Vaccaria perfoliata* Halacsy, Consp. Fl. Graec. 1: 189. 1900, nom. illegit.; non (Roxburgh ex Willdenow) Sweet, 1830. *Vaccaria perfoliata* Halacsy var. *typica* Halacsy, Consp. Fl. Graec. 1: 190. 1900, nom. inadmiss. *Vaccaria pyramidata* Medikus var. *parviflora* Hayek, Repert. Spec. Nov. Regni Veg. Beih. 30(1): 218. 1924, nom. inadmiss. *Vaccaria hispanica* (Miller) Rauschert var. *vaccaria* (Linnaeus) Greuter, Willdenowia 25: 139. 1995. *Vaccaria hispanica* (Miller) Rauschert subsp. *pyramidata* (Medikus) J. Holub, Preslia 70: 116. 1998.

Erect annual herb, to 8 dm; stem simple below, branched above, glabrous, glaucous. Leaves with the blade lanceolate to lance-ovate, 5–10 cm long, 2–4 cm wide, the apex acute, the base cordate or auriculate-clasping or that of the lower ones connate, the margin entire, the upper and lower surfaces glabrous, glaucous, the lower ones often short-petiolate. Flowers in an open, often flat-topped paniculiform cyme; lower bracts foliaceous, reduced upward; sepals connate into a tube 11–17 mm long, this strongly 5-ribbed and wing-angled, the lobes triangular, 2–3 mm long; petals pink, 18–22 mm long, the blade obovate, 5–8 mm long, the margin retuse, the claw usually exceeding the calyx; stamens exserted. Capsule 6–8 mm long, enclosed by the inflated, ovoid or flask-shaped, often purplish calyx; seeds globose, 2–3 mm long, the surface minutely tuberculate, reddish brown to black.

Disturbed sites. Rare; Gadsden and Miami-Dade Counties. Escaped from cultivation. Nearly throughout North America; South America; Europe, Africa, Asia, Australia, and Pacific Islands. Native to Europe, Africa, and Asia. Summer.

AMARANTHACEAE Juss., nom. cons. 1789. AMARANTH FAMILY

Annual or perennial herbs or subshrubs. Leaves alternate or opposite, simple, pinnate-veined, petiolate or epetiolate, estipulate. Flowers in axillary or terminal glomerate cymes, racemes, spikes, panicles or heads, bisexual or unisexual (plants monoecious, dioecious, or polygamous), each flower or flower cluster usually subtended by 1 bract and 2 bracteoles; perianth segments (1)2–5 or absent, free or basally connate, scarious, chartaceous, or fleshy; stamens 1–5 or absent, opposite the perianth segments, filaments distinct or basally connate, sometimes forming a cup or corolla-like tube, the anthers dorsifixed, 2- or 4-loculate, opening by longitudinal slits, introrse; ovary superior, 1- to 3(5)-carpellate, 1-loculate, the styles 1–3(5), free or connate, terminal, the stigmas elongate or capitate. Fruit an indehiscent, irregularly dehiscent, or circumscissile utricle or a nut, sometimes enclosed by a persistent perianth; seeds 1–many.

A family with about 175 genera and about 2,400 species; nearly cosmopolitan.

The Chenopodiaceae traditionally has been treated as distinct. However, recent studies provide strong support for placing it in the Amaranthaceae.

Chenopodiaceae Vent., nom. cons. (1799).

Selected references: Robertson and Clemants (2003); Welsh et al. (2003).

1. Perianth segments dry and papery; stamens slightly to completely connate.
    2. Leaves alternate.
        3. Fruit 1-seeded ....**Amaranthus**
        3. Fruit several-seeded ....Celosia
    2. Leaves opposite.
        4. Flowers in dense or spreading panicles .... Iresine
        4. Flowers in heads or spikes.
            5. Flowers in elongate, interrupted spikes; fruits deflexed..**Achyranthes**
            5. Flowers in short, congested spikes or heads; fruits erect or spreading.
                6. Flowers in spikes more than 2 times as long as broad...Froelichia
                6. Flowers in heads or spikes less than 1.5 times as long as broad.
                    7. Perianth segments villous; flower spikes subtended by 2 leaflike bracteoles ....
                    ....**Gomphrena**
                    7. Perianth segments not villous; flower spikes not subtended by 2 leaflike bracteoles.
                        8. Stigma capitate; staminodes present; leaves flat ....**Alternanthera**
                        8. Stigma with slender lobes; staminodes absent; leaves terete ....**Blutaparon**
1. Perianth segments membranaceous to fleshy; anthers free.
    9. Leaves reduced and scalelike, all opposite; flower spikes succulent, jointed.
        10. Perennial from a slender rhizome; all 3 flowers of the cluster inserted at or near the same level .... **Sarcocornia**
        10. Annual with a slender taproot; center flower of the cluster elevated above the 2 lateral ones ....
        ....Salicornia
    9. Leaves foliaceous, alternate (lower ones opposite in *Atriplex*); flower spikes not succulent or jointed.
        11. Leaves evidently spine-tipped.. **Salsola**
        11. Leaves merely acute.

12. Leaves linear-filiform ....**Suaeda**

12. Leaves not linear-filiform.

  13. Fruit enclosed by 2 stiff, dentate bracts, each crested on the surface by 2 pointed tu-
  bercles..**Atriplex**

  13. Fruit mostly enclosed by the perianth segments, these not crested.

    14. Leaves with resinous, sessile glands or glandular trichomes......................**Dysphania**

    14. Leaves glabrous or farinose.

      15. Perianth segments 3 or 4...**Oxybasis**

      15. Perianth segments 5.

        16. Young stems and leaves with the vesicular trichomes fully collapsed when dry;
        perianth segments with a prominent midvein on the inner surface ....
        ....**Chenopodiastrum**

        16. Young stems and leaves with the vesicular trichomes becoming cup-shaped
        when dry; perianth segments without a prominent midvein on the inner sur-
        face..**Chenopodium**

Achyranthes L., nom. cons. 1753. DEVIL'S HORSEWHIP

Annual herbs. Leaves opposite, simple, pinnate-veined, petiolate, estipulate. Flowers in terminal spikes or panicles, bisexual; bract 1, bracteoles 2; perianth segments 4–5, basally connate;
stamens 5, the filaments alternating with 5 pseudostaminodes, connate below into a cup, the
anthers 4-loculate; ovary 1-carpellate and -loculate, the style 1, the stigma capitate. Fruit an
indehiscent utricle enclosed by the indurate sepals; seed 1, smooth.

A genus of about 12 species; nearly cosmopolitan. [From the Greek *achyron*, chaff or husk,
and *anthos*, flower, in reference to the chaffy appearance of the mature inflorescence.]

Centrostachys Wall. (1824).

Selected reference: Robertson (2003b).

Achyranthes aspera L. [Rough or harsh.] DEVIL'S HORSEWHIP.

Erect or spreading annual or perennial herb, to 2 m; stem simple or branched, terete or obscurely quadrangular, whitish pilose. Leaves with the blade orbiculate, obovate-orbiculate to
broadly rhombic or elliptic, ovate to broadly ovate, 1–20 cm long, 2–6 mm wide, the apex
rounded or apiculate to acuminate, the base rounded to cuneate, the margin entire, the upper surface pilose-sericeous or glabrate, the lower surface pilose-sericeous, the petiole 3–15
mm long. Flowers in a spike to 3(4) dm long, 6–7 mm thick, the rachis pilose or villous; bract
broadly ovate to orbiculate, the midvein indurate and extended into a rigid spine as long as the
body of the bract or longer; bracteoles ovate, long-aristate, shorter than the sepals; perianth
segments narrowly lanceolate, 3–7 mm long, green, acuminate; pseudostaminodes with the
margin apically fimbriate. Fruit oblong, 2–4 mm long, membranaceous, the apex truncate;
seed ovoid, smooth.

Achyranthes aspera is a variable, pantropical species consisting of six varieties (Townsend,
1974), of which two occur in Florida. The nomenclature of our two taxa was long confused with

the name *A. indica* misapplied to *A. aspera* var. *aspera* (see Excluded Taxon below) and with *A. aspera* var. *aspera* misapplied to *A. aspera* var. *pubescens*.

1. Leaf blades orbiculate, obovate-orbiculate, or broadly rhombic, the apex rounded or apiculate; perianth segments usually 3–4(5) mm long .... var. **aspera**
1. Leaf blades elliptic, ovate, or broadly ovate, the apex acuminate; perianth segments (5)6–7 mm long.. .... var. **pubescens**

Achyranthes aspera L. var. **aspera** DEVIL'S HORSEWHIP.

> *Achyranthes aspera* Linnaeus, Sp. Pl. 204. 1753. *Stachyarpagophora aspera* (Linnaeus) M. Gómez de la Maza y Jiménez, Fl. Haban. 93. 1897. *Centrostachys aspera* (Linnaeus) Standley, J. Wash. Acad. Sci. 5: 75. 1915.
>
> *Achyranthes aspera* Linnaeus var. *indica* Linnaeus, Sp. Pl. 204. 1753. *Cadelaria indica* (Linnaeus) Rafinesque, Fl. Tellur. 3: 39. 1837 ("1836").
>
> *Achyranthes obtusifolia* Lamarck, Encycl. 1: 545. 1785. *Centrostachys indica* (Linnaeus) Standley, J. Wash. Acad. Sci. 5: 75. 1915. *Achyranthes aspera* Linnaeus var. *obtusifolia* (Lamarck) Suessenguth, Mitt. Bot. Staatssamml. München 1: 152. 1952.

Leaf blades orbiculate, obovate-orbiculate, or broadly rhombic, 1–4(6) cm long and wide, the apex rounded or apiculate, the base rounded. Perianth segments 3–4(5) mm long. Fruit 2–3 mm long.

Disturbed sites. Rare; central and southern peninsula, Franklin County. Nearly cosmopolitan. Native to Europe, Africa, Asia, and Australia. All year.

Achyranthes aspera L. var. **pubescens** (Moq.) C. C. Towns. [Hairy.] DEVIL'S HORSEWHIP.

> *Achyranthes fruticosa* Lamarck var. *pubescens* Moquin-Tandon in A. de Candolle, Prodr. 13(2): 314. 1849. *Achyranthes aspera* Linnaeus var. *pubescens* (Moquin-Tandon) C. C. Townsend., Kew Bull. 29: 473. 1974.

Leaf blades elliptic, ovate, or broadly ovate, 4–20 cm long, 2–5 cm wide, the apex acuminate, the base cuneate. Perianth segments 6–7 mm long. Fruit 3–4 mm long.

Disturbed sites. Rare; central and southern peninsula. Native to Europe, Africa, and Asia. All year.

EXCLUDED TAXON

> *Achyranthes indica* (Linnaeus) Miller—Misapplied by Correll and Johnston (1970), Small (1933), and Wunderlin (1982) to material of *A. aspera* var. *aspera*.

Alternanthera Forssk. 1775. JOYWEED

Annual or perennial herbs. Leaves opposite, simple, pinnate-veined, petiolate or epetiolate, estipulate. Flowers in terminal or axillary, sessile or pedunculate, capitate spikes, bisexual; bract 1, bracteoles 2; perianth segments 4–5, free; stamens (3)5, the filaments alternating with 5 pseudostaminodes, partly united into a tube or short cup, the anthers 2-loculate; ovary 1-carpellate and -loculate, the style 1, the stigma capitate. Fruit an indehiscent utricle; seed 1, smooth.

A genus of about 80 species; nearly cosmopolitan. [*Alternus*, alternate, and *anthera*, flower, in reference to the alternating stamens and pseudostaminodes.]

Selected reference: Clemants (2003a).

1. Inflorescence pedunculate.
    2. Leaves slightly succulent, the apex obtuse to rounded; bracts not keeled; perianth segments glabrous ....**A. philoxeroides**
    2. Leaves chartaceous, the apex acute to acuminate; bracts keeled; perianth segments pilose.
        3. Peduncle 3–6 cm long; bracts less than ½ as long as the perianth segments ....**A. flavescens**
        3. Peduncle 1–2 cm long; bracts slightly shorter than or equaling the perianth segments ....
        ....**A. brasiliana**
1. Inflorescence sessile or subsessile.
    4. Pseudostaminodes regularly fimbriate apically; anthers oblong.
        5. Perianth segments rigid, striate, glabrous; leaves succulent ....**A. maritima**
        5. Perianth segments soft, chartaceous, pilose; leaves chartaceous................................. **A. ficoidea**
    4. Pseudostaminodes entire to dentate; anthers globose.
        6. Perianth segments equal or subequal, glabrous or with simple trichomes; fruit bilaterally asymmetrical.
            7. Mature fruit expanded beyond the glabrous perianth segments.............................. **A. sessilis**
            7. Mature fruit contained within the slightly pubescent perianth segments.................................
            ....**A. paronichyoides**
        6. Perianth segments distinctly unequal, the trichomes barbed; fruit radially symmetrical.
            8. Perianth segments 5–7 mm long, strongly spinose-tipped, sparsely villous; pseudostaminode margin dentate.. **A. pungens**
            8. Perianth segments 3–5 mm long, only slightly spinose tipped, densely villous; pseudostaminode margin entire..**A. caracasana**

Alternanthera brasiliana (L.) Kuntze [Of Brazil.] BRAZILIAN JOYWEED.

Gomphrena brasiliana Linnaeus, Cent. Pl. 2: 13. 1756. *Gomphrena dentata* Moench, Suppl. Meth. 273. 1802, nom. illegit. *Philoxerus brasilianus* (Linnaeus) Smith, in Rees, Cycl. 27: Philoxeris no. 4. 1814. *Mogiphanes brasiliana* (Linnaeus) Martius, Nov. Gen. Sp. Pl. 2: 34. 1826. *Caraxeron brasilianus* (Linnaeus) Rafinesque, Fl. Tellur. 3: 38. 1837 ("1836"). *Telanthera brasiliana* (Linnaeus) Moquin-Tandon, in A. de Candolle, Prodr. 13(2): 382. 1849. *Alternanthera brasiliana* (Linnaeus) Kuntze, Revis. Gen. Pl. 2: 537. 1891. *Achyranthes brasiliana* (Linnaeus) Standley, J. Wash. Acad. Sci. 5: 74. 1915.

Erect annual or perennial herb or subshrub, to 6 dm; stem villous to glabrate. Leaves with the blade ovate to lanceolate, 1–7 cm long, 0.5–1 cm wide, chartaceous, the apex acuminate, the base cuneate, the margin entire, the upper and lower surfaces villous, the petiole 2–3 cm long. Flowers in a terminal or axillary pedunculate, globose head 0.7–1 cm long; bract and bracteoles keeled, slightly shorter than to equaling the perianth segments; perianth segments equal or subequal, lanceolate, 3–4 mm long, green to stramineous, the apex acuminate, pilose, the trichomes not barbed; stamens 5, the anthers oblong; pseudostaminodes ligulate, the margin fimbriate. Fruit ellipsoid, ca. 2 mm long, included within the sepals, brown; seed ovoid-oblong, ca. 1 mm long.

Wet, disturbed sites. Rare; Sarasota, Lee, and Broward Counties. Florida; West Indies, Mexico, Central America, and South America. Native to tropical America. Summer–fall.

Alternanthera caracasana Kunth [Of Caracas, Venezuela.] WASHERWOMAN.

Alternanthera caracasana Kunth, in Humboldt et al., Nov. Gen. Sp. 2: 206. 1818. *Telanthera caracasana* (Kunth) Moquin-Tandon, in A. de Candolle, Prodr. 13(2): 370. 1849.

Illecebrum peploides Willdenow ex Schultes, in Roemer & Schultes, Syst. Veg. 5: 517. 1820. *Alternanthera peploides* (Willdenow ex Schultes) Urban, Repert. Spec. Nov. Regni Veg. 15: 168. 1918. *Achyranthes peploides* (Willdenow ex Schultes) Britton & P. Wilson, Bot. Porto Rico 5: 279. 1924.

Prostrate to procumbent perennial herb; stem to 5 dm long, branched at the base, rooting at the nodes, pilose to glabrate. Leaves with the blade rhombic-ovate, ovate, or obovate to suborbicular, 0.5–2.5 cm long, 0.3–1.5 cm wide, the apex obtuse apiculate, the base cuneate, the margin entire, the upper and lower surfaces sparsely pilose, the petiole 1–5 mm long. Flowers in an axillary, ovoid to short-cylindric, sessile head; bract ⅔ as long as the bracteoles, the latter shorter than or subequal to the perianth segments, somewhat keeled, the apex long-attenuate, with a prominent excurrent midrib, dorsally pubescent; perianth segments 5, distinctly unequal, lanceolate, 3–5 mm long, aristate, 1-nerved, pilose with barbed trichomes on the back and margin; stamens 5, the anthers globose; pseudostaminodes subulate, subequaling the filaments, the margin entire. Fruit ovoid, with a slight wing just below the apex; seed ovate-orbicular, 1–1.5 mm long.

Disturbed sites. Rare; St. Johns and Escambia Counties. Not recently collected. Maryland south to Florida, west to California; West Indies, Mexico, Central America, and South America; Europe and Africa. Native to tropical America. Summer–fall.

Achyranthes repens (=*Alternanthera pungens*) was misapplied to this species by Small (1933).

Alternanthera ficoidea (L.) P. Beauv. [Fig-like, perhaps in reference to its resemblance to some species of fig marigold (*Mesembryanthemum* spp.) (Aizoaceae).] SLENDER JOYWEED.

Gomphrena ficoidea Linnaeus, Sp. Pl. 225. 1753, nom. cons. *Illecebrum ficoideum* (Linnaeus) Linnaeus, Sp. Pl., ed. 2. 300. 1762. *Achyranthes ficoidea* (Linnaeus) Lamarck, Encycl. 1: 548. 1785. *Paronychia ficoidea* (Linnaeus) Desfontaines, Tabl. École Bot. 44. 1804. *Alternanthera ficoidea* (Linnaeus) Palisot de Beauvois, Fl. Oware 2: 66. 1818. *Bucholzia ficoidea* (Linnaeus) Martius, Nov. Gen. Sp. Pl. 2: 52. 1826. *Steiremis ficoidea* (Linnaeus) Rafinesque, Fl. Tellur. 33: 41. 1837 ("1836"). *Telanthera ficoidea* (Linnaeus) Moquin-Tandon, in A. de Candolle, Prodr. 13(2): 363. 1849.

Alternanthera tenella Colla, Mem. Reale Accad. Sci. Torino 33: 131, t. 9. 1829.

Telanthera bettzickiana Regel, Index Sem. Hort. Petrop. 1862: 28. 1862. *Telanthera bettzickiana* Regel var. *typica* Regel, Index Sem. Hort. Petrop. 1869: 102. 1869, nom. inadmiss. *Alternanthera bettzickiana* (Regel) G. Nicholson, Ill. Dict. Gard. 1: 59. 1884. *Alternanthera bettzickiana* (Regel) G. Nicholson forma *typica* Voss, Vilm. Blumengärtn., ed. 3. 1: 869. 1896, nom. inadmiss. *Achyranthes bettzickiana* (Regel) Standley, in Britton, N. Amer. Fl. 21: 138. 1917. *Alternanthera ficoidea* (Linnaeus) Palisot de Beauvois var. *bettzickiana* (Regel) Backer, in Steenis, Fl. Males. 4(2): 93. 1949. *Alternanthera tenella* Colla var. *bettzickiana* (Regel) Veldkamp, Taxon 27: 313. 1978.

Erect ascending, decumbent, or prostrate annual or perennial herb or subshrub; stem to 1 m long, villous or gray-pubescent when young, glabrescent in age. Leaves with the blade rhombic-ovate to rhombic-obovate, 1–6 cm long, 0.5–2 cm wide, the apex acute to acuminate, the base cuneate, decurrent on the petiole, green or multicolored, the margin entire, the upper and lower surfaces glabrate, the petiole to 1 cm long. Flowers in an axillary or terminal, sessile, in globose, ovoid, or short-cylindric head 5–8 mm long; bract and bracteoles ovate to elliptic, 2–3 mm long, the apex acuminate, aristate, white-hyaline, pilose or glabrate; perianth segments unequal, lanceolate-oblong or ovate-oblong, 2–4 mm long, whitish or stramineous, the margin chartaceous, the apex acute to acuminate, mucronate, pilose; stamens 5, the anthers oblong; pseudostaminodes lanceolate, equaling or longer than the stamens, the margin fimbriate. Fruit ovoid, ca. 2 mm long, included within the sepals; seed ovoid, ca. 1 mm long.

Disturbed sites. Rare; Hillsborough, Sarasota, Lee, and Broward Counties. Escaped from cultivation. Florida and Texas; West Indies, Mexico, Central America, and South America. Native to tropical America. All year.

The cultivar with variously multicolored leaves, called "parrotleaf" or "calico plant," is the form encountered as an escape in Florida.

Alternanthera flavescens Kunth [Yellowish.] YELLOW JOYWEED.

Alternanthera flavescens Kunth, in Humboldt et al., Nov. Gen. Sp. 2: 207. 1818.

Mogiphanes ramosissima Martius, Nov. Gen. Sp. Pl. 2: 31, t. 130. 1826. *Telanthera ramosissima* (Martius) Moquin-Tandon, in A. de Candolle, Prodr. 13(2): 381. 1849. *Alternanthera ramosissima* (Martius) Chodat, Bull. Herb. Boissier, ser. 2. 3: 355. 1903. *Achyranthes ramosissima* (Martius) Standley, J. Wash. Acad. Sci. 5: 74. 1915.

Telanthera floridana Chapman, Fl. South. U.S. 383. 1860. *Alternanthera floridana* (Chapman) Small, Fl. S.E. U.S. 396, 1330. 1903. TYPE: FLORIDA: Monroe Co.: Key West, s.d., *Chapman s.n.* (isotype: GH).

Ascending to spreading or climbing perennial herb, to 1.5(4) m; stem with long, straight, appressed, tawny trichomes or glabrate. Leaves with the blade ovate to lanceolate or elliptic, 2–9 cm long, 0.5–3 cm wide, chartaceous, the apex acute to long-acuminate, the base rounded to cuneate, the margin entire or minutely toothed, the upper and lower surfaces sparsely strigose or glabrous, the petiole to 1 cm long. Flowers sessile or subsessile in a subglobose to cylindric, long-pedunculate spike; bract and bracteoles ovate to lanceolate, 1.5–2.5 mm long, mucronate, hyaline, sparsely to densely pilose; perianth segments equal, lanceolate or lanceolate-oblong to ovate, 4–5 mm long, the apex acute to acuminate, apiculate, scarious, pilose, trichomes not barbed; stamens 5, the anthers oblong; pseudostaminodes ligulate, exceeding the stamens, the margin fimbriate. Fruit ellipsoid, 2–3 mm long; seed ovoid, ca. 1 mm long.

Coastal hammocks. Frequent; Clay County, central and southern peninsula. Florida; West Indies, Mexico, Central America, and South America. All year.

Alternanthera maritima (Mart.) A. St. Hil. [Growing by the sea]. SEASIDE JOYWEED.

Bucholzia maritima Martius, Nov. Gen. Sp. Pl. 2: 50, t. 147. 1826. *Illecebrum maritimum* (Martius) Sprengel, Syst. Veg. 4(2): 103. 1827. *Alternanthera maritima* (Martius) A. Saint-Hilaire, Voy. Distr. Diam. 2: 437. 1833. *Alternanthera maritima* (Martius) A. Saint-Hilaire var. *parvifolia* A. Saint-Hilaire, Voy. Distr. Diam. 2: 438. 1833, nom. inadmiss. *Telanthera maritima* (Martius) Moquin-Tandon, in A. de Candolle, Prodr. 13(2): 364. 1849. *Telanthera maritima* (Martius) Moquin-Tandon var. *parvifolia* Moquin-Tandon, in A. de Candolle, Prodr. 13(2): 364. 1849, nom. inadmiss. *Achyranthes maritima* (Martius) Standley, J. Wash. Acad. Sci. 5: 73. 1915.

Procumbent or prostrate perennial herb; stem to 8 dm long, branched, lightly ridged or angled, rooting at the nodes, glabrous. Leaves with the blade oblong, oblanceolate, obovate, or oval, 2–6 cm long, 1–3 cm wide, succulent, the apex obtuse to acute, mucronate, the base cuneate, the margin entire, the upper and lower surfaces glabrous, sessile or subsessile. Flowers in an axillary, sessile, globose or ovoid head 0.6–1.2 mm long, stramineous; bract and bracteoles ovate, keeled, the midrib excurrent into a short awn, ½–⅔ as long as sepals; perianth segments 5, equal, ovate or lanceolate, glabrous, the apex acuminate; stamens 5, the anthers oblong; pseudostaminodes lanceolate, equaling or exceeding the filaments, cleft or fimbriate apically. Fruit obovoid to subglobose, ca. 2 mm long; seed subglobose, 1–2 mm long.

Coastal strands. Occasional; St. Lucie County southward. Florida; West Indies, South America; Africa. All year.

Alternanthera paronichyoides A. St. Hil. [Resembling *Paronychia* (Caryophyllaceae).] SMOOTH JOYWEED.

Alternanthera paronichyoides A. Saint-Hilaire, Voy. Distr. Diam. 2: 439. 1833.

Procumbent or prostrate perennial herb; stem to 4(8) dm long, sparingly pilose or glabrate. Leaves with the blade elliptic, ovate-rhombic, ovate, or obovate, 1–3 cm long, 0.5–1 cm wide, the apex acute to obtuse, the base cuneate, the margin entire, the upper and lower surfaces pilose to glabrate, the petiole 5–10 mm. Flowers in an axillary, globose to subcylindric, sessile head 8–12 mm long, white; bract and bracteoles lanceolate, shorter than the perianth segments; perianth segments 5, slightly equal, lanceolate, 3–4(5) mm long, white, the apex acute or acuminate, pilose at the base, the trichomes not barbed; stamens 5, the anthers globose; pseudostaminodes ligulate, subequaling or shorter than the filaments, entire or dentate at the apex. Fruit obcordate, 2–3 mm long, compressed; seed lenticular, 1–2 mm long.

Salt flats and disturbed sites. Rare; central and southern peninsula, Escambia County. Pennsylvania and New Jersey, North Carolina south to Florida west to Texas; West Indies, Mexico, Central America, and South America; Asia, Africa, and Pacific Islands. Native to tropical America. All year.

Alternanthera polygonoides (=*A. sessilis*) was misapplied to this species by various authors (e.g., Clewell, 1985; Correll and Johnston, 1970; Radford et al., 1964, 1968; Small, 1933 (as *Achyranthes polygonoides*); Wunderlin, 1982).

Alternanthera philoxeroides (Mart.) Griseb. [Resembling *Philoxerus* (Amaranthaceae).] ALLIGATORWEED.

Bucholzia philoxeroides Martius, Nova Acta Phys.-Med. Acad. Caes. Leop.-Carol. Nat. Cur. 13(1): 315. 1826. *Telanthera philoxeroides* (Martius) Moquin-Tandon, in A. de Candolle, Prodr. 13(2): 362. 1849. *Telanthera philoxeroides* (Martius) Moquin-Tandon var. *obtusifolia* Martius ex Moquin-Tandon, in A. de Candolle, Prodr. 13(2): 363. 1849, nom. inadmiss. *Alternanthera philoxeroides* (Martius) Grisebach, Abh. Königl. Ges. Wiss. Göttingen 24: 36. 1879. *Alternanthera philoxeroides* (Martius) Grisebach var. *obtusifolia* Hicken, Apuntes Hist. Nat. 2: 94. 1910, nom. inadmiss. *Achyranthes philoxeroides* (Martius) Standley, J. Wash. Acad. Sci. 5: 74. 1915.

Prostrate to decumbent perennial herb; stem to 5 m long, forming dense, tangled, often in float-
ing mats, rooting at the nodes, glabrous except for a narrow band of multicellular trichomes in
the leaf base, hollow and slightly flattened in age, rooting at the nodes. Leaves with the blade
narrowly elliptic, elliptic, or oblanceolate, 4–7(9) cm long, 1–2 cm wide, fleshy, the apex acute
to obtuse, mucronate, the base tapering to form a short petiole-like base, this dilating, clasping
the stem and adnate to the opposite one to form a narrow sheath, the margin hyaline, with
minute dentations which bear long trichomes, the upper and lower surfaces with scattered
trichomes when young, glabrous in age. Flowers in an axillary or terminal, subglobose, head
or short spike ca. 1.5 cm long, on peduncles to 6 cm long; bract and bracteoles 2–3 mm long;
perianth segments subequal, lanceolate, 5–6 mm long, white, glabrous, the apex acute; stamens
5, the anthers oblong; pseudostaminodes ligulate, fimbriate. Fruit and seed not seen.

Pond and lake margins, shallow streams, and swamps. Frequent; nearly throughout. Virginia south to Florida, west to Texas, also California; West Indies, Mexico, Central America,
and South America; Europe, Asia, Australia, and Pacific Islands. Native to South America. All
year.

Alternanthera philoxeroides apparently only reproduces vegetatively in North America as no
mature fruit or seed has been seen. It is listed as a Category II invasive species in Florida by the
Florida Exotic Pest Plant Council (FLEPPC, 2015). In Florida, its possession, cultivation, and
transport is prohibited by state law (Florida Administrative Code, Chapter 62C-52).

Alternanthera pungens Kunth [Pointed.] KHAKIWEED.

> *Achyranthes repens* Linnaeus, Sp. Pl. 205. 1753. *Illecebrum achyrantha* Linnaeus, Sp. Pl., ed. 2. 299.
> 1762, nom. illegit. *Achyranthes mucronata* Lamarck, Encycl. 2: 547. 1785, nom. illegit. *Achyranthes
> radicans* Cavanilles, Anales Ci. Nat. 3: 27. 1801, nom. illegit. *Paronychia achyrantha* Desfontaines,
> Tabl. École Bot. 44. 1804, nom. illegit. *Pityranthus crassifolius* Martius, Denkschr. Königl. Akad.
> Wiss. München 5: 179. 1817, nom. illegit. *Alternanthera achyrantha* R. Brown ex Sweet, Hort. Sub-
> urb. Lond. 48. 1818. *Alternanthera repens* (Linnaeus) Link, Enum. Hort. Berol. Alt. 1: 154. 1821; non
> J. F. Gmelin, 1791.
>
> *Alternanthera pungens* Kunth, in Humboldt et al., Nov. Gen. Sp. 2: 206. 1818. *Illecebrum pungens*
> (Kunth) Sprengel, Syst. Veg. 1: 820. 1824 ("1825"). *Telanthera pungens* (Kunth) Moquin-Tandon,
> in A. de Candolle, Prodr. 13(2): 371. 1849. *Alternanthera achyrantha* (Linnaeus) R. Brown ex Sweet
> var. *leiantha* Seubert, in Martius, Fl. Bras. 5(1): 183. 1875. *Achyranthes leiantha* (Seubert) Standley,
> J. Wash. Acad. Sci. 5: 73. 1915. *Alternanthera leiantha* (Seubert) Alain, Contr. Mus. Hist. Nat. "La
> Salle" Havana 9: 1. 1950, nom. illegit. *Alternanthera pungens* Kunth var. *leiantha* (Seubert) Sues-
> senguth, Mitt. Bot. Staatssamml. München 4: 103. 1952, nom. inadmiss.

Prostrate perennial herb; stem to 8 dm long, villous. Leaves with the blade ovate to obovate or
orbicular, 1–3.5 cm long, 1–2 cm wide, the apex rounded, mucronate, the base broadly cuneate,
the margin entire, the upper and lower surfaces appressed pilose when young, glabrate in age,
the petiole 2–5 mm long. Flowers in an axillary, sessile, globose to ovoid head 8–10 mm long;
bract and bracteoles lance-oblong, equaling the perianth segments, the apex attenuate to aristate; perianth segments, oblong or ovate-oblong, 5–7 mm long, unequal, stramineous, pilose,
the trichomes barbed, the apex acuminate to long-aristate; stamens 3–5, the anthers globose;

pseudostaminodes triangular, shorter than stamens, the margin remotely dentate. Fruit ovate, ca. 2 mm long, strongly compressed, truncate; seed lenticular, ca. 1.5 mm long.

Disturbed sites. Rare; Escambia County. Not collected since 1897. New York, Virginia, South Carolina south to Florida, west to Texas; West Indies, Mexico, and South America; Asia, Australia, and Pacific Islands. Native to South America. Summer–fall.

Alternanthera sessilis (L.) R. Br. ex DC. [Sessile, in reference to the flowers.]
SESSILE JOYWEED.

> *Gomphrena sessilis* Linnaeus, Sp. Pl. 225. 1753. *Illecebrum sessile* (Linnaeus) Linnaeus, Sp. Pl., ed. 2. 300. 1762. *Paronychia tetragona* Moench, Methodus 315. 1794, nom. illegit. *Paronychia sessilis* (Linnaeus) Desfontaine, Tabl. École Bot. 44. 1804. *Alternanthera sessilis* (Linnaeus) R. Brown ex de Candolle, Cat. Pl. Horti Monsp. 4, 77. 1813. *Allaganthera forskalii* Martius, Pl. Hort. Erlang. 69. 1814, nom. illegit.
>
> *Gomphrena polygonoides* Linnaeus, Sp. Pl. 225. 1753. *Illecebrum polygonoides* (Linnaeus) Linnaeus, Sp. Pl., ed. 2. 300. 1762. *Achyranthes polygonoides* (Linnaeus) Lamarck, Encycl. 1: 547. 1783. *Alternanthera polygonoides* (Linnaeus) R. Brown ex Sweet, Hort. Suburb. Lond. 48. 1818. *Bucholzia polygonoides* (Linnaeus) Martius, Nov. Gen. Sp. Pl. 2: 51. 1826. *Steiremis repens* Rafinesque, Fl. Tellur. 3: 41. 1837 ("1836"), nom. illegit. *Telanthera polygonoides* (Linnaeus) Moquin-Tandon, in A. de Candolle, Prodr. 13(2): 363. 1849.

Weakly erect or prostrate annual or perennial herb; stem to 6 dm long, rooting at the nodes, pubescent in lines between the nodes and with horizontal rows of trichomes at the nodes. Leaves with the blade elliptic to oblong or oblanceolate, 1–5 cm long, 0.5–2 cm wide, the apex acute or obtuse, the base narrowed and short petiole-like, the margin entire, the upper surface glabrous, the lower surface sparsely pubescent. Flowers in an axillary, subglobose or ovoid, sessile head 0.5–1 cm long; bract and bracteoles ovate, ca. 1 mm long, mucronate; perianth segments ovate to lanceolate, 2–4 mm long, white, the apex acuminate, sparsely pilose, the trichomes not barbed; stamens 3–5, the anthers globose; pseudostaminodes subulate, equal or shorter than the stamens, the margin laciniate. Fruit obcordate, compressed, exceeding the sepals; seed lenticular, ca. 1 mm long.

Wet, disturbed sites. Occasional; peninsula, central and western panhandle. Alabama, Georgia, Florida, and Louisiana; West Indies, Mexico, Central America, and South America; Africa, Asia, and Pacific Islands. Native to Asia. All year.

EXCLUDED TAXON

> *Alternanthera moquinii* (Webb ex Moquin-Tandon) Dusen—Misapplied by Small (1903, 1913a) to material of *A. flavescens*.

Amaranthus L. 1753. AMARANTH

Annual herbs. Leaves alternate, simple, pinnate-veined, petiolate, estipulate. Flowers in terminal and/or spikes or panicles, unisexual (plants dioecious or monoecious); bract 1, bracteoles 2. Staminate flowers with the perianth segments 3–5, free; stamens 3–5, free, the anthers 4-thecal;

pseudostaminodes absent. Carpellate flowers with the perianth segments (1–2)3–5, free or basally connate; ovary 1-carpellate and -loculate, the styles short or absent, the stigmas 2–3. Fruit a utricle, 2- to 3-beaked with the persistent styles, circumscissile, irregularly dehiscent, or indehiscent; seed 1.

A genus of about 70 species; cosmopolitan. [From the Greek *amaranthos*, unfading or not wilting, in reference to the long-lasting flowers.]

Acnida L. (1753).

Selected reference: Mosyakin and Robertson (2003).

1. Plants dioecious.
    2. Plant with carpellate flowers.
        3. Perianth segments regularly present and well developed (at least 1 mm long and with distinct midveins).
            4. Perianth segments 1 or 2, lanceolate to linear; fruit indehiscent ....**A. floridanus**
            4. Perianth segments 5, at least the inner ones spatulate; fruit circumscissile...........**A. palmeri**
        3. Perianth segments completely lacking or rudimentary (less than 1 mm long and without visible midveins).
            5. Seed ca. 1 mm long; fruit to 2 mm long .... **A. australis**
            5. Seed 2–3 mm long; fruit 2.5–4 mm long..**A. cannabinus**
    2. Plant with staminate flowers.
        6. Perianth segments with a strong midvein; bract with a strong midvein, ca. 4 mm long..............
            ....**A. palmeri**
        6. Perianth segments without a strong midvein; bract with a slender midvein, not more than 2 mm long.
            7. Bract and bracteoles less than 1 mm long, the midvein scarcely excurrent .....**A. cannabinus**
            7. Bract and bracteoles more than 1 mm long, the midvein conspicuously excurrent.
                8. Leaf blade usually lanceolate, more than 1 cm wide; inflorescence often branched above the uppermost leaves .... **A. australis**
                8. Leaf blade usually linear, to 1 cm wide; inflorescence unbranched above the leaves ....
                    ....**A. floridanus**
1. Plants monoecious, with staminate and carpellate flowers intermingled or in nearly separate inflorescences.
    9. Stems with paired axillary spines... **A. spinosus**
    9. Stems not spinose.
        10. Perianth segments of the carpellate flowers connate basally, spatulate, usually contracted into a narrow claw at the base.
            11. Fruit indehiscent; pedicel, bract, and bracteoles much thickened and indurate......................
                ....**A. crassipes**
            11. Fruit dehiscent; pedicel, bract, and bracteoles not thickened.......................**A. polygonoides**
        10. Perianth segments of the carpellate flowers free, oblong to obovate, not clawed at the base.
            12. Fruit indehiscent.
                13. Fruit conspicuously muricate; seed with shallow, scurfy verrucae on the reticulate pattern of the testa (with 50x lens), therefore somewhat dull ....**A. viridis**
                13. Fruit smooth or slightly muricate; seed without verrucae, glossy.
                    14. Fruit ellipsoid...**A. deflexus**
                    14. Fruit round to slightly pyriform ....**A. blitum**

12. Fruit dehiscent.

    15. Perianth segments of the carpellate flowers obtuse or truncate.

        16. Flowers in axillary clusters or glomerules.. **A. albus**

        16. Flowers in terminal or axillary erect spikes or panicles .... **A. retroflexus**

    15. Perianth segments of the carpellate flowers acute to acuminate.

        17. Bract and bracteoles conspicuously longer than the perianth segments ....

        ....**A. hybridus**

        17. Bract and bracteoles shorter than or subequaling the perianth segments.

            18. Apex of the perianth segments incurved against the fruit .... **A. dubius**

            18. Apex of the perianth segments straight or spreading away from the fruit..............

        .... **A. caudatus**

Amaranthus albus L. [White, in reference to its general appearance.] PROSTRATE PIGWEED; TUMBLEWEED.

Amaranthus albus Linnaeus, Syst. Nat., ed. 10. 1268. 1759. *Galliaria albida* Bubani, Fl. Pyren. 1: 185. 1897, nom. illegit.
Amaranthus leucanthus Rafinesque, Fl. Ludov. 32. 1817.

Erect, ascending, or prostrate annual herb; stem to 1 m long, much branched, glabrous, sparingly puberulent, or villous. Leaves with the blade elliptic to oblong or spatulate or obovate, 1–7 cm long, the apex rounded or mucronate-cuspidate, the base cuneate, the margin entire, the upper and lower surfaces glabrous, sparingly puberulent, or villous, the petiole 0.5–5 cm long. Flowers in an axillary glomerule, unisexual (plants monoecious). Staminate flowers usually few, intermixed with the carpellate; bract and bracteoles subulate to linear lanceolate, 2–4 mm long, rigid; perianth segments 3, oblong, cuspidate, scarious, green; stamens 3. Carpellate flowers with bract and bracteoles as in staminate; perianth segments 3, oblong to linear, slightly unequal, thin, green along the nerve, often tinged with red, the apex acute; style branches 3, erect. Fruit subglobose, longer than the persistent calyx, 1–2 mm long, rugose distally, circumscissile dehiscent; seed lenticular, 0.6–0.8 mm in diameter, mahogany-colored, lustrous.

Disturbed sites. Rare; Gadsden, Leon, and Wakulla Counties. Nearly throughout North America; West Indies, Mexico, Central America, and South America; Europe, Africa, Asia, and Australia. Native to the North American Great Plains. Spring.

Amaranthus australis (A. Gray) J. D. Sauer [Southern.] SOUTHERN AMARANTH.

Acnida australis A. Gray, Amer. Naturalist 10: 489. 1876. *Acnida cannabina* Linnaeus var. *australis* (A. Gray) Uline & W. L. Bray, Bot. Gaz. 20: 157. 1895. *Amaranthus australis* (A. Gray) J. D. Sauer, Madroño 13: 15. 1955. TYPE: FLORIDA: Franklin Co.: Apalachicola, s.d., *Chapman s.n.* (lectotype: GH). Lectotypified by Sauer (1955: 15).
Acnida cuspidata Bertero ex Sprengel, Syst. Veg. 3: 903. 1826. *Acnida cannabina* Linnaeus var. *cuspidata* (Bertero ex Sprengel) Moquin-Tandon, in A. de Candolle, Prodr. 13(2): 277. 1849.
Acnida alabamensis Standley, in Britton, N. Amer. Fl. 21: 121. 1917.

Erect annual herb, to 3(9) m; stem branched, glabrous. Leaves with the blade lanceolate to ovate, 10–20(30) cm long, 1–4 cm wide, the apex acute to acuminate, the base cuneate, the margin entire, the upper and lower surfaces glabrous, the petiole 5–20 cm long. Flowers in an elongate, simple or paniculate spike, unisexual (plants dioecious). Staminate flowers with the bract and bracteoles 1–2 mm long; perianth segments 5, 2.5–3 mm long, equal, 2–3 mm long, the apex subacute to mucronulate; stamens 5. Carpellate flowers with the bract and bracteoles 1–2 mm long; perianth segments absent; style branches 3–5, spreading. Fruit obovoid to turbinate, 2–3 mm long, with 3(5) longitudinal ridges corresponding to the style branches, smooth (slightly rugose in drying), indehiscent, stramineous; seed lenticular, ca. 1 mm wide, dark reddish brown to black, lustrous.

Freshwater and brackish marshes. Frequent; nearly throughout. Virginia south to Florida, west to Texas; West Indies, Mexico, Central America, and South America. All year.

Amaranthus blitum L. subsp. **emarginatus** (Salzm. ex Uline & W. L. Bray) Carretero et al. [Old generic name for some species of *Chenopodium* (Chenopodiaceae) including the "strawberry blite," from the Greek *bliton*, blite; emarginate, in reference to the leaves.] PURPLE AMARANTH.

> *Euxolus viridis* (Linnaeus) Moquin-Tandon var. *polygonoides* Moquin-Tandon, in A. de Candolle, Prodr. 13(2): 274. 1849. *Amaranthus emarginatis* Salzmann ex Uline & W. L. Bray, Bot Gaz. 19: 319. 1894. *Amaranthus ascendens* Loiseleur-Deslongchamps var. *polygonoides* (Moquin-Tandon) Thellung, Mém. Soc. Sci. Nat. Cherbourg 38: 215. 1912. *Amaranthus lividus* Linnaeus var. *polygonoides* (Moquin-Tandon) Thellung ex Druce, Bot Soc. Exch. Club Brit. Isles 5: 574. 1920. *Amaranthus lividus* Linnaeus subsp. *polygonoides* (Moquin-Tandon) Probst, Wolladventivfl. Mitteleur. 74. 1949. *Amaranthus ascendens* Loiseleur-Deslongchamps subsp. *polygonoides* (Moquin-Tandon) Priszter, Agrártud. Egyet. Kert-Szoeloegazdaságtud. Karának Evk. 2(2): 221. 1953. *Amaranthus blitum* Linnaeus subsp. *polygonoides* (Moquin-Tandon) Cerretero, Anales Jard. Bot. Madrid 41: 276. 1985. *Amaranthus blitum* Linnaeus subsp. *emarginatus* (Salzmann ex Uline & W. L. Bray) Carretera et al., Anales Jard. Bot. Madrid 44: 599. 1987. *Amaranthus blitum* Linnaeus var. *emarginatus* (Salzmann ex Uline & W. L. Bray) Lambinon, Bull. Soc. Échange Pl. Vasc. Eur. Occid. Bassin Médit. 24: 55. 1993.
> *Amaranthus lividus* Linnaeus forma *pseudogracilis* Thellung, in Ascherson & Graebner, Syn. Mitteleur. Fl. 5: 321. 1914. *Amaranthus emarginatus* Salzmann ex Uline & W. L. Bray subsp. *pseudogracilis* (Thellung) Hugin, Willdenowia 16: 463. 1987. *Amaranthus blitum* Linnaeus var. *pseudogracilis* (Thellung) Costea, in Costea et al. Sida 19: 981. 2001.

Erect, ascending, or prostrate annual herb; stem to 7 dm long, simple or branched, glabrous. Leaves with the blade ovate, rhombic-ovate, or obovate, 1–3(6) cm long, 0.5–3(4) cm wide, the apex retuse to broadly emarginate, mucronulate, the base long-cuneate, the margin entire, the upper and lower surfaces glabrous, the petiole equaling to 2 times as long as the blade. Flowers in a slender, erect or sometimes reflexed, terminal spike or panicle, the spike, rarely in an axillary cluster, unisexual (plants monoecious). Staminate flowers clustered at the tips of the spike; bract and bracteoles ovate, the apex acute, subequal to the perianth segments; perianth segments 3, ovate-elliptic, the apex acute, stamens 3. Carpellate flowers with the perianth segments 3, elliptic or spatulate, 1–2 mm long, unequal or subequal, the apex acute to mucronate; style branches 3, erect. Fruit subglobose to obovate, 1–2(3) mm long, indehiscent, smooth or faintly rugose at maturity; seed subglobose to broadly lenticular, ca. 1 mm long, dark brown to black, lustrous.

Disturbed sites. Frequent; peninsula, central and western panhandle. Nearly cosmopolitan. Native to Europe, Africa, and Australia. All year.

Amaranthus cannabinus (L.) J. D. Sauer [Resembling *Cannabis* (Cannabaceae).] TIDALMARSH AMARANTH.

Acnida cannabina Linnaeus Sp. Pl. 1027. 1753. *Amaranthus cannabinus* (Linnaeus) J. D. Sauer, Madroño 13: 11. 1955.
Acnida elliotii Rafinesque, New Fl. 1: 54. 1836). TYPE: "Carolina and Florida."

Erect annual herb, to 3 m; stem branched, glabrous. Leaves with the blade narrowly lanceolate to linear or ovate, to 20 cm long, to 4 cm wide, the apex acute to acuminate, the base cuneate, the margin entire, the upper and lower surfaces glabrous, the petiole about ½ the blade length. Flowers in a simple or paniculate spike, unisexual (plants dioecious). Staminate flowers with the bract and bracteoles 1–2 mm long; perianth segments 5, 2.5–3 mm long, unequal, the inner ones with the apex emarginate, the outer ones with the apex acute; stamens 3–5. Carpellate flowers with the bract and bracteoles 1–2 mm long; perianth segments absent or 1–2 and rudimentary; style branches 3–5, erect to slightly spreading. Fruit ovoid or obovoid, 2.5–4 mm long, angled, usually with 3–5 longitudinal ridges corresponding to style branches, indehiscent, slightly rugose when dry; seed lenticular, 2–3 mm long, dark reddish brown to black, lustrous.

Freshwater and brackish swamps and marshes. Rare; Duval County. Not recently collected. Maine south to Florida, west to Louisiana. All year.

Amaranthus caudatus L. [Tailed, in reference to the elongate inflorescence branches.] LOVE-LIES-BLEEDING.

Amaranthus caudatus Linnaeus, Sp. Pl. 990. 1753.

Erect annual herb, to 1.5(2.5) m; stem moderately branched or rarely simple, pubescent distally, glabrescent in age. Leaves with the blade rhombic-ovate, ovate, or elliptic to broadly lanceolate, 5–15(20) cm long, 2–10 cm wide, the apex acute to obtuse, mucronate, the base cuneate, the margin entire, the upper and lower surfaces glabrous, the petiole shorter than or equaling the blade length. Flowers in a terminal or axillary, drooping spike or panicle, usually red, purple, or white, rarely green or yellow, unisexual (plants monoecious). Staminate flowers mostly at the tip of the inflorescence; bract and bracteoles narrowly lanceolate; perianth segments (4)5; stamens 5. Carpellate flowers with the bract and bracteoles as in the staminate; perianth segments 5, spatulate-obovate or lanceolate-obovate, subequal, 1–2 mm long, the apex obtuse, slightly emarginate, subacute or mucronate; style branches 3, spreading or reflexed. Fruit ovoid to subglobose, 1–2 mm long, circumscissile dehiscent; seed lenticular to subglobose, ca. 1 mm long, dark brown to yellowish white, smooth.

Disturbed sites. Rare; Alachua County. Escaped from cultivation. Scattered nearly throughout the United States. Central America and South America. Native to South America. Spring–fall.

Amaranthus caudatus may have originated as a selection from the *A. hybridus* complex in cultivation in South America. It is commonly cultivated as an ornamental and to a lesser extent as a grain crop nearly worldwide. Its treatment here as a distinct species is tentative.

Amaranthus crassipes Schltdl. [Thickened, in reference to its general appearance.] SPREADING AMARANTH.

Amaranthus crassipes Schlechtendal, Linnaea 6: 757. 1831. *Scleropus crassipes* (Schlechtendal) Moquin-Tanden, in A. de Candolle, Prodr. 13(2): 271. 1849. *Euxolus crassipes* (Schlechtendal) Hieronymus Bol. Acad. Nac. Ci. 4: 13. 1881.

Prostrate or decumbent annual herb; stem to 6 dm long, usually much branched at the base, glabrous. Leaves with the blade suborbicular to ovate or obovate, 1–3 cm long, 0.5–2 cm wide, the apex rounded or retuse, the base cuneate, the margin entire, the upper and lower surfaces glabrous, the petiole to 4 cm long, about ½ as long as the blade. Flowers in a short, axillary cluster, unisexual (plants monoecious), the staminate intermixed with the carpellate. Staminate flowers with the bract and bracteoles ovate, keeled, indurate at maturity, the midrib prominent and slightly excurrent; perianth segments 5, elliptic to ovate, the apex acute, midrib prominent and slightly excurrent, subequal; stamens 3. Carpellate flowers with the bract and bracteoles as in the staminate; perianth segments 5, narrowly spatulate, 2–3 mm long, the apex rounded and apiculate, indurate and scarious at maturity; style branches 2(3), spreading. Fruit obovoid to compressed-obovoid, 1.5–2 mm long, the surface muricate above the middle, indehiscent; seed compressed obovoid to lenticular, 1–1.5 mm long, dark brown or black, lustrous.

Disturbed coastal sites. Rare; Monroe County keys. Not recently collected. Pennsylvania, South Carolina, Florida, and Alabama, Texas west to Arizona; West Indies, Mexico, and South America. Native to tropical America. Spring.

Amaranthus deflexus L. [Turned back, apparently in reference to the branches.] LARGEFRUIT AMARANTH.

Amaranthus deflexus Linnaeus, Mant. Pl. 2: 295. 1771. *Glomeraria deflexa* (Linnaeus) Cavanilles, Descr. Pl. 319, 617. 1803. *Euxolus deflexus* (Linnaeus) Rafinesque, Fl. Tellur. 3: 42. 1837 ("1836"). *Albersia deflexa* (Linnaeus) Fourreau, Ann. Soc. Linn. Lyon, ser. 2. 17: 142. 1869.

Prostrate to ascending annual or short-lived perennial herb; stem 0.5 m long, much branched near the base, puberulent in the upper part. Leaves with the blade rhombic-ovate to ovate or lanceolate, 1–3 cm long, 0.5–2 cm wide, the apex subacute to obtuse or retuse, rarely shallowly emarginate, mucronulate, the base subtruncate to cuneate, the margin entire, the upper and lower surfaces glabrous, the petiole ½ as long as to equaling the blade. Flowers in a dense, erect, terminal panicle 2–5 cm long, smaller clusters also present in the axil of the upper leaves, unisexual (plants monoecious). Staminate flowers at the tip of the inflorescence branches; bract and bracteoles deltoid-ovate to ovate-lanceolate, ca. ½ the length of the perianth segments; perianth segments 2–3, lanceolate, equal or subequal, the apex mucronate; stamens 2–3. Carpellate flowers with the bract and bracteoles as in the staminate; perianth segments 2–3, linear-elliptic to oblanceolate, equal or subequal, 1–2 mm long, the apex acute to obtuse, the styles branches 3, erect. Fruit ellipsoid, 2–3 mm long, ca. 1 mm wide, wrinkled or rugose when dry, indehiscent; seed obovate, ca. 1 mm long, black, lustrous.

Disturbed sites. Rare; Escambia County. Nearly cosmopolitan. Native to South America. Summer–fall.

Amaranthus dubius Mart. ex Thell. [Doubtful, in reference to its original identity.] SPLEEN AMARANTH.

Amaranthus dubius Martius ex Thellung, Fl. Adv. Montpellier 203. 1912.

Erect annual or short-lived perennial herb, to 1 m; stem simple or much branched, glabrous or sparsely pubescent. Leaves with the blades ovate to rhombic-ovate or elliptic, 3–12 cm long, 2–8 cm wide, the apex obtuse to acute or slightly retuse, mucronate, the base cuneate, the margin entire, the upper and lower surfaces glabrate, the petiole 2–9 cm long, equaling or longer than the blade. Flowers in a terminal erect or drooping panicle or axillary spike to 15 cm long, unisexual (plants monoecious). Staminate flowers at the tip of the inflorescence branches; bract and bracteoles lanceolate to ovate, less than 2 mm long, shorter than the perianth segments, the apex spinescent; perianth segments 5, equal or subequal, oblong-spatulate to oblong, 1–2 mm long, the apex acute, mucronate; stamens 5. Carpellate flowers with the bract and bracteoles as in the staminate; perianth segments 5, equal or subequal, oblong-spatulate to oblong, 1–2 mm long, the apex acute, mucronate; style branches 3, strongly spreading. Fruit ovoid to subglobose, 1–2 mm long, smooth or irregularly wrinkled, circumscissile dehiscent; seed subglobose to lenticular, ca. 1 mm long, dark reddish brown to black, lustrous.

Disturbed sites. Rare; Pinellas, Hillsborough, and Charlotte Counties, Monroe County keys. Nearly cosmopolitan. Native to tropical America. Spring.

Amaranthus dubius is morphologically and genetically closely related to *A. spinosus* and members of the *A. hybridus* complex and may have originated as a result of ancient hybridization (Mosyakin and Robertson, 2003).

Amaranthus floridanus (S. Watson) J. D. Sauer [Of Florida.] FLORIDA AMARANTH.

Acnida floridana S. Watson, Proc. Amer. Acad. Arts 17: 376. 1882. *Amaranthus floridanus* (S. Watson) J. D. Sauer, Madroño 13: 25. 1955. TYPE: FLORIDA: Monroe Co.: Key West, s.d., *Blodgett s.n.* (lectotype: GH). Lectotypified by Sauer (1955: 25).

Erect annual herb, to 1.5 m; stem simple or branched at the base, glabrous. Leaves with the blade linear to narrowly oblong, 1–10 cm long, 2–10 mm wide, the apex obtuse to rounded, the base cuneate, the margin entire, the upper and lower surfaces glabrous, the petiole 0.5–3 cm long, ca. ¼ the blade length. Flowers in a terminal spike or panicle, unisexual (plants dioecious). Staminate flowers with the bract and bracteoles triangular-lanceolate or ovate, 1–2 mm long; perianth segments 5, ovate, 1–2 mm long, equal; or subequal, the apex acute, mucronate; stamens 5. Carpellate flowers with bract and bracteoles as in the staminate; perianth segments 1–2(3), lanceolate to narrowly lanceolate, unequal, the apex acute to acuminate; style branches 3, spreading. Fruit subglobose to obovoid, 2–3 mm long, somewhat rugulose, with 3 longitudinal ridges; seed lenticular, ca. 1 mm long, smooth, reddish brown to black, lustrous.

Coastal dunes and beaches. Occasional; peninsula. Endemic. Spring–fall.

Amaranthus hybridus L. [Of hybrid origin.] SLIM AMARANTH; PIGWEED.

Amaranthus hybridus Linnaeus, Sp. Pl. 990. 1753. *Amaranthus retroflexus* Linnaeus var. *hybridus* (Linnaeus) A. Gray, Manual, ed. 5. 412. 1867. *Amaranthus hybridus* Linnaeus var. *typicus* Beck, in Reichenbach, Icon. Fl. Germ. Helv. 24: 175. 1908, nom. inadmiss. *Galliaria hybrida* (Linnaeus) Nieuwland, Amer. Midl. Naturalist 3: 278. 1914.

Amaranthus hypochondriacus Linnaeus, Sp. Pl. 991. 1753. *Amaranthus hybridus* Linnaeus var. *erythrostachys* Moquin-Tanden, in A. de Candolle, Prodr. 13(2): 259. 1849. *Amaranthus hybridus* Linnaeus var. *hypochondriacus* (Linnaeus) B. L. Robinson, Rhodora 10: 32. 1908, nom. illegit. *Amaranthus hybridus* Linnaeus subsp. *hypochondriacus* (Linnaeus) Thellung, Fl. Adv. Montpellier 204. 1912. *Amaranthus chlorostachys* Willdenow var. *erythrostachys* (Moquin-Tanden) Aellen, in Hegi, Ill. Fl. Mitt.-Eur., ed. 2. 3: 482. 1959.

Amaranthus cruentus Linnaeus, Syst. Nat., ed. 10. 1269. 1759. *Amaranthus paniculatus* Linnaeus var. *cruentus* (Linnaeus) Seubert, in Martius, Fl. Bras. 5(1): 238. 1875. *Amaranthus hybridus* Linnaeus subsp. *cruentus* (Linnaeus) Thellung, Fl. Adv. Montpellier 205. 1912.

Amaranthus chlorostachys Willdenow, Hist. Amaranth. 34, t. 10(19). 1790. *Amaranthus retroflexus* Linnaeus var. *chlorostachys* (Willdenow) A. Gray, Manual, ed. 5. 412. 1867. *Amaranthus hybridus* Linnaeus var. *chlorostachys* (Willdenow) Beck, in Reichenbach, Icon. Fl. Germ. Helv. 24: 175. 1908.

Erect or ascending annual herb, to 1.5(2.5) m; stem branched, glabrous or pubescent. Leaves with the blade ovate to rhombic-ovate or lanceolate, 4–15 cm long, 2–6 cm wide, the apex acute or rounded, mucronate, the base cuneate or rounded, the margin entire, the upper and lower surfaces pubescent or glabrous, the petiole ½ as long as to as long as the blade. Flowers in a terminal or axillary, erect or nodding spike, the terminal one to 2 dm long and often with numerous shorter branches at the base, unisexual (plants monoecious). Staminate flowers at the inflorescence tips; bract and bracteoles linear-lanceolate, 2–4 mm long, the apex spinescent; perianth segments 5, narrowly oblong to ovate, 2–3 mm long, subequal or unequal, the apex acuminate, aristate, 2–3 mm long; stamens (4)5. Carpellate flowers with the bract and bracteoles as in the staminate; perianth segments 5, lanceolate to linear-lanceolate, subequal or unequal, the acuminate, aristate; style branches 3, erect. Fruit ovoid, 2–3 mm long, base of the style branches connate and forming a short neck, circumscissile dehiscence smooth; seed lenticular-globose, ca. 1 cm long, dark brown to black, lustrous.

Disturbed sites. Frequent; nearly throughout. Nearly cosmopolitan. Native to the southwestern United States and tropical America. Summer–fall.

Amaranthus palmeri S. Watson [Commemorates Edward Palmer (1821–1911), English-born natural history collector, army surgeon, and ethnobotanist.] PALMER'S PIGWEED; CARELESSWEED.

Amaranthus palmeri S. Watson, Amer. Acad. Arts. 12: 274. 1877.

Erect annual herb, to 1.5(3) m; stem simple or branched at the base, glabrous or glabrescent. Leaves with the blade rhombic-ovate to rhombic-lanceolate or elliptic, 1–6 cm long, 1–4 cm wide, the apex acute, usually mucronate, the base cuneate, the margin entire, the upper and lower surfaces glabrous, the petiole subequaling the blade. Flowers in a dense to interrupted, simple spike or raceme 2–7 cm long, terminating the stem or a branch, often lax or drooping, often with some flowers in the lower axillary clusters, unisexual (plants dioecious). Staminate flowers with the bract and bracteoles ca. 4 mm long, the apex long-acuminate; perianth segments 5, 2–4 mm long, unequal, the apex acute to long-acuminate, mucronulate; stamens 5. Carpellate flowers with the bract and bracteoles 4–6 mm long, the apex acuminate, mucronulate; perianth segments 5, 2–4 mm long, the apex acuminate, mucronulate; style branches 2(3), spreading. Fruit obovoid to subglobose, 1.5–2 mm long, circumscissile dehiscent, smooth or slightly rugose; seed obovate, lenticular, dark reddish brown to black, ca. 1 mm long, smooth, lustrous.

Disturbed sites. Rare; northern counties. Ontario south to Florida, west to California; Mexico; Europe, Africa, Asia, and Australia. Native to southwestern United States and Mexico. Summer–fall.

Amaranthus polygonoides L. [Resembling *Polygonum* (Polygonaceae).] TROPICAL AMARANTH.

Amaranthus polygonoides Linnaeus, Pl. Jamaic. Pug. 27. 1759. *Roemeria polygonoides* (Linnaeus) Moench, Methodus 341. 1794. *Glomeraria polygonoides* (Linnaeus) Cavanillas, Elench. Pl. Hort. Matr. 17. 1803. *Amblogyna polygonoides* (Linnaeus) Rafinesque, Fl. Tellur. 3: 42. 1837 ("1836").

Erect to ascending or prostrate annual herb; stem to 0.5 m long, much branched from the base, usually pinkish and glaucous, pubescent or glabrous. Leaves with the blade ovate- to obovate-rhombic to ovate or lanceolate, 1–3 cm long, 0.5–2 cm wide, the apex rounded, obtuse to subtruncate or retuse, mucronate, the base cuneate, the margin entire, the upper surface glabrous, the lower surface sparsely pubescent or glabrous, the petiole to 2.5 cm long, subequaling the blade. Flowers in a dense axillary cluster, unisexual (plants monoecious). Staminate flowers intermixed with the carpellate; bract and bracteoles lanceolate to linear, acuminate, ca. 1 mm long; perianth segments (4)5, ovate to lanceolate, free; stamens 2–3. Carpellate flowers with the bract and bracteoles lanceolate to linear, ca. 1 mm long; perianth segments 5, thickened and connate in proximal ⅓, equal or subequal, spatulate, the apex obtuse or rounded, mucronate; style branches 3, erect to spreading. Fruit cylindric to narrowly turbinate, 2–3 mm long, rugose distally, indehiscent; seed ovoid, lenticular, ca. 1 mm long, dark brown or black, lustrous.

Disturbed coastal sites. Rare; Miami-Dade County, Monroe County keys. South Carolina, Florida, Texas, and New Mexico; West Indies, Mexico, Central America, and South America; Europe. Native to tropical America. Summer–fall.

Amaranthus retroflexus L. [Bent backward, in reference to the sometimes reflexed inflorescence tips.] RED PIGWEED; ROUGH PIGWEED.

Amaranthus retroflexus Linnaeus, Sp. Pl. 991. 1753. *Gaillaria scabra* Bubani, Fl. Pyren. 1: 87. 1897, nom. illegit. *Amaranthus retroflexus* Linnaeus var. *typicus* Fiori & Paoletti, Fl. Anal. Ital. 1: 321. 1898, nom. inadmiss. *Galliaria retroflexa* (Linnaeus) Nieuwland, Amer. Midl. Naturalist 3: 278. 1914. *Amaranthus retroflexus* Linnaeus var. *genuinus* Thellung ex Probst, Mitt. Naturf. Ges. Solothurn 6: 26. 1920, nom. inadmiss.

Erect or ascending annual herb, to 1.5(2) m; stem simple or distally branched, villous-puberulent, white or reddish striate. Leaves with the blade rhombic-ovate to ovate or lanceolate,

2–8(15) cm long, 1–7 cm wide, the apex acute or obtuse to slightly emarginate, mucronate, the base cuneate to rounded cuneate, the margin entire, the upper and lower surfaces villous, at least on the veins on the lower surface; the petiole ½ as long as or equaling the blade length. Flowers in a dense terminal and axillary spike or panicle, often in a dense cluster in the axil of the upper leaves (plants monoecious). Staminate flowers few at the inflorescence tips; bract and bracteoles lanceolate to subulate, 3.5–5 mm long, the apex acuminate, spinose, sparsely villous; perianth segments 5, spatulate to obovate, 2–4 mm long, unequal, the apex emarginate or obtuse, mucronate; stamens (3)4–5. Carpellate flowers with the bract and bracteoles lanceolate to subulate, 3.5–5 mm long, the apex acuminate, spinose, sparsely villous; perianth segments 5, spatulate to obovate, 2–4 mm long, unequal, the apex emarginate or obtuse, mucronate; style branches 3, erect to slightly spreading. Fruit obovate to elliptic, 2–3 mm long, smooth or slightly rugose, circumscissile dehiscent; seed subglobose or compressed lenticular, ca. 1 mm long, black, smooth, lustrous.

Disturbed sites. Rare; Glades and Lee Counties. Not recently collected. Nearly cosmopolitan. Native to central and eastern North America to the north of Florida. Summer.

Amaranthus spinosus L. [Spiny.] SPINY AMARANTH.

Amaranthus spinosus Linnaeus, Sp. Pl. 991. 1753. *Galliaria spinosa* (Linnaeus) Nieuwland, Amer. Midl. Naturalist 3: 278. 1914.

Erect or ascending annual herb, to 1 m; stem much branched, often reddish, glabrous or sparsely pubescent distally, most nodes with paired, divergent spines 5–10 mm long. Leaves with the blade ovate-lanceolate to ovate or rhombic-ovate, 3–10 cm long and wide, the apex obtuse or subobtuse to emarginate, mucronulate, the base broadly cuneate, the margin entire, the upper and lower surfaces glabrous to sparsely pubescent, the petiole ½ as long as to longer than the blade. Flowers in a terminal or axillary paniculate spike or an axillary subglobose cluster, the terminal spike often wholly or mainly staminate, the apex reflexed or nodding, the basal portion and the axillary clusters mostly carpellate. Staminate flowers with the bract and bracteoles lanceolate or subulate, 0.5–1 mm long; perianth segments 5, obovate or spatulate-lanceolate, 1–2 mm long, equal or subequal, the apex acute, mucronate or short-aristate; stamens 5. Carpellate flowers with the bract and bracteoles lanceolate or subulate, 0.5–1 mm long; perianth segments 5, obovate or spatulate-lanceolate, 1–2 mm long, equal or subequal, the apex acute, mucronate or short-aristate; styles 3, erect or spreading. Fruit ovoid or subglobose, 1–2(3) mm long, indehiscent or irregularly dehiscent, roughened above; seed lenticular or subglobose, ca. 1 mm long, black, lustrous.

Disturbed sites. Frequent; nearly throughout. Nearly cosmopolitan. Native to tropical America. All year.

Amaranthus viridis L. [Green, apparently in reference to its general appearance.] SLENDER AMARANTH.

Amaranthus viridis Linnaeus, Sp. Pl., ed. 2. 1405. 1763. *Amaranthus spicatus* Lamarck, Fl. Franc. 2: 192. 1779 ("1778"), nom. illegit. *Blitum viride* (Linnaeus) Moench, Methodus 359. 1794. *Glomeraria viridis* (Linnaeus) Cavanilles, Descr. Pl. 319. 1802. *Euxolus viridis* (Linnaeus) Moquin-Tandon, in A. de Candolle, Prodr. 13(2): 273. 1849. *Pyxidium retroflexum* Friche-Joset & Montandon, Syn. Jura 261. 1856, nom. illegit.

    Chenopodium caudatum Jacquin, Collectanea 2: 325. 1789. *Amaranthus gracilis* Desfontaines, Tabl. École Bot. 43. 1804. *Euxolus caudatus* (Jacquin) Moquin-Tandon, in A. de Candolle, Prodr. 13(2): 274. 1849.

Erect to decumbent annual or short-lived perennial herb; stem to 1 m long, glabrous or lightly pubescent. Leaves with the blade rhombic-ovate or ovate, 1–7 cm long, 0.5–5 cm wide, the apex obtuse, rounded, or emarginate, mucronate, the base acute to rounded, the margin entire, the upper and lower surfaces glabrous or lightly pubescent, sometimes with whitish or purplish markings in the center, the petiole 1–2 cm long. Flowers in a terminal or paniculate spike 10–20 cm long, with several ascending lateral branches nearly equaling the terminal one, unisexual (plants monoecious). Staminate flowers inconspicuous, mostly at the tip of the inflorescence; bract and bracteoles ovate to lanceolate, ca. 1 mm long; perianth segments 3, narrowly elliptic, obovate-elliptic, or spatulate, 1–2 mm long, subequal, the apex rounded to subacute, mucronate; stamens 3. Carpellate flowers with the bract and bracteoles ovate to lanceolate, ca. 1 mm long; perianth segments 3, narrowly elliptic, obovate-elliptic, or spatulate, 1–2 mm long, subequal, the apex rounded to subacute, mucronate; style branches 3, erect. Fruit ovoid to compressed-ovoid, ca. 1 mm long, rugose, indehiscent; seed subglobose to lenticular, ca. 1 mm long, black or brown, minutely puncticulate, dull.

    Disturbed sites. Frequent; peninsula, central and western panhandle. Nearly cosmopolitan. Native to South America. All year.

EXCLUDED TAXA

    Amaranthus blitoides S. Watson—Small (1933) gives the range as "nearly throughout the U.S.," which would presumably include Florida. Mosyakin and Robertson (2003) also report it for Florida. No Florida specimens known.

    Amaranthus blitum Linnaeus—Because infraspecific categories were not recognized, this taxon interpreted as the typical subspecies, was reported by Wunderlin (1982, 1998), Wilhelm (1984), and Mosyakin and Robertson (2003). All our material is of subsp. *emarginatus*.

    Amaranthus graecizans Linnaeus—Misapplied to material of *Amaranthus albus* by several authors, including Small (1933) and Radford et al. (1964, 1968).

    Amaranthus lividus Linnaeus—This synonym of the typical subspecies of *A. blitum* was reported by Chapman (1860, 1883, both as *Euxolus lividus* (Linnaeus) Moquin-Tandon) and Correll and Correll (1982). All Florida material is *A. blitum* subsp. *emarginatus*.

    Amaranthus powellii S. Watson—Reported by Mosyakin and Robertson (2003). No Florida material is known, but this species is easily misidentified as *A. hybridus*, so further study may well reveal a Florida specimen.

    Amaranthus tuberculatus (Moquin-Tandon) J. D. Sauer—Reported by Radford et al. (1964, 1968). Excluded from Florida by Sauer (1955). No Florida specimens known.

Atriplex L., nom. cons. 1753. SALTBUSH

Annual or perennial herbs or subshrubs. Leaves alternate or rarely the proximalmost opposite or subopposite, simple, pinnate-veined, petiolate or epetiolate, estipulate. Flowers in axillary or in terminal spikes or panicles, unisexual (plants monoecious). Staminate flowers ebracteate and ebracteolate; perianth segments 3–5, basally connate; stamens 3–5, free, the anthers 2-thecal; gynoecium rudimentary or lacking. Carpellate flowers ebracteate, 2-bracteolate; perianth segments absent; staminodes absent; ovary 1-carpellate and -loculate, the stigmas 2. Fruit a utricle enclosed by the fruiting bracteoles, the pericarp adnate to the seed, indehiscent; seed 1, vertical.

A genus of about 250 species; nearly cosmopolitan. [The ancient Latin name for *Atriplex hortensis* L., perhaps from *ater*, black, and *plexus*, braided or interwoven, the application unclear.]

Selected reference: Welsh (2003).

1. Leaves entire, oblong to elliptic or obovate, the base rounded to cuneate .... **A. pentandra**
1. Leaves irregularly sinuate-dentate, the base often subhastate.
    2. Plant erect, simple or few branched; leaves with Kranz anatomy; surface of the fruiting bracteoles tuberculate to nearly smooth.. **A. rosea**
    2. Plant decumbent, much branched; leaves without Kranz anatomy; surface of the fruiting bracteoles reticulate.. **A. tatarica**

Atriplex pentandra (Jacq.) Standl. [From the Greek *pentos*, five, and *andros*, male, in reference to the 5 perianth segments.] CRESTED SALTBUSH.

Axyris pentandra Jacquin, Select. Stirp. Amer. Hist. 244. 1763. *Atriplex pentandra* (Jacquin) Standley, in Britton, N. Amer. Fl. 21: 54. 1916. *Atriplex pentandra* (Jacquin) Standley subsp. *typica* H. M. Hall & Clements, Publ. Carnegie Inst. Wash. 326: 294. 1923, nom. inadmiss. *Obione pentandra* (Jacquin) Ulbrich, in Engler & Prantl, Nat. Pflanzenfam., ed. 2. 16C: 507. 1934, nom. illegit. *Atriplex cristata* Humboldt & Bonpland ex Willdenow, Sp. Pl. 4: 959. 1806. *Obione cristata* (Humboldt & Bonpland ex Willdenow) Moquin-Tandon, Chenop. Monogr. Enum. 73. 1840.
Atriplex mucronata Rafinesque, Amer. Monthly Mag. & Crit. Rev. 2: 119. 1817.
Atriplex arenaria Nuttall, Gen. N. Amer. Pl. 1: 198. 1818. *Obione arenaria* (Nuttall) Moquin-Tandon, Chenop. Monogr. Enum. 71. 1840. *Atriplex cristata* Humboldt & Bonpland ex Willdenow var. *arenaria* (Nuttall) Kuntze, Revis. Gen. Pl. 2: 546. 1891. *Atriplex pentandra* (Jacquin) Standley subsp. *arenaria* (Nuttall) H. M. Hall & Clements, Carnegie Inst. Wash. 326: 294. 1923.
Atriplex tampicensis Standley, in Britton, N. Amer. Fl. 21: 56. 1916.

Ascending or procumbent annual or perennial herb, sometimes suffruticose; stem to 8 dm long, much branched, finely furfuraceous when young, glabrate in age. Leaves alternate or the proximalmost opposite or subopposite, the blade oblong or oval to obovate or elliptic, 1–4 cm long, 3–15 mm wide, the apex rounded to acute, the base cuneate to rounded, the margin entire or with 1–2 teeth to repand-dentate or sinuate-dentate to undulate, the upper surface gray-green, glabrate, the lower surface densely white-furfuraceous, the petiole short or absent. Staminate flowers in a short, dense or interrupted, terminal spike or panicle 1–2 cm long; perianth segments 5; stamens 3. Carpellate flowers in an axillary fascicle; bracteoles sessile or subsessile, cuneate-orbicular, 3–6 mm long and wide, compressed, united at the base or to the middle, the apex rounded, the base truncate to cuneate, 3- to 5-dentate, the lobes keeled, the teeth subequal, the sides irregularly tuberculate or with 2 lateral dentate crests, appendaged or rarely not appendaged. Fruit and seed laterally compressed, 1–2 mm long, brown or reddish brown.

Salt marshes, beaches, and coastal dunes. Frequent; nearly throughout. New Hampshire south to Florida, west to Texas; West Indies, Mexico, and South America. All year.

Welsh (2003) treats *A. mucronata* as distinct from *A. pentandra*, but the two intergrade and are nearly sympatric throughout most of their range.

Atriplex rosea L. [In reference to the flowers which are sometimes red-tinged.] TUMBLING SALTWEED.

Atriplex rosea Linnaeus, Sp. Pl., ed. 2, 2: 1493. 1763.

Erect annual herb, to 1 m; stem simple or divaricately branched, scurfy to glabrate. Leaves alternate, the blade ovate to lanceolate, 1.5–8 cm long, 0.6–5 cm wide, the apex acute to obtuse, the base cuneate, the margin irregularly sinuate-dentate, often subhastately lobed or rarely entire, the upper and lower surfaces scurfy to glabrate. Staminate flowers in an axillary glomerule or an interrupted terminal spike; perianth segments 4–5. Carpellate flowers in an axillary glomerule; bracteoles deltoid, 3–6(10) mm long and wide, prominently 3- to 5-veined, sessile or short-stipitate, sometimes subhastately lobed at the base, dentate, the surfaces sharply tuberculate to nearly smooth. Fruit and seed laterally compressed, dimorphic, either 2–3 mm long and brown or 1–2 mm long and black.

Disturbed sites. Rare; Escambia County, based on a single collection (*Mohr s.n.*, US) from Pensacola (Hall and Clements, 1923). Nearly throughout North America; Mexico; Europe, Africa, Asia, and Pacific Islands. Native to Europe, Africa, and Asia. Summer–fall.

Atriplex tatarica L. [Of Tataria (Tatarstan), now a republic within the Russian Federation.] TATARIAN ORACHE.

Atriplex tatarica Linnaeus, Sp. Pl. 1053. 1753. *Schizotheca tatarica* (Linnaeus) Celakovský, Arch. Naturwiss. Landesdurchf. Böhmen 2: 149. 1873.

Erect annual herb, to 1(1.5) m; stem, much divaricately branched, sparsely scurfy or glabrous. Leaves alternate or the proximalmost opposite, the blade ovate to triangular, 1.5–5 cm long, 1–4 cm wide, the distalmost ones becoming linear or linear-oblong, entire, the apex acute to obtuse, the base cuneate to subhastate, the margin sinuate-dentate, the upper and lower surfaces sparsely scurfy or glabrous, the petiole long, becoming much reduced distally on the stem. Staminate flowers in a slender, usually interrupted terminal spike or panicle; perianth segments 5. Carpellate flowers in an axillary fascicle; bracteoles ovate-rhombic or subcordate, 4–8 mm long, 3–7 mm wide, sessile or short-stipitate, 3-veined, moderately compressed, the margin coarse dentate, the surfaces tuberculate or smooth. Fruit and seed lateral compressed, ca. 2 mm long, brown.

Disturbed sites. Rare; Escambia County, based on a single collection (*Curtiss 6865*, NY) from ballast ground near Pensacola (Fernald, 1902, as *A. lampa* (moquin-Tandon) Gillis ex small, misapplied]; Hall and Clements, 1923). New Hampshire, Massachusetts, New Jersey, Pennsylvania, Connecticut, Florida, and Alabama; South America; Europe, Africa, and Asia. Native to Europe, Africa, and Asia. Summer–fall.

EXCLUDED TAXA

Atriplex lampa (Moquin-Tandon) Gillies ex Small—Reported by Fernald (1902), Small (1933), Wilhelm (1984), Clewell (1985), and Wunderlin (1998), the name misapplied to material of *A. tatarica* L.

Atriplex patula Linnaeus—Reported by Small (1933). No Florida specimens known.

Blutaparon Raf. 1838.

Perennial or annual herbs. Leaves opposite, simple, pinnate-veined, epetiolate, estipulate. Flowers in subglobose or cylindrical, pedunculate spikes, solitary and terminal on the branches, short-pedicellate, bisexual; bract 1, bracteoles 2; perianth segments 5, basally connate, unequal; stamens 5, the filaments basally connate, the anthers 2-thecal; pseudostaminodes absent; ovary superior, 1-carpellate and -loculate, the stigmas 2. Fruit a utricle, indehiscent or irregularly splitting; seed 1.

A genus of 4 species; North America, West Indies, Mexico, Central America, Africa, Asia, and Pacific Islands. [Abridged from *Bulutulaparon*, an old Latin name, probably a corruption of the Latin *volutum laparum*, a loose twiner.]

Caraxeron Raf. (1836); *Philoxerus* R. Br. (1810).

Selected references: Clemants (2003d); Mears (1982).

Blutaparon vermiculare (L.) Mears [Wormlike, in reference to the leaf shape.]
SAMPHIRE; SILVERHEAD.

Gomphrena vermicularis Linnaeus, Sp. Pl. 224. 1753. *Illecebrum vermiculare* Linnaeus, Sp. Pl., ed. 2. 300. 1762, nom. illegit. *Philoxerus vermicularis* (Linnaeus) Palisot de Beauvois, Fl. Owaré 2: 65. 1814. *Philoxerus vermicularis* Smith, in Rees, Cycl. 27: Philoxerous no. 3. 1814, nom. illegit. *Achyranthes vermicularis* (Linnaeus) Elliott, Sketch Bot. S. Carolina 1: 310. 1821. *Caraxeron vermicularis* (Linnaeus) Rafinesque, Fl. Tellur. 3: 38. 1837 ("1836"). *Blutaparon repens* Rafinesque, New Fl. 4: 46. 1838 ("1836"), nom. illegit. *Iresine vermicularis* (Linnaeus) Moquin-Tandon, in A. de Candolle, Prodr. 13(2): 340. 1849. *Cruzeta vermicularis* (Linnaeus) M. Gómez de la Maza y Jiménez, Anales Inst. Segunda Enseñ. 2: 212. 1896. *Lithophila vermicularis* (Linnaeus) Uline, in Millspaugh, Publ. Field Columb. Mus., Bot. Ser. 2: 39. 1900, nom. illegit. *Blutaparon vermiculare* (Linnaeus) Mears, Taxon 31: 113. 1982.

Blutaparon breviflorum Rafinesque, New Fl. 4: 45. 1838 ("1836"). TYPE: FLORIDA: s.d., *Baldwin s.n.* (neotype: P-Durand). Neotypified by Mears (1982: 114).

Prostrate annual or perennial herb; stem to ca. 2 m long, much branched and mat-forming, succulent. Leaves with the blade linear to linear-oblong or oblong to clavate, subterete, 1.5–6 cm long, 2–12 mm wide, succulent, the apex acute or obtuse, the base tapering, joined at the base with opposite leaf and forming a narrow sheath, villous in the axils, otherwise glabrous, the margin entire. Flowers in a subglobose head to a cylindric spike 0.7–3 cm long, 0.5–1 cm wide, the rachis lanate; bract broadly ovate, chartaceous, 1-nerved, the apex acute or obtuse, glabrous, white; bracteoles ovate-oblong, slightly shorter than the perianth segments, the apex acute, glabrous, white; perianth segments basally thickened, 3–5 mm long, the apex obtuse, the outer ones elliptic oblong, glabrous, the inner ones lanceolate, usually lanate near the base; stamen filaments subulate. Fruit compressed, broadly ovoid; seed orbicular, ca. 1 mm long, dark brown, lustrous.

Tidal flats and coastal swales. Frequent; peninsula. Florida, Louisiana, and Texas; West Indies, Mexico, Central America, and South America; Africa and Pacific Islands. Native to North America, tropical America, and Africa. All year.

Celosia L. 1753. COCK'S COMB

Annual or perennial herbs. Leaves alternate, simple, pinnate-veined, petiolate, estipulate. Flowers in terminal and sometimes also axillary spikes or fascicles, pedunculate or sessile, bisexual; bract 1, bracteoles 2; perianth segments 5, free, subequal; stamens 5, the filaments basally connate, the anthers 4-thecal; pseudostaminodes usually absent, alternating with free portions of the filaments if present; ovary superior, 3-carpellate, 1-loculate, the style 1, the stigmas 3. Fruit a circumscissile utricle; seeds 3–many.

A genus of about 65 species; nearly cosmopolitan. [From the Greek *kelos*, to burn, in reference to the color and/or the appearance of the inflorescence, or from *kelis*, a bloodstain, in reference to the red spot often found on the leaves.]

Selected reference: Robertson (2003a).

1. Inflorescence usually a simple, terminal spike 15–20 mm wide (larger and crested in cultivated forms); perianth bright white, pink, or red (other colors in cultivated forms), 6–9 mm long....... **C. argentea**
1. Inflorescence a terminal or axillary panicle composed of few to numerous spikes 3–10 mm wide; perianth white or stramineous to dark brown, 3–6 mm long.
     2. Perianth segments ca. 5 mm long, firm, stramineous to dark brown, prominently and finely parallel-veined; style longer than the stigmas...**C. nitida**
     2. Perianth segments ca. 3 mm long, membranaceous, white, 1-veined; style shorter than the stigmas...**C. trigyna**

Celosia argentea L. [Silvery.] SILVER COCK'S COMB; CRESTED COCK'S COMB.

Celosia argentea Linnaeus, Sp. Pl. 205. 1753. *Celosia pallida* Salisbury, Prodr. Stirp. Chap. Allerton 145. 1796, nom. illegit. *Celosia argentea* Linnaeus var. *vera* Voss, Vilm. Blumengärtn., ed. 3, 1: 864. 1896, nom. inadmiss.

Celosia cristata Linnaeus, Sp. Pl. 205. 1753. *Celosia argentea* Linnaeus var. *cristata* (Linnaeus) Kuntze, Revis. Gen. Pl. 2: 541. 1891. *Celosia argentea* Linnaeus forma *cristata* (Linnaeus) Schinz, in Engler & Prantl, Nat. Pflanzenfam., ed. 2. 16C: 29. 1934.

Erect annual herb, to 1 m; stem simple or much branched, glabrous. Leaves with the blade oblong-lanceolate to linear, 8–15 cm long, 1–6 cm wide, the upper leaves reduced, the apex acute to long-acuminate, the base cuneate, the margin entire, the upper and lower surfaces glabrous, the petiole 1–3 cm long. Flowers in a dense cylindric or ovoid spike 13–20 mm wide or variously fasciated, crested, or plumose; bract and bracteoles lanceolate to deltoid, 3–5 mm long, hyaline, short-aristate with the excurrent midrib; perianth segments narrowly oblong-elliptic, 6–8 mm long, the apex acute to obtuse, mucronate with the excurrent midrib, 3-veined, bright white, pink, or red, sometimes other colors in cultivated forms; style 3–4 mm long. Fruit ovoid or subglobose, 3–4 mm long; seeds 3–10, lenticular, ca. 1.5 mm long, black, faintly reticulate or smooth, lustrous.

Disturbed sites. Rare; Leon and Escambia Counties. Escaped from cultivation. Native to Asia. Summer.

The cultivated cockscomb with a fasciated, crested, or plumose inflorescence has been variously treated as a distinct species, a variety, or a form. Khoshoo and Pal (1973) present convincing evidence that a wild form of *C. argentea* showing potential for fasciation gave rise to the cultivated cockscomb.

Celosia nitida Vahl [Shining, in reference to the leaves.] WEST INDIAN COCK'S COMB.

Celosia nitida Vahl, Symb. Bot. 3: 44. 1794.

Erect or clambering perennial herb or subshrub, to 2 m; stem branched from the base, glabrous. Leaves with the blade ovate to deltoid-ovate or triangular-lanceolate, 2–7 cm long, 1–4 cm wide, the apex obtuse to acuminate, the base obtuse to truncate, slightly decurrent, the margin entire, the upper and lower surfaces glabrous, the petiole 0.5–2 cm long, often with axillary fascicles of leaves. Flowers in a lax, terminal or axillary, solitary spike or panicle of spikes, 1–5 cm long, 7–10 mm wide; bract and bracteoles rounded-ovate, the apex obtuse to acute, ca. ¼ as long as the perianth segments, often ciliolate; perianth segments oblong to elliptic-lanceolate, 3–7 mm long, the apex acute to acuminate and mucronulate, dark brown, strongly parallel-veined; style ca. 1 mm long. Fruit globose-ovoid, 4–5 mm long; seeds 10–20, lenticular, ca. 1 mm long, faintly reticulate or smooth, black.

Hardwood hammocks. Occasional; central and southern peninsula. Florida and Texas; West Indies, Mexico, Central America, and South America. Summer–fall.

Celosia nitida is listed as endangered in Florida (Florida Administrative Code, Chapter 5B-40).

Celosia trigyna L. [From the Greek *tri*, three, and *gyne*, female, referring to the three stigmatic lobes.] WOOLFLOWER.

Celosia trigyna Linnaeus, Mant. Pl. 212. 1771.

Erect or scrambling annual herb, to 1 m; stem simple or branching from the base, glabrous or sparsely puberulent at the nodes. Leaves with the blade lanceolate to ovate, 2–8(11) cm long,

1–4(6) cm wide, unlobed or occasionally with a broad, obtuse lateral lobe near the base, the uppermost leaves somewhat reduced, the apex acute to acuminate, the base subcordate to truncate or cuneate, slightly decurrent on the petiole, the margin entire, the upper surface glabrous, the lower surface glabrous or sparsely puberulent near the base, the petiole to 5 cm long. Flowers in a subglobose cluster in an axillary or terminal, simple or branched, lax, spiciform panicle 6–30 cm long; bract and bracteoles ovate to ovate-elliptic, 1–2 mm long, scarious with a darker midrib, glabrous, the margin minutely erose-denticulate; perianth segments ovate-elliptic, 2–3 mm long, the apex short-mucronate with the excurrent midrib, glabrous, scarious, the margin denticulate, at least above; style less than 0.5 mm long. Fruit ovoid, 3–4 mm long; seeds 4–8, lenticular, ca. 1 mm long, faintly reticulate, black, lustrous.

Disturbed sites. Occasional; central peninsula, Okaloosa County. Florida; Africa and Asia. Native to Africa and Asia. Summer–fall.

Chenopodiastrum S. Fuentes et al. 2012.

Annual herbs. Leaves alternate, simple, pinnate-veined, the upper and lower surfaces farinose or glabrous, petiolate, estipulate. Flowers in clusters or glomerules in axillary cymes or panicles, bisexual; ebracteate and ebracteolate; perianth segments 5; stamens 5; ovary superior, 1-carpellate and -loculate, the stigmas 2. Fruit a utricle, the pericarp free from the seed; seed 1, horizontal.

A genus of about 7 species; nearly cosmopolitan. [*Chenopodium*, which it resembles, and *astrum*, incomplete likeness.]

Recent molecular phylogenetic studies (Fuentes-Bazan et al., 2012) show that *Chenopodium* s.l. is polyphyletic. In making it monophyletic, *Chenopodiastrum* is recognized (Fuentes-Bazan et al., 2012; Mosyakin, 2013).

Chenopodiastrum murale (L.) S. Fuentes et al. [Of walls.] NETTLELEAF GOOSEFOOT.

Chenopodium murale Linnaeus, Sp. Pl. 219. 1753. *Atriplex muralis* (Linnaeus) Crantz, Inst. Rei Herb. 1: 206. 1766. *Vulvaria trachisperma* Bubani, Fl. Pyren. 1: 177. 1897, nom. illegit. *Chenopodiastrum murale* (Linnaeus) S. Fuentes et al., Willdenowia 42: 14. 2012.

Erect annual herb, to 1 m; stem simple or commonly branching at the base, the proximal branches decumbent, glabrous or sparsely farinose. Leaves with the blade triangular, ovate, or rhombic-ovate, 1–8 cm long, 0.5–5 cm wide, chartaceous, the apex acute or acuminate, the base subtruncate to cuneate or rounded, the margin coarsely sinuate dentate, the upper surface glabrous, the lower surface glabrate or densely farinose, the petiole 1–2.5 cm long. Flowers a cluster or glomerule, sessile, in a lax or dense axillary or terminal cyme or panicle; perianth segments with the lobes oblong, obtuse, keeled, farinose; stamens 5; stigmas 2. Fruit ovoid, 1–2 mm long, enclosed by the persistent calyx, pericarp minutely rugulose or smooth, adherent to the seed; seed horizontal, minutely rugose to smooth, black.

Disturbed sites. Occasional; peninsula, Leon County. Nearly cosmopolitan. Native to Europe, Africa, and Asia. Spring–summer.

Chenopodium L. 1753. GOOSEFOOT

Annual or perennial herbs, rarely suffrutescent. Leaves alternate, simple, pinnate-veined, the upper and lower surfaces farinose or glabrous, petiolate or epetiolate, estipulate. Flowers in axillary or terminal spikes or glomerules, bisexual or rarely unisexual (plant polygamous); ebracteate, ebracteolate; perianth segments 5, persistent; stamens 1–5; ovary superior, 1-carpellate, 1-locular, the styles 2–5. Fruit a utricle, the pericarp free or adherent to the seed, sometimes fleshy; seed 1, horizontal or vertical.

A genus of about 100 species; nearly cosmopolitan. [From the Greek *chen*, goose, and *podion*, foot, referring to the leaf shape in some species.]

Selected references: Clemants & Mosyakin (2003b).

1. Perianth segments rounded, conforming to the shape of the seed .... **C. vulvaria**
1. Perianth segments keeled or hooded (perianth pentagonal or star-shaped when viewed from above).
   2. Fruit pericarp and seed reticulate..**C. berlandieri**
   2. Fruit pericarp and seed smooth to tuberculate-roughened .... **C. album**

Chenopodium album L. [White.] LAMB'S-QUARTERS.

Chenopodium album Linnaeus, Sp. Pl. 219. 1753. *Atriplex alba* (Linnaeus) Crantz, Inst. Rei Herb. 1: 206. 1766. *Chenopodium candicans* Lamarck, Fl. Franc. 3: 248. 1799 ("1778"), nom. illegit. *Chenopodium leiospermum* de Candolle, in de Candolle & Lamarck, Fl. Franc. 3: 390. 1805, nom. illegit. *Chenopodium leiospermum* de Candolle var. *album* (Linnaeus) Becker, Fl. Frankfurt 171. 1827. *Chenopodium album* Linnaeus var. *spicatum* Wimmer & Grabowski, Fl. Siles. 1: 236. 1827, nom. inadmiss. *Chenopodium album* Linnaeus var. *paniculatum* Kunth, Fl. Berol. 2: 150. 1838, nom. inadmiss. *Chenopodium album* Linnaeus var. *vulgare* Patze et al., Fl. Preuss. 162. 1848, nom. inadmiss. *Chenopodium album* Linnaeus var. *commune* Moquin-Tandon, in A. de Candolle, Prodr. 13(2): 71. 1849, nom. inadmiss. *Chenopodium album* Linnaeus var. *albofarinosum* Sonder, Fl. Hamburg. 143. 1850, nom. inadmiss. *Anserina candicans* Friche-Joset & Montandon, Syn. Fl. Jura 262. 1856, nom. illegit. *Chenopodium album* Linnaeus var. *melanocarpum* Alefeld, Landw. Fl. 276. 1866, nom. inadmiss. *Chenopodium album* Linnaeus var. *candicans* Syme, in Smith, Engl. Bot., ed. 3B. 8: 8. 1868, nom. inadmiss. *Chenopodium album* Linnaeus var. *typicum* Beck, Fl. Nieder-Oesterreich 332. 1890, nom. inadmiss. *Vulvaria albescens* Bubani, Fl. Pyren. 1: 176. 1897, nom. illegit. *Chenopodium album* Linnaeus forma *linneanum* Beck, in Reichenbach, Icon. Fl. Germ. Helv. 24: 104. 1907, nom. inadmiss. *Botrys alba* (Linnaeus) Nieuwland, Amer. Midl. Naturalist 3: 276. 1914. *Chenopodium album* Linnaeus var. *eualbum* A. Ludwig, in Schinz & Thellung, Fl. Schweiz., ed. 3. 2: 95. 1914, nom. inadmiss. *Chenopodium album* L. forma *spicatum* A. Ludwig, in Schinz & Tellung, Fl. Schweiz., ed. 3. 2: 95. 1914, nom. inadmiss. *Chenopodium album* Linnaeus subsp. *eualbum* Aellen, Amer. Midl. Naturalist 30: 68. 1943, nom. inadmiss. *Chenopodium album* Linnaeus var. *polymorphum* Aellen, Amer. Midl. Naturalist 30: 68. 1943, nom. inadmiss.

Chenopodium viride Linnaeus, Sp. Pl. 319. 1753. *Atriplex viridis* (Linnaeus) Crantz, Inst. Rei Herb. 1: 206. 1766. *Chenopodium album* Linnaeus var. *viride* (Linnaeus) Pursh, Fl. Amer. Sept. 198. 1814. *Chenopodium album* Linnaeus var. *cymigerum* W.D.J. Koch, Syn. Fl. Germ. Helv. 606. 1837, nom. illegit. *Chenopodium album* Linnaeus forma *cymigerum* A. Ludwig, in Schinz & Tellung, Fl. Schweiz., ed. 3. 2: 95. 1914. *Chenopodium album* Linnaeus subvar. *cymigerum* (A. Ludwig) Arlt & Jüttersonke, in Jüttersonke & Arlt, Feddes Repert. 100: 20. 1989. *Chenopodium album* Linnaeus subforma *cymigerum* (A. Ludwig) Arlt & Jüttersonke, in Jüttersonke & Arlt, Feddes Repert. 100: 20. 1989.

Chenopodium lanceolatum Muhlenberg ex Willdenow, Enum. Pl. 291. 1809. *Chenopodium viride* Linnaeus forma *lanceolatum* (Muhlenberg ex Willdenow) Petermann, Fl. Lips. Excurs. 201. 1838. *Chenopodium album* Linnaeus var. *lanceolatum* (Muhlenberg ex Willdenow) Cosson & Germain de Saint-Pierre, Fl. Descr. Anal. Paris 451. 1845. *Chenopodium album* Linnaeus subsp. *lanceolatum* (Muhlenberg ex Willdenow) Murr, in Urban & Graebner, Festschr. Ascherson 219. 1904. *Chenopodium album* Linnaeus forma *lanceolatum* (Muhlenberg ex Willdenow) A. Ludwig, in Schinz & Thellung, Fl. Schweiz. ed. 3. 2: 95. 1914. *Chenopodium album* Linnaeus subforma *paucidens* Arlt & Jüttersonke, in Jüttersonke & Arlt, Feddes Repert. 100: 19. 1989.

Chenopodium album Linnaeus var. *microphyllum* Boenninghausen, Prodr. Fl. Monast. Westphal. 77. 1824.

Chenopodium giganteum D. Don, Prodr. Fl. Nepal. 75. 1825.

Chenopodium album Linnaeus subsp. *amaranthicolor* H. J. Coste & A. Reynier, Bull. Herb. Boissier, ser. 2. 5: 979. 1905. *Chenopodium amaranthicolor* (H. J. Coste & A. Reynier) H. Coste & A. Reynier, Bull. Soc. Bot. France 54: 178. 1907.

Chenopodium missouriense Aellen, Bot. Not. 1928: 206. 1928. *Chenopodium album* Linnaeus var. *missouriense* (Aellen) Bassett & Crompton, Canad. J. Bot. 60: 603. 1982.

Chenopodium missouriense Aellen var. *bushianum* Aellen, Repert. Spec. Nov. Regni Veg. 26: 156. 1929.

Chenopodium standleyanum Aellen, Repert. Spec. Nov. Regni Veg. 26: 153. 1929.

Erect to sprawling annual herb, to 1.5 m; stem simple below the inflorescence or paniculate branched above, farinose or glabrous. Leaves with the blade rhombic-ovate to lanceolate, (2)3–5(8) cm long, (1)2–3(4) cm wide, the apex acute to rounded, apiculate, the base rounded to cuneate, often shallowly 3-lobed, the margin irregularly and coarsely sinuate-dentate to entire, the upper surface glabrate, the lower surface sparsely to densely farinose, the petiole 1–3 cm long. Flowers in a glomerule in an axillary or terminal, ascending or spreading paniculate spike to 30 cm long; perianth segments 5, the lobes ovate, free nearly to the base, ca. 1 mm long, moderately to densely farinose, keeled; stamens 5; stigmas 2. Fruit 1–1.5 mm long, the pericarp lightly roughened, smooth, or tuberculate, free or adherent to the seed; seed horizontal, nearly smooth to minutely pitted, black, lustrous.

Disturbed sites. Frequent; nearly throughout. Nearly cosmopolitan. Native of Europe, Africa, and Asia. Spring–summer.

Chenopodium album is a highly polymorphic, cosmopolitan, weedy species, which has been variously interpreted. Aellen and Just (1943) recognized two subspecies, five varieties, and four forms on a worldwide basis. Others have treated it as a "dust bin" species and swept several other closely related ones into it (e.g., *C. berlandieri*).

Chenopodium giganteum, *C. missouriense*, and *C. standleyanum* are recognized as distinct species in several recent treatments (e.g., Clemants and Mosyakin, 2003b). However, we could find no consistent differences to separate any of them from *C. album*.

Chenopodium berlandieri Moq. [Commemorates the discoverer, Jean Louis Berlandier (1805–1851), Belgian explorer and plant collector in North America and Mexico.] PITSEED GOOSEFOOT.

Chenopodium berlandieri Moquin-Tandon, Chenop. Monogr. Enum. 23. 1840. *Chenopodium album* Linnaeus var. *berlandieri* (Moquin-Tandon) Mackenzie & Bush, Man. Fl. Jackson County 80. 1902. *Botrys berlandieri* (Moquin-Tandon) Nieuwland, Amer. Midl. Naturalist 3: 276. 1914.

Chenopodium berlandieri Moquin-Tandon subsp. *euberlandieri* Aellen, Repert. Spec. Nov. Regni Veg. 26: 62. 1929, nom. inadmiss. *Chenopodium berlandieri* Moquin-Tandon var. *typicum* Aellen, Repert. Spec. Nov. Regni Veg. 26: 52. 1929, nom. inadmiss.

Chenopodium boscianum Moquin-Tandon, Chenop. Monogr. Enum. 21. 1840. *Botrys bosciana* (Moquin-Tandon) Nieuwland, Amer. Midl. Naturalist 3: 275. 1914. *Chenopodium berlandieri* Moquin-Tandon subsp. *boscianum* (Moquin-Tandon) Aellen, Amer. Midl. Naturalist 30: 71. 1943. *Chenopodium berlandieri* Moquin-Tandon var. *boscianum* (Moquin-Tandon) H. Wahl, Bartonia 27: 42. 1954.

Chenopodium zschackei Murr, Deutsche Bot. Monatsschr. 19: 39. 1901. *Chenopodium berlandieri* Moquin-Tandon subsp. *zschackei* (Murr) A. Zobel, Verz. Anhalt Phan 3: 70. 1909.

Chenopodium album Linnaeus subsp. *collinsii* Murr, Bull. Herb. Boissier, ser. 2. 4: 990, t. 5(1). 1904. *Chenopodium bushianum* Aellen, Repert. Spec. Nov. Regni Veg. 26: 63. 1929. *Chenopodium berlandieri* Moquin-Tandon var. *bushianum* (Aellen) Cronquist, in Gleason & Cronquist, Man. Vasc. Pl. N.E. U.S., ed. 2. 863. 1991.

Chenopodium macrocalycium Aellen, Repert Nov. Spec. Regni Veg. 26: 119. 1929. *Chenopodium berlandieri* Moquin-Tandon var. *macrocalycium* (Aellen) Cronquist, in Gleason & Cronquist, Man. Vasc. Pl. N.E. U.S., ed. 2. 863. 1991.

Erect or ascending annual herb, to 1.5 m; stem usually simple at the base, much branched above, farinose or glabrate. Leaves with the blade rhombic-triangular or ovate-lanceolate, 2–4 cm long, (0.5)1–3(4) cm wide, the apex obtuse to acute, often mucronulate, the base broadly cuneate or rounded, the margin irregularly sinuate, sinuate-serrate, or serrate-dentate, rarely 3-lobed at the base, the upper and lower surfaces moderately to densely farinose, glabrate in age, the petiole 1–6 cm long. Flowers in a glomerule clustered into a dense to interrupted panicle; perianth segments 5, the lobes ovate, free nearly to the base, moderately to densely farinose, sharply keeled; stamens 5; stigmas 2. Fruit 1–2 mm long, the pericarp alveolate, adherent to seed except at style base; seed horizontal, alveolate, black, lustrous.

Disturbed areas. Occasional; peninsula, central and western panhandle. Nearly throughout North America; Mexico; Europe. Native to North America and Mexico. Summer.

Chenopodium berlandieri is here considered as comprising a single, widespread, weedy, and highly variable taxon. The species is often broken up into several weak varieties with most of the plants in the lower southeastern coastal plain referred to var. *boscianum*. However, assignment of many collections to a variety often is arbitrary.

Chenopodium vulvaria L. [Vulva, external female genitalia, in reference to the characteristic odor of the crushed leaves.] STINKING GOOSEFOOT.

Chenopodium vulvaria Linnaeus, Sp. Pl. 220. 1753. *Atriplex vulvaria* (Linnaeus) Crantz, Inst. Rei Herb. 1: 207. 1766. *Vulvaria vulgaris* Bubani, Fl. Pyren. 1: 175. 1897.

Erect or ascending annual herb, to 4 dm; stem branched from the base, farinose. Leaves with the blade rhombic to broadly ovate, 0.5–1.5 cm long and wide, the apex obtuse or acute, the base rounded to broadly cuneate, the margin entire, the upper surface glabrescent, the lower surface densely farinose, the petiole 0.8–0.9 cm long. Flowers in a small glomerule in a compact leafy spike, this terminal or in the axil of the upper leaves; perianth segments 5, connate into a short tube, the lobes deltate, rounded abaxially, farinose; stamens 5; stigmas 2. Fruit ovoid, ca. 1 cm long, enclosed by persistent calyx, the pericarp adherent to seed, slightly puncticulate; seed vertical, the surface slightly puncticulate with shallow striations, black.

Disturbed sites. Rare; Escambia County. Not recently collected. Quebec south to Maryland, Ohio, and Indiana, also Florida, Alabama, Texas, Oregon, and California; South America; Europe, Africa, Asia, Australia, and Pacific Islands. Native of Europe, Africa, and Asia. Summer–fall.

EXCLUDED TAXA

Chenopodium ficifolium Smith—Reported by Clemants and Mosyakin (2003b), perhaps based on *Baker 1579* (NY). (See *C. serotinum* below.)

Chenopodium paganum Reichenbach—Reported by Small (1933), who apparently misapplied this name to material of *C. album*.

Chenopodium pratericola Rydberg—Reported by Clemants and Mosyakin (2003b). Material of the collection that serves as the basis of this report (*Genelle & Fleming 1277*) and seen at USF is here referred to *C. berlandieri*.

Chenopodium serotinum Linnaeus—Reported by Wahl (1954) based on *Baker 1579* (NY). However, Uotila (1977) has shown that the type of *C. serotinum* is an immature *Atriplex* species. The Baker specimen has been questionably referred to *C. ficifolium* Smith by Hugh D. Wilson. However, there is some doubt as to its identity; it may represent *C. berlandieri*.

Chenopodium strictum Roth—Reported by Clemants and Mosyakin (2003b). No Florida specimens known; probably misapplied to material of *C. album* s.l.

Dysphania R. Br. 1810.

Annual or perennial herbs. Leaves alternate, simple, pinnate-veined, the upper and lower surfaces with resinous, sessile glands or glandular trichomes, petiolate or epetiolate, estipulate. Flowers in clusters or glomerules in axillary cymes or panicles, bisexual or unisexual (plants monoecious); ebracteate, ebracteolate; perianth segments 3–5; stamens 1–5; ovary superior, 1-carpellate and -locular, the stigmas 1–5. Fruit a utricle, adherent or free from the seed; seed 1, horizontal or vertical.

A genus of about 32 species; nearly cosmopolitan. [From the Greek *dysphanis*, obscure, in reference to the inconspicuous flowers.]

Recent molecular phylogenetic studies (Fuentes-Bazan et al., 2012) show that *Chenopodium* s.l. is polyphyletic. In making it monophyletic, *Dysphania* is recognized (Fuentes-Bazan et al. 2012; Mosyakin, 2013).

Ambrina Moq. (1840); *Roubieva* Moq. (1834).

Selected reference: Clemants and Mosyakin (2003a).

1. Perianth segments free nearly to the base... **D. pumilo**
1. Perianth segments connate ½ or more their length.
    2. Plant prostrate; perianth segments connate for about ½ their length ....**D. multifida**
    2. Plant erect or ascending; perianth segments connate nearly their length............... **D. ambrosioides**

Dysphania ambrosioides (L.) Mosyakin & Clemants [Resembling *Ambrosia* (Asteraceae).] MEXICAN TEA.

Chenopodium ambrosioides Linnaeus, Sp. Pl. 219. 1753. *Atriplex ambrosioides* (Linnaeus) Crantz, Inst. Rei Herb. 1: 207. 1766. *Orthosporum ambrosioides* (Linnaeus) Kosteletzky, Allg. Med.-Pharm. Fl. 4: 1433. 1835. *Ambrina ambrosioides* (Linnaeus) Spach, Hist. Nat. Vég. 5: 197. 1836. *Chenopodium ambrosioides* Linnaeus var. *genuinum* Willkomm, Prodr. Fl. Hispan. 1: 271. 1861, nom. inadmiss. *Vulvaria ambrosioides* (Linnaeus) Bubani, Fl. Pyren. 1: 178. 1897. *Chenopodium ambrosioides* L. var. *typicum* Spegazzini, Anales Soc. Ci. Argent. 53: 282. 1902; Anales Mus. Nac. Hist. Nat. Buenos Aires 7: 137. 1902, nom. inadmiss. *Blitum ambrosioides* (Linnaeus) Beck, in Reichenbach, Icon. Fl. Germ. Helv. 24: 118. 1908. *Botrys ambrosioides* (Linnaeus) Nieuwland, Amer. Midl. Naturalist 3: 275. 1914. *Chenopodium ambrosioides* Linnaeus subsp. *euambrosioides* Aellen, Repert. Spec. Nov. Regni Veg. 26: 34. 1929, nom. inadmiss. *Chenopodium ambrosioides* Linnaeus forma *genuinum* Aellen, Repert. Spec. Nov. Regni Veg. 26: 34. 1929, nom. inadmiss. *Teloxys ambrosioides* (Linnaeus) W. A. Weber, Phytologia 58: 477. 1985. *Dysphania ambrosioides* (Linnaeus) Mosyakin & Clemants, Ukrayins'k. Bot. Zhurn., ser. 2. 59: 382. 2002.

Chenopodium anthelminticum Linnaeus, Sp. Pl. 220. 1753. *Atriplex anthelmintica* (Linnaeus) Crantz, Inst. Rei Herb. 1: 207. 1766. *Orthosporum anthelminticum* (Linnaeus) Kosteletzky, Allg. Med.-Pharm. Fl. 4: 1433. 1835. *Ambrina anthelmintica* (Linnaeus) Spach, Hist. Nat. Vég. 5: 298. 1836. *Roubieva anthelmintica* (Linnaeus) Hooker & Arnott, Bot. Beechey Voy. 387. 1840. *Chenopodium ambrosioides* L. var. *anthelminticum* (Linnaeus) A. Gray, Manual, ed. 2. 364. 1856. *Botrys anthelmintica* (Linnaeus) Nieuwland, Amer. Midl. Naturalist 3: 275. 1914. *Ambrina ambrosioides* (Linnaeus) Spach var. *anthelmintica* (Linnaeus) Moldenke, Phytologia 1: 274. 1938. *Dysphania anthelmintica* (Linnaeus) Mosyakin & Clemants, Ukrayins'k. Bot. Zhurn., ser. 2. 59: 382. 2002.

Erect or ascending annual or short-lived perennial herb, to 1.5 m; stem simple or much branched above, glabrous or glandular puberulent below, usually glandular-pubescent above. Leaves with the blade oblong or ovate to lanceolate, 2–8(12) cm long, 1–4(6) cm wide, the apex obtuse to attenuate, the base cuneate, the margin serrate, dentate or sinuate to shallowly sinuate-pinnatifid, the upper and lower surfaces densely yellow-glandular to glabrate, the petiole ca. 1 cm long, the upper leaf blades reduced, narrower, and becoming entire, sessile, and sometimes bract-like. Flowers in a sessile glomerule (rarely solitary), in a spike 3–7 cm long; perianth segments (3)4–5, the lobes rounded-ovate, obtuse, connate to ca. ½ their length, ca. 1 mm long, glabrous or short villous, sparsely glandular to glabrous; stamens 4 or 5; stigmas 3. Fruit ca. 1 mm long, the pericarp rugulate to smooth, free from the seed; seed horizontal (rarely vertical), rugose or smooth, dark brown or black.

Disturbed sites. Common; nearly throughout. Nearly cosmopolitan. Native to North America in the southwestern U.S., Mexico, Central America, and South America. All year.

Dysphania anthelmintica essentially differs from *D. ambroisioides* in lacking reduced bract-like leaves subtending the glomerules. However, this character is variable and cannot reliably be used to distinguish taxa.

Dysphania multifida (L.) Mosyakin & Clemants [Much-divided, in reference to the leaf dissection.] CUTLEAF GOOSEFOOT.

Chenopodium multifidum Linnaeus, Sp. Pl. 220. 1753. *Roubieva multifida* (Linnaeus) Moquin-Tandon, Ann. Sci. Nat. Bot., ser. 2. 1: 292. 1834. *Orthosporum multifidum* (Linnaeus) Kosteletzky, Allg.

Med.-Pharm. Fl. 4: 1434. 1835. *Ambrina pinnatisecta* Spach, Hist. Nat. Vég. 5: 296. 1836, nom. illegit. *Chenopodium multifidum* Linnaeus forma *typicum* Aellen, Amer. Midl. Naturalist 30: 49. 1943, nom. inadmiss. *Teloxys multifida* (Linnaeus) W. A. Weber, Phytologia 58: 478. 1985. *Dysphania multifida* (Linnaeus) Mosyakin & Clemants, Ukrayins'k. Bot. Zhurn., ser. 2. 59: 382. 2002.

Prostrate perennial herb; stem to 6 dm long, villous-puberulent. Leaves with the blade narrowly oblong, 1–4 cm long, 0.5–2 cm wide, deeply and irregularly pinnatilobate, the upper ones much reduced and becoming bract-like, the apex acute to obtuse, the base cuneate, the upper surface obscurely puberulent, the lower surface finely resinous-glandular, the petiole absent or indistinguishable from the blade. Flowers in a small axillary glomerule; perianth segments 5, the lobes obovate, 1.5–2.5 mm long, raised reticulate-veined; stamens 5; styles 2–5, basally connate. Fruit loosely but completely enclosed by the persistent calyx, pericarp thin, free; seed vertical, obovate, ca. 1 mm long, thick-lenticular, smooth.

Disturbed sites. Rare; Walton County. Not recently collected. Massachusetts and New York south to Florida, west to Alabama, also Indiana, Oregon, and California; West Indies, Mexico, Central America, and South America; Europe, Africa, Asia, and Australia. Native to South America. Summer.

Dysphania pumilo (R. Br.) Mosyakin & Clemants [Dwarf.] RIDGED GOOSEFOOT.

Chenopodium pumilo R. Brown, Prodr. 407. 1910. *Ambrina pumilo* (R. Brown) Moquin-Tandon, Monogr. Chenop. Enum. 42. 1840. *Blitum pumilo* (R. Brown) Moquin-Tandon, in A. de Candolle, Prodr. 13(2): 82. 1849. *Teloxys pumilo* (R. Brown) W. A. Weber, Phytologia 58: 478. 1985. *Dysphania pumilo* (R. Brown) Mosyakin & Clemants, Ukrayins'k. Bot. Zhurn., ser. 2. 59: 382. 2002.

Spreading to prostrate annual herb; stem, to 4 dm long, much branched, with uniseriate trichomes and sessile or stipitate-glandular trichomes. Leaves with the blade elliptic to ovate, 1–3 cm long, 0.5–1.5 cm wide, the upper ones 3–5 mm long and bract-like, the apex obtuse, the base cuneate, the margin sinuate-dentate, the upper surface glabrous, the lower surface densely glandular-pilose, the petiole 0.5–1.5 cm long. Flowers in a small glomerule aggregated into a small axillary glomerule on a short spike to 1 cm long; perianth segments 5, elliptic to narrowly oblong, glandular-pilosulose; stamens absent or 1; stigmas 2. Fruit ovoid, the pericarp adherent, slightly rugose; seeds vertical, lenticular, 0.5 mm long, smooth, reddish brown.

Disturbed sites. Rare; Hillsborough County. Massachusetts south to Florida, west to Wisconsin, Kansas, Oklahoma, and Texas, also British Columbia, south to Arizona; Mexico; Europe, Africa, Asia, Australia, and Pacific Islands. Native to Australia. Summer–fall.

EXCLUDED TAXA

Dysphania botrys (Linnaeus) Mosyakin & Clemants—Reported by Correll & Johnston (1970, as *Chenopodium botrys* Linnaeus) as "naturalized as a weed throughout U.S. & Can." which presumably includes Florida. No Florida specimens known.

Froelichia Moench 1794. SNAKECOTTON

Annual herbs. Leaves opposite, simple, pinnate-veined, petiolate, estipulate. Flowers in pedunculate spikes, bisexual; ebracteate, bracteoles 2; perianth segments 5; stamens 5, the filaments connate to form an elongate tube; pseudostaminodes 5, ligulate; ovary superior, 1-carpellate and -loculate, the style 1, the stigma capitate, irregularly lobed. Fruit a utricle, enclosed by the persistent calyx with facial wings, spines, and tubercles.

A genus of about 16 species; North America, West Indies, Mexico, Central America, South America, and Australia. [In honor of Joseph Aloys von Froelich (1766–1841), German physician and botanist.]

Selected reference: McCauley (2003).

Froelichia floridana (Nutt.) Moq. [Of Florida.] COTTONWEED; PLAINS SNAKECOTTON.

> *Oplotheca floridana* Nuttall, Gen. N. Amer. Pl. 2: 79. 1818. *Gomphrena floridana* (Nuttall) Sprengel, Syst. Veg. 1: 824. 1824 ("1825"). *Froelichia floridana* (Nuttall) Moquin-Tandon, in A. de Candolle, Prodr. 13(2): 420. 1849.
>
> *Froelichia floridana* (Nuttall) Moquin-Tandon var. *pallescens* Moquin-Tandon, in A. de Candolle, Prodr. 13(2): 421. 1849. TYPE: FLORIDA: *Nuttall s.n.*

Erect, ascending, or procumbent annual herb; stem to 1.5 m long, simple or sparsely branched from the base or above, puberulent or tomentulose with short, often viscid whitish or brownish trichomes. Leaves with the blade oblanceolate to elliptic-lanceolate, 4–11 cm long, 1–3 cm wide, the apex obtuse to acute, the base attenuate to cuneate, the margin entire, the upper surface canescent to subscabrous, the lower surface sericeous-tomentose, the petiole ca. 1 cm long. Flowers in a dense, branched spike 1–10 cm long, whitish; bracteoles acuminate or cuspidate, fuscous, stramineous, or blackish, lanate; perianth connate to about the middle into a tube, greenish white or pinkish, lanate, the lobes lanceolate, the tube 2-lipped. Fruit flask-shaped, ca. 4–5 mm long and wide, with irregular dentate lateral wings, both surfaces with spines or tubercles; seed obovoid, 1–2 mm long, brown.

Sandhills, dry hammocks, and disturbed sites. Common; nearly throughout. New York south to Florida, west to North Dakota, South Dakota, Nebraska, Colorado, and New Mexico; West Indies; Australia. Native to North America and West Indies. Summer–fall.

Gomphrena L. 1753. GLOBE AMARANTH

Annual or perennial herbs. Leaves opposite, simple, pinnate-veined, sessile or short-petiolate or epetiolate, estipulate. Flowers in solitary, sessile terminal or axillary heads, subtended by an involucre of sessile leaves, bisexual; ebracteate, bracteoles 2; perianth segments 5; stamens 5, the filaments connate into a tube; pseudostaminodes absent; ovary superior, 1-carpellate and -loculate, the style 1, the stigmas 2. Fruit an indehiscent utricle, enclosed by the persistent calyx and staminal tube; seed 1.

A genus of about 100 species; North America, West Indies, Mexico, Central America, South America, Africa, Asia, Australia, and Pacific Islands. [Old Latin name for a type of amaranth.] Selected references: Clemants (2003b); Mears (1980).

Gomphrena serrata L. [Saw-edged.] PROSTRATE GLOBE AMARANTH; ARRASA CON TODO.

> *Gomphrena serrata* Linnaeus, Sp. Pl. 224. 1753. *Xeraea serrata* (Linnaeus) Kuntze, Revis. Gen. Pl. 2: 545. 1891.
> *Gomphrena decumbens* Jacquin, Pl. Hort. Schoenbr. 4: 41, t. 482. 1804. *Xeraea decumbens* (Jacquin) Kuntze, Revis. Gen. Pl. 2: 545. 1891. *Amaranthoides decumbens* (Jacquin) M. Gómez de la Maza y Jiménez, Anales Inst. Segunda Enseñ. 2: 313. 1896.
> *Gomphrena dispersa* Standley, Contr. U.S. Natl. Herb. 18: 91. 1916.

Prostrate, procumbent, or decumbent annual or perennial herb; stem to 1 m long, much branched, rooting at the nodes, appressed-pilose. Leaves with the blade oblong to obovate, 1.5–7.5 cm long, 0.5–2.5 cm wide, the apex rounded or obtuse, the base narrowly cuneate, the margin entire, the upper surface pilose-sericeous or glabrate, the lower surface pilose-sericeous, sessile or the petiole to 6 mm long. Flowers in a solitary, terminal or axillary, subglobose to short-cylindric head 9–13 mm in diameter, white, subtended by 2–3 sessile leaves shorter than the spike; bracteoles 5–6 mm long, white, with a denticulate or laciniate crest along the keel; perianth subequaling the bracteoles, densely lanate, the lobes oblong-linear, the apex acuminate, often denticulate; staminal tube short, densely lanate. Fruit ovoid, ca. 2 mm long; seed ovoid, ca. 1 mm long, reddish brown, lustrous.

Dry, disturbed sites. Common; nearly throughout. Maryland, Virginia, Georgia, Florida, Louisiana, and Texas; West Indies, Mexico, Central America, and South America; Asia and Pacific Islands. Native to North America, West Indies, Mexico, and Central America. All year.

Iresine P. Browne 1756. BLOODLEAF

Annual or perennial herbs. Leaves opposite, simple, pinnate-veined, petiolate, estipulate. Flowers in terminal panicles, bisexual or unisexual (plants polygamous or dioecious); bract 1, bracteoles 2; perianth basally connate and 5-lobed or free; stamens 5, the filaments basally connate; pseudostaminodes absent or obsolete; Ovary superior, 1-carpellate and -loculate, the style 1, the stigmas 2–3. Fruit an indehiscent utricle; seed 1.

A genus of about 70 species; North America, West Indies, Mexico, Central America, Africa, and Asia. [From the Greek *eiresione*, a harvest garland wound with wool, in reference to the long trichomes on the flowers and seeds of some species.]

Selected reference: Clemants (2003c).

1. Perianth segments of the carpellate flowers 3-veined, usually obtuse, exceeding the fruit; plant annual or short-lived perennial, not rhizomatous..**I. diffusa**
1. Perianth segments of the carpellate flowers faintly 1-veined, acute, not exceeding the fruit; plant a rhizomatous perennial .... **I. rhizomatosa**

Iresine diffusa Humb. & Bonpl. ex Willd. [Diffuse, in reference to the inflorescence.] JUBA'S BUSH.

Celosia paniculata Linnaeus, Sp. Pl. 206. 1753, nom. cons. *Iresine celosia* Linnaeus, Syst. Nat., ed. 10. 1291. 1759, nom. illegit. *Iresine celosioides* Linnaeus, Sp. Pl., ed. 2. 1456. 1763, nom. illegit. *Lophoxera paniculata* (Linnaeus) Rafinesque, Fl. Tellur. 3: 42. 1837 ("1836"). *Xerandra celosioides* Rafinesque, Fl. Tellur. 3: 43. 1837 ("1836"), nom. illegit. *Gonufas paniculata* (Linnaeus) Rafinesque, Sylva Tellur. 124. 1838. *Iresine paniculata* (Linnaeus) Kuntze, Revis. Gen. Pl. 2: 542. 1891; non Poiret, 1814.

Iresine diffusa Humboldt & Bonpland ex Willdenow, Sp. Pl. 4: 765. 1806.

Iresine canescens Humboldt & Bonpland ex Willdenow, Sp. Pl. 4: 765. 1806. *Trommsdorffia canescens* (Humboldt & Bonpland ex Willdenow) Martius, Nov. Gen. Sp. Pl. 2: 42. 1826.

Iresine paniculata (Linnaeus) Kuntze var. *floridana* Uline & W. L. Bray, Bot. Gaz. 21: 353. 1896. TYPE: FLORIDA: St. Johns Co.: Anastasia Island, 1875, *Reynolds s.n.* (holotype: US; isotype: US).

Erect or ascending annual or short-lived perennial herb, to 3 m; stem much branched, glabrous or sparsely villous. Leaves with the blade ovate to lanceolate, 3–14 cm long, 1.5–7 cm wide, the apex acute to acuminate, the base broadly cuneate to truncate, slightly decurrent, the margin entire, the upper and lower surfaces glabrous or sparsely villous, minutely pellucid-dotted, the petiole to 6.5 cm long. Flowers usually in a large and much-branched panicle 10–40 cm long, the branches slightly villous, the spikelets alternate, opposite, or verticillate, sessile or pedunculate; bract and bracteoles ovate to ovate-orbicular, ⅓–⅔ as long as the perianth, white to stramineous; perianth 0.5–1 mm long, deeply 5-lobed, the lobes oblong, conspicuously 3-nerved, white to stramineous, the apex obtuse to rounded, the staminate segments glabrous, the carpellate ones 3-veined, villous. Fruit ovoid, 0.6–0.8 mm long, enclosed within the persistent calyx; seed broadly obovoid to suborbicular, 0.5–0.7 mm long, smooth, dark red, lustrous.

Hammocks and disturbed sites. Common; peninsula, central panhandle. Florida, Louisiana, and Texas; West Indies, Mexico, Central America, and South America. All year.

Iresine rhizomatosa Standl. [Bearing rhizomes.] ROOTSTOCK BLOODLEAF.

Iresine rhizomatosa Standley, Proc. Biol. Soc. Wash. 28: 172. 1915.

Erect, stoloniferous perennial herb, to 1 m; stem mostly unbranched to the inflorescence, glabrous or sparsely puberulent to villous. Leaves with the blade lanceolate to ovate-lanceolate to elliptic, 6–15 cm long, 2–7 cm wide, the apex acute to long-acuminate, the base cuneate, decurrent, the margin entire, the upper and lower surfaces finely puberulent or glabrous, the petiole 1–4 cm long. Flowers in a panicle 7–30 cm long; bract and bracteoles ovate, shorter than the perianth, silvery white; perianth 1–1.5 mm long, deeply 5-lobed, the lobes ovate, 1-nerved, silvery white to stramineous, the staminate segments glabrous, the carpellate ones faintly 1-vined, subtended by a ring of woolly trichomes. Fruit ovoid, 1–1.5 mm long, enclosed in the persistent calyx; seed obovoid, 0.5–1 mm long, smooth, reddish brown, lustrous.

Floodplain forests and adjacent mesic bluff forests. Occasional; northern peninsula, west to central panhandle. Pennsylvania south to Florida, west to Kansas, Oklahoma, and Texas. Summer–fall.

Oxybasis Kar. & Kir. 1841.

Annual herbs. Leaves alternate, simple, pinnate-veined, the upper and lower surfaces farinose or glabrous, petiolate, estipulate. Flowers in glomerules in axillary spikes or panicles, bisexual or unisexual (plants polygamous); ebracteate, ebracteolate; perianth segments 3–4, free nearly to the base; stamen 1; ovary superior, 1-carpellate and -loculate, the stigmas 2; Fruit a utricle, the pericarp free from the seed; seed 1, horizontal.

A genus of about 10 species; nearly cosmopolitan. [From the Greek *oxys*, sharply, and *basis*, base, in reference to the branches arising abruptly from the base.]

Recent molecular phylogenetic studies (Fuentes-Bazan et al., 2012) show that *Chenopodium* s.l. is polyphyletic. In making it monophyletic, *Oxybasis* is now recognized (Fuentes-Bazan et al, 2012; Mosyakin, 2013).

Oxybasis glauca (L.) S. Fuentes et al. [Covered with a white waxy coating.] OAKLEAF GOOSEFOOT.

> *Chenopodium glaucum* Linnaeus, Sp. Pl. 220. 1753. *Atriplex glauca* (Linnaeus) Crantz, Inst. Rei Herb. 1: 207. 1766. *Blitum glaucum* (Linnaeus) W.D.J. Koch, Syn. Fl. Germ. Helv. 608. 1837. *Orthosporum glaucum* (Linnaeus) Petermann, Fl. Bienitz 94. 1841. *Orthospermum glaucum* (Linnaeus) Opiz, Seznam 71. 1852. *Agathophytum glaucum* (Linnaeus) Fuss, Fl. Transsilv. 553. 1866. *Botrys glauca* (Linnaeus) Nieuwland, Amer. Midl. Naturalist 3: 275. 1914. *Chenopodium glaucum* Linnaeus subsp. *euglaucum* Aellen, Repert. Spec. Nov. Regni Veg. 24: 45. 1929, nom. inadmiss. *Oxybasis glauca* (Linnaeus) S. Fuentes et al., Willdenowia 42: 15. 2012.

Erect or prostrate annual herb, to 4 dm; stem, solitary or much branched from the base, farinose. Leaves with the blade oblong to ovate or lanceolate, 0.5–4 cm long, 0.4–2 cm wide, becoming reduced, elliptic and bract-like upward and into the lower half of the inflorescence, the apex obtuse, the base cuneate, the margin entire to undulate or sinuate-dentate, the upper surface glabrous, the lower surface densely farinose, glaucous, the petiole to 1 cm long. Flowers in a small glomerule in an axillary spike or rarely in a terminal spike or panicle to 4 cm long; perianth segments 3–4, free nearly to the base, the lobes ovate-oblong, the apex obtuse and rounded, glabrous; stamen 1; stigmas 2. Fruit ca. 1 mm long, the pericarp smooth, separable from the seed; seed horizontal, reticulate-punctate to nearly smooth, reddish brown.

Disturbed sites. Rare; Miami-Dade County. Nearly throughout North America; Europe and Asia. Native to Europe and Asia. Summer–fall.

Salicornia L. 1753. GLASSWORT

Annual herbs. Leaves opposite, simple, pinnate-veined, epetiolate, estipulate. Flowers in terminal spikes, bisexual or unisexual (plants polygamous); perianth segments connate nearly to the apex, 3(4)-lobed; stamens 1–2; Ovary superior, 1-carpellate and -loculate, the style 1, the stigmas 2. Fruit an indehiscent utricle; seed 1, vertical.

A genus of about 10 species; North America, West Indies, Mexico, Europe, Africa, and Asia. [*Sal*, salt, and *cornu*, a horn, saline plants with hornlike branches.]

Salicornia is a genus with various and conflicting treatments, none of which are completely satisfactory. The treatment here must be considered tentative.

Selected reference: Ball (2003a).

Salicornia bigelovii Torrey [Commemorates Jacob Bigelow (1787–1879), American botanist.] ANNUAL GLASSWORT; DWARF GLASSWORT.

Salicornia mucronata Bigelow, Fl. Boston., ed. 3. 2. 1840; non Lagasca y Segura, 1817. *Salicornia bigelovii* Torrey, in Emory, Rep. U.S. Mex. Bound. 2(1): 184. 1858 "1859."

Erect annual herb, to 6 dm; stem stout, jointed, succulent, green or suffused with red pigment, with few to many stout, spreading or ascending branches, the joints 7–25 mm long, 2–3 mm wide. Leaves with the blade ovate to triangular-ovate, 2–4 mm long, basally connate, the apex acute, mucronate, the margin entire, the upper and lower surfaces glabrous. Flowers in a spike 2–10 cm long, 4–6 mm wide, the 3-flowered cymes embedded in and adherent to the internode tissue, the 2 lateral ones contiguous below the central one, the central one 2–3 mm long, slightly higher and larger than the lateral ones, the flowers separated by persistent flaps of internodal tissue. Fruit ellipsoid, ca. 2 mm long; seed ellipsoid, 1.5–2 mm long, puberulent, yellowish brown to nearly black.

Salt marshes and salt flats. Occasional; peninsula, west to central panhandle. Maine south to Florida, west to New Mexico, also California; West Indies and Mexico. Summer–fall.

EXCLUDED TAXA

Salicornia europaea Linnaeus—Reported by Small (1903, 1913a, 1913b, 1933, all as *S. herbacea* Linnaeus) and Gleason and Cronquist (1991). Excluded from Florida by Godfrey and Wooten (1981). Two specimens from Monroe County (*Small & Carter 3124*, NY; *Small s.n.*, NY) may represent this species, but material is insufficient for positive identification.

Salicornia virginica Linnaeus—Reported by Radford et al. (1964, 1968), Correll and Johnston (1970), Long and Lakela (1971), Godfrey and Wooten (1981), Wunderlin (1982), Correll and Correll (1982), and Clewell (1985). This name, a synonym of the annual *S. europaea* Linnaeus has been misapplied to our material of *Sarcocornia ambigua*.

Salsola L. 1753. SALTWORT

Annual herbs. Leaves alternate, simple, pinnate-veined, epetiolate, estipulate. Flowers solitary in the axils of bracts or reduced distal leaves and forming spikes, bisexual; bract 1, bracteoles 2; perianth segments 5, basally connate; stamens 5, inserted at the edge of a minute lobed disk; Ovary superior, 1-carpellate and -loculate, the style 1, the stigmas 2. Fruit an indehiscent utricle, enclosed by the persistent calyx, horizontally ridged or winged; seed 1, horizontal.

A genus of about 130 species; North America, Mexico, Central America, South America, Europe, Africa, and Asia. [*Salsus*, salty, referring to its saline or alkaline habitat.]

Selected references: Mosyakin (1996, 2003); Rilke (1999).

Salsola kali L. subsp. **pontica** (Pall.) Mosyakin [Old generic name from the Persian, a large carpet; *ponticus*, of or pertaining to the Black Sea and adjacent regions.] PRICKLY RUSSIAN THISTLE.

> *Salsola kali* Linnaeus var. *pontica* Pallas, Ill. Pl. 37, t. 29(2). 1803. *Salsola pontica* (Pallas) Iljin, in Komarov, Fl. URSS 6: 212. 1936. *Salsola kali* Linnaeus subsp. *pontica* (Pallas) Mosyakin, Ann. Missouri Bot. Gard. 83: 389. 1996.

Erect or ascending annual herb, to 8 dm; stem branched from the base, coarsely hirsute to papillose or glabrous. Leaves with the blade linear to filiform, 2–8 cm long, 1–2 mm wide, reduced upward, the apex spinose, the base broad, the margin entire to denticulate, hyaline, the upper and lower surfaces coarsely hirsute to papillose or glabrous. Flowers sessile, in a short, erect, interrupted spike 1–7(10) cm long; bract and bracteoles ovate to narrowly deltoid, 3–8 mm long, spreading or recurved, glabrous or pubescent at base, the apex strongly spinulose, the margin entire to denticulate-crenulate, the bracteoles connate at the base; perianth segments ovate to oval or oblong below the median protuberance, narrowly deltoid above the protuberance, 2.5–3.5 cm long, the midvein obscure, the apex acute, the margin entire to crenulate. Fruit obovoid to conical, 3–10 mm long, 4–6(7) mm wide, with a scarious, winged rim; seeds ca. 1.5 mm long, smooth, blackish, lustrous.

Beaches. Occasional; east coastal peninsula, central coastal panhandle. Massachusetts and New York south to Florida, west to Texas, also Oregon and California; Mexico; Europe, Africa, and Asia. Native to Europe, Africa, and Asia. Summer–fall.

EXCLUDED TAXON

> *Salsola kali* Linnaeus—The typical subspecies was reported by Small (1903, 1913a, 1933), Radford et al. (1964, 1968), Long and Lakela (1971), Wunderlin (1982), and Clewell (1985). All Florida material is subsp. *pontica*.

Sarcocornia A. J. Scott 1978. SWAMPFIRE

Shrubs. Leaves opposite, simple, pinnate-veined, petiolate, estipulate. Flowers in terminal and lateral spikes, bisexual or unisexual (plants polygamous), some stems terminated by an inflorescence, others vegetative; ebracteate, ebracteolate; perianth segments connate nearly to the apex; stamens 1–2; ovary superior, 1-carpellate and -loculate, the style 1, the stigmas 2–3. Fruit an indehiscent utricle; seed 1, vertical.

A genus of about 15 species; North America, West Indies, Mexico, South America, Europe, Africa, Asia, and Australia. [From the Greek *Sarco*, fleshy, and the latin *cornu*, horn, in reference to the fleshy stems and hornlike branches.]

Selected references: Alonso and Crespo (2008); Ball (2003b); Scott (1977).

Sarcocornia ambigua (Michx.) M. A. Alonso & M. B. Crespo [Ambiguous, as to its identity.] PERENNIAL GLASSWORT; VIRGINIA GLASSWORT.

> *Salicornia ambigua* Michaux, Fl. Bor.-Amer. 1: 2. 1803. *Arthrocnemum ambiguum* (Michaux) Moquin-Tandon, Chenop. Monogr. Enum. 112. 1840. *Sarcocornia ambigua* (Michaux) M. A. Alonso & M.

B. Crespo, Ann. Bot. Fenn. 45: 247. 2008. TYPE: "Carolina ad Floridam," s.d., *Michaux s.n.* (lectotype: P). EPITYPE: FLORIDA: Monroe Co.: N end of Big Pine Key, 15 Sep 1982, *Correll & Correll 54061* (epitype: USF). Lecto- and epitypified by Alonso and Crespo (2008: 247–48).

Erect, procumbent, or prostrate shrub, to 7 dm; stem freely rooting at the nodes and often forming mats, glabrous, succulent. Leaves with the blade triangular, 2–4 mm long and wide, the apex acute, the margin scarious, the upper and lower surfaces glabrous, the petiole connate at the base, decurrent, eventually deciduous. Flowers in groups of 3 in a spike 1–6 cm long, 2.5–5 mm thick, the spike usually solitary and terminal, sometimes pedunculate and axillary, the central flower 1–3 mm long, scarcely surpassing the lateral ones. Fruit obovate, 1–2 mm long, the pericarp membranous; seed 1–1.5 mm long, densely covered with slender curved or hooked trichomes 1–2 mm long.

Salt marshes and salt flats. Frequent; nearly throughout. New Hampshire south to Florida, west to Louisiana, also Alaska, Washington, and Oregon; West Indies, Mexico, and South America; Europe, Africa, and Asia. Native to North America, West Indies, Mexico, Europe, Africa, and Asia. Summer–fall.

EXCLUDED TAXA

Sarcocornia perennis (Miller) S. J. Scott—Reported by Small (1933) and Wunderlin and Hansen (2003), misapplied to our material of *S. ambigua*. Alonso and Crespo (2008) confine this name to a Mediterranean species.

Sarcocornia pacifica (Standley) A. J. Scott—Reported by Ball (2003b), the name misapplied to our material of *S. ambigua*.

Suaeda Forssk. ex J. F. Gmel. 1776. SEEPWEED

Annual or perennial herbs. Leaves alternate, simple, pinnate-veined, epetiolate, estipulate. Flowers in 1- to 3-flowered dichasial cymes and forming axillary clusters, each cluster subtended by a bract-like leaf, bisexual; ebracteate, ebracteolate; perianth segments 5, basally connate, unequal; stamens 5; ovary superior, 1-carpellate and -loculate, the style 1, the stigmas 3. Fruit an indehiscent utricle; seed 1, horizontal.

A genus of about 110 species, nearly cosmopolitan. [From the Arabic *suaed*, black, in reference to the seed color of some species.]

Dondia Adans. (1763).

Selected references: Ferren and Schenk (2003).

Suaeda linearis (Elliott) Moq. [Linear, in reference to the leaves.] SEA BLITE; ANNUAL SEEPWEED.

Salsola linearis Elliott, Sketch Bot. S. Carolina 1: 332. 1817. *Suaeda linearis* (Elliott) Moquin-Tandon, Chenop. Monogr. Enum. 130. 1840. *Chenopodina linearis* (Elliott) Moquin-Tandon, in A. de Candolle, Prodr. 13(2): 164. 1849. *Dondia linearis* (Elliott) A. Heller, Cat. N. Amer. Pl. 3. 1898.

Erect or ascending annual or perennial herb, to 9 dm; stem usually much branched from the base, glabrous, often glaucous. Leaves with the blade narrowly linear, 1–5 cm long, 1–2 mm wide, subterete, the apex acute, the margin entire, the upper and lower surfaces glabrous, usually glaucous, sessile. Flowers in a short, loose dichasial cyme; each cluster subtended by a bract-like reduced leaf 2–7 mm long; perianth ca. 2 mm long and wide, the lower 2 segments rounded on the back, the upper 3 larger and distally hooded or keeled. Fruit 1–3 mm long; seed black or brown, finely reticulate, lustrous or dull.

Salt marshes, coastal strands, and salt flats. Frequent; peninsula, west to central panhandle. Maine south to Florida, west to Texas; West Indies and Mexico. Spring–fall.

EXCLUDED TAXON

Suaeda maritima (Linnaeus) Dumortier—Reported by Chapman (1860, 1883, both as *Chenopodina maritima* (Linnaeus) Moquin-Tandon), Long and Lakela (1971), Hopkins and Blackwell (1977), Godfrey and Wooten (1981), Clewell (1985), and Gleason and Cronquist (1991), who misapplied the name to material of *S. linearis*.

EXCLUDED GENUS

Chamissoa altissima (Jacquin) Kunth—Reported (questionably) by Standley (1917). No Florida specimens known.

Literature Cited

Adema, F. 1991. *Cupaniopsis* Radlk. (Sapindaceae): Monograph. Leiden Bot. Ser. 15: 1–190.

Aellen, P., and T. Just. 1943. Key and synopsis of the American species of the genus *Chenopodium* L. Amer. Midl. Naturalist 30: 47–76.

Alexander, S. N., L. C. Hayek, and A. Weeks. 2012. A subspecific revision of North American saltmarsh mallow *Kosteletzkya pentacarpos* (L.) Ledeb. (Malvaceae). Castanea 77: 106–22.

Alonso, M. A., and M. B. Crespo. 2008. Taxonomic and nomenclatural notes on South American taxa of *Sarcocornia* (Chenopodiaceae). Ann. Bot. Fenn. 45: 241–54.

Al-Shehbaz, I. A. 1985. The genera of Brassicaceae (Cruciferae; Brassicaceae) in the southeastern United States. J. Arnold Arbor. 65: 279–351.

———. 1986. The genera of the Lepidieae (Cruciferae; Brassicaceae) in the southeastern United States. J. Arnold Arbor. 67: 265–311.

———. 1988. The genera of Sisymbrieae (Cruciferae); Brassicaceae) in the southeastern United States. J. Arnold Arbor. 69: 213–37.

———. 2010a. Brassicaceae. *In*: Flora of North America Editorial Committee. Flora of North America North of Mexico. 7: 224–48. New York/Oxford: Oxford University Press.

———. 2010b. *Capsella*. *In*: Flora of North America Editorial Committee. Flora of North America North of Mexico. 7: 453–54. New York/Oxford: Oxford University Press.

———. 2010c. *Nasturtium*. *In*: Flora of North America Editorial Committee. Flora of North America North of Mexico. 7: 489–92. New York/Oxford: Oxford University Press.

———. 2010d. *Planodes*. *In*: Flora of North America Editorial Committee. Flora of North America North of Mexico. 7: 492–93. New York/Oxford: Oxford University Press.

———. 2010e. *Rorippa*. *In*: Flora of North America Editorial Committee. Flora of North America North of Mexico. 7: 493–506. New York/Oxford: Oxford University Press.

———. 2010f. *Erysimum*. *In*: Flora of North America Editorial Committee. Flora of North America North of Mexico. 7: 534–45. New York/Oxford: Oxford University Press.

———. 2010g. *Sisymbrium*. *In*: Flora of North America Editorial Committee. Flora of North America North of Mexico. 7: 667–71. New York/Oxford: Oxford University Press.

———. 2010h. *Warea*. *In*: Flora of North America Editorial Committee. Flora of North America North of Mexico. 7: 742–44. New York/Oxford: Oxford University Press.

———. 2010i. *Thlaspi*. *In*: Flora of North America Editorial Committee. Flora of North America North of Mexico. 7: 745–46. New York/Oxford: Oxford University Press.

Al-Shehbaz, I. A., and J. F. Gaskin. 2010. *Lepidium*. *In*: Flora of North America Editorial Committee. Flora of North America North of Mexico. 7: 570–95. New York/Oxford: Oxford University Press.

Al-Shehbaz, I. A., K. Marhold, and J. Lihová. 2010. *Cardamine*. *In*: Flora of North America Editorial Committee. Flora of North America North of Mexico. 7: 464–84. New York/Oxford: Oxford University Press.

Al-Shehbaz, I. A., and R. C. Rollins. 1988. A reconsideration of *Cardamine curvisiliqua* and *C. gambellii* as species of *Rorippa* (Cruciferae). J. Arnold Arbor. 69: 65–71.

Al-Shehbaz, I. A., and M. D. Windham. 2010. *Boechera. In*: Flora of North America Editorial Committee. Flora of North America North of Mexico. 7: 348–412. New York/Oxford: Oxford University Press.

Al-Shehbaz, I. A., M. D. Windham, and R. Elven. 2010. *Draba. In*: Flora of North America Editorial Committee. Flora of North America North of Mexico. 7: 269–347. New York/Oxford: Oxford University Press.

Anderson, L. C. 1991. *Paronychia chartacea* ssp. *minima* (Caryophyllaceae): A new subspecies of a rare Florida endemic. Sida 14: 435–41.

Arrington, J. M. 2004. Systematics of the Cistaceae. PhD dissertation. Duke University, Durham, NC.

Bailey, L. H. 1962. A revision of the genus *Ptelea* (Rutaceae). Brittonia 14: 1–45.

Bailey, L. H., E. Z. Bailey, and Liberty Hyde Bailey Hortorium. 1976. Hortus Third: A Concise Dictionary of Plants Cultivated in the United States and Canada. New York: MacMillan.

Ball, P. W. 2003a. *Salicornia. In*: Flora of North America Editorial Committee. Flora of North America North of Mexico. 4: 382–84. New York/Oxford: Oxford University Press.

———. 2003b. *Sarcocornia. In*: Flora of North America Editorial Committee. Flora of North America North of Mexico. 4: 384–87. New York/Oxford: Oxford University Press.

Bates, D. M. 2015. *Abelmoschus. In*: Flora of North America Editorial Committee. Flora of North America North of Mexico. 6: 219–20. New York/Oxford: Oxford University Press.

Baum, B. R. 1967. Introduced and naturalized tamarisks in the United States and Canada [Tamaricaceae]. Baileya 15: 19–25.

———. 1978. The Genus *Tamarix*. Jerusalem: Israel Academy of Science and Humanities.

Bayer, C., M. W. Chase, and M. F. Fay. 1998. Muntingiaceae, a new family of dicotyledons with malvalean affinities. Taxon 47: 37–42.

Bayer, R. J., D. J. Mabberley, C. Morton, C. H. Miller, I. K. Sharma, B. E. Pfeil, S. Rich, R. Hitchcock, and S. Sykes. 2009. A molecular phylogeny of the orange subfamily (Rutaceae: Auranthioideae) using nine cpDNA sequences. Amer. J. Bot. 96: 668–85.

Blanchard, O. J. 2008. Innovations in *Hibiscus* and *Kosteletzkya* (Malvaceae, Hibisceae). Novon 18: 4–8.

———. 2015a. *Hibiscus. In*: Flora of North America Editorial Committee. Flora of North America North of Mexico. 6: 252–67. New York/Oxford: Oxford University Press.

———. 2015b. *Kosteletzkya. In*: Flora of North America Editorial Committee. Flora of North America North of Mexico. 6: 272–74.

Bogel, A. L. 1997. Leitneriaceae. *In*: Flora of North America Editorial Committee. Flora of North America North of Mexico. 3: 414–15. New York/Oxford: Oxford University Press.

Borgen, L. 2010. *Lobularia. In*: Flora of North America Editorial Committee. Flora of North America North of Mexico. 7: 597–98. New York/Oxford: Oxford University Press.

Brandbyge, J. 1986. A revision of the genus *Triplaris* (Polygonaceae). Nord. J. Bot. 6: 545–70.

Breteler, F. J. 2003. The African genus *Sorindeia* (Anacardiaceae): A synoptic revision. Adansonia, sér. 3. 25(1): 93–113.

Brizicky, G. K. 1962a. The genera of Rutaceae in the southeastern United States. J. Arnold Arbor. 43: 1–22.

———. 1962b. The genera of Simaroubaceae and Burseraceae in the southeastern United states. J. Arnold Arbor. 43: 173–86.

———. 1962c. The genera of Anacardiaceae in the southeastern United states. J. Arnold Arbor. 43: 359–75.

———. 1963. The genera of Sapindales in the southeastern United States. J. Arnold Arbor. 44: 462–501.

———. 1964. The genera of Cistaceae in the southeastern United States. J. Arnold Arbor. 45: 346–57.

———. 1965. The genera of Tiliaceae and Elaeocarpaceae in the southeastern United States. J. Arnold Arbor. 46: 286–307.

———. 1966. The genera of Sterculiaceae in the southeastern United States. J. Arnold Arbor. 47: 60–74.

Channell, R. B., and C. W. James. 1964. Nomenclatural and taxonomic corrections in *Warea* (Cruciferae). Rhodora 66: 18–26.

Chapman, A. W. 1860. Flora of the Southern United States. New York: Ivison, Phinney & Co.

———. 1883. Flora of the Southern United States, ed. 2. New York: Ivison, Blakeman, Taylor & Co.

———. 1897. Flora of the Southern United States, ed. 3. New York: American Book Co.

Chaudhri, M. N. 1968. A Revision of the Paronychiinae. Tilburg: Drukkerij J. Gianotten N.V.

Clemants, S. E. 2003a. *Alternanthera*. *In*: Flora of North America Editorial Committee. Flora of North America North of Mexico. 4: 447–51. New York/Oxford: Oxford University Press.

———. 2003b. *Gomphrena*. *In*: Flora of North America Editorial Committee. Flora of North America North of Mexico. 4: 451–54. New York/Oxford: Oxford University Press.

———. 2003c. *Iresine*. *In*: Flora of North America Editorial Committee. Flora of North America North of Mexico. 4: 454–56. New York/Oxford: Oxford University Press.

———. 2003d. *Blutaparon*. *In*: Flora of North America Editorial Committee. Flora of North America North of Mexico. 4: 456. New York/Oxford: Oxford University Press.

Clemants, S. E., and S. L. Mosyakin. 2003a. *Dysphania*. *In*: Flora of North America Editorial Committee. Flora of North America North of Mexico. 4: 267–74. New York/Oxford: Oxford University Press.

———. 2003b. *Chenopodium*. *In*: Flora of North America Editorial Committee. Flora of North America North of Mexico. 4: 275–99. New York/Oxford: Oxford University Press.

Clewell, A. F. 1985. A Guide to the Vascular Plants of the Florida Panhandle. Tallahassee: University Presses of Florida/Florida State University Press.

Cochrane, T. S., and H. H. Iltis. 2014. Studies in the Cleomaceae VII: Five new combinations in *Corynandra*, an earlier name for *Arivela*. Novon 23: 21–26.

Core, E. L. 1939. A taxonomic revision of the genus *Siphonychia*. J. Elisha Mitchell Sci. Soc. 55: 339–45.

———. 1941. The North American species of *Paronychia*. Amer. Midl. Naturalist 26: 369–97.

Correll, D. S., and H. B. Correll. 1982. Flora of the Bahama Archipelago. Vaduz: J. Cramer Verlag.

Correll, D. S., and M. C. Johnston. 1970. Manual of the Vascular Plants of Texas. Renner: Texas Research Foundation.

Costea, M., F. J. Tardif, and H. R. Hinds. 2005. *Polygonum*. *In*: Flora of North America Editorial Committee. Flora of North America North of Mexico. 5: 547–71. New York/Oxford: Oxford University Press.

Crins, W. J. 1989. The Tamaricaceae in the southeastern United States. J. Arnold Arbor. 70: 403–25.

Cristóbal, C. L. 1960. Revisión del género "*Ayenia*" (Sterculiaceae). Opera Lilloana 4: 1–230.

Crow, G. E. 1978. A taxonomic revision of *Sagina* (Caryophyllaceae) in North America. Rhodora 80: 1–91.

———. 2005. *Sagina*. *In*: Flora of North America Editorial Committee. Flora of North America North of Mexico. 5: 140–47. New York/Oxford: Oxford University Press.

Daoud, H. S., and R. L. Wilbur. 1965. A revision of the North American species of *Helianthemum* (Cistaceae). Rhodora 67: 63–82, 201–16, 255–312.

DeFilips, R. A. 1968. A revision of *Ximenia* [Plum.] L. (Olacaceae). 129 pp. PhD dissertation. Southern Illinois University, Carbondale.

DeLaney, K. R. 2010. *Paronychia discoveryi* (Caryophyllaceae), a new species from Florida. Bot. Explor. 4: 69–98.

Der, J. P., and D. L. Nickrent. 2008. A molecular phylogeny of Santalaceae (Santalales). Syst. Bot. 33: 107–16.

Dillenberger, M. S., and J. W. Kadereit. 2014. Maximum polyphyly: Multiple origins and delimitation with plesiomorphic characters require a new circumscription of *Minuartia* (Caryophyllaceae). Taxon 63: 64–88.

Dorr, L. J. 1990. A revision of the North American genus *Callirhoë* (Malvaceae). Mem. New York Bot. Gard. 56: 1–76.

———. 2015a. *Firmiana*. *In*: Flora of North America Editorial Committee. Flora of North America North of Mexico. 6: 190–91. New York/Oxford: Oxford University Press.

———. 2015b. *Ayenia*. *In*: Flora of North America Editorial Committee. Flora of North America North of Mexico. 6: 202–7. New York/Oxford: Oxford University Press.

———. 2015c. *Callirhoe*. *In*: Flora of North America Editorial Committee. Flora of North America North of Mexico. 6: 240–45. New York/Oxford: Oxford University Press.

Fassett, N. C. 1949. The variation of *Polygonum punctatum*. Brittonia 6: 369–93.

Fernald, M. L. 1902. Some little known plants from Florida and Georgia. Bot. Gaz. 33: 154–57.

———. 1950. Gray's Manual of Botany, ed. 8. New York: American Book Co.

Fernando, E. S., and C. J. Quinn. 1995. Picramniaceae, a new family, and a recircumscription of Simaroubaceae. Taxon 44: 177–81.

Ferren, W. R., and H. J. Schenk. 2003. *Suaeda*. *In*: Flora of North America Editorial Committee. Flora of North America North of Mexico. 4: 390–98. New York/Oxford: Oxford University Press.

Floden, A., and L. I. Nevling. 2015. *Dirca*. *In*: Flora of North America Editorial Committee. Flora of North America North of Mexico. 6: 381–83. New York/Oxford: Oxford University Press.

Florida Exotic Pest Plant Council (FLEPPC). 2015. Florida Exotic Pest Plant Council's 2015 List of Invasive Plant Species. (http://fleppc.org).

Freeman, C. C. 2005a. *Antigonon*. *In*: Flora of North America Editorial Committee. Flora of North America North of Mexico. 5: 481. New York/Oxford: Oxford University Press.

———. 2005b. *Coccoloba*. *In*: Flora of North America Editorial Committee. Flora of North America North of Mexico. 5: 483–84. New York/Oxford: Oxford University Press.

———. 2005c. *Emex*. *In*: Flora of North America Editorial Committee. Flora of North America North of Mexico. 5: 487–88. New York/Oxford: Oxford University Press.

———. 2005d. *Polygonella*. *In*: Flora of North America Editorial Committee. Flora of North America North of Mexico. 5: 534–40. New York/Oxford: Oxford University Press.

Freeman, C. C., and H. R. Hinds. 2005. *Fallopia*. *In*: Flora of North America Editorial Committee. Flora of North America North of Mexico. 5: 541–46. New York/Oxford: Oxford University Press.

Freeman, C. C., and J. L. Reveal. 2005. Polygonaceae. *In*: Flora of North America Editorial Committee. Flora of North America North of Mexico. 5: 216–18. New York/Oxford: Oxford University Press.

Fryxell, P. A. 1969. The genus *Cienfuegosia* Cav. (Malvaceae). Ann. Missouri Bot. Gard. 56: 179–250.

———. 1985. Sidus sidarum–V. The North and Central American species of *Sida*. Sida 11: 62–91.

———. 1987. Revision of the genus *Anoda* (Malvaceae). Aliso 11: 485–522.

———. 1988. Malvaceae of Mexico. Syst. Bot. Monogr. 25: 1–522.

———. 1992. A revised taxonomic interpretation of *Gossypium* L. (Malvaceae). Rheedea 2: 108–65.

———. 1997. The American genera of Malvaceae–II. Brittonia 49: 204–69.

———. 1999. *Pavonia* Cavanilles (Malvaceae). Fl. Neotropica Monogr. 76: 1–284.

———. 2001. *Talipariti* (Malvaceae), a segregate from *Hibiscus*. Contr. Univ. Michigan Herb. 23: 225–70.

Fryxell, P. A., and S. R. Hill. 2015a. *Abutilon*. *In*: Flora of North America Editorial Committee. Flora of North America North of Mexico. 6: 221–27. New York/Oxford: Oxford University Press.

———. 2015b. *Anoda*. *In*: Flora of North America Editorial Committee. Flora of North America North of Mexico. 6: 234–37. New York/Oxford: Oxford University Press.

———. 2015c. *Cienfuegosia*. *In*: Flora of North America Editorial Committee. Flora of North America North of Mexico. 6: 245–46. New York/Oxford: Oxford University Press.

———. 2015d. *Gossypium*. *In*: Flora of North America Editorial Committee. Flora of North America North of Mexico. 6: 250–51. New York/Oxford: Oxford University Press.

———. 2015e. *Malachra*. *In*: Flora of North America Editorial Committee. Flora of North America North of Mexico. 6: 279. New York/Oxford: Oxford University Press.

———. 2015f. *Pavonia*. *In*: Flora of North America Editorial Committee. Flora of North America North of Mexico. 6: 305–8. New York/Oxford: Oxford University Press.

————. 2015g. *Sida*. *In*: Flora of North America Editorial Committee. Flora of North America North of Mexico. 6: 310–19. New York/Oxford: Oxford University Press.

————. 2015h. *Talipariti*. *In*: Flora of North America Editorial Committee. Flora of North America North of Mexico. 6: 370–71. New York/Oxford: Oxford University Press.

————. 2015i. *Thespesia*. *In*: Flora of North America Editorial Committee. Flora of North America North of Mexico. 6: 372. New York/Oxford: Oxford University Press.

Fuentes-Bazan, S., P. Uotila, and T. Borsch. 2012. A novel phylegeny-based generic classification for *chenopodium* sensu lato, and a tribal arrangement of Chenopodioideae (Chenopodiaceae). Willdenowia 42: 5–24.

Galasso, G., E. Banfi, F. De Mattia, F. Grassi, S. Sgorbati, and M. Labra. 2009. Molecular phylogeny of *Polygonum* L. s.l. (Polygonoideae, Polygonaceae), focusing on European taxa; preliminary results and systematic considerations based on *rbcL* plastid sequence data. Atti Soc. Ital. Sci. Nat. Mus. Civico Storia Nat. Milano 150(1): 113–48.

Gaskin, J. F. 2015. *Tamaricaceae*. *In*: Flora of North America Editorial Committee. Flora of North America North of Mexico. 6: 413–17. New York/Oxford: Oxford University Press.

Gillis, W. T. 1971. The systematics and ecology of poison-ivy and the poison-oaks (*Toxicodendron*, Anacardiaceae). Rhodora 73: 72–159, 161–237, 370–443, 465–540.

Gleason, H. A., and A. Cronquist. 1991. Manual of the Vascular Plants of Northeastern United States and Adjacent Canada, ed. 2. Bronx: New York Botanical Garden.

Godfrey, R. K. 1988. Trees, Shrubs, and Woody Vines of Northern Florida and Adjacent Georgia and Alabama. Athens: University of Georgia Press.

Godfrey, R. K., and J. W. Wooten. 1981. Aquatic and Wetland Plants of Southeastern United States: Dicotyledons. Athens: University of Georgia Press.

Goldberg, A. 1967. The genus *Melochia* L. (Sterculiaceae). Contr. U.S. Natl. Herb. 34: 191–363.

————. 2015. *Melochia*. *In*: Flora of North America Editorial Committee. Flora of North America North of Mexico. 6: 210–12. New York/Oxford: Oxford University Press.

Goodson, B. E., and I. A. Al-Shehbaz. 2010. *Descurainia*. *In*: Flora of North America Editorial Committee. Flora of North America North of Mexico. 7: 518–29. New York/Oxford: Oxford University Press.

Graham, S. A. 1964a. The genera of Lythraceae in the southeastern United States. J. Arnold Arbor. 45: 235–50.

————. 1964b. The genera of Rhizophoraceae and Combretaceae in the southeastern United States. J. Arnold Arbor. 45: 285–301.

Graham, S. A., and C. E. Wood. 1965. The genera of Polygonaceae in the southeastern United States. J. Arnold Arbor. 46: 91–113.

Guzmán, B., and P. Vargas. 2009. Historical biogeography and character evolution of Cistaceae (Malvales) based on analysis of plastid *rbcL* and *trnL* and *trnL-trnF* sequences. Organ. Divers. Evol. 9: 83–99.

Hall, H. M., and F. E. Clements. 1923. The phylogenetic method in taxonomy: The North American species of *Artemisia*, *Chrysothamnus*, and *Atriplex*. Publ. Carnegie Inst. Wash. 326: 1–255.

Haraldson, K. 1978. Anatomy and taxonomy in Polygonaceae subfam. Polygonoideae Meisn. emend. Jaretzky. Acta Univ. Upsal., Symb. Bot. Upsal. 22: 1–93.

Hardin, J. 1957. A revision of the American Hippocastanaceae II. Brittonia 9: 173–95.

Hartman, R. L., and R. K. Rabeler. 2005a. *Spergula*. *In*: Flora of North America Editorial Committee. Flora of North America North of Mexico. 5: 14–16. New York/Oxford: Oxford University Press.

————. 2005b. *Spergularia*. *In*: Flora of North America Editorial Committee. Flora of North America North of Mexico. 5: 16–23. New York/Oxford: Oxford University Press.

Hartman, R. L., R. K. Rabeler, and F. H. Utech. 2005. *Arenaria*. *In*: Flora of North America Editorial Committee. Flora of North America North of Mexico. 5: 51–56. New York/Oxford: Oxford University Press.

Hartman, R. L., J. W. Thieret, and R. K. Rabeler. 2005. *Paronychia*. *In*: Flora of North America Editorial

Committee. Flora of North America North of Mexico. 5: 30–43. New York/Oxford: Oxford University Press.

Hill, S. R. 1982. A monograph of the genus *Malvastrum* A. Gray (Malvaceae: Malveae). Rhodora 84: 1–83, 159–264, 317–409.

———. 2015a. *Malva*. In: Flora of North America Editorial Committee. Flora of North America North of Mexico. 6: 286–93. New York/Oxford: Oxford University Press.

———. 2015b. *Malvastrum*. In: Flora of North America Editorial Committee. Flora of North America North of Mexico. 6: 293–98. New York/Oxford: Oxford University Press.

———. 2015c. *Modiola*. In: Flora of North America Editorial Committee. Flora of North America North of Mexico. 6: 303–4. New York/Oxford: Oxford University Press.

———. 2015d. *Urena*. In: Flora of North America Editorial Committee. Flora of North America North of Mexico. 6: 373. New York/Oxford: Oxford University Press.

Hinds, H. R., and C. C. Freeman. 2005a. *Fagopyrum*. In: Flora of North America Editorial Committee. Flora of North America North of Mexico. 5: 572–73. New York/Oxford: Oxford University Press.

———. 2005b. *Persicaria*. In: Flora of North America Editorial Committee. Flora of North America North of Mexico. 5: 574–94. New York/Oxford: Oxford University Press.

Hitchcock, C. L., and B. Maguire. 1947. A revision of the North American species of *Silene*. Univ. Washington Publ. 13: 1–73.

Hodgdon, A. R. 1938. A taxonomic study of *Lechea*. Rhodora 40: 29–69, 87–131.

Holmes, W. C. 2005. *Brunnichia*. In: Flora of North America Editorial Committee. Flora of North America North of Mexico. 5: 482–83. New York/Oxford: Oxford University Press.

———. 2010. Caricaceae. In: Flora of North America Editorial Committee. Flora of North America North of Mexico. 7: 170–71. New York/Oxford: Oxford University Press.

Hopkins, C. O., and W. H. Blackwell. 1977. Synopsis of *Suaeda* (Chenopodiaceae) in North America. Sida 7: 147–73.

Horton, J. H. 1972. Studies of the southeastern United States flora. IV. Polygonaceae. J. Elisha Mitchell Sci. Soc. 88: 92–102.

Iltis, H. H., and T. S. Cochrane. 2007. Studies in the Cleomaceae V: A new genus and ten new combinations for the Flora of North America. Novon 17: 447–51.

Iltis, H. H., and X. Cornejo. 2010. Studies in Capparaceae XXVIII: The *Quadrella cyanophallophora* complex. J. Bot. Res. Inst. Texas 4: 93–115.

James, C. W. 1956. A revision of *Rhexia* (Melastomataceae). Brittonia 8: 201–30.

Johnston, M. C. 1957. *Phoradendron serotinum* for *P. flavescens* (Loranthaceae): Nomenclature corrections. Southw. Naturalist 2: 45–47.

Jones, G. N. 1968. The taxonomy of American species of linden (*Tilia*). Illinois Biol. Monogr. 39: 1–156.

Judd, W. S. 1983. The taxonomic status of *Stipulicida filiformis* (Caryophyllaceae). Sida 10: 33–36.

Judd, W. S., E. R. Bécquer, J. D. Skean, and L. C. Majure. 2014. Taxonomic studies in the Miconieae (Melastomataceae). XII. Revision of *Miconia* sect. *Miconiastrum*, with emphasis on the *Miconia bicolor* complex. J. Bot. Res. Inst. Texas 8: 457–91.

Khoshoo, T. N., and M. Pal. 1973. The probable origin and relationships of the garden cockcomb. J. Linn. Soc., Bot. 66: 127–41.

Kostermans, A.J.G.H. 1957. The genus *Firmiana* (Sterculiaceae). Pengum. Balai Besar Penjel. Kehut. Indonesia 54: 3–33.

Kral, R., and P. E. Bostick. 1969. The genus *Rhexia* (Melastomatacee). Sida 3: 387–440.

Krapovickas, A. 1996. La identidad de *Wissadula amplissima* (Malvaceae). Bonplandia 9: 89–94.

———. 2003. *Sida* sección *Distichifolia* (Monteiro) Krapov. comb. nov., stat. nov. (Malvaceae-Malveae). Bonplandia (Corrientes) 12: 83–121.

Kriebel, R. 2008. Systematics and Biogeography of the Neotropical Genus *Acisanthera* (Melastomataceae). MS thesis. San Francisco State University, San Francisco, CA.

Kuijt, J. 1982. The Viscaceae in the southeastern United States. J. Arnold Arbor. 63: 401–10.

Kuntze, C.E.O. 1891. Revisio Generum Plantarum. 2: 375–1011. Leipzig: Arthur Felix.

Kurz, H., and R. K. Godfrey. 1962. Trees of North Florida. Gainesville: University of Florida Press.

La Duke, J. 2015. *Herissantia*. *In*: Flora of North America Editorial Committee. Flora of North America North of Mexico. 6: 251–52. New York/Oxford: Oxford University Press.

Landrum, L. R. 1986. *Pimenta*. Fl. Neotropica Monogr. 45: 72–115.

Larsen, G. E. 1986. Caryophyllaceae. *In*: R. L. McGregor, T. M. Barkley, R. E. Brooks, and E. K. Schofield. Flora of the Great Plains. 192–214. Lawrence: University Press of Kansas.

Leenhouts, P. W. 1971. A revision of *Dimocarpus* (Sapinadaceae). Blumea 19: 113–31.

———. 1983. Notes on the extra-Australian species of *Dodonaea* (Sapindaceae). Blumea 28: 271–89.

Lemke, D. E. 2015. *Lechea*. *In*: Flora of North America Editorial Committee. Flora of North America North of Mexico. 6: 389–97. New York/Oxford: Oxford University Press.

Long, R. W., and O. Lakela. 1971. A Flora of Tropical Florida. Coral Gables: University of Miami Press.

Luteyn, J. L. 1976. Revision of *Limonium* (Plumbaginaceae) in eastern North America. Brittonia 28: 303–17.

———. 1990. The Plumbaginaceae in the flora of the southeastern United States. Sida 14: 169–78.

Mabberley, D. J. 1997. A classification for edible *Citrus* (Rutaceae). Telopea 7: 167–72.

———. 1998. Australian Citreae with notes on other Aurantioideae (Rutaceae). Telopea 7: 333–44.

Malécot, V., and D. L. Nickrent. 2008. Molecular phylogenetic relationships of Olacaceae and related Santalales. Syst. Bot. 33: 97–106.

Martínez-Laborde, J. B. 2010. *Diplotaxis*. *In*: Flora of North America Editorial Committee. Flora of North America North of Mexico. 7: 432–33. New York/Oxford: Oxford University Press.

Maurin, O., M. W. Chase, M. Jordaan, and M. van der Bank. 2010. Phylogenetic relationships of Combretaceae inferred from nuclear and plastid DNA sequence data: Implications for generic classification. Bot. J. Linn. Soc. 162: 453–76.

McCauley, R. A. 2003. *Froelichia*. *In*: Flora of North America Editorial Committee. Flora of North America North of Mexico. 4: 443–47. New York/Oxford: Oxford University Press.

McNeill, J. 1980. The delimitation of *Arenaria* (Caryophyllaceae) and related genera in North America, with 11 new combinations in *Minuartia*. Rhodora 82: 495–502.

Mears, J. A. 1980. The Linnaean species of *Gomphrena* L. (Amaranthaceae). Taxon 29: 85–95.

———. 1982. A summary of *Blutaparon* Rafinesque including species earlier known as *Philoxerus* R. Brown (Amaranthaceae). Taxon 31: 111–17.

Mellichamp, T. L. 2015. Droseraceae. *In*: Flora of North America Editorial Committee. Flora of North America North of Mexico. 6: 418–25. New York/Oxford: Oxford University Press.

Mendenhall, M. G., and P. A. Fryxell. 2015. *Malvaviscus*. *In*: Flora of North America Editorial Committee. Flora of North America North of Mexico. 6: 298–300. New York/Oxford: Oxford University Press.

Menzel, M. Y., P. A. Fryxell, and F. D. Wilson. 1983. Relationships among New World species of *Hibiscus* section Furcaria (Malvaceae). Brittonia 35: 204–21.

Merrill, E. D. 1949. Index Rafinesquianus. Jamaica Plain: Arnold Arboretum of Harvard University.

Mertens, T. R., and P. H. Raven. 1965. Taxonomy of *Polygonum* section *Polygonum* (*Avicularia*) in North America. Madroño 18: 85–92.

Meyer, F. G. 1976. A revision of the genus *Koelreuteria* (Sapindaceae). J. Arnold Arbor. 57: 129–66.

Meyer, K. 2001. Revision of the southeast Asian genus *Melastoma* (Melastomataceae). Blumea 46: 351–98.

Miller, N. G. 1990. The genera of Meliaceae in the southeastern United States. J. Arnold Arbor. 71: 453–86.

Mohr, C. 1901. Plant life of Alabama. Contr. U.S. Nat. Herb. 6: 1–921.

Moldenke, H. N. 1944. A contribution to our knowledge of the wild and cultivated flora of Florida–I. Amer. Midl. Naturalist 32: 529–90.

Morton, J. K. 2005a. *Cerastium*. *In*: Flora of North America Editorial Committee. Flora of North America North of Mexico. 5: 74–93. New York/Oxford: Oxford University Press.

———. 2005b. *Stellaria*. *In*: Flora of North America Editorial Committee. Flora of North America North of Mexico. 5: 96–114. New York/Oxford: Oxford University Press.

———. 2005c. *Silene*. *In*: Flora of North America Editorial Committee. Flora of North America North of Mexico. 5: 166–214. New York/Oxford: Oxford University Press.

Mosyakin, S. L. 1996. A taxonomic synopsis of the genus *Salsola* (Chenopodiaceae) in North America. Ann. Missouri Bot. Garden. 83: 387–95.

———. 2003. *Salsola*. *In*: Flora of North America Editorial Committee. Flora of North America North of Mexico. 4: 398–403. New York/Oxford: Oxford University Press.

———. 2005. *Rumex*. *In*: Flora of North America Editorial Committee. Flora of North America North of Mexico. 5: 498–533. New York/Oxford: Oxford University Press.

———. 2013. New nomenclatural combinations in *Blitum*, *Oxybasis*, *Chenopodiastrum*, and *Lipandra* (Chenopodiaceae). Phytoneuron 2013-56: 1–8.

Mosyakin, S. L., and K. R. Robertson. 2003. *Amaranthus*. *In*: Flora of North America Editorial Committee. Flora of North America North of Mexico. 4: 410–35. New York/Oxford: Oxford University Press.

Munz, P. A. 1938. A revision of the genus *Gaura*. Bull. Torrey Bot. Club 65: 105–22, 211–28.

———. 1944. Studies in Onagraceae–XIII. The American species of *Ludwigia*. Bull. Torrey Bot. Club 71: 152–65.

Nakayama, H., K. Fukushima, T. Fukuda, J. Yokoyama, and S. Kimura. 2014. Molecular phylogeny determined using DNA inferred a new Phylogenetic relationship of *Rorippa aquatica* (Eaton) E. J. Palmer & Stermark (Brassicaceae)—lake cress. Amer J. Plant Sci. 5: 48–54.

Nesom, G. L. 2015a. *Corchorus*. *In*: Flora of North America Editorial Committee. Flora of North America North of Mexico. 6: 197–200. New York/Oxford: Oxford University Press.

———. 2015b. *Triumfetta*. *In*: Flora of North America Editorial Committee. Flora of North America North of Mexico. 6: 200–202. New York/Oxford: Oxford University Press.

Nesom, G. L., and J. T. Kartesz. 2000. Observations on the *Ludwigia uruguayensis* complex (Onagraceae) in the United States. Castanea 65: 123–25.

Nevling, L. I. 1962. The Thymelaeaceae in the southeastern United States. J. Arnold Arbor. 63: 428–34.

Olson, M. E. 2010. Moringaceae. *In*: Flora of North America Editorial Committee. Flora of North America North of Mexico. 7: 167–69. New York/Oxford: Oxford University Press.

Peng, C.-I. 1986. A new combination in *Ludwigia* sect. *Microcarpium* (Onagraceae). Ann. Missouri Bot. Gard. 73: 490.

———. 1989. The systematics and evolution of *Ludwigia* sect. *Microcarpium* (Onagraceae). Ann. Missouri Bot. Gard. 76: 221–302.

Peng, C.-I, C. L. Schmidt, P. C. Hoch, and P. H. Raven. 2005. Systematics and evolution of *Ludwigia* section *Dantia* (Onagraceae). Ann. Missouri Bot. Gard. 92: 307–59.

Poindexter, D. B., K. E. Bennett, and A. S. Weakley. 2014. A morphologically based taxonomic reevaluation of the genus *Stipulicida* (Caryophyllaceae), with comments on rank. J. Bot. Res. Inst. Texas 8: 419–30.

Porter, D. M. 1976. *Zanthoxylum* (Rutaceae) in North America North of Mexico. Brittonia 28: 443–47.

Pringle, J. S. 2005. *Gypsophila*. *In*: Flora of North America Editorial Committee. Flora of North America North of Mexico. 5: 153–56. New York/Oxford: Oxford University Press.

Rabeler, R. K., and R. L. Hartman. 2005a. *Drymaria*. *In*: Flora of North America Editorial Committee. Flora of North America North of Mexico. 5: 9–14. New York/Oxford: Oxford University Press.

———. 2005b. *Dianthus*. *In*: Flora of North America Editorial Committee. Flora of North America North of Mexico. 5: 159–62. New York/Oxford: Oxford University Press.

Rabeler, R. K., R. L. Hartman, and F. H. Utech. 2005. *Minuartia*. *In*: Flora of North America Editorial Committee. Flora of North America North of Mexico. 5: 116–36. New York/Oxford: Oxford University Press.

Rabeler, R. K., and J. W. Thieret. 1988. Comments on the Caryophyllaceae of the southeastern United States. Sida 13: 149–56.

Radford, A. E., H. E. Ahles, and C. R. Bell. 1964. Guide to the Vascular Flora of the Carolinas. Chapel Hill: University of North Carolina Press.

———. 1968. Manual of the Vascular Flora of the Carolinas. Chapel Hill: University of North Carolina Press.

Ramamoorthy, T. P., and E. Zardini. 1987. The systematics and evolution of *Ludwigia* sect. *Myrtocarpus* sensu lato (Onagraceae). Monogr. Syst. Bot. Missouri Bot. Gard. 19: 1–120.

Raven, P. H. 1963. The Old World species of *Ludwigia* (including *Jussiaea*), with a synopsis of the genus (Onagraceae). Reinwardtia 6: 327–427.

Reveal, J. L. 2005. *Eriogonum. In:* Flora of North America Editorial Committee. Flora of North America North of Mexico. 5: 221–430. New York/Oxford: Oxford University Press.

Reveal, J. L., and M. C. Johnston. 1989. A new classification in *Phorodendron* (Viscaceae). Taxon 38: 107–8.

Rilke, S. 1999. Revision de Sektion *Salsola* s.l. de Gattung *Salsola* (Chenopodiaceae). Bibliotheca Botanica (Stuttgart) 149: 1–190.

Robertson, K. R. 1982. The genera of Olacaceae in the southeastern United States. J. Arnold Arbor. 63: 387–99.

———. 2003a. *Celosia. In:* Flora of North America Editorial Committee. Flora of North America North of Mexico. 4: 407–9. New York/Oxford: Oxford University Press.

———. 2003b. *Achyranthus. In:* Flora of North America Editorial Committee. Flora of North America North of Mexico. 4: 435–37. New York/Oxford: Oxford University Press.

Robertson, K. R., and S. E. Clemants. 2003. Amaranthaceae. *In:* Flora of North America Editorial Committee. Flora of North America North of Mexico. 4: 405–6. New York/Oxford: Oxford University Press.

Rodman, J. E. 1974. Systematics and evolution of the genus *Cakile* (Cruciferae). Contr. Gray Herb. 205: 3–146.

———. 2010. *Cakile. In:* Flora of North America Editorial Committee. Flora of North America North of Mexico. 7: 424–28. New York/Oxford: Oxford University Press.

Rogers, G. K. 1982. The Bataceae in the southeastern United States. J. Arnold Arbor. 63: 375–86.

Rollins, R. C. 1978. Watercress in Florida. Rhodora 80: 147–53.

Ronse Decraene, L.-P., and J. R. Akeroyd. 1988. Generic limits in *Polygonum* and related genera (Polygonaceae) on the basis of floral characters. Bot. J. Linn. Soc. 98: 321–71.

Ronse Decraene, L.-P., S. P. Hong, and E. F. Smets. 2004. What is the taxonomic status of *Polygonella*? Ann. Missouri Bot. Gard. 91: 320–45.

Rossbach, R. P. 1940. *Spergularia* in North and South America. Rhodora 42: 57–83, 105–43, 158–93, 203–15.

Salywon, A. 2003. A monograph of *Mosiera* (Myrtaceae). PhD dissertation. Arizona State University, Tempe.

Sanchez, A., T. M. Schuster, J. M. Burke, and K. A. Kron. 2011. Taxonomy of Polygonoideae (Polygonaceae): A new tribal classification. Taxon 60: 151–60.

Sargent, C. S. 1926. Manual of the Trees of North America, ed. 2. Boston and New York: Houghton Mifflin Co.

Sauer, J. D. 1955. Revision of the dioecious amaranths. Madroño 13: 5–46.

Saunders, J. G. 2015. *Waltheria. In:* Flora of North America Editorial Committee. Flora of North America North of Mexico. 6: 212–15. New York/Oxford: Oxford University Press.

Schrader, J. A., and W. R. Graves. 2011. Taxonomy of *Leitneria* (Simaroubaceae) resolved by ISSR, ITS, and morphometric characterization. Castanea 76: 313–38.

Schulz, O. E. 1923. *Cakile.* In: Engler, A. Das Pflanzenreich 4(Heft 84): 18–28.

Schuster, T. M., J. L. Reveal, M. J. Bayly, and K. A. Kron. 2015. An updated molecular phylogeny of

Polygonaideae (Polygonaceae): Relationships of *Oxygonium*, *Pteroxygonium*, and *Rumex*, and a new circumscription of *Koenigia*. Taxon 64: 1188–208.

Scott, A. J. 1977. Reinstatement and revision of Salicorniaceae J. Agardh (Caryophyllales). Bot. J. Linn. Soc. London 75: 357–74.

Shinners, L. A. 1962a. *Drosera* (Droseraceae) in the southeastern United States: An interim report. Sida 1: 53–59.

———. 1962b. *Siphonychia* transferred to *Paronychia* (Caryophyllaceae). Sida 1: 101–3.

Siedo, S. J. 1999. A taxonomic treatment of *Sida* sect. *Ellipticifolae* (Malvaceae). Lundellia 2: 100–127.

Small, J. K. 1903. Flora of the Southeastern United States. New York: Published by the author.

———. 1913a. Flora of the Southeastern United States, 2nd ed. New York: Published by the author.

———. 1913b. Flora of Miami. New York: Published by the author.

———. 1913c. Florida Trees. New York: Published by the author.

———. 1913d. Flora of the Florida Keys. New York: Published by the author.

———. 1913e. Shrubs of Florida. New York: Published by the author.

———. 1933. Manual of the Southeastern Flora. New York: Published by the author.

Smith, A. R. 2005a. *Limonium*. *In*: Flora of North America Editorial Committee. Flora of North America North of Mexico. 5: 606–10. New York/Oxford: Oxford University Press.

———. 2005b. *Plumbago*. *In*: Flora of North America Editorial Committee. Flora of North America North of Mexico. 5: 610–11. New York/Oxford: Oxford University Press.

Sorrie, B. A. 2015. *Crocanthemum*. *In*: Flora of North America Editorial Committee. Flora of North America North of Mexico. 6: 400–408. New York/Oxford: Oxford University Press.

Spaulding, D. D. 2013. Key to the pinweeds (*Lechea*, Cistaceae) of Alabama and adjacent states. Phytoneuron 2013-99; 1–15.

Spongberg, S. A. 1971. The staphyleaceae in the southeastern United States. J. Arnold Arbor. 52: 196–203.

Stace, C. E. 2010. Combretaceae. Fl. Neotrop. Monogr. 107: 1–369.

Standley, P. C. 1917. Amaranthaceae. *In*: N. L. Britton et al., eds. 1905+. North American Flora. 21(2): 95–169. New York: New York Botanical Garden.

Stone, B. C. 1985. A conspectus of the genus *Glycosmis* Correa: Studies in Malesian Rutaceae, III. Proc. Acad. Nat. Sci. Philadelphia 137(2): 1–27.

Strother, J. L. 2015a. Muntingiaceae. *In*: Flora of North America Editorial Committee. Flora of North America North of Mexico. 6: 185–86. New York/Oxford: Oxford University Press.

———. 2015b. *Tilia*. *In*: Flora of North America Editorial Committee. Flora of North America North of Mexico. 6: 194–96. New York/Oxford: Oxford University Press.

Styles, B. T., and F. White. 1991. Meliaceae. Fl. Trop. E. Africa 18: 1–67.

Su, H.-J., Hu, J.-M., F. E. Anderson, J. P. Der, and D. L. Nickrent. 2015. Phylogenetic relationships of Santalales with insights into the origins of holoparasitic Balanophoraceae. Taxon 64: 491–506.

Swanson, A., and R. K. Rabeler. 2005. *Stipulicida*. *In*: Flora of North America Editorial Committee. Flora of North America North of Mexico. 5: 27–28. New York/Oxford: Oxford University Press.

Swingle, W. T. 1944. The history, botany and breeding. Volume 1. *In*: H. J. Webber and L. D. Batchelor. The Citrus Industry. Berkeley: University of California Press.

Thieret, J. W. 2005. *Argostemma*. *In*: Flora of North America Editorial Committee. Flora of North America North of Mexico. 5: 214–15. New York/Oxford: Oxford University Press.

Thieret, J. W., and R. K. Rabeler. 2005a. *Polycarpaea*. *In*: Flora of North America Editorial Committee. Flora of North America North of Mexico. 5: 23–25. New York/Oxford: Oxford University Press.

———. 2005b. *Polycarpon*. *In*: Flora of North America Editorial Committee. Flora of North America North of Mexico. 5: 25–26. New York/Oxford: Oxford University Press.

———. 2005c. *Scleranthus*. *In*: Flora of North America Editorial Committee. Flora of North America North of Mexico. 5: 149–53. New York/Oxford: Oxford University Press.

———. 2005d. *Vaccaria*. *In*: Flora of North America Editorial Committee. Flora of North America North of Mexico. 5: 156. New York/Oxford: Oxford University Press.

——. 2005e. *Saponaria. In*: Flora of North America Editorial Committee. Flora of North America North of Mexico. 5: 157–58. New York/Oxford University Press.

Thorne, R. F. 2010. Bataceae. *In*: Flora of North America Editorial Committee. Flora of North America North of Mexico. 7: 186–88. New York/Oxford: Oxford University Press.

Townsend, C. C. 1974. Notes on Amaranthaceae: 2. Kew Bull. 29: 461–75.

Tucker, G. C. 2010a. Capparaceae. *In*: Flora of North America Editorial Committee. Flora of North America North of Mexico. 7: 194–98. New York/Oxford: Oxford University Press.

——. 2010b. *Polanisia. In*: Flora of North America Editorial Committee. Flora of North America North of Mexico. 7: 201–4. New York/Oxford: Oxford University Press.

——. 2010c. *Cleome. In*: Flora of North America Editorial Committee. Flora of North America North of Mexico. 7: 215–16. New York/Oxford: Oxford University Press.

——. 2010d. *Cleoserrata. In*: Flora of North America Editorial Committee. Flora of North America North of Mexico. 7: 216–18. New York/Oxford: Oxford University Press.

——. 2010e. *Arivela. In*: Flora of North America Editorial Committee. Flora of North America North of Mexico. 7: 221–22. New York/Oxford: Oxford University Press.

——. 2010f. *Gynandropsis. In*: Flora of North America Editorial Committee. Flora of North America North of Mexico. 7: 222–23. New York/Oxford: Oxford University Press.

Tucker, G. C., and H. H. Iltis. 2010. *Tarenya. In*: Flora of North America Editorial Committee. Flora of North America North of Mexico. 7: 218–19. New York/Oxford: Oxford University Press.

Tucker, G. C., and S. S. Vanderpool. 2010. Cleomaceae. *In*: Flora of North America Editorial Committee. Flora of North America North of Mexico. 7: 199–201. New York/Oxford: Oxford University Press.

Turner, B. L., and M. G. Mendenhall. 1993. A revision of *Malvaviscus* (Malvaceae). Ann. Missouri Bot. Gard. 80: 439–57.

Uotila, P. 1977. *Atriplex prostrata* ssp. *policinia*, and *Chenopodium serotinum*. Ann. Bot. Fenn. 14: 197–98.

USDA, NRCS. 2015. The PLANTS database. National Plant Data Team, Greensboro, NC. (http://plants.usda.gov).

Wagner, W. L., P. C. Hoch, and P. H. Raven. 2007. Revised classification of the Onagraceae. Syst. Bot. Monogr. 83: 1–240.

Wahl, H. A. 1954. A preliminary study of the genus *Chenopodium* in North America. Bartonia 27: 1–46.

Ward, D. B. 1977. Keys to the Florida of Florida–2, *Paronychia* (Caryophyllaceae). Phytologia 35: 414–18.

——. 2004. *Acer floridanum*: The correct name of the Florida maple. Castanea 69: 230–32.

——. 2008. Thomas Walter typification project, V: Neotypes and epitypes for 63 Walter Names, of genera D through Z. J. Bot. Res. Inst. Tex. 2: 475–86.

——. 2011. Native or not: Studies in problematic species, No. 6. Papaya, *Carica papaya* (Caricaceae). Palmetto 28(1): 8–11.

Warwick, S. I. 2010a. *Brassica. In*: Flora of North America Editorial Committee. Flora of North America North of Mexico. 7: 419–24. New York/Oxford: Oxford University Press.

——. 2010b. *Erucastrum. In*: Flora of North America Editorial Committee. Flora of North America North of Mexico. 7: 435. New York/Oxford: Oxford University Press.

——. 2010c. *Raphanus. In*: Flora of North America Editorial Committee. Flora of North America North of Mexico. 7: 438–40. New York/Oxford: Oxford University Press.

——. 2010d. *Sinapis. In*: Flora of North America Editorial Committee. Flora of North America North of Mexico. 7: 441–43. New York/Oxford: Oxford University Press.

——. 2010e. *Conringia. In*: Flora of North America Editorial Committee. Flora of North America North of Mexico. 7: 517. New York/Oxford: Oxford University Press.

Welsh, S. L. 2003. *Atriplex. In*: Flora of North America Editorial Committee. Flora of North America North of Mexico. 4: 322–81. New York/Oxford: Oxford University Press.

Welsh, S. L., C. W. Crompton, and S. E. Clemants. 2003. Chenopodiaceae. *In*: Flora of North America Editorial Committee. Flora of North America North of Mexico. 4: 258–61. New York/Oxford: Oxford University Press.

Whetstone, R. D. 1983. The Sterculiaceae in the flora of the southeastern United States. Sida 10: 15–23.

Wilbur, R. L., and H. S. Daoud. 1961. The genus *Lechea* (Cistaceae) in the southeastern United States. Rhodora 63: 103–18.

Wilhelm, G. S. 1984. Vascular Flora of the Pensacola Region. PhD dissertation. Southern Illinois University, Carbondale.

Wilson, P. 1911. *Zanthoxylum*. *In*: N. L. Britton. North American Flora. 25: 177–99. New York: New York Botanical Garden.

Wood, C. E. 1960. The genera of Sarraceniaceae and Droseraceae in the southeastern United States. J. Arnold Arbor. 40: 94–112.

Wunderlin, R. P. 1981. *Polygonella polygama* in Florida. Florida Scientist 44: 78–80.

———. 1982. Guide to the Vascular Plants of Central Florida. Gainesville: University Presses of Florida/ University of South Florida Press.

———. 1998. Guide to the Vascular Plants of Florida. Gainesville: University Press of Florida.

Wunderlin, R. P., and B. F. Hansen. 2003. Guide to the Vascular Plants of Florida. 2nd ed. Gainesville: University Press of Florida.

———. 2011. Guide to the Vascular Plants of Florida. 3rd ed. Gainesville: University Press of Florida.

Wurdack, J. J., and R. Kral. 1982. The genus of Melastomataceae in the southeastern United States. J. Arnold Arbor. 63: 429–39.

Wynne, F. E. 1944. *Drosera* in eastern North America. Bull. Torrey Bot. Club 71: 166–74.

Zhang, D., and D. J. Mabberly. 2008. *Citrus*. *In*: Z. Y. Wu, P. H. Raven, and D. Y. Hong (eds.) Flora of China. 11: 90–96. Beijing: Science Press/St. Louis: Missouri Botanical Garden Press.

Index to Common Names

Index to Scientific Names

Richard P. Wunderlin, professor emeritus of biology at the University of South Florida, is the author of *Guide to the Vascular Plants of Central Florida* and *Guide to the Vascular Plants of Florida*, and coauthor of *Guide to the Vascular Plants of Florida*, second edition, *Guide to the Vascular Plants of Florida*, third edition, and the *Atlas of Florida Plants* website (www.florida.plantatlas.usf.edu).

Bruce F. Hansen, curator emeritus of biology at the University of South Florida Herbarium, is coauthor of *Guide to the Vascular Plants of Florida*, second edition, *Guide to the Vascular Plants of Florida*, third edition, and the *Atlas of Florida Plants* website (www.florida.plantatlas.usf.edu).

Alan R. Franck, curator of the University of South Florida Herbarium, is coauthor of the *Atlas of Florida Plants* website (www.florida.plantatlas.usf.edu).

CPSIA information can be obtained
at www.ICGtesting.com
Printed in the USA
LVHW01*0421030518
575796LV00009B/23/P